ELSEVIER'S DICTIONARY OF CHEMISTRY

ELSEVIER'S DICTIONARY OF CHEMISTRY

INCLUDING TERMS FROM BIOCHEMISTRY

in

English, French, Spanish, Italian and German

compiled by

A.F. DORIAN

ELSEVIER

AMSTERDAM/OXFORD/NEW YORK 1983

ELSEVIER SCIENCE PUBLISHERS B.V.
1 Molenwerf
P.O. Box 211, 1000 AE Amsterdam, The Netherlands

Distributors for the United States and Canada:

ELSEVIER SCIENCE PUBLISHING COMPANY INC.
52, Vanderbilt Avenue
New York, N.Y. 10017

ISBN 0-444-42230-7

Printed in The Netherlands

Electronic data processing:
Büro für Satztechnik W. Meyer KG
Weissensberg, W. Germany

PREFACE

It is now an established fact that chemistry constitutes a body of knowledge without which no scientific work in this field can proceed unfettered. The author has attempted to include in the present dictionary of chemistry a large number of terms specifically relating to biochemistry which provides a link between organic chemistry, physiology, genetics and medicine. It is hoped that this will render this multilingual dictionary a more useful and supple work of reference.

Research workers, scientists, students, translators and technical editors will certainly not find it complete, or even near-complete, this being, in our times, an almost impossible task. The author's hope is that it may prove of use to all those who consult it to speed up and ease their work.

A.F. Dorian

Abbreviations:

f	français
e	español
i	italiano
d	deutsch

adj	adjective
GB	English usage
US	American usage
v	verb

BASIC TABLE

A

1 Abderhalden reaction
f réaction d'Abderhalden
e reacción de Abderhalden
i reazione di Abderhalden
d Abderhalden-Reaktion

2 Abegg's rule
f règle d'Abegg
e regla de Abegg
i regola di Abegg
d Abeggsche Regel

3 abelite
f abélite
e abolito
i abelite
d Abelit

4 abelmosk
f abelmosch; ambrette
e abelmoscho; ambarina
i abelmosco; ambretta
d Abelmoschus; Bisamkörner

5 Abel test
f essai d'Abel
e prueba Abel
i prova di Abel
d Abel-Probe

6 abietate
f abiétate
e abietinato
i abictato; abietinato
d abietinsaures Salz

7 abietene
f abiétène
e abieteno
i abietene
d Abietin

8 abietic acid
f acide abiétique
e ácido abiético; ácido abietínico
i acido abietico
d Abietinsäure

9 abiuret
f abiuret
e abiureto
i abiuret
d Abiuret

10 ablastin
f ablastine

e ablastina
i ablastina
d Ablastin

11 ablative compounds
f plastiques ablatifs
e compuestos ablativos
i composti ablativi
d ablative Kunststoffe

12 aborticide; abortifacient
f abortif
e abortivo
i abortivo
d Abtreibungsmittel

* **abrasin oil** → 8568

13 abrasives; abradants
f abrasifs
e abrasivos
i abrasivi
d Abreibungsmittel; Schleifmittel

14 abrotine
f abrotine
e abrotina
i abrotina
d Abrotin

* **absinthe oil** → 8863

15 absinthin
f absinthine
e absintina
i absintina
d Absinthin; Bitterstoff im Absinth

16 absolute alcohol
f alcool absolu
e alcohol absoluto
i alcool assoluto
d absoluter Alkohol

17 absolute electrode potential
f tension absolue d'une électrode
e tensión absoluta de un electrodo
i potenziale assoluto di elettrodo
d absolutes Elektrodenpotential

18 absolute pressure
f pression absolue
e presión absoluta
i pressione assoluta
d absoluter Druck

19 absolute scale
f échelle absolue
e escala absoluta

i scala assoluta
d absolute Skala

20 absolute temperature
f température absolue
e temperatura absoluta
i temperatura assoluta
d absolute Temperatur

21 absolute zero
f zéro absolu
e cero absoluto
i zero assoluto
d absolute Null

22 absorbability
f absorbabilité
e absorbibilidad
i assorbabilità
d Absorbierbarkeit; Aufsaugfähigkeit

23 absorbent
f absorbant
e absorbente; absorbedor
i assorbente; colonna di assorbimento
d Absorbens; Absorber

24 absorber
f absorbeur
e absorbedor
i assorbitore
d Absorber; Absorptionsapparat

25 absorptance
f capacité d'absorption
e absorptancia
i fattore di assorbimento; capacità di assorbimento
d Absorptionsvermögen

26 absorptiometer
f absorptiomètre
e absorciómetro
i assorbimetro
d Absorptiometer; Absorptionsmesser

27 absorption
f absorption
e absorción
i assorbimento
d Absorption; Aufnahme (bei Geräusch)

28 absorption analysis
f analyse par absorption
e análisis por absorción
i analisi per assorbimento
d Absorptionsanalyse

29 absorption band of a scintillating material
f bande d'absorption d'une matière scintillante
e banda de absorción de un material de centelleo
i banda d'assorbimento d'un materiale scintillante
d Absorptionsbande eines szintillierenden Materials

30 absorption cell
f cellule d'absorption
e cuba de absorción
i cellula di assorbimento
d Absorptionszelle

31 absorption curve
f courbe d'absorption
e curva de absorción
i curva di assorbimento
d Absorptionskurve

32 absorption extraction
f extraction par absorption
e extracción por absorción
i estrazione per assorbimento
d Absorptionsextraktion

33 absorption factor
f facteur d'absorption
e factor de absorción
i fattore di assorbimento
d Absorptionsfaktor

34 absorption hygrometer
f hygromètre d'absorption
e higrómetro de absorción
i igrometro di assorbimento
d Absorptionshygrometer

35 absorption oils
f huiles d'absorption
e aceites absorbentes
i oli di assorbimento; oli di lavaggio
d Absorptionsöle

36 absorption pipette
f pipette d'absorption
e pipeta de absorción
i pipetta di assorbimento
d Absorptionspipette

37 absorption spectrophotometry
f spectrophotométrie d'absorption
e espectrofotometría de absorción
i spettrofotometria di assorbimento
d Absorptionsspektrophotometrie

38 absorption tower
f tour d'absorption; colonne d'absorption
e torre de absorción
i colonna di assorbimento
d Absorptionsturm

39 absorptive power
f pouvoir absorbant
e poder absorbente
i potere assorbente
d Absorptionsvermögen

40 abundance ratio
f richesse relative en isotopes
e abundancia relativa isotópica
i abbondanza relativa isotopica
d Isotopenverhältnis

41 acacatechin
f acacatechine
e acacatechina
i acacatechina
d Akakatekin

42 acacetin
f acacétine
e acacetina
i acacetina
d Akazetin

43 acaciin
f acaciine
e acaciina
i acaciina
d Akaziin

44 acaricide
f acaricide
e acaricida
i acaricida
d Acarizid

45 accelerated filtration
f filtration accélérée
e filtración acelerada
i filtrazione accelerata
d beschleunigte Filtrierung

46 accelerator
f accélérateur
e acelerante
i accelerante
d Beschleuniger

47 acclimatization
f accoutumance
e aclimatización
i acclimatazione
d Akklimatisierung

48 acenaphthene
f acénaphtène
e acenafteno
i acenaftene
d Acenaphthen

49 acenaphthenequinone
f acénaphtène-quinone
e acenaftenoquinona
i acenaftenechinone
d Acenaphthenchinon

50 acetal
f acétal; diéthylacétate
e acetal
i acetale; azina acetaldeide
d Acetal

51 acetaldehydase
f acétaldéhydase
e acetaldehidasa
i acetaldeidasi
d Acetaldehydase

52 acetaldehyde; acetic aldehyde
f acétaldéhyde; aldéhyde acétique
e acetaldehído; aldehído acético
i acetaldeide; aldeide acetica
d Acetaldehyd

53 acetamide
f acétamide
e acetamida
i acetammide
d Acetamid

54 acetamidine
f acétamidine
e acetamidina
i acetamidina
d Acetamidin

55 acetanilide
f acétanilide
e acetanilida
i acetanilida
d Acetanilid

56 acetate
f acétate
e acetato
i acetato
d Acetat; Azetat

57 acetate film
f pellicule d'acétocellulose
e película de acetato celulósico
i pellicola all'acetato di cellulosa
d Celluloseacetatfilm

58 acetic acid
 f acide acétique
 e ácido acético
 i acido acetico
 d Essigsäure

* **acetic aldehyde** → 52

59 acetic anhydride; acetic oxide
 f anhydride acétique
 e anhídrido acético
 i anidride acetica
 d Essigsäureanhydrid

60 acetic bacteria
 f acétobacter; bactéries acétiques
 e bacterias acéticas
 i batteri acetici
 d Acetobakterien

61 acetic ether
 f éther acétique
 e éter acético
 i etere acetico
 d Essigäther

62 acetic fermentation
 f fermentation acétique
 e fermentación acética
 i fermentazione acetica
 d Essigsäuregärung

* **acetic oxide** → 59

63 acetifier
 f acétifiant
 e acetificante
 i acetificante
 d Schnellsäurer

64 acetin
 f acétine
 e acetina
 i acetina
 d Acetin

65 acetoacetanilide
 f acéto-acétanilide
 e acetoacetanilida
 i acetoacetoanilide
 d Acetoacetanilid

66 aceto-acetic ether
 f éther acéto-acétique
 e éter acetoacético
 i etere acetoacetico
 d Acetessigäther

67 acetobacter
 f acétobacter
 e acetobacteria
 i acetobacter
 d Acetobakterie

68 acetobromal; diethylbromoacetamine
 f acétobromal
 e acetobromal
 i dietilbromoacetamina
 d Diäthylbromacetamin

69 acetobromide; n-bromoacetamide
 f acéto-bromure
 e acetobromuro; n-bromoacetamida
 i n-bromoacetammide
 d Acetobromid

70 acetoin
 f acétoïne
 e acetoína
 i acetoino
 d Acetoin

71 acetol
 f acétol
 e acetolo
 i acetolo
 d Acetol

72 acetolysis
 f acétolyse
 e acetolisis
 i acetolisi
 d Acetolyse

73 acetometer
 f acétomètre
 e acetómetro
 i acetometro
 d Acetometer

74 acetonaphthone
 f acétonaphtone
 e acetonaftona
 i acetonaftone
 d Acetonaphthon

75 acetone
 f acétone
 e acetona
 i acetone
 d Aceton

76 acetone bodies
 f corps cétoniques
 e cuerpos acétonicos
 i corpi acetonici; corpi chetonici
 d Acetonkörper; Ketonkörper

77 acetone oils
f huiles d'acétone
e aceites de acetona
i oli di acetone
d Acetonöle

78 acetophenone
f acétophénone
e acetofenona
i acetofenone
d Acetophenon

79 acetoxime
f acétoxime
e acetoxima
i acetossima
d Acetoxim

80 acetoxyl group
f groupe acétoxyle
e grupo acetoxilo
i gruppo acetossile
d Acetoxylgruppe

81 acetylacetone
f acétylacétone
e acetilacetona
i acetilacetone
d Acetylaceton

82 acetylation
f acétylation
e acetilación
i acetilazione
d Acetylierung

83 acetylator
f acétylateur
e acetilador
i acetilatore
d Acetylierungsapparat

84 acetylbenzoyl peroxide
f peroxyde d'acétylbenzoyle
e peróxido de acetilbenzoílo
i perossido di acetilbenzoile
d Acetylbenzoylperoxyd

85 acetyl bromide
f bromure d'acétyle
e bromuro de acetilo
i bromuro di acetile
d Acetylbromid

86 acetylcellulose
f acétocellulose
e acetilcelulosa
i acetato di cellulosa
d Acetylcellulose

87 acetyl chloride
f chlorure d'acétyle
e cloruro de acetilo
i cloruro di acetile
d Acetylchlorid

88 acetylcholine
f acétylcholine
e acetilcolina
i acetilcolina
d Acetylcholin; Vagusstoff

89 acetylene
f acétylène
e acetileno
i acetilene
d Acetylen

90 acetylene black
f noir d'acétylène
e negro de acetileno
i nero di acetilene
d Acetylenruß

91 acetylene generator
f générateur d'acétylène
e generador de acetileno
i generatore di acetilene
d Acetylenentwickler; Acetylenerzeuger

92 acetylene tetrabromide
f tétrabromure d'acétylène
e tetrabromuro de acetileno
i tetrabromuro di acetile
d Acetylentetrabromid

93 acetyl group
f groupe acétyle
e grupo acetilo
i gruppo acetile
d Acetylgruppe

94 acetyl iodide
i iodure d'acétyle
e yoduro de acetilo
i ioduro di acetile
d Acetyljodid

95 acetylisoeugenol
f acétylisoeugénol
e acetilisoeugenol
i acetilisoeugenolo
d Acetylisoeugenol

96 acetylphenylhydrazine
f acétylphénylhydrazine
e acetilfenilhidracina
i fenilidrazina di acetile
d Acetylphenylhydrazin

97 acetylsalicylic acid
f acide acétylsalicylique
e ácido acetilsalicílico
i acido acetilsalicilico
d Acetylsalicylsäure

98 acetyl value
f indice d'acétyle
e índice de acetilo
i numero di acetile
d Acetylzahl

99 achrodextrine
f achrodextrine
e acrodextrina
i acrodestrina
d Akrodextrin

100 achromatic
f achromatique
e acromático
i acromatico
d achromatisch

101 achromatic figure
f fuseau achromatique
e huso acromático
i fuso acromatico
d achromatische Spindel; achromatische
   Figur

102 achromia
f achromie
e acromia
i acromia
d Farblosigkeit

103 acid
f acide
e ácido
i acido
d Säure

104 acid amides
f amides d'acides
e amidas de ácidos
i ammidi acidi
d Säureamide

105 acid azides
f azides acides
e azidas de ácidos
i azoturi acidi
d Säureazide

106 acid chlorides
f chlorures d'acides
e cloruros de ácidos
i acicloruri
d Säurechloride

107 acid dipping
f décapage au bain acidulé
e inmersión en ácido
i decapaggio acido
d Eintauchen in Säure

108 acid dyes
f colorants acides
e colorantes ácidos
i tinte acide
d Säurefarbstoffe

109 acid egg; acid elevator
f monte-acides
e monta-ácidos; bomba para trasvase de
   ácidos
i montaliquidi
d Druckbehälter, Druckbirne

110 acid esters
f esters acides
e ésteres ácidos
i esteri acidi
d Säureester

111 acid-fast
f résistant aux acides
e ácidorresistente
i resistente all'acido
d säurebeständig; säurefest

112 acid fixing bath
f fixage acide
e baño de fijador ácido
i bagno di fissazione acida
d Säurefixierbad

113 acid heat test
f essai de rechauffement par acides
e prueba acidotérmica
i prova di riscaldamento con acido
d Säurehitzungsprobe

114 acid hydrazides
f hydrazides acides
e hidracidas ácidas
i idrazidi acide
d Säurehydrazide

115 acidic
f acidifiant
e acidificante
i acidificante
d säurebildend

116 acidiferous
f acidifère
e acidifero
i acidifero
d säurehaltig

117 acidifier
f acidifiant
e acidificante
i acidificante
d Säurebildner

118 acidify v
f acidifier
e acidificar
i acidificare
d ansäuern

" **acidimeter — 125**

119 acidimetry
f acidimétrie
e acidimetria
i acidimetria
d Acidimetrie; Säuremessung

120 acidiphile
f acidophile
e acidófilo
i acidofilo
d Azidophil

121 acidity
f acidité
e acidez
i acidità
d Säure

122 acidity coefficient
f coefficient d'acidité
e coeficiente de acidez
i coefficiente di acidità
d Säurekoeffizient

123 acidize v
f acidifier; traiter à l'acide
e acidificar; cargar con ácido
i acidificare
d säuern

124 acidolysis
f acidolyse
e acidólisis
i idrolisi acida
d Acidolyse

125 acidometer; acidimeter
f acidimètre
e acidímetro; pesa-ácidos

i acidometro
d Säuremesser

126 acidophilic
f acidophile
e acidófilo
i acidofilo
d azidophil

127 acid polishing
f polissage par l'attaque à l'acide
e pulimento al ácido
i lucidatura all'acido
d Säurepolieren

128 acid radical
f radical acide
e radical ácido
i radicale acido
d Säureradikal

129 acid refractory
f réfractaire aux acides
e refractario ácido
i refrattario all'acido
d säurefeste Auskleidung

130 acid room
f dépôt des acides
e depósito de los ácidos
i deposito degli acidi
d Säurenraum

131 acid salts
f sels acides
e sales ácidas
i sali acidi
d saure Salze

132 acid sludge
f boues acides
e fangos ácidos
i fango acido
d Sauerschlamm

133 acid solution
f solution acide
e solución ácida
i soluzione acida
d Säurelösung

134 acid spray
f pulvérisation d'acide
e pulverización de ácido
i spruzzo acido
d Säuresprühen

135 acidulate v
f aciduler

e acidular
i acidulare
d ansäuern

136 acid value
f indice d'acide; valeur de l'acidité
e índice de acidez
i numero di acidità
d Säurezahl

137 aconitic acid
f acide aconitique
e ácido aconítico
i acido aconítico
d Akonitinsäure

138 aconitine
f aconitine
e aconitina
i aconitina
d Akonitin

139 acridine
f acridine
e acridina
i acridina
d Acridin

140 acridine dye
f colorant acridinique
e colorante acrídico
i colorante acridico
d Acridinfarbstoff

141 acridone
f acridone
e acridona
i acridone
d Acridon

142 acriflavine
f acriflavine
e acriflavina
i acriflavina
d Acriflavin

143 acrolein; acrylaldehyde; propenal
f acroléine
e acroleína
i acroleina
d Acrolein

144 acrylamide
f acrylamide
e acrilamida
i acrilammide
d Acrylamid

* **acrylate resins** → 147

145 acrylic acid
f acide acrylique
e ácido acrílico
i acido acrilico
d Acrylsäure

146 acrylic fibre
f fibre acrylique
e fibra acrílica
i fibra acrilica
d Acrylfaser

147 acrylic resins; acrylate resins
f résines acryliques
e resinas acrílicas
i resine acriliche
d Acrylharze

148 acrylic rubber
f caoutchouc acrylique
e caucho acrílico
i gomma acrilica
d Acrylkautschuk

149 acrylonitrile
f acrylonitrile; nitrile acrylique
e acrilonitrilo
i acrilonitrile
d Acrylnitril

150 acrylonitrile-butadiene-styrene resins
f résines acrylonitrile-butadiène-styrène
e resinas de acrilonitrilo-butadieno-estireno
i resine di acrilonitrile-butadiene-stirene
d Acrylnitril-Butadien-Styrolharze

* **ACTH** → 200

151 actinic radiation
f radiation actinique
e radiación actínica
i radiazione attinica
d aktinische Strahlung

152 actinides; actinoids
f actinides
e actinidos
i attinidi
d Aktiniden

153 actinium
f actinium
e actinio
i attinio
d Actinium

154 actinium series; actinium family
f famille de l'actinium

e serie del actino
i serie dell'attinio
d Aktiniumreihe

155 actino-chemistry
f chimie actinique
e actinoquímica
i attinochimica
d Aktinochemie; Strahlenchemie

* **actinoids** → **152**

156 actinolite
f actinolite
e actinolita
i actinolite
d Actinolith; Strahleisen

157 actinology
f actinologie
e actinología
i attinologia
d Aktinologie

158 actinometer
f actinomètre
e actinómetro
i attinometro
d Aktinometer

159 actinon
f actinon
e actinón
i attinon
d Aktinon

160 activated alumina
f alumine activée
e alúmina activada
i allumina attivata
d aktivierte Tonerde

161 activated carbon; active carbon
f charbon actif
e carbón activado
i carbone attivo
d Aktivkohle

162 activated molecule
f molécule activée
e molécula activada
i molecola attivata
d angeregtes Molekül

163 activated water
f eau activée
e agua activada
i acqua attivata
d aktiviertes Wasser

164 activation analysis
f analyse par activation
e análisis por activación
i analisi per attivazione
d Aktivierungsanalyse

165 activator
f activant; agent d'activation
e activador
i attivatore
d Aktivator; Aktivierungsmittel

* **active carbon** → **161**

166 active element
f élément actif; élément radioactif
e elemento activo; elemento radioactivo
i elemento attivo; elemento radioattivo
d aktives Element; radioaktives Element

167 active hydrogen
f hydrogène actif
e hidrógeno activo
i idrogeno attivo
d aktiver Wasserstoff

168 active mass
f masse active
e masa activa
i massa attiva
d aktive Masse

169 active material
f matière active
e materia activa
i materiale attivo
d aktive Masse; wirksame Masse

170 active nitrogen
f azote actif
e nitrógeno activo
i azoto attivo
d aktiver Stickstoff

171 active sampling equipment
f installation d'échantillonnage actif
e equipo de muestreo activo
i impianto di prelievo di campioni attivi
d Einrichtung zur aktiven Probennahme

172 activity
f activité
e actividad
i attività
d Aktivität

173 activity coefficient
f coefficient d'activité
e coeficiente de actividad

i coefficiente d'attività
d Aktivitätskoeffizient

174 activity spectrum
f spectre d'activité
e espectro de actividad
i spettro di attività
d Aktivitätsspektrum

175 acylation
f acylation
e acilación
i acilazione
d Einführung eines Säureradikals

176 acyl group
f groupe acyle
e grupo acilo
i gruppo acile
d Acylgruppe

177 acyloin
f acyloïne
e aciloina
i aciloina
d Acyloin

178 acyloxy-
Prefix relating to radicals derived from
the removal of hydrogen from oxygen in
organic acids.
f acyloxy-
e aciloxi-
i acilossi-
d acyloxy-

179 adaptation
f adaptation
e adaptación
i adattamento
d Anpassung; Adaptation

180 adaptive modification
f modification d'adaptation
e modificación adaptiva
i modificazione adattiva
d adaptive Modifikation

181 addition agent
f agent d'addition
e agente de adición
i agente di addizione
d Zusatz

182 adenine
f adénine
e adenina
i adenina
d Adenin

183 adenochrome
f adénochrome
e adenocromo
i adenocromo
d Adenochrom

184 adenosine
f adénosine
e adenosina
i adenosina
d Adenosin; Adeninribose

185 adenosine phosphate
f phosphate d'adénosine
e fosfato de adenosina; adenosinfosfato
i acido adenofosforico; adenofosfato
d Adenosinphosphat;
Adenosinphosphorsäure

186 adenosine triphosphate
f triphosphate d'adénosine
e trifosfato de adenosina
i trifosfato di adenosina
d Adenosintriphosphat

187 adenylic acid
f acide adénylique
e ácido adenílico
i acido adenilico
d Adenylsäure

188 adiabatic
f adiabatique
e adiabático
i adiabatico
d adiabatisch

189 adiactinic
f adiactinique
e adiactínico
i adiattinico
d aktinisch undurchlässig

190 adion
f adion
e adión
i adione
d Adion

191 adipic acid
f acide adipique
e ácido adípico
i acido adipico
d Adipinsäure

192 adipocere
f adipocère
e adipocira

i adipocera
d Adipozere; Leichenwachs

193 adipoid
f lipide; corps gras
e lípido
i lipide
d Lipid; Fettkörper

194 adiponitrile
f adiponitrile
e adiponitrilo
i adiponitrile
d Adiponitril

195 adjective dye
f colorant adjectif
e colorante adjetivo
i colorante aggettivo
d adjektiver Farbstoff

196 admixture
f mélange additionnel
e agregado a la mezcla
i aggiunta in mescolatura
d Beimischung

* **adrenalin** → 3076

197 adrenergic
f adrénergique
e adrenérgico
i adrenergico
d adrenergisch

198 adrenochrome
f adrénochrome
e adrenocromo
i adrenocromo
d Adrenochrom

199 adrenocortical
f corticosurrénal
e adrenocortical
i corticosurrenale
d adrenokortikal

200 adrenocorticotropic hormone; ACTH
f hormone adrénocorticotropique
e hormona adrenocorticotrópica
i ormone adrenocorticotropo; ormone
   adrenotropo
d adrenokortikotropisches Hormon

201 adrenosterone
f adrénostérone
e adrenosterona
i adrenosterone
d Adrenosteron

202 adsorbability
f adsorbabilité
e adsorbabilidad
i adsorbabilità
d Adsorptionsfähigkeit

203 adsorbate
f substance adsorbée
e adsorbido
i adsorbato; sostanza adsorbita
d adsorbierte Substanz

204 adsorbent
f agent adsorbant
e adsorbente
i adsorbente
d Adsorptionsmittel

205 adsorption
f adsorption
e adsorción
i adsorbimento
d Adsorption

206 adsorption cycle
f cycle d'adsorption
e ciclo de adsórcion
i ciclo di adsorbimento
d Adsorptionsvorgang

207 adsorption indicator
f indicateur d'adsorption
e indicador de adsorción
i indicatore di adsorbimento
d Adsorptionsindikator

208 adsorption stage
f étage d'adsorption
e etapa de adsorción
i stadio d'adsorbimento
d Adsorptionsstufe

209 adulteration
f falsification
e adulteración
i sofisticazione; adulterazione
d Verfälschung

210 adventitious
f adventice
e adventicio
i avventizio
d zusätzlich

211 aerate *v*
f aérer; battre à mousse
e ventilar; batir a espuma
i areare; montare a schiuma
d belüften; Schaum schlagen

212 **aerobe**
 f aérobie
 e célula aeróbica; organismo aerobio
 i cella aerobica; organismo aerobico
 d Aerobe; Aerobier

213 **aerobic fermentation**
 f fermentation aérobie
 e fermentación aeróbica
 i fermentazione aerobica
 d aerobe Gärung

214 **aerobiology**
 f aérobiologie
 e aerobiología
 i aerobiologia
 d Aerobiologie

215 **aerobioscope**
 f aérobioscope
 e aerobioscopio
 i aerobioscopio
 d Aerobioskop

216 **aerogenic**
 f aérogène
 e aerógeno
 i aerogeno
 d gasbildend

217 **aeroscope**
 f aéroscope
 e aeroscopio
 i aeroscopio
 d Aeroskop

218 **aerosol**
 f aérosol
 e aerosol
 i aerosol
 d Aerosol

219 **aerosol propellants**
 f propulseurs d'aérosol
 e propulsantes de aerosol
 i propellenti per aerosol
 d Aerosoltreibstoffe

220 **aerothermodynamics**
 f aérothermodynamique
 e aerotermodinámica
 i aerotermodinamica
 d Aerothermodynamik

221 **affination**
 f affinage
 e refinación
 i affinazione
 d Affination

222 **after-bake**
 f après cuisson; étuvage
 e post-cocer
 i ricottura
 d Nachhärten

223 **after-blow**
 f sursoufflage
 e sobresoplado; continuación del soplado
   para eliminar el fósforo
 i postsoffiatura
 d Nachblasen

224 **after-chrome** *v*
 f postchromer
 e postchromar
 i postchromare
 d nachverchromen

225 **after-cooler**
 f post-refroidisseur
 e postrefrigerador; postrefrigerante
 i postrefrigeratore
 d Nachkühler

226 **after-tack**
 f persistance d'adhérence
 e persistencia de adherencia
 i persistenza della colla
 d Klebevermögen

227 **agar-agar**
 f agar-agar
 e agar-agar
 i agar-agar
 d Agar-agar

228 **agar slant**
 f gélose inclinée
 e agar inclinado
 i inclinato agarico
 d Agarschrägfläche

229 **agate mortar**
 f mortier d'agate
 e mortero de ágata
 i mortaio d'agata
 d Achatmörser

230 **ageing**
 f vieillissement
 e envejecimiento
 i invecchiamento
 d Alterung

231 **ageing oven**
 f étuve de vieillissement
 e estufa de envejecimiento

i stufa di invecchiamento
d Alterungsschrank

232 ageing resistance
f résistance au vieillissement
e resistencia al envejecimiento
i resistenza all'invecchiamento
d Alterungsbeständigkeit

233 ageing test
f essai de vieillissement
e ensayo de envejecimiento
i prova di invecchiamento
d Alterungsprüfung

234 agenesis
f agénèse
e agénesis
i agenesl
d Agenese

235 age resister
f antivieillisseur
e agente resistente al envejecimiento
i antinvecchiante
d Alterungsschutzmittel

236 agglomerate
f agglomérat
e aglomerado
i agglomerato
d Agglomerat

237 agglutination
f agglutination
e aglutinación
i agglutinazione
d Agglutination

238 agglutinin
f agglutinine
e aglutinina
i agglutinina
d Agglutinin

239 agglutlnogen
f agglutinogène
e aglutinógeno
i agglutinogeno
d Agglutinogen

240 aggregate
f agrégat
e agregado
i aggregato
d Aggregat

241 aggressin
f agressine

e agresina
i aggressina
d Aggressin; Angriffsstoff

242 aggressinogen
f agent d'agression
e agresinógeno
i fattore stressante
d Aggressinbildner

243 agitation
f agitation
e agitación
i agitazione
d Rühren

244 agitation pan
f bac à agitation
e recipiente de agitación
i marmitta dell'agitatore
d Rührpfanne

245 agitator
f agitateur
e agitador mecánico
i agitatore
d Rührwerk

246 aglycosuric
f aglycosurique
e aglicosúrico
i aglicosurico
d glucosefrei; zuckerfrei

247 agonisis; certation
f certation
e agonisis
i certatio
d Zertation

248 agranulocyte
f agranulocyte
e agranulocito
i agranulocite
d Agranulozyt

249 air blowing
f soufflage
e sopladura con aira
i getto d'aria
d Windfrische

250 air brush
f pistolet à peinture
e pulverizador por aire
i spruzzatore per vernice
d Spritzpistole; Farbzerstäubungsbürste

* **air cell** → 253

251 air circulation
 f circulation de l'air
 e circulación de aire
 i circolazione d'aria
 d Luftzirkulation

252 air cooling
 f refroidissement par air
 e refrigeración por aire
 i raffreddamento ad aria
 d Luftkühlung

253 air-depolarized cell; air cell *(US)*
 f pile à dépolarisation par l'air
 e pila de despolarización por aire
 i pila depolarizzata ad aria
 d Luftsauerstoffelement

254 air-drying lacquer
 f vernis séchant à l'air
 e barniz secante al aire
 i lacca che si asciuga all'aria
 d lufttrocknender Lack

255 air-elutriation
 f élution pneumatique; séparation par courant d'air
 e neumoelutriación; elutriación por corriente de aire
 i elutriazione ad aria
 d Auswaschung mittels Luftstrom

256 air equivalent
 f équivalent-air; équivalent en air
 e equivalente en aire
 i equivalente in aria
 d Luftäquivalent

257 air factor
 f facteur d'air
 e coeficiente de aire
 i coefficiente d'aria
 d Luftfaktor

* **air-humidity indicator → 4254**

258 air-lift; air-lift pump
 f pompe élévatoire
 e bomba elevadora de líquidos por aire
 i pompa a compressione d'aria
 d Druckluftheber

259 air lock
 f labyrinthe d'entrée
 e laberinto de entrada
 i labirinto d'ingresso
 d Eintrittsschleuse

260 air oven
 f étuve à l'air chaud
 e estufa de aire caliente
 i forno ad aria calda
 d Heißluftofen

261 air oven ageing
 f essai de vieillissement à l'air chaud
 e ensayo de envejecimiento con aire caliente
 i prova d'invecchiamento con aria calda
 d Heißluft-Alterungsprobe

262 air-pump
 f pompe à air
 e bomba de aire
 i pompa pneumatica
 d Luftpumpe

263 air receiver
 f réservoir d'air
 e depósito de aire comprimido
 i serbatoio d'aria compressa; polomone d'aria compressa
 d Luftkessel

264 air separation
 f séparation de l'air
 e separación del aire
 i separazione dall'aria
 d Luftabscheidung

* **ajacine → 2389**

265 akaryocyte
 f érythrocyte sans noyau
 e acariocito
 i emazia nucleopriva
 d kernlose Zelle; rote Zelle

266 alabandite
 f alabandite; alabandine
 e alabandita
 i alabandina
 d Alabandin; Manganblende

267 alabaster
 f albâtre
 e alabastro
 i alabastro
 d Alabaster

268 albedo
 f albédo
 e albedo
 i albedine
 d Albedo

269 Alberger salt process
f procédé de production de sel pur
e proceso para la producción de sal pura
i processo per la produzione di sale puro
d Alberger Salzverfahren

270 albertite
f albertite
e albertita
i albertite
d Albertit

271 albumen
f albumen; albumine de l'œuf
e albumen; clara del uevo
i albume; bianco dell'uovo
d Albumin; Eiweiß; Eiklar

272 albumin
f albumine
e albúmina
i albumina
d Albumin

273 albuminates
f albuminates
c albuminados
i albuminati
d Albuminate

274 albuminoids; scleroproteins
f albuminoïdes
e albuminoides
i albuminoidi
d Albuminoide

275 albumoses
f albumoses
e albumosas
i albumosi
d Albumosen

276 alcohol
f alcool
e alcohol
i alcole; alcool
d Alkohol

277 alcoholature
f alcoolature
e alcoholatura
i alcoolatura
d Alkoholatur

278 alcohol content
f teneur d'alcool
e centenido de alcohol
i contenuto di alcool
d Alkoholgehalt

279 alcoholic fermentation
f fermentation alcoolique
e fermentación alcohólica
i fermentazione alcoolica
d Alkoholgärung

280 alcoholic potash
f potasse alcoolique
e potasa alcohólica
i potassa alcoolica
d Alkoholpottasche

281 alcoholization
f alcoolisation
e alcoholización
i alcoolizzazione
d Alkoholisierung

282 alcoholometer
f alcoolomètre
e alcoholímetro
i alcoolometro
d Alkoholmeter

283 alcoholometry
f alcoolométrie
e alcoholimetría
i alcoolometria
d Alkoholometrie

284 alcosol
f alcoosol
e alcosol
i alcosol
d Alkohollösung

285 aldehydase
f aldéhydase
e aldehidasa
i aldeidasi
d Aldehydase

286 aldehyde acids
f acides-aldéhydes
e ácidos aldehídicos
i acidi aldeidici
d Aldehydsäuren

287 aldehyde ammonia
f aldéhydate d'ammoniaque
e aldehidato amónico
i aldeide ammoniaca
d Aldehydammoniak

288 aldehyde resins
f résines aldéhydiques
e resinas aldehídicas
i resine aldeidiche
d Aldehydharze

289 aldehydes
f aldéhydes
e aldehídos
i aldeidi
d Aldehyde

290 aldimines
f aldimines
e aldiminas
i aldimine
d Aldimine

291 aldohexoses
f aldohexoses
e aldohexosas
i aldoesosi
d Aldohexosen

292 aldoketenes
f aldocétènes
e aldocetenas
i aldochetoni
d Aldoketene

293 aldol
f aldol
e aldol
i aldolo
d Aldol

294 aldolase
f aldolase
e aldolasa
i aldolasi
d Aldolase

295 aldoses
f aldoses
e aldosas
i aldosi
d Aldosen

296 aldosterone
f aldostérone
e aldosterona
i aldosterone
d Aldosteron

297 aldoximes
f aldoximes
e aldoximas
i aldossime
d Aldoxime

298 aldrin
f aldrine
e aldrina
i aldrina
d Aldrin

299 aletocyte
f cellule migratrice
e célula emigrante
i cellula migrante
d Wanderzelle

300 aleurone
f aleurone
e aleurona
i aleurone
d Klebermehl

301 alexin
f alexine
e alexina
i alessina
d Alexin

302 Alfin catalyst
f catalyseur Alfin
e catalizador de Alfin
i catalizzatore di Alfin
d Alfin-Katalysator

303 alga
f algue
e alga
i alga
d Alge

304 algesiogenic
f algogène
e algesiógeno
i algogeno
d schmerzauslösend

305 alginates
f alginates
e alginatos
i alginati
d Alginate

306 alginic acid
f acide alginique
e ácido algínico
i acido alginico
d Alginsäure

307 algodonite
f algodonite
e algodonita
i algodonite
d Algodonit

308 alible
f alibile
e asimilable
i assimilabile
d assimilierbar

309 alicyclic
f alicyclique
e alicíclico
i aliciclico
d alizyklisch

310 alicyclic compounds
f composés alicycliques
e compuestos alicíclicos
i composti aliciclici
d alizyklische Verbindungen

311 alicyclic hydrogenation
f hydrogénation alicyclique
e hidrogenación alicíclica
i idrogenazione aliciclica
d alizyklische Hydrierung

312 aliesterase
f aliestérase
e aliesterasa
i aliesterasi
d Aliesterase

313 aliphatic
f aliphatique
e alifático
i alifatico
d aliphatisch

314 aliphatic acid
f acide aliphatique; acide gras
e ácido alifático
i acido alifatico
d Fettsäure

315 aliphatic compounds
f composés aliphatiques
e compuestos alifáticos
i composti alifatici
d aliphatische Verbindungen;
   Fettverbindungen

316 alite
f alite
e alita
i alite
d Alit

317 alizarin
f alizarine
e alizarina
i alizarine; alizarone
d Alizarin

318 alizarin black
f noir d'alizarine
e negro de alizarina

i nero di alizarina
d Alizarinschwarz

319 alizarin brown
f marron d'alizarin
e marrón de alizarina
i antragallolo
d Alizarinmarron

320 alkali
f alcali
e álcali
i alcali
d Alkali

321 alkali cellulose
f alcali-cellulose
e álcalicelulosa; celulosa alcalina
i alcali-cellulosa
d Alkalicellulose

322 alkali content
f teneur en alcali
e contenido en álcali
i contenuto alcalino
d Alkaligehalt

323 alkali metals
f métaux alcalins
e metales alcalinos
i metalli alcalini
d Alkalimetalle

324 alkalimeter
f alcalimètre
e alcalímetro
i alcalimetro
d Alkalimeter; Laugenmesser

325 alkalimetry
f alcalimétrie
e alcalimetría
i alcalimetria
d Laugenmessung

326 alkaline cleaning
f dégraissage alcalin
e desengrase alcalino
i sgrassaggio alcalino
d Reinigung mit alkalischen Lösungsmitteln

327 alkaline hydrolysis
f hydrolyse alcaline
e hidrólisis alcalina
i idrolisi alcalina
d Alkalihydrolyse

328 alkaline-soak cleaning
f dégraissage alcalin
e desengrase alcalino
i sgrassagio alcalino
d Reinigung mit alkalischen Lösungsmitteln

329 alkaline steeping agent
f bain alcalin
e agente de enriar alcalino
i imbibitore alcalino
d alkalisches Tränkmittel

330 alkaline storage battery
f accumulateur alcalin
e acumulador alcalino
i accumulatore alcalino
d alkalischer Akkumulator

331 alkalinity
f alcalinité; basicité
e alcalinidad; basicidad
i alcalinità; basicità
d Alkalinität; Basizität

332 alkali-refined linseed oil
f huile de lin raffinée à l'alcali
e aceite de lino refinado por proceso
alcalínico
i olio di lino raffinato con processo
alcalinico
d alkalisiertes Leinölraffinat

333 alkali reserve
f réserve alcaline
e reserva alcalina
i riserva alcalina
d Alkalireserve

334 alkalize v
f alcaliser
e alcalinizar
i alcalinizzare
d alkalisieren

335 alkaloids
f alcaloïdes
e alcaloides
i alcaloidi
d Alkaloide

336 alkalosis
f alcalose
e alcalosis
i alcalosi
d Alkalose

337 alkane
f alcane

e alcano
i alcano
d Alkan

338 alkanesulphonic acid
f acide alcanosulphonique
e ácido alcanosulfónico
i acido alcanosolfonico
d Alkansulfonsäure

339 alkyd resins; glyptal resins
f résines alkydes
e resinas alquídicas
i resine alchidi
d Alkydharze

340 alkyl
f alcoyle
e alcohilo; alkilo
i alcoile; alchile
d Alkyl

341 alkylaryl phosphate
f phosphate d'alkylaryle
e fosfato de alkilarilo
i alchilarilfosfato
d Alkylarylphosphat

342 alkylaryl sulphonates
f sulfonates d'alkylaryl
e sulfonatos de alkilarilo
i solfonati di alchilarile
d Alkylarylsulfonate

343 alkylate
f produit d'alcoylation
e alcohilato; alkilato
i alchilato
d Alkylat

344 alkylation
f alkylation; alcoylation
e alcohilación; alkilización
i alchilazione
d Alkylierung; Alkylation

345 alkylene
f alcène
e alquileno
i alchilene
d Alkylen

* alkyl sulphides → 8241

346 Allan cell
f cellule d'Allan
e célula Allán; elemento Allán
i cella elettrolitica Allan
d Allan-Zelle

347 allanite
 f allanite
 e alanita
 i allanite
 d Allanit

348 allantoin
 f allantoïne
 e alantoína
 i allantoina
 d Allantoin

349 allele
 f allèle
 e alelo
 i allelo
 d Allel

350 allelism
 f allélisme
 e alelismo
 i allelismo
 d Allelie

351 allene; propadiene
 f allene; propadiène
 e aleno
 i allene
 d Allen

352 allethrin
 f alléthrine
 e aletrina
 i alettrina
 d Allethrin

353 allobar
 f allobare
 e alobara
 i allobara
 d Allobar

354 allogamy
 f allogamie; fécondation croisée
 e alogamia; fecundación cruzada
 i allogamia
 d Allogamie; Fremdbestäubung

355 allogene
 f allogène
 e alógeno
 i allogeno
 d Allogen; rezessives Allel

356 allogenetic
 f allogénétique
 e alogenético
 i allogenetico
 d allogenetisch

357 alloiogenesis
 f alloiogénèse
 e aloiogénesis
 i alloiogenesi
 d Alloiogenesis

358 alloisomerism
 f allomérie
 e aloisomerismo
 i alloisomeria
 d Alloisomerie

359 allomeric
 f allomérique
 e alomérico
 i allomerico
 d allomerisch

360 allomorphic
 f allomorphe
 e alomorfo
 i allomorfo
 d allomorph

361 allomorphism
 f allomorphisme
 e alomorfismo
 i allomorfismo
 d Allomorphismus

362 allomorphosis
 f allomorphose
 e alomórfosis
 i allomorfosi
 d Allomorphose

363 allo-ocimene
 f allo-ocimène
 e alocimeno
 i alloocimene
 d Allo-ocimen

364 allopatric
 f allopatrique
 e alopátrico
 i allopatrico
 d allopatrisch

365 allose
 f allose
 e alosa
 i allosio
 d Allose

366 allosome
 f allosome
 e alosoma
 i allosoma
 d Allosom

367 **allosteric**
 f allostérique
 e aloestérico
 i allosterico
 d allosterisch

368 **allotrope**
 f allotrope
 e alotrópico
 i allotropo
 d allotropisch

369 **allotropy**
 f allotropie
 e alotropía
 i allotropia
 d Allotropie

370 **alloxazine**
 f alloxazine
 e aloxacina
 i allossazina
 d Alloxazin

371 **alloy**
 f alliage
 e aleación
 i lega
 d Metallegierung

372 **alloy plate**
 f dépôt d'alliage
 e depósito de aleación
 i deposito di leghe
 d Legierungsüberzug

373 **allyl alcohol**
 f alcool allylique
 e alcohol alílico
 i alcool allilico
 d Allylalkohol

374 **allyl amine**
 f allyle-amine
 e alilamina
 i allilammina
 d Allylamin

375 **allyl bromide**
 f bromure d'allyle
 e bromuro de alilo
 i bromuro di allile
 d Allylbromid

376 **allyl caproate**
 f caproate d'allyle
 e caproato de alilo
 i caproato di allile
 d Allylcaproat

377 **allyl chloride**
 f chlorure allylique
 e cloruro de alilo
 i cloruro allilico
 d Allylchlorid

378 **allyl cyanide**
 f cyanure d'allyle
 e cianuro de alilo
 i cianuro di allile
 d Allylcyanid

379 **allyl ester**
 f ester allylique
 e éster alílico
 i estere allilico
 d Allylester

380 **allylidene diacetate**
 f allylidène diacétate
 e diacetato de alilideno
 i diacetato di allilidene
 d Allylidendiacetat

381 **allyl isothiocyanate**
 f isothiocyanate d'allyle
 e isotiocianato de alilo
 i isotiocianato di allile
 d Allylisothiocyanat

382 **allyl resins**
 f résines allyliques
 e resinas alílicas
 i resine alliliche
 d Allylharze

383 **allylthiourea**
 f thiourée d'allyle
 e aliltiourea
 i tiourea allilica
 d Allylthioharnstoff

384 **almandine; almandite**
 f almandine; alamandine
 e almandina
 i almandina
 d Almandin; Karfunkel; roter Granat

385 **Almen-Nylander test**
 f essai Almen-Nylander
 e prueba Almen-Nylander
 i prova di Almen-Nylander
 d Almen-Nylander-Probe

386 **aloin**
 f aloïne
 e aloína
 i aloina
 d Aloin

387 alpha
 f alpha
 e alfa
 i alfa
 d Alpha

388 alpha-cellulose
 f alpha-cellulose
 e alfacelulosa
 i alfa-cellulosa
 d Alphacellulose

389 alpha-iron
 f fer alpha
 e hierro alfa
 i ferro alfa
 d Alpha-Eisen

390 alpha-particle binding energy
 f énergie de liaison de particules alpha
 e energía de enlace de partículas alfa
 i energia di legame di particelle alfa
 d Bindungsenergie von Alphateilchen

391 alpha-particle spectrum
 f spectre de particules alpha
 e espectro de partículas alfa
 i spettro di particelle alfa
 d Spektrum von Alphateilchen

392 alpha-radiation
 f rayonnement alpha
 e radiación alfa
 i radiazione alfa
 d Alphastrahlung

393 alpha-radiation spectrometer
 f spectromètre alpha
 e espectrómetro alfa
 i spettrometro alfa
 d Alphaspektrometer; Alphastrahl
 spektrometer

394 alphatron gauge
 f alphatron
 e alfatrón
 i alfatrone
 d Alphatron

395 alpha-uranium
 f uranium alpha
 e uranio alfa
 i uranio alfa
 d Alpha-Uran

396 alphyls
 f alphyles
 e alfilos

 i alfili
 d Alphyle

 * **Altmann's granule → 946**

397 altrose
 f altrose
 e altrosa
 i altrosio
 d Altrose

398 aludel
 f aludel
 e vasija periforme con dos cuellos
 i vaso a forma di pera per collegamenti in
 serie
 d Aludel

399 alum
 f alun
 e alumbre
 i allume
 d Alaun

400 alumina
 f alumine
 e alúmina
 i allumina
 d Tonerde

401 alumina-blanc fixe
 f blanc fixe d'alumine
 e blanco fijo de alúmina
 i bianco fisso di allumina
 d Aluminium-blanc fixe

402 alumina trihydrate
 f trihydrate d'alumine
 e trihidrato de alúmina
 i ortoidrossido alluminio; allumina
 triidrato
 d Aluminiumhydrat

403 aluminite
 f aluminite
 e aluminita
 i alluminite
 d Aluminit

404 aluminium
 f aluminium
 e aluminio
 i alluminio
 d Aluminium

405 aluminium acetate
 f acétate d'aluminium
 e acetato alumínico

i acetato di alluminio
d Aluminiumacetat; essigsaure Tonerde

406 aluminium ammonium sulphate
f sulfate double d'aluminium et
d'ammonium
e sulfato alumínico-amónico
i solfato ammonico di alluminio
d Aluminiumammoniumsulfat

407 aluminium borohydride
f borohydrure d'aluminium
e borohidruro alumínico
i boroidruro di alluminio
d Aluminiumborhydrid

408 aluminium bromide
f bromure d'aluminium
e bromuro alumínico
i bromuro di alluminio
d Aluminiumbromid

409 aluminium carbide
f carbure d'aluminium
e carburo alumínico
i carburo di alluminio
d Aluminiumcarbid

410 aluminium chloride
f chlorure d'aluminium
e cloruro alumínico
i cloruro di alluminio
d Aluminiumchlorid

411 aluminium distearate
f distéarate d'aluminium
e distearato alumínico
i distearato di alluminio
d Aluminiumdistearat

412 aluminium hydroxide
f hydroxyde d'aluminium
e hidróxido de aluminio
i idrossido di alluminio
d Aluminiumhydroxyd

413 aluminium hydroxystearate
f hydroxystéarate d'aluminium
e hidroxiestearato alumínico
i idrossistearato di alluminio
d Aluminiumhydroxystearat

414 aluminium isopropylate
f isopropylate d'aluminium
e isopropilato alumínico
i isopropilato di alluminio
d Aluminiumisopropylat

415 aluminium metaphosphate
f métaphosphate d'aluminium
e metafosfato alumínico
i metafosfato di alluminio
d Aluminiummetaphosphat

416 aluminium monostearate
f monostéarate d'aluminium
e monostearato alumínico
i monostearato di alluminio
d Aluminiummonostearat

417 aluminium nitrate
f nitrate d'aluminium
e nitrato alumínico
i nitrato di alluminio
d Aluminiumnitrat

418 aluminium oleate
f oléate d'aluminium
e aleato alumínico
i oleato di alluminio
d Aluminiumoleat

419 aluminium oxide
f oxyde d'aluminium
e óxido alumínico
i ossido di alluminio
d Aluminiumoxyd

420 aluminium palmitate
f palmitate d'aluminium
e palmitato alumínico
i palmitato di alluminio
d Aluminiumpalmitat

421 aluminium potassium sulphate
f sulfate double d'aluminium et de
potassium
e sulfato alumínico-potásico
i solfato di potassio d'aluminio
d Aluminiumkaliumsulfat

422 aluminium silicate
f silicate d'aluminium
e silicato alumínico
i silicato di alluminio
d Aluminiumsilikat

423 aluminium sodium sulphate
f sulfate double d'aluminium et de sodium
e sulfato alumínico-sódico
i solfato di sodio di alluminio
d Aluminiumnatriumsulfat

424 aluminium sulphate
f sulfate d'aluminium
e sulfato alumínico

i solfato di alluminio
d Aluminiumsulfat

425 aluminium triformate
f triformiate d'aluminium
e triformiato alumínico
i triformato d'alluminio
d Aluminiumtriformiat

426 aluminothermic process
f aluminothermie
e proceso aluminotérmico
i alluminotermia
d Aluminothermie; Thermitverfahren

427 aluminous cement
f ciment alumineux
e cemento aluminoso
i cemento alluminoso
d Schmelzzement

428 alunite
t alunite
e alunita
i alunite
d Alunit; Alaunstein

429 alunogen
f alunogène
e alunógeno
i alunogeno
d Federalaun; Haarsalz

430 amalgam
f amalgame
e amalgama
i amalgama
d Amalgam

431 amalgamation process
f amalgamation
e amalgamación
i processo di amalgamazzione
d Amalgamieren

432 amber
f ambre
e ámbar
i ambra
d Amber; Bernstein

433 ambergris
f ambre gris
e ámbar gris
i ambra grigia
d grauer Amber

434 ambient temperature
f température ambiante

e temperatura ambiente
i temperatura ambiente
d Raumtemperatur; umgebende
Temperatur

435 ameba; amoeba
f amibe
e amiba
i ameba
d Amibe

436 ameiosis
f améiose
e ameiosis
i ameiosi
d Ameiose

437 americium
t americium
e americio
i americio
d Americium

* **amethocaine hydrochloride → 8099**

438 amicron
f amicron
e amicrón
i amicrone
d Amikron

439 amictic
f amictique
e amíctico
i amittico
d amiktisch

440 amide
f amide
e amida
i ammide
d Amid

441 amidines
f amidine
e amidinas
i ammidine
d Amidine

442 amido group
f groupe amido
e grupo amido
i gruppo ammido
d Amidogruppe

443 amine
f amine
e amina

i ammina
d Amin

444 amino acids
f aminoacides; acides aminés
e aminoácidos
i amminoacidi
d Aminosäuren

* **aminocordine** → 5732

* **aminodimethyl benzene** → 8885

445 amino group
f groupe amino
e grupo amino
i gruppo ammino
d Aminogruppe

446 amino resins; amino plastics
f résines aminiques
e resinas amínicas
i resine amminiche
d Aminoharze

* **amithiozone** → 8253

447 amitosis
f amitose
e amitosis
i amitosi
d Amitose

448 amixia
f amixie; amixis
e amixia; amixis
i amissia; amixi
d Amixie; Amixis

449 ammines
f ammines
e amminas
i ammine
d Ammine

450 ammonia
f ammoniac; gaz ammoniac
e amoníaco
i ammoniaca
d Ammoniak

451 ammonia burner
f brûleur d'ammoniac
e quemador de amonio
i bruciatore per ammoniaca
d Ammoniakbrenner

452 ammoniacal latex
f latex ammonié
e látex amoniacal
i lattice ammoniacale
d ammoniakaler Latex

453 ammonia liquor; ammoniacal liquor; ammonia water *(US)*
f eau ammoniacale
e agua amoniacal
i ammoniaca
d wässerige Ammoniaklösung

* **ammonia-soda process** → 7651

454 ammonia synthesis
f synthèse de l'ammoniac
e síntesis del amoníaco
i sintesi dell'ammoniaca
d Ammoniaksynthese

* **ammonia, total** ~ → 8346

* **ammonia water** → 453

455 ammonium
f ammonium
e amonio
i ammonio
d Ammonium

456 ammonium acetate
f acétate d'ammonium
e acetato amónico
i acetato di ammonio
d Ammoniumacetat

457 ammonium alginate
f alginate d'ammonium
e alginato de amonio
i alginato di ammonio
d Ammoniumalginat

458 ammonium benzoate
f benzoate d'ammonium
e benzoato amónico
i benzoato di ammonio
d Ammoniumbenzoat

459 ammonium bifluoride
f bifluorure d'ammonium
e bifluoruro amónico
i bifluoruro di ammonio
d Ammoniumbifluorid

460 ammonium carbamate
f carbamate d'ammonium
e carbamato amónico
i carbammato di ammonio
d Ammoniumcarbamat

461 ammonium carbonate
f carbonate d'ammonium
e carbonato amónico
i carbonato di ammonio
d Ammoniumcarbonat

462 ammonium chloride; sal ammoniac
f chlorure d'ammonium; sel ammoniac
e cloruro amónico; sal amoníaco
i cloruro di ammonio
d Ammoniumchloride; Salmiak

463 ammonium chromate
f chromate d'ammonium
e cromato amónico
i cromato di ammonio
d Ammoniumchromat

464 ammonium dichromate
f bichromate d'ammonium
e dicromato amónico
i bicromato di ammonio
d Ammoniumdichromat

465 ammonium diuranate
f diuranate d'ammonium
e diuranato de amonio
i diuranato d'ammonio
d Ammoniumdiuranat

466 ammonium fluoride
f fluorure d'ammonium
e fluoruro amónico
i fluoruro di ammonio
d Ammoniumfluorid

467 ammonium hydroxide
f ammoniaque; hydroxyde d'ammonium
e hidróxido amónico
i idrossido di ammonio
d Ammoniumhydroxyd

468 ammonium laurate
f laurate d'ammonium
e laurato de amonio
i laurato di ammonio
d Ammoniumlaurat

469 ammonium linoleate
f linoléate d'ammonium
e linoleato amónico
i linoleato di ammonio
d Ammoniumlinoleat

470 ammonium metavanadate
f métavanadate d'ammonium
e metavanadato amónico
i metavadanato di ammonio
d Ammoniummetavanadat

471 ammonium molybdate
f molybdate d'ammonium
e molibdato amónico
i molibdato di ammonio
d Ammoniummolybdat

472 ammonium nitrate
f nitrate d'ammonium
e nitrato amónico
i nitrato di ammonio
d Ammoniumnitrat

473 ammonium oleate
f oléate d'ammonium
e oleato de amonio
i oleato di ammonio
d Ammoniumoleat

474 ammonium perchlorate
f perchlorate d'ammonium
e perclorato amónico
i perclorato di ammonio
d Ammoniumperchlorat

475 ammonium persulphate
f persulfate d'ammonium
e persulfato amónico
i persolfato di ammonio
d Ammoniumpersulfat

* **ammonium phosphate, dibasic ~
→ 2496**

* **ammonium phosphate,
monobasic ~ → 5565**

476 ammonium picrate
f picrate d'ammonium
e picrato amónico
i picrato di ammonio
d Ammoniumpikrat

477 ammonium stearate
f stéarate d'ammonium
e estearato amónico
i stearato di ammonio
d Ammoniumstearat

478 ammonium sulphate
f sulfate d'ammonium
e sulfato de amonio
i solfato di ammonio
d Ammoniumsulfat

479 ammonium sulphide
f sulfure d'ammonium
e sulfuro amónico
i solfuro di ammonio
d Ammoniumsulfid

480 ammonium thiocyanate
f thiocyanate d'ammonium
e tiocianato amónico
i tiocianato di ammonio
d Ammoniumthiocyanat

481 ammonium thiosulphate
f thiosulfate d'ammonium
e tiosulfato amónico
i tiosolfato di ammonio
d Ammoniumthiosulfat

482 ammonolysis
f ammonolyse
e amonólisis
i ammonolisi
d Ammonolyse

* **amoeba** → 435

483 amorphous
f amorphe
e amorfo
i amorfo
d amorph; gestaltlos

484 amphiaster
f amphiaster
e anfiaster
i amfiaster
d Amphiaster

485 amphiboles
f amphiboles
e anfíboles
i anfiboli
d Amphibole

486 amphibolite
f amphibolite
e amfibolita
i anfibolite
d Amphibolit

487 amphidiploidy
f amphidiploïdie
e anfidiploidia; doble diploidia
i amfidiploidismo; anfidiploidismo
d Amphidiploidie

488 amphigony
f amphigonie
e anfigonia
i anfigenia; anfigonia
d Amphigonie

489 amphimixia
f amphimixie
e anfimixis

i amfimissi; anfimixi
d Amphimixis; Amphimixie

490 amphogeny
f amphogénie
e anfogenia
i amfogenia; anfogenia
d Amphogenie

491 ampholyte
f ampholyte
e anfólito; electrólito anfotérico
i anfolito
d Ampholyte

492 ampholytoid
f ampholytoïde
e anfolitoide
i anfolitoide
d Ampholytoid

493 amphoteric
f amphotère
e anfótero
i anfoterico
d amphoter

494 amphoteric ion
f ion amphotérique
e ión anfotérico
i ione amfoterico
d Zwitterion

495 amyl
f amyle
e amilo
i amile
d Amyl

496 amyl acetate; pear oil
f acétate d'amyle; essence de poires
e acetato de amilo; esencia de peras
i acetato di amile
d Amylacetat; Birnenöl

497 amyl alcohol
f alcool amylique
e alcohol amílico
i alcool amilico
d Amylalkohol

498 amylase
f amylase
e amilasa
i amilasi
d Amylase

499 amyl chloride mixture
f mélange de chlorures d'amyle

e mezcla de cloruros de amilo
i miscela di cloruro di amile
d Amylchloridmischung

500 amyl citrate
f citrate d'amyle
e citrato de amilo
i citrato di amile
d Amylcitrat

501 amylene; pentene
f amylène
e amileno
i amilene
d Amylen

502 amyl ether
f éther amylique
e éter amílico
i etere amilico
d Amyläther

503 amyl formate
f formiate d'amyle
e formiato de amilo
i formiato di amile
d Amylformiat; Ameisensäureamylester

504 amyl group
f groupe amyle
e grupo amilo
i gruppo amilico
d Amylgruppe

505 amyl lactate
f lactate d'amyle
e lactato de amilo
i lattato di amile
d Amyllactat

506 amyl laurate
f laurate d'amyle
e laurato de amilo
i laurato di amile
d Amyllaurat

507 amyl mercaptan
f mercaptan amylique
e amilmercaptano
i mercaptano di amile
d Amylmercaptan

508 amyl naphthalene
f naphtalène d'amyle
e amilnaftaleno
i amilnaftalene
d Amylnaphthalin

509 amyl nitrite
f nitrite d'amyle
e nitrito de amilo
i nitrito di amile
d Amylnitrit

510 amylo fermentation
f fermentation amylique
e fermentación amílica
i fermentazione amilica
d Amylalkoholgärung

511 amyloid
f amyloïde
e amiloide
i amiloide
d Amyloid

512 amyloidosis
f amyloïdose
e amiloidosis
i amiloidosi
d Amyloidose

513 amyl oleate
f oléate d'amyle
e oleato de amilo
i oleato di amile
d Amyloleat

514 amylolysis
f amylolyse
e amilólisis
i amilolisi
d Stärkespaltung

515 amylopsin
f amylopsine
e amilopsina
i amilopsina
d Amylopsin

516 amylose
f amylose
e amilosa
i amilosio
d Amylose

517 amyl oxalate
f oxalate d'amyle
e oxalato de amilo
i ossalato di amile
d Amyloxalat

518 amyl propionate
f propionate d'amyle
e propionato de amilo
i propionato di amile
d Amylpropionat

519 amyl salicylate
f salicylate d'amyle
e salicilato de amilo
i salicilato di amile
d Amylsalicylat

520 amyl stearate
f stéarate d'amyle
e estearato de amilo
i stearato di amile
d Amylstearat

521 amyl tartrate
f tartrate d'amyle
e tartrato de amilo
i tartrato di amile
d Amyltartrat

522 ana-
Prefix meaning a condensed double
aromatic nucleus substituted in the 1.5
positions.
f ana-
e ana-
i ana-
d Ana-

523 anabasine
f anabasine
e anabasina
i anabasina
d Anabasin

524 anachromasis
f anachromasie
e anacromasia
i anacromasi
d Anachromasis

525 anaerobe
f anaérobe
e anaerobio; anaeróbico
i anaerobio
d Anaerobe

526 analcime; analcite
f analcime; analcite
e analcima; analcita
i analcime; analcite
d Analcim

527 analgesic
f analgésique
e analgésico
i analgesico
d Schmerzlinderungsmittel

528 analysis
f analyse

e análisis
i analisi
d Analyse

529 analytical balance
f balance de laboratoire; balance de
précision
e balanza para laboratorio
i bilancia di precisione
d Analysenwaage

530 anaphase
f anaphase
e anafase
i anafase
d Anaphase

531 anaphase movement
f mouvement anaphasique
e movimiento anafásico
i movimento anafasico
d Anaphasebewegung

532 anaphoresis
f anaphorèse
e anafóresis
i anaforesi
d Anaphorese

533 anchor agitator
f agitateur à palette en U
e mezclador con paletas en forma de ancla
i agitatore ad ancora
d Ankerrührwerk

534 andalusite
f andalousite
e andalucita
i andalusite
d Andalusit; Hartspat

535 androsterone
f androstérone
e androsterona
i androsterone
d Androsteron

536 anemometer
f anémomètre
e anemómetro
i anemometro
d Anemometer

537 aneroid manometer
f manomètre anéroïde
e manómetro aneroide
i manometro aneroide
d Aneroidmanometer; Dosenmanometer

538 **anethole**
f anéthole
e anetol
i anetolo
d Anethol

539 **aneuploid**
f aneuploíde
e aneuploide
i aneuploide
d Aneuploid

540 **aneuploidy**
f aneuploídie
e aneuploidia
i aneuploidismo
d Aneuploidie

541 **angioblast**
f angioblaste
e angioblasto
i angioblasto
d Angioblast

542 **anglesite**
f anglésite
e anglesita
i anglesite
d Anglesit; Bleisulfat; Vitriolbleierz

543 **Angola copal**
f angola; copal fossile
e copal de Angola
i coppale d'Angola
d Kopalin

544 **Angus Smith process**
f procédé Angus Smith
e proceso de Angus Smith
i processo di Angus Smith
d Angus-Smith-Verfahren

545 **anhydrides**
f anhydrides
e anhídridos
i anidridi
d Anhydride

546 **anhydrite**
f anhydrite
e anhidrita; sulfato de calcio anhidro
i anidrite
d Anhydrit; wasserfreier Gips; Karstenit

547 **anhydrite process**
f procédé anhydritique
e proceso anhidrítico
i processo anidritico
d Anhydrit-Verfahren

548 **anhydrous**
f anhydre
e anhidro
i anidro
d wasserfrei

549 **anhydrous ammonia**
f ammoniaque anhydre
e amoníaco anhidro
i ammoniaca anidra
d wasserfreies Ammoniak

550 **anhydrous borax**
f borax anhydre
e bórax anhidro
i borace anidro
d wasserfreier Borax

551 **anhydrous hydrochlorination of
    zirconium bearing fuels**
f hydrochloruration anhydrique de
    combustibles zirconifères
e hidrocloruración anhídrido de
    combustibles zirconiferos
i idroclorurazione anidrica di combustibili
    zirconiferi
d wasserfreie Hydrochlorierung zirkon-
    haltiger Brennstoffe

552 **anhydrous salt**
f sel anhydre
e sal anhidra
i sale anidro
d wasserfreies Salz

553 **anhydrous sodium carbonate; soda ash**
f carbonate de sodium
e carbonato de sodio
i carbonato di sodio; soda
d Natriumcarbonat; Sodaasche

554 **aniline**
f aniline
e anilina
i anilina
d Anilin

555 **aniline point**
f point d'aniline
e punto de anilina
i punto di anilina
d Anilinpunkt

556 **animal black; animal charcoal; spodium**
f noir animal; charbon animal; charbon
    d'os
e negro animal; carbón animal
i nero animale; carbone animale

d Knochenschwarz; tierische Kohle;
Beinschwarz; Knochenkohle; Spodium

557 animal-sized
f encollé à la colle animale
e encolado con gelatina
i incollato a gelatina
d behandelt mit Tierleim

558 anion
f anion; ion négatif
e anión; ión negativo
i anione; ione negativo
d Anion; negatives Ion

559 anise oil
f essence d'anis
e aceite de anís
i olio di semi d'anice
d Anisöl

560 anisidines
f anisidines
e anisidinas
i anisidine
d Anisidine

561 anisole
f anisol
e anisol
i anisolo
d Anisol

562 anisomeric
f anisomère
e anisomérico
i anisomerico
d anisomer

563 anisotonic
f anisotonique
e anisotónico
i anisotonico
d anisotonisch

564 anisotropic
f anisotrope
e anisotrópico
i anisotropico
d anisotrop

565 ankerite
f ankérite; spath brunissant
e anquerita
i ancherite
d Ankerit; Braunspat

566 annabergite; nickel bloom
f annabergite; nickelochre

e anabergita; eflorescencias de níquel
i annabergite
d Nickelblüte

567 annatto
f annatto
e anato; achiote
i arnotto
d Annatto-Farbstoff

568 annealing
f recuit; annealing
e recocido
i ricottura
d Ausglühen; Frischen; Tempern

569 annidation
f annidation
e anidación
i annidazione
d Annidation; Einnischung

570 anode
f anode
e ánodo
i anodo
d Anode

571 anode efficiency
f rendement anodique
e rendimiento anódico
i rendimento anodico
d anodische Stromausbeute

572 anode layer
f couche anodique
e capa anódica
i strato anodico
d anodische Schicht

573 anode scrap
f résidu anodique
e residuo anódico
i anodo residuo
d Anodenrest

574 anode slime
f boues métalliques à la surface de l'anode
e fango anódico
i fango anodico
d Anodenschlamm

575 anodic cleaning
f dégraissage anodique
e desengrase anódico
i sgrassaggio anodico
d anodische Reinigung

576 anodic inhibitors
f inhibiteurs anodiques
e inhibidores anódicos
i inibitori anodici
d anodische Inhibitoren

577 anodic passivation
f passivation anodique
e pasivación anódica
i passivazione anodica
d anodische Passivierung

578 anodic reaction
f réaction anodique
e reacción anódica
i reazione anodica
d Anodenreaktion

579 anodizing
f traitement anodique; oxydation anodique
e anodizado; anodización
i ossidazione anodica
d Eloxierung

580 anolyte
f anolyte
e anólito
i anolito
d Anolyt

581 anomalous valence
f valence anomale
e valencia anomala
i valenza anomala
d anomale Valenz

582 anorthosite
f anorthosite
e anortosita
i anortosite
d Anorthosit; Labradorfels

583 anthophyllite
f anthophyllite; antholite
e antofilita
i antofillite
d Anthophyllit

584 anthracene
f anthracène
e antraceno
i antracene
d Anthrazen

585 anthraflavine
f anthraflavine
e antraflavina
i antraflavina
d Anthraflavin

586 anthranilic acid
f acide anthranilique
e ácido antranílico
i antranilico
d Anthranilsäure

587 anthranol
f anthranol
e antranol
i antranolo
d Anthranol

588 anthraquinone
f anthraquinone
e antraquinona
i antrachinone
d Anthrachinon

590 anthrarobin
f anthrarobine
e antrarrobina
i antrarobina
d Anthrarobin

590 antialbumoses
f antialbumoses
e antialbumosas
i antialbumose
d Antialbumosen

591 antialdoximes
f antialdoximes
e antialdoximas
i antialdossime
d Antialdoxime

592 antibaryon
f antibaryon
e antibariona
i antibariona
d Antibaryon

593 antibiosis
f antibiose
e antibiosis
i antibiosi
d Antibiose

594 antibiotics
f antibiotiques
e antibióticos
i antibiotici
d Antibiotika

595 antibody
f anticorps
e anticuerpo
i anticorpo
d Immunkörper

596 anticatalyst
f anticatalyseur
e anticatalizador
i anticatalizzatore
d Antikatalysator

597 antichlor
f antichlore
e anticloro
i anticloro
d Antichlor

598 anticoagulant
f anticoagulant
e anticoagulante
i anticoagulante
d Antigerinnungsmittel

599 anticoagulating action
f action anticoagulante
e acción anticoagulante
i azione anticoagulante
d Antikoaguliervorgang

600 anticondensation paint
f peinture anticondensation
e pintura anticondensación
i vernice a prova di condensazione
d Antikondensationsfarbe

601 antidetonant
f antidétonant
e antidetonante
i antidetonante
d Klopffestigkeitsmittel

602 antidiazo compounds
f antidiazotates
e compuestos antidiazo
i composti anti-diazo
d Antidiazotate

603 antidote
f antidote
e antídoto
i antidoto
d Gegengift; Gegenmittel

604 antienzyme
f antienzyme
e antienzima
i antienzima
d Antienzym

605 antifoam; antifoaming agent
f agent antimousse
e agente antiespuma; agente antiespumante
i agente antischiuma; agente antischiumante
d Antischaummittel

606 antifouling composition
f enduit préservatif
e composición antivegetativa
i pittura sottomarina antivegetativa
d Bodenanstrich; Antifouling-Komposition

607 antigelling agent
f agent antigélifiant
e agente contra la formación de gel
i agente antigelizzazione
d Gelverhütungsmittel

608 antihistamine
f antihistamine
e antihistamina
i antiistaminico
d Antihistaminstoff

609 antihormone
f antihormone
e antihormona; chalona
i antiormone
d Antihormon; Gegenhormon

610 antiknock additives
f antidétonants
e aditivos antidetonantes
i antidetonanti
d Antiklopfmittel

611 antimonial
f antimonié
e antimonial
i antimoniale
d antimonartig

612 antimonial lead
f plomb antimonial; plomb aigre
e plomo antimoniado
i piombo antimoniale; piombo indurito
d Antimonblei

613 antimoniates
f antimoniates
e antimoniatos
i antimoniati
d Antimoniate

614 antimonite; stibnite
f stibine; stibnite
e antimonita; estibnita
i antimonite; stibnite
d Antimonglanz; Grauspießglanzerz; Schwefelspießglanz; Stibnit

615 antimony
f antimoine
e antimonio
i antimonio
d Antimon

616 antimony orange
f orange d'antimoine
e anaranjado de antimonio
i arancio di antimonio
d Antimonorange

617 antimony oxide
f oxyde d'antimoine
e óxido de antimonio
i ossido di antimonio
d Antimonoxyd

618 antimony oxychloride
f oxychlorure d'antimoine
e oxicloruro de antimonio
i ossicloruro di antimonio
d Antimonoxychlorid

019 antimony pentachloride
f pentachloruro d'antimoine
e pentacloruro de antimonio
i pentacloruro di antimonio
d Antimonpentachlorid

620 antimony pentasulphide
f pentasulfure d'antimoine
e pentasulfuro de antimonio
i pentasolfuro di antimonio
d Antimonpentasulfid

021 antimony potassium tartrate
f tartrate antimonico-potassique
e tartrato de antimonio y potasio
i tartrato di potassio e antimonio
d Antimonkaliumtartrat

622 antimony trichloride
f trichlorure d'antimoine
e tricloruro de antimonio
i tricloruro di antimonio
d Antimontrichlorid

623 antimony trioxide
f trioxyde d'antimoine
e trióxido de antimonio
i triossido di antimonio
d Antimontrioxyd

624 antimony trisulphide
f trisulfure d'antimoine
e trisulfuro de antimonio
i trisolfuro di antimonio
d Antimontrisulfid

625 antimony vermilion
f vermillon d'antimoine
e bermellón de antimonio
i vermiglione di antimonio
d Kermes

626 antimony white
f blanc d'antimoine
e blanco de antimonio
i bianco di antimonio
d Antimonweiß

627 antimycin
f antimycine
e antimicina
i antimicina
d Antimycin

628 antinoise paint
f peinture d'insonorisation
e pintura contra ruidos
i pittura antirombo
d geräuschdämpfende Anstrichfarbe

629 antioxidant
f antioxydant
e antioxidante
i antiossidante
d Antioxydant; Antioxydationsmittel

630 antiozonant
f agent antiozone
e antiozonizante; agente antiozono
i agente antiozono
d Antiozonisator; Ozonschutzmittel

031 antipyrine
f antipyrine
e antipirina
i antipirina
d Antipyrin

632 antiseptics
f antiseptiques
e antisépticos
i antisettici
d Antiseptika

633 antishatter composition
f verre ne produisant pas d'éclat
e composición inastillable
i composizione infrangibile
d splitterfreies Glas; Sigla-Glas

634 antiskinning agent
f agent antipeaux
e agente contra la formación de films
i agente antiformazione di pelle
d Hautbildungsverhinderungsmittel

635 antislip paint
f peinture antidérapante
e pintura antideslizante
i pittura antisdrucciolevole
d rutschfeste Farbe

636 antistatic cleaning
f nettoyage antistatique
e depuración antiestática
i pulitura ad aria ionizzata
d Antistatikreinigung

637 antistatic solution
f solution antistatique
e solución antiestática
i soluzione antistatica
d Antistatiklösung

638 antitracking varnish
f vernis antitrace
e pintura contra formación de sendas
   eléctricas
i pittura antitraccia
d Kriechspur-Schutzanstrich

639 apatite
f apatite
e apatita
i apatite
d Apatit

640 apochromatic
f apochromatique
e apocromático
i apocromatico
d apochromatisch

641 apocrine
f apocrine
e apocrino
i apocrino
d apokrin; absondernd

642 apoenzyme
f apoferment
e apoenzima
i apofermento
d Zwischenferment

643 apogamy
f apogamie
e apogamia
i apogamia
d Apogamie

644 apomorphine
f apomorphine
e apomorfina

i apomorfina
d Apomorphin

645 apparent absorption
f absorption apparente
e absorción aparente
i assorbimento apparente
d scheinbare Absorption

646 apple-jack *(US)*
f calvados; eau-de-vie de cidre
e aguardiente de manzana; sidra
   fermentada
i sidro ad alto grado alcoolico
d Apfelschnaps

647 aqua fortis
f eau forte
e agua fuerte
i acquaforte
d Gelbbrennsäure; Ätze

648 aqua regia
f eau régale
e agua regia
i acqua regia
d Königswasser; Goldscheidewasser

649 aqueous desizing
f désensimage aqueux
e tratamiento acuoso; desensimado por
   agua
i sbozzimatura
d Wasserwäsche

650 aqueous dispersion
f dispersion aqueuse
e dispersión acuosa
i dispersione acquosa
d wässerige Dispersion

651 aqueous medium
f milieu aqueux
e medio acuoso
i mezzo acquoso
d wässeriges Medium

652 arabinose
f arabinose
e arabinosa
i arabinosio
d Arabinose

653 arachidic acid
f acide arachidique
e ácido araquídico
i acido arachidico
d Arachinsäure

654 arachis oil
f huile d'arachides
e aceite de cacahuete
i olio di arachide
d Arachisöl; Erdnußöl

655 aragonite
f aragonite
e aragonito
i aragonite
d Aragonit; Schalenkalk

656 arborescent
f arborescent
e arborescente
i a forma d'albero
d baumähnlich; dendritisch

657 arc process
f procédé à l'arc voltaïque
e tratamiento por arco voltaico
i processo all'arco voltaico
d Lichtbogenverfahren

658 area density
f densité superficielle
e densidad superficial
i densità superficiale
d Oberflächendichte

659 areopyknometer
f aréopycnomètre
e areopicnómetro
i areopicnometro
d Aräopyknometer

660 argentite
f argentite
e argentita
i argentite
d Glanzsilber; Silberglanzerz

661 argillaceous
f argileux
e arcilloso
i argilloso
d tonartig; tonig

662 arginine
f arginine
e arginina
i arginina
d Arginin

663 argol
f argol
e argol
i argol
d Weinstein

664 argon
f argon
e argón
i argon
d Argon

665 arm *(of a balance)*
f bras
e brazo
i braccio
d Arm

666 aromatic compounds
f composés aromatiques
e compuestos aromáticos
i composti aromatici
d aromatische Verbindungen

667 arsanilic acid
f acide arsanilique
e ácido arsanílico
i acido arsanilico
d Arsanilsäure

668 arsen-fast
f arséno-résistant
e arsenorresistente
i arsenoresistente
d arsenfest

669 arsenic
f arsenic
e arsénico
i arsenico
d Arsen

670 arsenical copper
f cuivre arsénical
e cobre arsenical
i rame arsenicale
d Arsenkupfer

671 arsenical pyrite; arsenopyrite; mispickel
f arsénopyrite; mispickel
e arsenopirita; hierro arsenical; mispickel
i arsenopirite; mispickel
d Arsenkies; Mispickel

672 arsenic disulphide
f bisulfure d'arsenic
e disulfuro de arsénico
i bisolfuro di arsenico
d Arsendisulfid

673 arsenic pentasulphide
f pentasulfure d'arsenic
e pentasulfuro de arsénico
i pentasolfuro di arsenico
d Arsenpentasulfid

674 arsenic trichloride
f trichlorure d'arsenic
e tricloruro de arsénico
i tricloruro di arsenico
d Arsentrichlorid

675 arsenic trioxide
f trioxyde d'arsenic
e trióxido de arsénico
i triossido di arsenico
d Arsentrioxyd

* **arsenopyrite → 671**

676 arsine
f arsine
e arsina; arsenamina
i arsina
d Arsin

677 arsphenamine
f arsphénamine
e arsfenamina
i salvarsan
d Dioxydiamino-arsenobenzol; Salvarsan

678 artificial atmosphere furnace
f four à atmosphère artificielle
e horno de atmósfera controlada
i forno ad atmosfera controllata
d Schutzgasofen

679 artificial silk
f soie artificielle
e seda artificial
i seta artificiale
d Kunstseide

680 artificial transmutation of elements
f transmutation artificielle d'éléments
e transmutación artificial de elementos
i trasmutazione artificiale d'elementi
d künstliche Transmutation von Elementen

681 aryl compounds
f composés aryliques
e compuestos arílicos
i composti arilici
d Arylverbindungen

682 asbestos
f amiante; asbeste
e amianto; asbesto
i amianto; asbesto
d Asbest

683 asbolane; asbolite; earthy cobalt
f asbolane; asbonite; cobalt oxydé noir
e asbolana; asbolita; cobalto oxidado negro
i asbolite
d Asbolit; Erdkobalt

684 ascaridole
f ascaridol
e ascaridol
i ascaridolo
d Ascaridol

685 ascorbic acid; vitamin C
f acide ascorbique
e ácido ascórbico
i acido ascorbico
d Ascorbinsäure

686 aseptic
f aseptique
e aséptico
i aseptico
d aseptisch

687 ash content
f taux de cendres; teneur en cendres
e contenido en cenizas
i tenore in ceneri
d Aschengehalt

688 ashless filter paper
f papier-filtre sans cendres
e papel de filtro sin cenizas
i carta da filtro a minima percentuale di ceneri
d aschfreies Filterpapier

689 asphalt
f asphalte
e asfalto
i asfalto
d Asphalt

690 asphaltite
f asphaltite
e asfaltita
i asfaltite
d Asphaltit

691 aspirator
f aspirateur
e aspirador
i aspiratore
d Aspirator; Luftsauger

692 assay balance
f balance d'essai
e balanza de ensayo
i bilancia d'assaggio
d Versuchswaage

693 assayer
f essayeur
e ensayador
i assaggiatore
d Prüfer

694 assay furnace
f four d'essai
e horno de ensayo
i fornace per assaggio
d Versuchsofen

695 assaying
f analyse des minerais
e ensayo; análisis
i prova; analisi
d Auswerten; Erzprobe

696 association
f association
e asociación
i associazione
d Assoziation; Bindung

697 astatine
f astate
e astatino
i astatinio
d Astatin

698 aster
f aster
e áster
i aster
d Aster

699 astrocyte; spider cell
f astrocyte; cellule étoilée
e astrocito; célula estrellada
i astrocita; cellula aracnoiforme
d Astrozyt; Spinnenzelle

700 asymmetric system
f système asymétrique; système triclinique
e sistema asimétrico; sistema triclínico
i sistema triclino
d triklinisches System

701 atacamite
f atacamite
e atacamita
i atacamite
d Atakamit

702 atactic
f atactique
e atáctico
i atattico
d ataktisch

703 atmolysis
f atmolyse
e atmólisis
i atmolisi
d Atmolyse

704 atomic absorption coefficient
f coefficient d'absorption atomique
e coeficiente de absorción atómica
i coefficiente d'assorbimento atomico
d atomarer Absorptionskoeffizient

705 atomic arrangement
f disposition atomique
e configuración atómica
i disposizione atomica
d räumliche Anordnung der Atome

706 atomic distance
f distance des atomes dans la molécule
e distancia de los atomos en la molécula
i distanza degli atomi nella molecola
d Atomabstand im Molekül

707 atomic heat
f chaleur atomique
e calor atómico
i calore atomico
d Atomwärme

708 atomic hydrogen welding
f soudure à l'hydrogène atomique
e soldadura oxhídrica con soplete de hidrógeno atómico
i saldatura all'idrogeno atomico
d atomare Wasserstoffschweißung

709 atomicity
f atomicité; valence
e atomicidad
i atomicità
d Wertigkeit; Valenz

710 atomic migration
f migration atomique
e migración atómica
i migrazione atomica
d Atomwanderung

711 atomic refraction
f réfraction atomique
e refracción atómica
i rifrazione atomica
d Atomrefraktion

712 atomic scale
f échelle des poids atomiques
e escala de los pesos atómicos

i scala dei pesi atomici
d Atomgewichtsskala

713 atomic volume
f volume atomique
e volumen atómico
i volume atomico
d Atomvolumen

714 atomic weight
f poids atomique
e peso atómico
i peso atomico
d Atomgewicht

715 atomic weight unit
f unité de poids atomique
e unidad de peso atómico
i unità di peso atomico
d Einheit des Atomgewichts

716 atropine
f atropine
e atropina
i atropina
d Atropin

717 attapulgite
f attapulgite
e atapulguita
i argilla del Texas
d Attapulgit

718 attemperator
f serpentin; réfrigérant de cuve de
  fermentation
e serpentín de refrigeración
i serpentina refrigerante
d Gärbottichkühler

719 attritus
f durain
e dureno
i durame
d Mattkohle

720 auric oxide; auric acid
f oxyde aurique
e óxido áurico; ácido áurico
i acido aurico
d Goldsäure

721 aurines
f aurines
e aurinas
i aurine
d Aurine

722 aurous
f aureux
e auroso
i auroso
d Gold-

723 austenite
f austénite
e austenita
i austenite
d Austenit

724 autocatalysis
f autocatalyse
e autocatálisis
i autocatalisi
d Autokatalyse

725 autoclave
f autoclave
e autoclave
i autoclave
d Autoklav

726 autolysis
f autolyse
e autólisis
i autolisi
d Autolyse

727 autoploidy
f autoploïdie
e autoploidia
i autoploidismo
d Autoploidie

728 autoradiolysis
f autoradiolyse
e autorradiólisis
i autoradiolisi
d Autoradiolyse

* autosome → 3279

729 autotetraploidy
f autotétraploïdie
e autotetraploidia
i autotetraploidismo
d Autotetraploidie

730 autoxidation
f autoxydation
e autooxidación
i autossidazione
d Autooxydation

731 autunite
f autunite
e autunita

i autunite
d Autunit; Kalkuranglimmer

732 auxochromes
f auxochromes
e auxocromos
i auxocromi
d Auxochrome

733 auxocyte
f auxocyte
e auxocito
i auxocito
d Auxocyte; Auxozyte

734 avidin
f avidine
e avidina
i avidina
d Avidin

735 Avogadro constant; Avogadro number
f nombre d'Avogadro
e número de Avogadro
i numero di Avogadro
d Avogadrosche Zahl

736 Avogadro's hypothesis
f hypothèse d'Avogadro
e hipótesis de Avogadro
i ipotesi di Avogadro
d Avogadrosche Hypothese

737 azaserine
f azasérine
e azaserina
i azaserina
d Azaserin

738 azelaic acid
f acide azélaïque
e ácido azelaico
i acido azelaico
d Azelainsäure

739 azeotrope
f azéotrope
e azeotropo
i azeotropo
d Azeotrop

740 azeotropic distillation
f distillation azéotropique
e destilación azeotrópica
i distillazione azeotropica
d azeotropische Destillation

741 azeotropic mixture
f azéotrope

e mezcla azeotrópica
i azeotropo
d Azeotrop

742 azides
f azides; azothydrure
e azidas
i azoturi
d Azide

743 azimino compounds
f composés azimino
e compuestos azimínicos
i azimino-composti
d Aziminoverbindungen

744 azines
f azines
e azinas
i azine
d Azine

745 azobenzene
f azobenzène
e azobenceno
i azobenzene
d Azobenzol

746 azo dyes
f colorants azoïques
e colorantes azoicos
i azocoloranti
d Azofarbstoffe

747 azo group
f groupe azoïque
e grupo azo
i gruppo azoico
d Azogruppe

748 azoimide
f acide azothydrique
e azoimida
i azoimide
d Azoimid

749 azophenine
f azophénine
e azofenina
i azofenina
d Azophenin

* azote → 5766

750 azotometer
f azotomètre
e azotómetro
i azotometro
d Azotometer

* **azurite** · 1694

B

751 bacillus
f bacille
e bacilo
i bacillo
d Bacillus

752 bacitracin
f bacitracine
e bacitracina
i bacitracina
d Bacitracin

753 backwash
f lavage par inversion de courant
e lavado a contracorriente
i lavaggio a inversione di flusso;
   desoliatura
d zurücklaufende Strömung

754 backwash cascade
f cascade de réextraction
e cascada de recxtracción
i cascata di riestrazione
d Rückwaschkaskade

755 backwashing
f réextraction
e reextracción
i riestrazione
d Rückwaschung

756 bacterial coagulation
f coagulation bactérienne
e coagulación bacteriana
i coagulazione batterica
d bakterielle Koagulation

757 bacteria-propagation tank
f vase clos de propagation des bactéries
e depósito para la propagación de bacterias
i serbatoio per la propagazione dei batteri
d Bakterienfortpflanzungsbehälter

758 bactericide
f substance bactéricide
e bactericida
i battericida
d Bakterizid

759 bacteriocin
f bactériocine
e bacteriocina
i batteriocina
d Bacteriocin; Bakteriozin

760 bacteriogenic
f bactériogène
e bacteriógeno
i batteriogeno
d bakteriogen

761 bacteriolysin
f bactériolysine
e bacteriolisina
i batteriolisina
d Bakteriolysin

762 bacteriolysis
f bactériolyse
e bacteriólisis
i batteriolisi
d Bakteriolyse

763 bacteriophage
f bactériophage
e bacteriofago
i batteriofago
d Bakteriophage; Phage

764 bacteriostasis
f bactériostase
e bacterióstasis
i batteriostasi
d Bakteriostase

765 bacteriostat
f bactériostatique
e bacteriostato
i batteriostato
d Bakteriostat

766 bacteriostatic
f bactériostatique
e bacteriostático
i batteriostatico
d bakteriostatisch

767 bacterium
f bactérie
e bacteria
i batterio
d Bakterie

768 baculiform
f bacilliforme; en forme de bâtonnet
e baculiforme
i baculiforme
d stabförmig; stäbchenförmig

769 baddeleyite
f oxyde de zirconium pur
e badeleyita
i ossido di zirconio naturale
d Baddeleyit

770 baffle plate
 f chicane; plaque de déviation
 e chicana; placa deflectora
 i deflettore
 d Prallplatte

771 bagasse
 f bagasse
 e bagazo
 i bagasse esaurite
 d Bagasse

772 bag-type construction
 f montage habillé
 e montaje tipo saco
 i montaggio a catodo rivestito
 d Wickelbauweise; gewickelter Beutel

773 baking press
 f presse à blocs
 e prensa de cocimiento
 i pressa a cottura
 d Kochpresse

* baking soda → 7495

774 balance desiccator
 f récipient à chlorure de calcium
 e copa para cloruro de calcio
 i piattello di bilancia per cloruro di calcio
 d Waageeinsatz für Chlorcalcium

775 balance room
 f salle de pesée; salle de balances
 e sala de las balanzas
 i sala delle bilancie
 d Wägezimmer

776 balata
 f balata
 e balata
 i balata
 d Balata

* ballas → 1130

777 ball burnishing
 f brunissage à la bille
 e bruñido con bola
 i brunitura a sfera
 d Kugelpolierung

778 ball clay
 f argile; figuline
 e arcilla grasa
 i argilla grassa
 d Bindeton

779 ball viscosimeter
 f viscosimètre à chute de bille
 e viscosímetro de bola
 i viscosimetro a sfera
 d Kugelfallviskosimeter

780 balm; balsam
 f baume
 e bálsamo
 i balsamo
 d Balsam

781 bar agitator
 f agitateur à barres
 e agitador de barras
 i agitatore ad asta
 d Stangenrührer

782 barbital
 f barbital
 e barbital
 i acido dietilbarbiturico
 d Barbital

783 barbitone; veronal; diethyl-malonylurea
 f véronal
 e barbitona; veronal
 i barbitone; veronale
 d Diäthylmalonylharnstoff; Veronal

784 barbiturate
 f barbiturique
 e barbitúrico
 i barbiturico
 d Barbiturpräparat

785 barbituric acid
 f acide barbiturique
 e ácido barbitúrico
 i acido barbiturico
 d Barbitursäure

786 barilla
 f barille
 e barrilla
 i residuo di ceneri di soda
 d Barilla

787 barium
 f baryum
 e bario
 i bario
 d Barium

788 barium acetate
 f acétate de baryte
 e acetato bárico
 i acetato di bario
 d Bariumacetat; essigsaurer Baryt

789 barium carbonate
f carbonate de baryum
e carbonato de bario
i carbonato di bario
d Bariumkarbonat; kohlensaures Barium

790 barium chlorate
f chlorate de baryum
e clorato bárico
i clorato di bario
d Bariumchlorat

791 barium chloride
f chlorure de baryum
e cloruro bárico
i cloruro di bario
d Bariumchlorid; Chlorbarium

792 barium chromate
f chromate de baryum
e cromato bárico
i cromato di bario
d Bariumchromat

793 barium citrate
f citrate de baryum
e citrato de bario
i citrato di bario
d Bariumcitrat

794 barium concrete
f béton au baryum
e hormigón de bario
i conglomerato di bario
d Bariumbeton

795 barium fluosilicate
f fluosilicate de baryum
e fluosilicato bárico
i fluosilicato di bario
d Bariumfluosilikat

796 barium hydroxide
f monohydrate de baryum
e hidróxido bárico
i idrossido di bario
d Bariumhydroxyd

797 barium manganate
f manganate de baryum
e manganato bárico
i manganato di bario
d Bariummanganat

798 barium mixer
f mélangeur de boisson d'épreuve du
   baryum
e mezclador de papilla barítica

i mescolatore di pasto opaco
d Bariumbreimischer

799 barium nitrate
f nitrate de baryum
e nitrato bárico
i nitrato di bario
d Bariumnitrat

800 barium oxide
f oxyde de baryum
e óxido de bario; barita
i ossido di bario
d Bariumoxyd; Baryt

801 barium peroxide
f peroxyde de baryum
e peróxido de bario
i perossido di bario
d Bariumperoxyd

802 barium plaster
f plâtre baryté
e mortero de bario; argamasa de bario
i intonaco al bario
d Bariumweißkalk

803 barium stearate
f stéarate de baryum
e estearato bárico
i stearato di bario
d Bariumstearat

804 barium sulphate
f sulfate de baryum
e sulfato de bario
i solfato di bario
d Bariumsulfat; schwefelsaures Barium

805 barium sulphide
f sulfure de baryum
e sulfuro bárico
i solfuro di bario
d Bariumsulfid

806 barium thiosulphate
f thiosulfate de baryum
e tiosulfato bárico
i tiosolfato di bario
d Bariumthiosulfat

807 barley
f orge
e cebada
i orzo
d Gerste

808 barm
f levure de bière; levain

   e giste; jiste
   i lievito di birra
   d Zeug; Stellhefe

809 barodiffusion
   f barodiffusion
   e barodifusión
   i barodiffusione
   d Barodiffusion

810 barometric leg
   f tube barométrique
   e tubo barométrico
   i tubo barometrico
   d Fallwasserrohr

811 barrel
   f barrel
   e barrel
   i barrel
   d Barrel

812 barrel attemperator
   f serpentin de fût intérieur
   e enfriador de barril
   i serpentina refrigerante
   d Faßkühler

813 barrel mixer
   f mélangeur à tonneau
   e mezclador de tambor giratorio
   i mescolatore a tamburo
   d Trommelmischer

814 barrel plating
   f galvanoplastie au tambour
   e galvanoplastia en tambor
   i galvanoplastica in tamburo
   d Trommelgalvanisierung

815 barthrin
   f barthrine
   e bartrina
   i bartrina
   d Barthrin

816 baryta; heavy spar
   f baryte
   e barita
   i barite
   d Baryt; Schwerspat

817 basal metabolic rate
   f métabolisme de base
   e metabolismo basal
   i metabolismo basale
   d Grundumsatzwert

818 basalte; black Egyptian ware
   f basalte
   e basalto; gres sin vidriar negro y de grano fino
   i basalto
   d Basalt

819 base
   Substance which dissolves in water with formation of hydroxyl ions, and reacts with an acid to produce a salt, without the evolution of a gas.
   f base
   e base
   i base
   d Base

820 base
   Solid ingredient, usually a metallic oxide, in a coating preparation.
   f base
   e base
   i base
   d Grundstoff

821 base bullion
   f plomb d'œuvre
   e plombo en bruto
   i piombo impuro
   d Rohblei

822 base coat
   f couche de fond
   e capa de base; capa de primera mano
   i mano di fondo
   d Grundstrich

823 base material
   f matière première
   e material básico
   i materiale base
   d Grundstoff

824 base metals
   f métaux vils
   e metales viles
   i metalli vili
   d Grundmetalle

825 BASF process
   f procédé BASF
   e proceso BASF
   i processo BASF
   d BASF-Verfahren

826 basicity
   f basicité
   e basicidad

i basicità
d Basität

827 basic lead carbonate
f carbonate de plomb basique
e carbonato básico de plomo
i carbonato basico di piombo
d basisches Bleicarbonat

828 basic lead sulphate
f sulfate basique de plomb
e sulfato básico de plomo
i solfato di piombo basico
d basisches Bleisulfat

829 basic oxide
f oxyde basique
e óxido básico
i ossido basico
d basisches Oxyd

830 basic refractory
f matériau réfractaire basique
e refractario alto en óxidos básicos
i materiale refrattario basico
d basisches feuerfestes Material

831 basic salts
f sels basiques
e sales básicas
i sali basici
d basische Salze

832 basic slag
f scorie basique; scorie Thomas
e escorias Thomas
i scoria Thomas
d Thomasmehl; basische Schlacke

833 basic substance
f substance de base
e substancia clave
i sostanza di base
d Ausgangsstoff

834 basket *(sugar)*
f panier
e rejilla de centrifugación
i paniere di centrifugazione
d Korb

835 basocyte; basophilic cell
f cellule basophile
e basocito; leucocito básófilo
i leucocita basofilo; mastleucocita
d Mastzelle; Basozyt

836 basophil
f basophile

e basófilo
i basofilo
d Basophil

837 basophile; basophilous
f basophile
e basófilo
i basofilo
d basophil

838 basophile granule
f granule basophile
e gránulo basófilo
i granulo basofilo
d basophiles Körnchen

* **basophilic cell** → 835

* **basophilous** → 837

839 bastnaesite
f bastnaésite
e bastnesita
i bastnaesite
d Bastnäsit

840 batch
f fournée; charge; lot
e lote; partida; tanda
i lotto; partita
d Partie; Charge

841 batch distillation
f distillation intermittente
e destilación por tandas
i distillazione a partite
d Postendestillation

842 batching unit
f unité de fournée
c medidor de cargas
i dispositivo dosatore a peso
d Füllvorrichtung

843 batch mixer
f mélangeur en discontinu
e mezclador para tratamiento por lotes
i mescolatore in serie
d Satzmischer

844 batch *v* **off**
f enlever du malaxeur (en petites quantités)
e tirar de la batidora (en pequeñas cantidades)
i estarre dal mescolatore (in piccoli lotti)
d von der Walze schneiden (in "Puppen")

845 batch processing
 f traitement intermittent
 e proceso en grupos; proceso en lotes
 i processo discontinuo
 d Postenverfahren

846 batchwise operation
 f opération discontinue
 e operación por lotes
 i operazione a lotti
 d Satzbetrieb

847 bath
 f bain
 e baño
 i bagno
 d Bad

848 batonet
 f pseudochromosome
 e seudocromosoma
 i pseudocromosoma; falso cromosoma
 d Pseudochromosom

849 battery acid
 f acide pour accumulateur
 e ácido de baterias
 i acido solforico
 d Akkumulatorsäure

850 Baudouin reaction
 f réaction de Baudouin
 e reacción de Baudouin
 i reazione di Baudouin
 d Baudouin-Reaktion

851 Baumé scale
 f échelle Baumé
 e escala Baumé
 i scala di Baumé
 d Baumé-Skala

852 bauxite
 f bauxite
 e bauxita
 i bauxite
 d Bauxit

853 Bayer process
 f procédé de Bayer
 e método Bayer
 i processo Bayer
 d Bayer-Verfahren

854 beaker
 f bécher; vase à précipitation
 e cubilete
 i bicchiere
 d Becherglas

855 beam *(of a balance)*
 f fléau de balance
 e brazo de balanza
 i bilanciere
 d Waagebalken

856 beating machine
 f fouetteuse
 e batidora mecánica
 i abbattitore
 d Schlagmaschine

857 bebeerine
 f bébirine
 e bebeerina; bebirina
 i bebeerina
 d Bebeerin

858 Beckmann rearrangement
 f transposition de Beckmann
 e transposición de Beckmann
 i trasformazione molecolare di Beckmann
 d Beckmannsche Umlagerung der
   Ketoxime

859 Beckmann thermometer
 f thermomètre de Beckmann
 e termómetro de Beckmann
 i termometro di Beckmann
 d Beckmann-Thermometer

860 Becquerel rays
 f rayons de Becquerel
 e rayos de Becquerel
 i raggi di Becquerel
 d Becquerel-Strahlen

861 beer most
 f moût de bière
 e mosto de cerveza
 i mosto di birra
 d Bierwürze

862 beer still
 f alambic pour la distillation de la bière
 e alambique de cerveza
 i distillatore di birra
 d Bierdestillierapparat

863 beer well
 f cuve à liquide fermenté
 e cuba para líquidos fermentados
 i serbatoio di liquidi fermentati
 d Bierbottich

864 beeswax
 f cire d'abeilles
 e cera de abejas

i cera d'api
d Bienenwachs

865 beet sugar
f sucre de betteraves
e azúcar de remolacha
i zucchero di barbatietola
d Rübenzucker

866 behenic acid
f acide béhénique
e ácido behénico
i acido beenico
d Behensäure

867 belladonna
f belladone
e belladona
i belladonna
d Belladonna; Tollkirsche

868 bell-jar
f cloche en verre
e campana de vidrio
i campana di vetro
d Glasglocke

* bell metal ore → 7750

869 belt dressing
f apprêt de courroi
e aprestos para correas
i appretto di cinghia
d Riemenappretur

870 belt-dryer
f sécheur pour bandes transporteuses
e secadora por correa
i essicatore a nastro
d Bandtrockner

871 belt feed
f alimentateur à courroie
e alimentador a correa
i alimentatore a nastro trasportatore
d Gurtzuführer

872 Bemelmans' reclaiming process
f procédé de régénération Bemelmans
e procedimiento de regeneración de Bemelmans
i processo di rigenerazione Bemelmans
d Bemelmans-Regenerierverfahren

873 Benedict solution
f solution de Benedict
e solución de Benedict
i soluzione di Benedict
d Benedict-Lösung

874 bentonite
f bentonite
e bentonita
i bentonite
d Bentonit

875 benzaldehyde
f benzaldéhyde; aldéhyde benzoïque
e benzaldehído
i benzaldeide
d Benzaldehyd

876 benzalkonium
f benzalconium
e benzalconio
i benzalconio
d Benzalkonium

877 benzamide
f benzamide
e benzamida
i benzamide
d Benzamid

878 benzamine; betacaine; beta-eucaine
f benzamine
e benzamina
i benzamina
d Benzamin

879 benzanilide
f benzanilide
e benzanilida
i benzanilide
d Benzanilid

880 benzedrine
f benzédrine
e benzedrina
i benzedrina
d Benzedrin

881 benzene; benzol; benzole
f benzène; benzol
e benceno; benzol
i benzene; benzolo
d Benzol

882 benzene hexachloride
f hexachlorure de benzène
e hexacloruro de benceno
i estacloruro di benzene
d Benzolhexachlorid

883 benzenephosphonic acid
f acide benzène-phosphonique
e ácido bencenofosfónico
i acido benzenofosfonico
d Benzolphosphonsäure

884 **benzene phosphorus dichloride**
f dichlorophosphorobenzène
c dicloruro de benceno fosforoso
i benzenefosforodicloruro
d Benzolphosphordichlorid

885 **benzene phosphorus oxydichloride**
f oxydichlorophosphorobenzène
e oxidicloruro de benceno fosforoso
i benzenefosforoossidicloruro
d Benzolphosphoroxydichlorid

886 **benzene ring**
f anneau de benzène
e anillo de benceno
i formula del benzene
d Benzolring

887 **benzene series**
f séries benzénique
e serie bencénica
i serie benzenica
d Benzolreihe

888 **benzenesulphinic acid**
f acide benzène-sulfonique
e ácido bencenosulfónico
i acido benzensolfonico
d Benzolsulfonsäure

889 **benzenesulphonbutylamide**
f benzène-sulfonbutylamide
e bencenosulfonbutilamida
i benzenesolfonbutilammide
d Benzolsulfonbutylamid

* **benzhexol hydrochloride** → 8480

890 **benzidine**
f benzidine
e bencidina
i benzidina
d Benzidin

891 **benzidine yellow**
f jaune de benzidine
e amarillo de bencidina
i giallo di benzidina
d Benzidingelb

892 **benzine**
f benzine
e bencina
i benzina
d Benzin

893 **benzoguanimine**
f benzoguanimine
e benzoguanimina

i benzoguanimina
d Benzoguanimin

894 **benzoic acid**
f acide benzoïque
e ácido benzoico
i acido benzoico
d Benzoesäure

895 **benzoin gum**
f benjoin
e goma benjuí
i gomma benzoe
d Benzoeharz

* **benzol** → 881

896 **benzol black**
f noir de benzol
e negro de benceno
i nero di benzolo
d Benzolschwarz

* **benzole** → 881

897 **benzol scrubber**
f scrubber à benzol
e lavador de benzol
i gorgogliatore di lavaggio al benzolo
d Benzolskrubber

898 **benzonitrile**
f nitrile benzoïque
e benzonitrilo
i benzonitrile
d Benzonitril

899 **benzophenone**
f benzophénone
e benzofenona
i benzofenone
d Benzophenon

900 **benzopurpurin**
f rouge pour coton; benzopurpurine
e benzopurpurina
i benzoporporina
d Benzopurpurin

901 **benzotrichloride**
f benzotrichlorure
e benzotricloruro
i cloruro di benzenile
d Benzotrichlorid

902 **benzotrifluoride**
f trifluorure de benzényle; benzotrifluorure
e benzotrifluoruro

i benzotrifluoro
d Benzotrifluorid

903 benzoyl peroxide
f peroxyde de benzoyle
e peróxido de benzoílo
i perossido di benzoile
d Benzoylperoxyd

904 benzyl acetate
f acétate de benzyle
e acetato de bencilo
i benzilacetato
d Benzylacetat

905 benzyl alcohol
f alcool benzylique
e alcohol bencílico
i alcool di benzile
d Benzylalkohol

906 benzylaniline
f benzylaniline
e bencilanilina
i benzilanilina
d Benzylanilin

907 benzyl benzoate
f benzylbenzoate
e benzoato de bencilo
i benzilbenzoato
d Benzylbenzoat

908 benzyl bromide
f bromure de benzyle
e bromuro de bencilo
i bromuro di benzile
d Benzylbromid

909 benzyl butyrate
f butyrate de benzyle
e butirato de bencilo
i butirrato di benzile
d Benzylbutyrat

910 benzyl cellulose
f benzylcellulose
e celulosa de bencilo
i cellulosa benzilica
d Benzylcellulose

911 benzyl chloride
f chlorure de benzyle
e cloruro de bencilo
i cloruro di benzile
d Benzylchlorid

912 benzyl cinnamate
f cinnamate de benzyle

e cinamato de bencilo
i cinnamato di benzile
d Benzylcinnamat

913 benzyl dichloride
f dichlorure de benzyle
e dicloruro de bencilo
i bicloruro di benzile
d Benzyldichlorid

914 benzyl formate
f benzylformiate
e formiato de bencilo
i benzil formiato
d Benzylformiat

915 benzylidene acetone
f benzylidène-acétone
e bencilidenacetona
i acetone di benzilidene
d Benzylidenaceton

916 benzyl isoeugenol
f isoeugénol de benzyle
e bencilisoeugenol
i isoeugenol di benzile
d Benzylisoeugenol

917 benzyl salicylate
f salicylate de benzyle
e salicilato de bencilo
i salicilato di benzile
d Benzylsalicylat; Salicylsäurebenzylester

918 benzyl thiocyanate
f thiocyanate de benzyle
e tiocianato de bencilo
i tiocianato di benzile
d Benzylthiocyanat

919 benzyltrimethylammonium chloride
f chlorure benzyltriméthylammonique
e cloruro de benciltrimetilamonio
i cloruro di benziltrimetilammonio
d Benzyltrimethylammoniumchlorid

920 berberamine
f berbéramine
e berberamina
i berberamina
d Berberamin

921 berkelium
f berkélium
e berquelio
i berchelio
d Berkelium

922 **Berlin blue**
  f bleu de Prusse
  e azul Berlín
  i azzurro di Berlino
  d Berlinerblau

923 **Berlin red**
  f rouge de Prusse
  e rojo Berlín
  i rosso di Berlino
  d Berlinerrot

924 **beryl**
  f béryl
  e berilo
  i berillo
  d Beryll

925 **beryllium**
  f béryllium
  e berilio
  i berillio
  d Beryllium

926 **beryllium metaphosphate**
  f métaphosphate de béryllium
  e metafosfato de berilio
  i metafosfato di berillio
  d Berylliummetaphosphat

927 **beryllium oxide**
  f oxyde de béryllium
  e óxido de berilio
  i ossido di berillio
  d Berylliumoxyd

  * **betacaine** → 878

  * **beta-eucaine** → 878

928 **betafite**
  f bétafite
  e betafita
  i betafite
  d Betafit

929 **Bettendorf's reagent**
  f réactif de Bettendorf
  e reactivo de Bettendorf
  i reagente di Bettendorf
  d Bettendorfsche Lösung

930 **Betterton-Kroll process**
  f procédé Betterton-Kroll
  e proceso Betterton-Kroll
  i processo di Betterton-Kroll
  d Betterton-Kroll-Verfahren

  * **bi-** → 2457

931 **Biazzi process**
  f procédé Biazzi
  e procedimiento Biazzi
  i processo di Biazzi
  d Biazzi-Verfahren

932 **bibasic**
  f bibasique
  e bibásico
  i bibasico
  d zweibasisch

933 **bibcock; bib-cock**
  f robinet à bec courbe
  e grifo de boca curva
  i rubinetto da cucinas
  d Ausflußhahn

934 **bicarbonates**
  f bicarbonates
  e bicarbonatos
  i bicarbonati
  d Bicarbonate

935 **Bicheroux process**
  f procédé Bicheroux
  e procedimiento Bicheroux
  i processo di Bicheroux
  d Bicheroux-Prozeß

936 **bichloride**
  f bichlorure
  e bicloruro
  i bicloruro
  d Bichlorid

937 **bichromate cell**
  f pile au bichromate de potassium
  e pila de bicromato potásico
  i pila bicromato di potassio
  d Bichromatzelle

938 **bichromated gelatine**
  f gélatine bichromatée
  e gelatina bicromatada
  i gelatina al bicromato
  d Chromgelatine

939 **bifunctional structural unit**
  f motif structural bifoncionnel
  e unidad estructural bifuncional
  i unità di struttura bifunzionale
  d bifunktionelle Struktureinheit

940 **bigeneric**
  f bigénérique
  e bigenérico
  i bigenerico
  d bigenerisch

941 bile acid
 f acide biliaire
 e ácido biliar
 i acido biliare
 d Gallensäure

942 bile pigment
 f pigment biliaire
 e pigmento biliar
 i pigmento biliare
 d Gallfarbstoff

943 bimetallic instrument
 f appareil bimétallique
 e aparato bimetálico
 i strumento bimetallico
 d Bimetallinstrument

944 binding agent; binder
 f liant
 e aglutinante; aglomerante
 i agente legante; cemento
 d Bindemittel

945 bioassay
 f essai biologique
 e dosaje biológico
 i saggio biologico
 d biologische Auswertung

946 bioblast; Altmann's granule
 f bioblaste
 c bioblasto
 i bioblasto
 d Bioblast

947 biochemical oxygen demand; B.O.D.
 f demande biochimique d'oxygène
 e demanda bioquímica de oxígeno
 i fabbisogno biochimico di ossigeno
 d biochemischer Sauerstoffverbrauch

948 biochemistry
 f biochimie
 e bioquimica
 i biochimica
 d Biochemie

949 biochore
 f biochore
 e biocora
 i biocora
 d Biochore

950 biogene
 f biogène
 e biógeno
 i biogeno
 d Biogen

951 biogenesis; biogeny
 f biogénèse; biogénie
 e biogénesis; biogenia
 i biogenesi; biogenia
 d Biogenese; Entstehung des Lebens

952 biogenetic
 f biogénétique
 e biogenético
 i biogenetico
 d biogenetisch

 * **biogeny** → 951

953 biological half-life
 f demi-vie biologique
 e período biólogico
 i tempo biologico di dimezzamento
 d biologische Halbwertzeit

954 biological hole
 f cavité pour matières biologiques
 e cavidad por materiales biológicos
 i cavità per materiali biologici
 d Hohlraum für biologisches Material

955 biological isolation
 f isolement biologique
 e aislamiento biológico
 i isolamento biologico
 d biologische Isolierung

956 biologic false positive reaction
 f réaction sérologique faussement positive
 e reacción serológica falsamente positiva
 i reazione serologica falsamente positiva
 d falsch-positive serologische Reaktion

957 bioluminescence
 f bioluminescence
 e bioluminiscencia
 i bioluminescenza
 d Biolumineszenz

958 biometry; biometrics
 f biométrie
 e biometría
 i biometria
 d Biometrie

959 bionics
 f bionique
 e biónica
 i bionica
 d Bionik

960 biophotometer
 f biophotomètre
 e biofotómetro

i biofotometro
d Biophotometer

961 biophysics
f biophysique
e biofisica
i biofisica
d Biophysik

962 bioplasm
f bioplasme
e bioplasma
i bioplasma
d Bioplasma

963 biopolymer
f biopolymère
e biopolimero
i biopolimero
d Biopolymer

964 biopsy
f biopsie
e biopsia
i biopsia
d Biopsie

965 biorhythm
f biorythme
e biorritmo
i bioritmo
d Biorhythmus

966 biostratigraphy
f biostratigraphie
e biostratigrafía
i biostratigrafia
d Biostratigraphie

967 biosynthesis
f biosynthèse
e biosíntesis
i biosintesi
d Biosynthese

968 biotic
f biotique
e biótico
i biotico
d biotisch

969 biotic potential
f potentiel biotique
e potencial biótico
i potenziale biotico
d biotisches Potential

970 biotin; vitamin H
f biotine

e biotina
i biotina
d Biotin

971 biotite
f biotite
e biotita
i biotite
d Magnesiaglimmer

972 biotope
f biotope
e biotopo
i biotopo
d Biotop

973 biotype
f biotype
e biotipo
i biotipo
d Biotyp

974 Birkeland-Eyde furnace
f four Birkeland-Eyde
e horno de Birkeland-Eyde
i forno di Birkeland e Eyde
d Birkeland-Eyde-Ofen

975 Bischof process
f procédé Bischof
e proceso Bischof
i processo di Bischof
d Bischof-Verfahren

976 bisethylxanthogen
f xanthogène de biséthyle
e xantogenato de bisetilo
i xantogenato di bisetile
d Diäthylxanthogen

977 bishydroxycoumarin
f bishydroxycoumarine
e bishidroxicumarina
i bisidrossicumarina
d Dihydrocumarin

978 bismanol
f bismanol
e bismanol
i bismanolo
d Bismanol

979 Bismarck brown
f brun Bismarck
e pardo de Bismarck
i bruno di Bismarck
d Bismarckbraun

* **bismite** → **986**

980 bismuth
  f bismuth
  e bismuto
  i bismuto
  d Wismut

981 bismuth ammonium citrate
  f citrate de bismuth-ammonium
  e citrato de bismuto y amonio
  i citrato di bismuto-ammonio
  d Wismutammoniumcitrat

982 bismuth chromate
  f chromate de bismuth
  e cromato de bismuto
  i cromato di bismuto
  d Wismutchromat

983 bismuth ethyl camphorate
  f éthylcamphorate de bismuth
  e etilcanforato de bismuto
  i etilcanforato di bismuto
  d Wismutäthylcamphorat

984 bismuthinite; bismuth glance
  f bismuthine
  e bismutinita; bismutina
  i bismutinite
  d Wismutinit

985 bismuth nitrate
  f nitrate de bismuth
  e nitrato de bismuto
  i nitrato di bismuto
  d Wismutnitrat

986 bismuth ochre; bismite
  f bismite
  e bismita; ocre de bismuto
  i bismite
  d Wismutocker

987 bismuth oxychloride; flake white
  f oxychlorure de bismuth; blanc de céruse
  e oxicloruro de bismuto; albayalde puro;
    blanco de España
  i ossicloruro di bismuto; bianco fiocco
  d Wismutoxychlorid

988 bismuth sodium tartrate
  f tartrate double de sodium et de bismuth
  e tartrato de bismuto y sodio
  i tartrato di bismuto-sodio
  d Wismutnatriumtartrat

989 bismuth subcarbonate
  f sous-carbonate de bismuth
  e subcarbonato de bismuto

  i subcarbonato di bismuto
  d Wismutsubcarbonat

990 bismuth subgallate
  f sous-gallate de bismuth
  e subgalato de bismuto
  i subgallato di bismuto
  d Wismutsubgallat

991 bismuth subnitrate
  f sous-nitrate de bismuth
  e subnitrato de bismuto
  i subnitrato di bismuto
  d Wismutsubnitrat

992 bismuth tannate
  f tannate de bismuth
  e tanato de bismuto
  i tannato di bismuto
  d Wismuttannat

993 bismuth telluride
  f tellurure de bismuth
  e telururo de bismuto
  i tellururo di bismuto
  d Wismuttellurid

994 bismutite
  f bismuthite
  e bismutita
  i bismutite
  d Wismutit; Wismutspat

995 bisphenol
  f diphénol; bisphénol
  e bisfenol
  i difenolo
  d Biphenol; Diphenol

996 bistre
  f bistre
  e bistre; bistro
  i bistro
  d Bister

997 bistrichlorosilyl ethane
  f bistrichlorosilylétane
  e bistriclorosililetano
  i bistriclorosililetano
  d Bistrichlorsilyläthan

998 bisulphates
  f bisulfates
  e bisulfatos
  i bisolfati
  d Bisulfate

999 bisulphites
  f bisulfites

e bisulfitos
i bisolfiti
d Bisulfite

1000 bithionol
f bithionol
e bitionol
i bitionolo
d Bithionol

1001 bitter almond
f amande amère
e almendra amarga
i mandorla amara
d Bittermandel

1002 bittering power
f pouvoir d'amertume
e poder de amargor
i valore amaro
d Bitterwert

1003 bittern
f eau mère
e agua madre (salinas)
i acque madri
d Salzmutterlauge

* **bitter spar** → **2841**

1004 bitumen
f bitume
e bitumen; asfalto; betún
i bitume
d Bitumen

1005 bituminous coal
f houille bitumineuse; charbon bitumineux
e hulla bituminosa; carbón bituminoso
i carbone bituminoso
d bituminöse Kohle; Fettkohle; Steinkohle

1006 biuret
f biuret
e biureto
i biureto
d Biuret

1007 black body
f corps noir; radiateur intégral
e cuerpo negro
i corpo nero
d schwarzer Strahler

1008 black-body radiation
f rayonnement de corps noir
e radiación del cuerpo negro
i irraggiamento del corpo nero
d schwarze Strahlung

1009 black-body temperature
f température du corps noir
e temperatura del cuerpo negro
i temperature del corpo nero
d Schwarztemperatur

1010 black copper oxide
f cuproxyde noir
e óxido negro de cobre
i biossido di rame
d schwarzes Kupferoxyd

1011 black cyanide
f cyanure noir
e cianuro negro
i cianuro nero
d schwarzes Cyanid

1012 black-damp
f mofette; anhydride carbonique
e mofeta; anhídrido carbónico
i biossido di carbonio
d Nachschwaden

1013 black diamond; carbonado
f diamant noir; carbonate
e diamante negro; carbonado
i diamante nero; carbonado
d schwarzer Diamant; Karbonado

* **black Egyptian ware** → **818**

* **black jack** → **1041**

1014 Black Japan
f vernis d'asphalte
e barniz semitransparente de secado rápido
i vernice nera semitrasparente
d Asphaltlack

1015 black lead
f graphite; mine de plomb
e plomo negro
i grafite
d Graphitschwärze

1016 black malt
f malt torréfié
e malta colorante
i malto torrefatto
d Röstmalz; Farbmalz

1017 black opal
f opale noire
e ópalo de tinte negro
i opale nero
d schwarzer Opal

1018 blackstrap molasses
  f molasses noires
  e melazas de mieles pobres
  i melasse di olio emulsionato nero
  d Restmolasse

1019 black tellurium; nagyagite
  f tellure noir; nagyagite
  e telurio negro; nagiagita
  i tellurio nero; nagyagite
  d Tellurglanz; Nagyagit

1020 black uranium oxide
  f oxyde noir d'uranium
  e óxido negro de uranio
  i ossido nero d'uranio
  d schwarzes Uranoxyd

1021 Blagden's law
  f loi de Blagden
  e ley de Blagden
  i legge di Blagden
  d Blagdensches Gesetz

1022 blanc de Chine
  f blanc de Chine
  e vidriado blanco brillante
  i bianco di Cina
  d Zinkweiß

1023 blanc fixe; constant white
  f blanc fixe
  e blanco fijo
  i bianco fisso; bianco di barite
  d Blanc fixe; Bariumsulfat

1024 bland
  f anodin
  e no irritante
  i non irritante
  d bland

1025 blastema
  f blastème
  e blastema
  i blastema
  d Keimstoff

1026 blast furnace gas
  f gaz de haut-fourneau
  e gas de alto horno
  i gas d'altoforno
  d Gichtgas; Hochofengas

1027 blasting agent
  f explosif
  e explosivo
  i agente esplosivo
  d Sprengstoff

1028 blasting gelatine
  f gélatine explosive
  e gelatina explosiva
  i gelatina esplosiva; gelatina di nitro-
      glicerina
  d Sprenggelatine

1029 blasting powders
  f poudres de mine
  e pólvoras explosivas
  i polvere da mina
  d Sprengpulver

**1030 blastocoele; segmentation cavity; cavity
      of a blastula**
  f cavité de segmentation
  e cavidad de segmentación
  i cavità di segmentazione
  d Blastozöl

1031 blastocyte
  f blastocyte
  c blastocito
  i blastocito
  d Blastocyte; Blastozyte

1032 blastomere
  f blastomère
  e blastomero
  i blastomero
  d Furchungszelle

1033 bleach *v*
  f blanchir
  e blanquear; descolorar
  i candeggiare
  d bleichen

1034 bleacher
  f agent de blanchiment
  e blanqueador; descoloreador
  i candeggiante
  d Bleichmittel; Bleicher

1035 bleaching; decoloration
  f blanchiment
  e blanqueo
  i candeggio; bianca
  d Bleichen, Entfärbung

1036 bleaching assistants
  f auxiliaires de blanchiment
  e auxiliares de blanqueo
  i sostanze ausiliarie del candeggio
  d Hilfsmittel zum Bleichen

**1037 bleaching powder; chlorinated lime;
      chloride of lime**
  f chlorure de chaux; chaux chlorée; poudre

à blanchir
e cal clorada; cloruro de cal
i cloruro di calcio
d Bleichkalk; Chlorkalk; Bleichpulver

1038 bleeding *(of paint and adhesives)*
f pouvoir de saigner
e corrimiento del color
i sanguinamento
d Durchschlagen

1039 bleed through
f infiltration
e penetración del adhesivo
i essudazione
d Durchschlagen

1040 blend *v*
f mélanger; combiner
e mezclar
i mescolare
d mischen

**1041 blende; black jack; zinc blende;
sphalerite**
f blende; sphalérite; zinc sulfuré
e blenda; blenda oscura; esfalerita
i blenda; solfuro di zinco; sfalerite
d Zinkblende; Blende; Sphalerit

1042 blender
f mélangeuse
e máquina mezcladora
i miscelatore
d Mischer

1043 blending
f mélange
e mezclado
i mescolanza
d Mischung

1044 blepharoplast
f blépharoplaste
e blefaroplasto
i blefaroplasto
d Basalkörperchen

1045 blister packing
f emballage en cloque; emballage "blister"
e empaque de cubierta transparente
i imballaggio "blister"; imballaggio in
recipienti di film transparenti
d tiefgezogene Klarsichtverpackung

1046 blocking medium
f moyen de blocage
e medio de bloqueo

i mezzo di bloccaggio
d Sperrmedium

1047 block pattern effect
f effet de combinaison en bloc
e efecto de combinación en bloque
i effetto di combinazione in blocco
d Sperrmustereffekt

1048 block polymer
f polymère en masse
e bloque polimero
i polimero in massa
d Blockpolymer

1049 blood albumin
f albumine du sang
e albúmina de sangre
i albumina del sangue
d Blutalbumin

1050 blood black
f charbon de sang
e carbón sangre
i nero sangue
d Blutschwarz

1051 blood corpuscles
f globules sanguins
e glóbulos sanguíneos
i corpuscoli sanguigni
d Blutkörperchen

* bloom → 2932

1052 blooming
f turbidité
e turbieza
i velamento torbidità
d Trübung; Wolkenbildung

1053 bloom oil
f huile de colophane
e aceite de colofonia
i olio di colofonia; olio di resina
d Harzöl

1054 blow case
f monte-jus
e montajugos
i montaliquidi
d Druckbirne

1055 blower
f soufflante
e soplante
i soffiatore
d Gebläse

1056 blowing agents
 f agents gonflants
 e agentes esponjantes; agentes expansivos
 i agenti soffianti; agenti rigonfianti
 d Blähmittel; Treibmittel

1057 blowing-down
 f purge de chaudière
 e purga de calderas
 i spurgo di caldaia
 d Ausblasen

1058 blown asphalt
 f asphalte soufflé
 e asfalto soplado; asfalto oxidado
 i asfalto soffiato
 d geblasener Asphalt

1059 blown bitumen
 f bitume soufflè
 e betun soplado
 i bitume soffiato
 d geblasenes Bitumen

1060 blown linseed oil
 f huile de lin soufflée
 e aceite de linaza soplado
 i olio di semi di lino soffiato
 d gesprühtes Leinöl

1061 blown oils
 f huiles soufflées
 e aceites soplados
 i oli soffiati
 d gesprühte Öle

1062 blowpipe
 f chalumeau
 e caña de soplar
 i cannello ferruminatorio
 d Brenner

1063 blowpipe test
 f essai au chalumeau
 e ensayo con el soplete
 i analisi al cannello ferruminatorio
 d Brennerprobe

1064 blubber
 f blanc de baleine
 e grasa de ballena
 i grasso di balena
 d Walspeck

1065 blue asbestos; crocidolite
 f amiante bleue; crocidolite
 e crocidolita
 i asbesto blu; crocidolite
 d blauer Asbest; Krokydolith

1066 blue dip
 f bain bleu
 e baño azul
 i bagno azzurro
 d Blaubrenne

1067 blue frit
 f fritte bleue
 e frita azul
 i fritta azzurra (per smalti)
 d blaue Fritte

1068 blue gas; bluewater gas
 f gaz bleu; gaz à l'eau
 e gas azul; gas de agua
 i gas blu; gas d'acqua
 d Blauwassergas; Wassergas

1009 blueing
 f bleuissement
 e pavonado en azul
 i azzurraggio
 d bläuen

 * **blueprinting** → 2267

1070 blue verditer
 f bleu de cuivre
 e verditer azul
 i azzurro di Brema
 d Kupferblau

1071 blue vitriol
 f vitriol bleu
 e vitriolo azul
 i vetriolo azzurro
 d Blaustein; Kupfervitriol

 * **bluewater gas** → 1068

1072 blunge *v*
 f gâcher, tremper
 e amasar
 i impastare (argilla)
 d mit Wasser vermischen (Ton); anfeuchten

1073 blushing
 f opalescence; turbidité
 e turbieza
 i torbidità
 d Trübung; Mattierung

 * **boart** → 1130

1074 boat varnish
 f vernis marine
 e pintura impermeabilizante
 i vernice impermeabilizzante
 d Bootslack

* **B.O.D.** → 947

1075 body *v*
f épaissir; donner du corps
e dar cuerpo; aumentar la viscosidad
i incorporare
d verdicken

1076 body *(paint)*
f corps; consistance
e cuerpo; densidad; consistencia
i densità
d Zähflüssigkeit

1077 body colours *(paint)*
f couleurs opaques
e colores consistentes
i colori molto consistenti
d Deckfarben

1078 bodying speed
f vitesse d'accroissement de la viscosité
e velocidad de incrementado de viscosidad
i velocità di indurimento
d Geschwindigkeit des Anwachsens der
  Viskosität

1079 bodying up
f épaississement
e espesado de cuerpo; incrementado de
  cuerpo
i addensamento
d Verdicken

1080 boil *v* **down**
f concentrer; réduire par ébullition
e concentrar por ebullición
i concentrare
d verkochen

1081 boiled linseed oil
f huile de lin cuite
e aceite de linaza cocido
i olio di lino cotto
d Leinölfirnis

1082 boiled oil
f huile cuite
e aceite cocido; aceite hervido
i olio cotto
d Leinölfirnis

1083 boiler
f chaudière
e caldera
i caldaia
d Dampfkessel

1084 boiler efficiency
f rendement de chaudière
e rendimiento de la caldera
i rendimento di caldaia
d Kesselwirkungsgrad

1085 boiler scale
f tartre
e incrustación de calderas
i incrostatura di caldaia
d Kesselniederschlag;
  Kesselsteinablagerung

1086 boiling
f ébullition
e ebullición
i ebollizione
d Kochen; Sieden

1087 boiling fermentation
f fermentation bouleuse
e fermentación tumultuosa
i fermentazione in ebollizione
d kochende Gärung

1088 boiling flask
f ballon
e matraz
i ballone
d Kochkolben

1089 boiling plate
f plaque chauffante
e placa hirviente
i piastra di fornello elettrico
d Kochplatte

1090 boiling point
f point d'ébullition
e punto de ebullición
i punto di ebollizione
d Siedepunkt

1091 bold
f en pièces
e de componentes grandes
i in pezzi
d grobkörnig

1092 bole
f bol
e tazón
i bolo
d große Bille

1093 bolometer
f bolomètre
e bolómetro

i bolometro
d Bolometer

1094 bolting cloth
f toile pour blutage; toile à tamis
e estameña; tela para tamices
i garza per imbozzimatura
d Müllergaze

1095 bomb calorimeter
f bombe calorimétrique; calorimètre de Berthelot
e calorímetro de combustión
i calorimetro con bomba calorimetrica
d kalorimetrische Bombe

1096 bond
Adhesion between a coating and the surface to which it is applied.
f liaison
e ligazón; vínculo; unión
i agglomerante
d Bindung; Verklebung

1097 bond
The valency link between two or more atoms in a molecule.
f liaison
e ligazón
i legame
d Bindung

1098 bond direction
f direction de la valence
e dirección de la valencia
i direzione della valenza
d Valenzrichtung

1099 bonding
f collage
e ligazón; pegamiento
i collegamento
d Binden

1100 bonding agent
f liant; ciment
e agente de liga
i agente legante
d Bindemittel

1101 bonding orbital
f orbite électronique commune
e órbita electrónica común
i orbita elettronica comune
d gemeinsame Elektronenbahn

1102 bond moment
f moment de liaison
e momento de enlace

i momento di legame
d Bindungsmoment

1103 bond orbital
f orbite de liaison
e órbita de enlace
i orbita di legame
d Bindungsbahn

1104 bond strength
f force de liaison
e fuerza de enlace
i forza di legame
d Bindekraft

1105 bone ash
f cendre d'os
e ceniza de huesos
i solfato di calcio; cenere d'ossa
d Knochenasche

1106 bone black
f noir d'os; noir animal
e carbón de huesos; carbón animal
i nero d'ossa
d Knochenkohle; Knochenschwarz

1107 bone charcoal
f charbon d'os
e carbón de huesos
i carbone di ossa
d Knochenkohle

1108 bone china
f porcelaine phosphatée
e porcelana fosfatada
i porcellana dura
d feines Steingut

1109 bone crusher
f broyeur pour os
e triturador para huesos
i frantoio per ossa
d Knochenzerkleinerungsmaschine

1110 bone marrow; marrow
f moelle osseuse
e médula ósea
i midollo osseo
d Knochenmark

1111 bone oil; Dippel's oil
f huile d'os; huile de Dippel
e aceite de huesos; aceite animal; aceite de Dippel
i olio di ossa; olio di Dippel
d Knochenöl; Dippelsches Öl

1112 bone seeker
f substance ostéophile
e substancia osteófila
i sostanza osseofila
d Knochensucher

1113 booster-expander
f machine d'expansion à compresseur
e máquina de expansión con compresor
i macchina ad espansione con compressore
d Expansionsmaschine mit Kompressor

1114 borax
f borax
e bórax
i borace
d Borax

1115 borax pentahydrate
f borax pentahydrate
e bórax pentahidratado
i pentraidrato di borace
d Boraxpentahydrat

1116 Bordeaux mixture
f bouillie bordelaise
e caldo bordelés; calco de Burdeos
i miscela di Bordeaux
d Bordeaux-Brühe

* **Bordeaux turpentine** → 3655

1117 boric acid
f acide borique
e ácido bórico
i acido borico
d Borsäure

1118 boric oxide
f anhydride borique
e óxido bórico
i anidride borica
d Boroxyd

1119 borneol
f bornéol
e borneol
i borneolo
d Borneol

1120 bornite
f cuivre panaché
e bornita
i bornite
d Bornit; Buntkupferkies

1121 bornyl acetate
f bornylacétate
e acetato de bornilo

i bornil acetato
d Bornylacetat

1122 bornylisovaleriate
f isovalérianate de bornyle
e isovalerianato de bornilo
i isovalerianato di bornile
d Bornylisovalerianat

1123 boron
f bore
e boro
i boro
d Bor

1124 boron carbide
f carbure de bore
e carburo de boro
i carburo di boro
d Borkarbid

* **boron fluoride** → 1129

1125 boron hydride; diborane
f hydrure de bore; diborane
e hidruro de boro; diborano
i idruro di boro; diborano
d Borhydrid; Diboran

1126 boron nitride
f nitrure de bore
e nitruro de boro
i nitruro di boro
d Bornitrid

1127 boron tribromide
f tribromure de bore
e tribromuro de boro
i tribromuro di boro
d Bortribromid

1128 boron trichloride
f trichlorure de bore
e tricloruro de boro
i tricloruro di boro
d Bortrichlorid

1129 boron trifluoride; boron fluoride
f trifluorure de bore
e trifluoruro de boro
i trifluoruro di boro
d Bortrifluorid

1130 bort; ballas; bortz; boart
f boort
e bort; diamante industrial; diamante negro
i diamante industriale; diamante bort
d Bort

1131 Bosch-Meiser process
 f procédé Bosch-Meiser
 e método de Bosch-Meiser
 i processo di Bosch-Meiser
 d Bosch-Meiser-Verfahren

1132 bottle
 f bouteille
 e botella
 i bombola
 d Flasche

1133 bottle-blowing machine
 f soufflante pour bouteilles
 e máquina para soplar botellas
 i soffiatrice per bottiglie
 d Flaschenblasmaschine

1134 bottleneck
 f goulot d'embouteillage
 e embotellamiento
 i strozzatura
 d Engpaß

1135 bottle washer
 f laveuse de bouteilles; machine à rincer
   les bouteilles
 e lavadora de botellas; máquina para lavar
   botellas
 i lavatrice per bottiglie
 d Flaschenspülmaschine

1136 bottling
 f embouteillage
 e embotellado
 i imbottigliamento
 d in Flaschen füllen

1137 bottling machine
 f soutireuse à bouteilles
 e llenadora de botellas
 i imbottigliatrice
 d Flaschenfüllmaschine

1138 bottling plant
 f installation pour l'embouteillage
 e planta de embotellado; instalación de
   embotellado
 i impianto di imbottigliamento
 d Flaschenabfüllanlage

1139 bottling tank
 f tank de soutirage
 e tanque a presión
 i serbatoio a pressione
 d Drucktank

1140 bottoms of distillation
 f résidus de distillation

 e residuos de destilación
 i residui di distillazione
 d Destillationsrückstände

1141 bottom yeast
 f levure basse
 e levadura de fermentación baja
 i lievito per bassa fermentazione
 d untergärige Hefe

1142 "bouncing putty"
 f mastic à forte résilience
 e masilla botadora; goma de silicona
 i mastice resiliente
 d "hüpfender Kitt"; Silikonkitt

1143 bound
 f fixé; lié
 e fijado
 i fissato
 d gebunden

1144 boundary layer
 f couche limite
 e capa límite
 i strato limite
 d Grenzschicht

1145 boundary potential
 f potentiel de limite
 e potencial de límite
 i potenziale di contatto
 d Grenzflächenpotential

1146 bound rubber
 f caoutchouc lié
 e caucho ligado
 i gomma legata
 d gebundener Kautschuk

1147 bound sulphur
 f soufre combiné
 e azufre combinado
 i zolfo combinato
 d gebundener Schwefel

 * **bournite** → 3098

1148 bournonite; wheel ore
 f bournonite
 e bournonita
 i bournonite
 d Bournonit; Bleifahlerz; Rädelerz

1149 Boyle's law
 f loi de Boyle
 e ley de Boyle
 i legge di Boyle
 d Boylesches Gesetz

1150 brachymeiosis
f brachyméiose
e braquimeiosis
i brachimeiosi
d Brachymeiosis

1151 brackish
f saumâtre
e salmastro
i salmastro
d brockig; halbsalzig

1152 bran
f son
e salvado
i crusca
d Kleie

1153 branched
f ramifié
e ramificado
i ramificato
d vernetzt

1154 branched chromosome
f chromosome fourchu
e cromosoma ramificado
i cromosoma ramificato
d verzweigtes Chromosom

1155 brass
f laiton
e latón
i ottone
d Messing

* **Brazil wax** → 1471

1156 breaking
f floculation
e separación de una emulsión
i dissociazione
d Entmischung

1157 break point
f point de rupture
e punto de ruptura
i punto di rottura
d Durchbruchpunkt

1158 Bremen blue
f bleu de Brême
e azul de Brema; azul brema
i azzurro di Brema
d Bremerblau

1159 Bremen green
f vert de Brême
e verde de Brema; verde brema
i verde di Brema
d Bremergrün

1160 brewery
f brasserie
e fábrica de cerveza
i fabbrica di birra; birrifico
d Brauerei

1161 brewing liquor
f eau de brassage
e agua de cocimiento
i acqua (adatta) per barra
d Brauwasser

1162 Brewster process
f procédé Brewster
e proceso Brewster
i processo di Brewster
d Brewster-Verfahren

1163 brightness to MgO
f brillance comparée à MgO
e brillantez comparada con óxido de magnesio puro
i splendore in confronto a ossido di magnesio puro
d Glanz im Vergleich zu Magnesiumoxyd

1164 bright stock
f brightstock
e lubricante de petróleo residual
i olio lubrificante (di alta viscosità)
d Brightstock

1165 brilliance
Clearness of varnish, lacquer etc.
f brillance
e brillantez
i lucentezza
d Glanz; Hochglanz

1166 brilliance
Clarity and brightness of colour.
f brillance
e resplandor
i splendore
d Leuchtkraft

1167 brine
f saumure
e salmuera
i soluzione salina
d Salzlösung

1168 Brinell hardness
f dureté Brinell
e dureza Brinell

i durezza Brinell
d Brinell-Härte

1169 briquetting machine
f machine pour briquettes
e briqueteadora
i pressa per mattonelle
d Brikettierpresse

1170 brittleness
f fragilité
e fragilidad
i fragilità
d Brüchigkeit

* **brittle silver ore** → 7791

1171 broad-beam absorption
f absorption de faisceau large
e absorción de haz ancho
i assorbimento di fascio largo
d Breitbündelabsorption

1172 broad-beam attenuation
f atténuation de faisceau large
e atenuación de haz ancho
i attenuazione di fascio largo
d Breitbündelschwächung

1173 broad-spectrum antibiotic
f antibiotique à large spectre d'action
e antibiótico de amplio espectro
i antibiotico a largo spettro
d Breitband-Antibiotikum

1174 brochantite
f brocantite
e brocantita
i brocantite
d Brockantit

1175 broken white
f blanchâtre
o blanco quebrado; blanco interrumpido
(por otros pigmentos)
i biancastro
d gebrochenes weiß; weißlich

1176 bromides
f bromures
e bromuros
i bromuri
d Bromide

1177 bromine
f brome
e bromo
i bromo
d Brom

1178 bromine pentafluoride
f pentafluorure de brome
e pentafluoruro de bromo
i pentafluoruro di bromo
d Brompentafluorid

1179 bromine thermal value
f valeur thermique du brome
e valor térmico del bromo
i valore termico del bromo
d Bromzahl

1180 bromine water
f eau de brome
e agua de bromo
i acqua di bromo
d Bromwasser

* **n-bromoacetamide** → 69

1181 bromoacetone
f bromacétone
e bromoacetona
i bromoacetone
d Bromaceton

1182 bromobenzene
f bromobenzène
e bromobenceno
i bromobenzene
d Brombenzol

1183 bromochlorodimethylhydantoin
f bromochlorodiméthylhydantoïne
e bromoclorodimetilhidantoína
i bromoclorobimetilidantoina
d Bromchlordimethylhydantoin

1184 bromochloroethane
f bromochloroéthane
e bromocloroetano
i bromoclorometano
d Bromchloräthan

1185 bromocresol purple
f pourpre de bromocrésol; rouge de
bromocrésol
e púrpura de bromocresol
i porpora di bromocresolo
d Bromkresolpurpur

1186 bromodiethylacetylurea
f bromodiéthylacétylurée
e bromodietilacetilurea
i bromodietilacetilurea
d Bromdiäthylacetylharnstoff

1187 bromoform
f bromoforme

   e bromoformo
   i bromoformio
   d Bromoform

1188 bromomethyl ethyl ketone
   f bromométhyléthylcétone
   e bromometiletilcetona
   i bromometiletilchetone
   d Brommethyläthylketon

1189 bromophenol blue
   f bleu de bromophénol
   e azul de bromofenol
   i azzurro di bromofenolo
   d Bromphenolblau

1190 bromophosgene
   f bromosphosgène
   e bromofosgeno
   i bromofosgene
   d Bromophosgen

1191 bromostyrol
   f bromostyrol
   e bromoestirol
   i bromostirolo
   d Bromstyrol

1192 bromothymol blue
   f bleu de bromothymol
   e azul de bromotimol
   i azzurro di bromotimolo
   d Bromthymolblau

1193 bromotrifluoromethane
   f bromotrifluorométhane
   e bromotrifluorometano
   i bromotrifluorometano
   d Bromtrifluormethan

1194 bromyrite
   f bromyrite
   e bromirita
   i bromirite
   d Bromspat

1195 bronze
   f bronze
   e bronce
   i bronzo
   d Bronze

   * **brown coal** → 4871

1196 brown crepe
   f crêpe brun
   e crepe oscuro
   i crepe bruna
   d dunkler Krepp

1197 Brownian movement
   f mouvement brownien
   e movimiento browniano
   i movimento browniano
   d Brownsche Bewegung

1198 brucine
   f brucine
   e brucina
   i brucina
   d Brucin

1199 brucite
   f brucite
   e brucita
   i brucite
   d Brucit

1200 Brunswick black
   f vernis à l'asphalte
   e negro Brunswick; barniz opaco con base de asfalto
   i vernice nera semitrasparente
   d Braunschweigerschwarz

1201 Brunswick blue
   f bleu de Brunswick
   e azul Brunswick
   i blu di ferro
   d Braunschweigerblau

1202 brushability
   f facilité d'application
   e facilidad de aplicación con brocha
   i facilità di applicazione
   d Streichfähigkeit

1203 brush binder
   f colle à pinceau
   e cola para brochas
   i colla per pennelli
   d Bürstenklebemittel

1204 brushing paint
   f peinture au pinceau
   e pintura para aplicarla con brocha
   i vernice applicabile con pennelli
   d Streichfarbe

1205 brush plating
   f dépôt au tampon
   e depósito con almohadillas
   i deposito con tampone
   d Bürstenplattierung

1206 bubble
   f bulle
   e burbuja

i bolla
d Blase

1207 bubble cap
f cloche (de barbottage)
e casquete de burbujeo
i campana di gorgogliamento
d Glocke

1208 bubble gauge
f débitmètre à bulles de gaz
e flujómetro de burbujas de gas
i flussometro a bolle di gas
d Gasblasenströmungsmesser

1209 bubble plate
f plateau de barbotage
e plato de burbujeo
i piatto a campanella
d Glockenboden

1210 bubble plate column
f colonne à plateaux de barbotage
e columna de burbujeo
i colonna di gorgogliamento
d Glockenbodenkolonne

1211 bubble tray
f plateau de barbotage
e plato de burbujeo
i vassoio di gorgogliamento
d Glockenboden

1212 bubble tray column
f colonne à plateaux de barbotage
e columna de burbujeo
i colonna di gorgogliamento
d Glockenbodenkolonne

1213 bubbling
f barbotage
e burbujeo
i gorgogliamento
d Durchblasen

1214 buffer battery
f batterie-tampon
e batería compensadora
i batteria tampone
d Pufferbatterie

1215 buffer coating
f enduit-tampon
e barniz tamponador
i vernice tampone
d Zwischenanstrich; Schutzanstrich

1216 buffer salts
f sels tampons

e sales tampones
i tampone
d Puffersubstanz

1217 buffer solution
f solution tampon
e solución tampón
i soluzione tampone
d Pufferlösung

1218 buhr-stone mill
f broyeur triturateur
e molino triturador
i macina
d Mahlgang

1219 bulbocapnine
f bulbocapnine
e bulbocapnina
i bulbocapnina
d Bulbocapnin

1220 bulb tube
f tube à boules
e tubo de bolas
i tubo a bulbo
d Kugelröhre

1221 bulk density
f masse volumique apparente
e densidad volumétrica
i densità in mucchio
d Schüttdichte

1222 bulk factor
f facteur de contraction
e relación de la densidad de una pieza
moldeada a la densidad del polvo
constituyente
i fattore di massa
d Füllkonstante; Verdichtungsgrad

1223 bulk polymerization
f polimérisation en masse
e polimerización en masa
i polimerizzazione di massa
d Blockpolymerisation

1224 buna
f buna
e caucho buna; goma sintética
i buna; gomma sintetica
d Buna

1225 bung
f bouchon
e tapón
i tappo
d Zapfen

1226 **bunghole**
f bonde
e agujero del tapón
i cocchiume
d Zapfenloch; Spundloch

1227 **bunker**
f soute
e depósito
i bunker
d Sammelbehälter

1228 **Bunsen burner**
f bec Bunsen
e mechero Bunsen
i becco Bunsen
d Bunsenbrenner

1229 **Bunsen cell**
f pile Bunsen
e elemento de Bunsen
i pila di Bunsen
d Bunsenelement

1230 **buoyancy**
f flottabilité
e flotabilidad
i galleggiabilità
d Auftrieb; Schwimmvermögen

1231 **burette**
f burette
e bureta
i buretta
d Bürette; Meßrohr

1232 **Burgundy pitch**
f poix de Bourgogne
e pez de Borgoña
i pece di Borgondia
d Burgunderpech

1233 **burning rate**
f vitesse de combustion
e velocidad de combustión
i velocità di combustione
d Brenngeschwindigkeit

1234 **burnt sienna**
f terre de Sienne brulée
e tierra de Siena calcinada
i terra di Siena bruciata
d gebrannte Sienna

1235 **burnt umber**
f terre d'ombre brûlée
e sombra quemada
i terra di ombra bruciata
d gebrannte Umbra

1236 **busulphan**
f busulphan
e busolfan
i busolfan
d Busulfan

1237 **butabarbital sodium**
f sodium-butabarbital
e butabarbital sódico
i sodio-butabarbital
d Butabarbitalnatrium

1238 **butacaine sulphate**
f sulfate de butacaïne
e sulfato de butacaína
i solfato di butacaina
d Butacainsulfat

1239 **butadiene**
f butadiène
e butadieno
i butadiene
d Butadien

1240 **butane**
f butane
e butano
i butano
d Butan

1241 **butene**
f butène
e buteno
i butilene
d Buten

1242 **butonate**
f butonate
e butonato
i butonato
d Butonat

1243 **butopyronoxyl**
f butopyronoxyl
e butopironoxilo
i butopironossil
d Butopyronoxyl

1244 **butoxyethyl laurate**
f laurate butoxyéthylique
e laurato de butoxietilo
i butossietillaurato
d Butoxyäthyllaurat

1245 **butoxyethyl oleate**
f oléate butoxyéthylique
e oleato de butoxietilo
i butossietiloleato
d Butoxyäthyloleat

1246 butoxyethylstearate
 f stéarate butoxyéthylique
 e estearato de butoxietilo
 i butossietilstearato
 d Butoxyäthylstearat

1247 butyl acetoacetate
 f acéto-acétate de butyle
 e acetoacetato de butilo
 i butilacetoacetato
 d Butylacetoacetat

1248 butyl acetoxystearate
 f acétoxystéarate de butyle
 e acetoxiestearato de butilo
 i butilacetossistearato
 d Butylacetoxystearat

1249 butyl acetyl ricinoleate
 f butylacétyl ricinoléate
 o acetilricinoleato de butilo
 i butilacetilricinoleato
 d Butylacetylricinoleat

1250 butyl acrylate
 f butylacrylate
 e acrilato de butilo
 i butilacrilato
 d Butylacrylat

1251 butyl benzenesulphonamide
 f sulfamide de butylbenzène
 e butilbencenosulfonamida
 i butilbenzenesolfonammide
 d Butylbenzolsulfonamid

1252 butyl benzoate
 f benzoate de butyle
 e benzoato de butilo
 i butilbenzoato
 d Butylbenzoat

1253 butyl benzyl phthalate
 f phtalate de butylbenzyle
 e ftalato de butilbencilo
 i butilbenzilftalato
 d Butylbenzylphthalat

1254 butyl benzyl sebacate
 f sébacate de butylbenzyle
 e sebacato de butilbencilo
 i butilbenzilsebacato
 d Butylbenzylsebacat

1255 butyl chloral
 f chloral de butyle
 e butilcloral
 i cloralio di butile
 d Butylchloral

1256 butyl cyclohexyl phthalate
 f butylcyclohexyl phtalate
 e ftalato de butilciclohexilo
 i butilcicloessilftalato
 d Butylcyclohexylphthalat

1257 butyl decyl phthalate
 f phtalate de butyle et de décyle
 e ftalato de butildecilo
 i butildecilftalato
 d Butyldecylphthalat

1258 butyl diglycol carbonate
 f carbonate double de butyle et de diglycol
 e carbonato de butilo y diglicol
 i carbonato di butildiglicol
 d Butyldiglykolcarbonat

1259 butylene glycol
 f butylène glycol
 e butilenglicol
 i butileneglicol
 d Butylenglykol

1260 butylene oxide
 f oxyde de butylène
 e óxido de butileno
 i ossido di butilene
 d Butylenoxyd

1261 butylene oxide mixtures
 f mélanges à base d'oxyde de butylène
 e mezclas de óxidos de butileno
 i miscela di ossido di butilene
 d Butylenoxydmischungen

1262 butyle oleate
 f oléate de butyle
 e oleato de butilo
 i oleato di butile
 d Butyloleat; Ölsäurebutylester

1263 butyl epoxy stearate
 f butylépoxystéarate
 e estearato de butilo epóxido
 i butilepossistearato
 d Butylepoxystearat

1264 butyl ether
 f éther butylique
 e éter butílico
 i etere butilico
 d Butyläther

1265 butyl-ethylhexyl phthalate
 f phtalate de butyle et d'éthylhexyle
 e ftalato de butiletilhexilo
 i butiletilesilftalato
 d Butyläthylhexylphthalat

1266 butyl formate
f formiate de butyle
e formiato de butilo
i formiato di butile
d Butylformiat; Ameisensäurebutylester

1267 butyl isodecyl phthalate
f phtalate de butylisodécyle
e ftalato de butilisodecilo
i butilisodecilftalato
d Butylisodecylphthalat

1268 butyl isohexyl phthalate
f phtalate de butyle et d'isohexyle
e ftalato de butilisohexilo
i butilisoesilftalato
d Butylisohexylphthalat

1269 butyl lactate
f lactate de butyle
e lactato de butilo
i lattato di butile
d Butyllactat

1270 butyl laurate
f laurate de butyle
e laurato de butilo
i laurato di butile
d Butyllaurat

1271 butyllithium
f butyllithium
e butil-litio
i litio di butile
d Butyllithium

1272 butyl methacrylate
f butylméthacrylate
e metacrilato de butilo
i butilmetacrilato
d Butylmetacrylat

1273 butyl myristate
f myristate de butyle
e miristato de butilo
i miristato di butile
d Butylmyristat

1274 butyl octyl phthalate
f butyloctyl phtalate
e ftalato de butiloctilo
i butilottilftalato
d Butyloctylphthalat

1275 butyl perbenzoate
f perbenzoate de butyle
e perbenzoato de butilo
i perbenzoato di butile
d Butylperbenzoat

1276 butyl permaleic acid
f acide butylpermaléique
e ácido butilpermaleico
i acido butilpermaleico
d Butylpermaleinsäure

1277 butyl perphthalic acid
f acide butylperphtalique
e ácido butilperftálico
i acido butilperftalico
d Butylperphthalsäure

1278 butylphosphoric acid
f acide butylphosphorique
e ácido butilfosfórico
i acido butilfosforico
d Butylphosphorsäure

1279 butyl phthalyl butyl glycolate
f butylphtalylbutylglycolate; glycolate de
   butyle et phtalylbutyle
e butilftalilglicolato de butilo
i butilftalilbutilglicolato
d Butylphthalylbutylglykolat

1280 butyl ricinoleate
f ricinoléate de butyle
e ricinoleato de butilo
i ricinoleato di butile
d Butylricinoleat

1281 butyl rubber
f butylcaoutchouc
e caucho butílico
i gomma butilica
d Butylkautschuk

1282 butyl stearamide
f stéaramide de butyle
e butilestearamida
i butilstearammide
d Butylstearamid

1283 butyl stearate
f stéarate de butyle
e estearato de butilo
i stearato di butile
d Butylstearat

1284 butynediol
f butynediol
e butinodiol
i butinediolo
d Butyndiol

1285 butynol
f butynol
e butinol

i etanolo etinilico
d Butynol

1286 butyraldehyde
  f butyraldéhyde
  e aldehído butírico
  i aldeide butirrica
  d Butyraldehyd

1287 butyric acid
  f acide butyrique
  e ácido butírico
  i acido butirrico
  d Buttersäure

1288 butyrolactone
  f butyrolactone
  e butirolactona
  i lattone butirrico
  d Butyrolacton

1289 butyronitrile
  f butyronitril
  e butironitrilo
  i nitrile butirrico
  d Butyronitril

1290 butyroyl chloride
  f chlorure de butyroyle
  e cloruro de butiroílo
  i cloruro di butirroile
  d Butyroylchlorid

1291 by-pass valve
  f vanne de dérivation
  e válvula de desvío
  i valvola di sorpasso
  d Umlaufventil

1292 by-product
  f produit secondaire; sous-produit
  e subproducto; producto secundario
  i prodotto secondario, sottoprodotto
  d Nebenprodukt

C

1293 Cabot process
f procédé Cabot
e proceso Cabot
i processo Cabot
d Cabot-Verfahren

1294 cacao beans
f grains de cacao
e frutos del cacao
i semi di cacao
d Kakaobohnen

1295 cacao butter
f beurre de cacao
e manteca de cacao
i burro di cacao
d Kakaobutter

1296 cacodylic acid
f acide cacodylique
e ácido cacodílico
i acido cacodilico
d Kakodylsäure; Cacodylsäure

1297 cade oil
f huile de cade
e aceite de cada
i olio di cade
d Kadeöl; Kadinöl

1298 cadmium
f cadmium
e cadmio
i cadmio
d Cadmium

1299 cadmium acetate
f acétate de cadmium
e acetato de cadmio
i acetato di cadmio
d Cadmiumacetat

1300 cadmium bromide
f bromure de cadmium
e bromuro de cadmio
i bromuro di cadmio
d Cadmiumbromid

1301 cadmium chloride
f chlorure de cadmium
e cloruro de cadmio
i cloruro di cadmio
d Cadmiumchlorid

1302 cadmium copper
f cuivre de cadmium
e cobre de cadmio
i rame al cadmio
d Cadmiumkupfer

1303 cadmium green
f vert de cadmium
e verde cadmio
i verde cadmio
d Cadmiumgrün

1304 cadmium hydroxide
f hydroxyde de cadmium
e hidróxido de cadmio
i idrossido di cadmio
d Cadmiumhydroxyd

1305 cadmium iodide
f iodure de cadmium
e yoduro de cadmio
i ioduro di cadmio
d Cadmiumjodid

1306 cadmium-nickel storage battery
f accumulateur au cadmium-nickel
e acumulador de cadmio-níquel
i accumulatore a nichel-cadmio
d Nickel-Cadmium-Akkumulator

1307 cadmium nitrate
f nitrate de cadmium
e nitrato de cadmio
i nitrato di cadmio
d Cadmiumnitrat

1308 cadmium oxide
f oxyde de cadmium
e óxido de cadmio
i ossido di cadmio
d Cadmiumoxyd

1309 cadmium red
f rouge de cadmium
e rojo de cadmio
i rosso di cadmio
d Cadmiumrot

1310 cadmium ricinoleate
f ricinoléate de cadmium
e ricinoleato cádmico
i ricinoleato di cadmio
d Cadmiumricinoleat

1311 cadmium selenide
f séléniure de cadmium
e seleniuro de cadmio
i seleniuro di cadmio
d Cadmiumselenid

1312 cadmium sulphate
 f sulfate de cadmium
 e sulfato de cadmio
 i solfato di cadmio
 d Cadmiumsulfat

1313 cadmium sulphide
 f sulfure de cadmium
 e sulfuro de cadmio
 i solfuro di cadmio
 d Cadmiumsulfid

1314 cadmium test
 f essai au cadmium
 e prueba al cadmio
 i prova al cadmio
 d Cadmiumprobe

1315 cadmium tungstate
 f tungstate de cadmium
 e tungstato de cadmio
 i tungstato di cadmio
 d Cadmiumwolframat

1316 cadmium yellow
 f jaune de cadmium
 e amarillo de cadmio
 i giallo di cadmio
 d Cadmiumgelb

1317 caesium; cesium
 f césium
 e cesio
 i cesio
 d Zäsium; Cäsium

1318 caesium cell
 f cellule au césium
 e célula de cesio
 i cellula fotoelettrica al cesio
 d Zäsiumzelle

1319 caesium oxygen cell
 f cellule au césium en atmosphère
   d'oxygène
 e célula de cesio en atmósfera de oxígeno
 i cellula fotoelettrica al cesio in atmosfera
   di ossigeno
 d Zäsiumsauerstoffzelle

1320 caffeine
 f caféine
 e cafeína
 i caffeina
 d Coffein; Koffein

1321 cajeput oil
 f essence de cajeput
 e aceite de cayeput

 i olio di cajeput
 d Kajuputöl

1322 cake *v*
 f se grumeler
 e agrumarse
 i raggrumarsi
 d sich klumpen

1323 calamine
 f calamine
 e calamina
 i calamina
 d Galmei; Zinkspat

1324 calandria
 f élément chauffant
 e elemento de calefacción
 i elemento riscaldatore
 d Erhitzer

1325 calaverite
 f calavérite
 e calaverita
 i calaverite
 d Calaverit; Tellurgold

1326 calcareous
 f calcaire
 e calcareo
 i calcareo
 d kalkhaltig

1327 calcareous water
 f eau calcaire
 e agua calcárea
 i acqua calcarea
 d kalkhaltiges Wasser

1328 calcination
 f calcination
 e calcinación
 i calcinazione
 d Calcinierung; Kalzination

1329 calcine *v*
 f calciner
 e calcinar
 i calcinare
 d kalzinieren

1330 calcined magnesia
 f magnésie calcinée
 e magnesia calcinada
 i magnesia calcinata
 d gebrannte Magnesia

1331 calcite; calc-spar
 f calcite

e calcita
i calcite
d Calcit; Kalkspat

1332 **calcium**
f calcium
e calcio
i calcio
d Calcium; Kalzium

1333 **calcium acetate**
f acétate de calcium
e acetato cálcico
i acetato di calcio
d Calciumacetat

1334 **calcium acetylsalicylate**
f acétylsalicylate de calcium
e acetilsalicilato cálcico
i acetilsalicilato di calcio
d Calciumacetylsalicylat

1335 **calcium acrylate**
f acrylate de calcium
e acrilato cálcico
i acrilato di calcio
d Calciumacrylat

1336 **calcium alginate**
f alginate de calcium
e alginato de calcio
i alginato di calcio
d Calciumalginat

1337 **calcium aluminate**
f aluminate de calcium
e aluminato de calcio
i alluminato di calcio
d Calciumaluminat

1338 **calcium ammonium nitrate**
f nitrate de calcium et d'ammonium
e nitrato cálcico amónico
i nitrato di ammonio e di calcio
d Calciumammoniumnitrat

1339 **calcium arsenate**
f arséniate de calcium
e arseniato cálcico
i arseniato di calcio
d Calciumarsenat

1340 **calcium arsenite**
f arsénite de calcium
e arsenito cálcico
i arsenito di calcio
d Calciumarsenit

1341 **calcium bisulphite**
f bisulfite de calcium
e bisulfito de calcio
i bisolfito di calcio
d Calciumbisulfit

1342 **calcium bromide**
f bromure de calcium
e bromuro cálcico
i bromuro di calcio
d Calciumbromid

1343 **calcium carbide**
f carbure de calcium
e carburo de calcio
i carburo di calcio
d Calciumcarbid

1344 **calcium carbonate**
f carbonate de calcium
e carbonato de calcio
i carbonato di calcio
d Calciumcarbonat; kohlensaurer Kalk

1345 **calcium chlorate**
f chlorate de chaux; chlorate de calcium
e clorato de calcio
i clorato di calcio
d Calciumchlorat

1346 **calcium chloride**
f chlorure de calcium
e cloruro de calcio
i cloruro di calcio
d Calciumchlorid; Chlorcalcium

1347 **calcium chloride cylinder**
f tube à chlorure de calcium
e tubo para cloruro de calcio
i tubo per cloruro di calcio
d Chlorcalciumzylinder

1348 **calcium chloride tube**
f tube en U pour chlorure de calcium
e tubo en U para cloruro de calcio
i tubo per cloruro di calcio
d Chlorcalciumröhre

1349 **calcium chromate**
f chromate de calcium
e cromato cálcico
i cromato di calcio
d Calciumchromat

1350 **calcium citrate**
f citrate de calcium
e citrato cálcico
i citrato di calcio
d Calciumcitrat

1351 calcium cyanamide
 f cyanamide calcique
 e cianamida cálcica
 i calciocianamide
 d Calciumcyanamid

1352 calcium cyanide
 f cyanure de calcium
 e cianuro cálcico
 i cianuro di calcio
 d Calciumcyanid

1353 calcium dehydroacetate
 f déhydroacétate de calcium
 e dehidroacetato de calcio
 i deidroacetato di calcio
 d Calciumdehydroacetat

1354 calcium fluoride
 f fluorure de calcium
 e fluoruro cálcico
 i fluoruro di calcio
 d Calciumfluorid; Fluorcalcium

1355 calcium gluconate
 f gluconate de calcium
 e gluconato cálcico
 i gluconato di calcio
 d Calciumgluconat

1356 calcium glycerophosphate
 f glycérophosphate de calcium
 e glicerofosfato cálcico
 i glicerofosfato di calcio
 d Calciumglycerinphosphat

1357 calcium hydride
 f hydrure de calcium
 e hidruro cálcico
 i idruro di calcio
 d Calciumhydrid

1358 calcium hydroxide
 f hydroxyde de calcium; chaux éteinte
 e hidróxido cálcico
 i idrossido di calcio
 d Calciumhydroxyd; gelöschter Kalk

1359 calcium hypochlorite
 f hypochlorite de calcium
 e hipoclorito cálcico
 i ipoclorito di calcio
 d Calciumhypochlorit; unterchlorsaurer Kalk

1360 calcium hypophosphite
 f hypophosphite de calcium
 e hipofosfito cálcico

 i ipofosfito di calcio
 d Calciumhypophosphit

1361 calcium iodide
 f iodure de calcium
 e yoduro cálcico
 i ioduro di calcio
 d Calciumjodid

1362 calcium lactate
 f lactate de calcium
 e lactato cálcico
 i lattato di calcio
 d Calciumlactat; milchsaurer Kalk

1363 calcium laevulinate
 f lévulinate de calcium
 e levulinato cálcico
 i levulinato di calcio
 d Calciumlävulinat; lävulinsaurer Kalk

1364 calcium linoleate
 f linoléate de calcium
 e linoleato cálcico
 i linoleato di calcio
 d Calciumlinoleat

1365 calcium magnesium chloride
 f chlorure de magnésium et de calcium
 e cloruro cálcico magnésico
 i cloruro di magnesio e calcio
 d Calciummagnesiumchlorid

1366 calcium naphthenate
 f naphténate de calcium
 e naftenato cálcico
 i naftenato di calcio
 d Calciumnaphthenat

1367 calcium nitrate
 f nitrate de calcium; nitrate de chaux
 e nitrato cálcico
 i nitrato di calcio
 d Calciumnitrat; Kalksalpeter

1368 calcium oxalate
 f oxalate de calcium
 e oxalato cálcico
 i ossalato di calcio
 d Calciumoxalat; oxalsaurer Kalk

1369 calcium oxide; quicklime
 f oxyde de calcium; chaux vive
 e óxido cálcico; cal viva
 i ossido di calcio; calce viva
 d Calciumoxyd; gebrannter Kalk; Ätzkalk

1370 calcium palmitate
 f palmitate de calcium

e palmitato cálcico
i palmitato di calcio
d Calciumpalmitat

1371 calcium pantothenate
f pantothénate de calcium
e pantotenato cálcico
i pantotenato di calcio
d Calciumpantothenat

1372 calcium permanganate
f permanganate de calcium; permanganate de chaux
e permanganato cálcico
i permanganato di calcio
d Calciumpermanganat

1373 calcium peroxide
f peroxyde de calcium
e peróxido cálcico
i perossido di calcio
d Calciumperoxyd

* **calcium phosphate, dibasic** ~
→ 2497

* **calcium phosphate, monobasic** ~
→ 5566

* **calcium phosphate, tribasic** ~
→ 8393

1374 calcium phosphide
f phosphure de calcium
e fosfuro cálcico
i fosfuro di calcio
d Calciumphosphid; Phosphorcalcium

1375 calcium plumbate
f plombate de calcium
e plombato cálcico
i piombato di calcio
d Calciumplumbat

1376 calcium pyrophosphate
f pyrophosphate de calcium
e pirofosfato cálcico
i pirofosfato di calcio
d Calciumpyrophosphat

1377 calcium resinate
f résinate de calcium
e resinato cálcico
i resinato di calcio
d Calciumresinat

1378 calcium ricinoleate
f ricinoléate de calcium
e ricinoleato cálcico

i ricinoleato di calcio
d Calciumricinoleat

1379 calcium salicylate
f salicylate de calcium
e salicilato cálcico
i salicilato di calcio
d Calciumsalicylat

1380 calcium silicate
f silicate de calcium
e silicato de calcio
i silicato di calcio
d Calciumsilikat

1381 calcium stannate
f stannate de calcium
e estannato cálcico
i stannato di calcio
d Calciumstannat

1382 calcium stearate
f stéarate de calcium
e estearato de calcio
i stearato di calcio
d Calciumstearat

1383 calcium sulphamate
f sulfamate de calcium
e sulfamato de calcio
i solfamato di calcio
d Calciumsulfamat

1384 calcium sulphate
f sulfate de calcium; plâtre
e sulfato de calcio
i solfato di calcio
d Calciumsulfat; Gips

1385 calcium sulphide
f sulfure de calcium
e sulfuro cálcico
i solfuro di calcio
d Calciumsulfid

1386 calcium sulphite
f sulfite de calcium
e sulfito cálcico
i solfito di calcio
d schwefligsaurer Kalk; Calciumsulfit

1387 calcium thiocyanate
f thiocyanate de calcium
e tiocianato cálcico
i tiocianato di calcio
d Calciumthiocyanat

1388 calcium tungstate
f tungstate de calcium

e tungstato de calcio
i tungstato di calcio
d Calciumwolframat; wolframsaures
  Calcium

1389 calcium uranyl carbonate
f carbonate uranylique de calcium
e carbonato uranílico de calcio
i carbonato uranilico di calcio
d Calciumuranylkarbonat

* **calc-spar** → **1331**

1390 calender
f calandre
e calandra
i calandra
d Kalander

1391 calendered coating
f revêtement à la calandre
e revestimiento con la calandra
i rivestimento per calandra
d Kalanderauftrag

1392 calibrate v
f calibrer
e calibrar
i calibrare
d kalibrieren

1393 caliche
f caliche
e caliche
i caliche
d Caliche

1394 californium
f californium
e californio
i californio
d Californium

1395 caliper gauge
f calibre à coulisse
e calibre plano
i calibro piatto
d Rachenlehre; Tasterlehre

1396 Callan cell
f élément Callan
e elemento de Callan; pila de Callan
i pila di Callan
d Callanelement

1397 Callaud cell
f élément Callaud
e elemento de Callaud

i pila di Callaud
d Callaudelement

1398 calomel; horn quicksilver
f calomel; chlorure mercureux
e calomelano; cloruro mercurioso
i calomelano
d Kalomel; Hornquecksilber;
  Merkurochlorid

1399 calomel half-cell; calomel electrode
f électrode au calomel; demi-cellule au
  calomel
e electrodo de calomelanos; semicelda de
  calomelanos
i elettrodo al calomelano
d Kalomelelektrode; Kalomelhalbzelle

1400 calorie
f calorie
e caloría
i caloria
d Kalorie

1401 calorific value
f pouvoir calorifique
c poder calorífico
i potere calorifico
d Heizwert

1402 calorimeter
f calorimètre
e calorímetro
i calorimetro
d Kalorimeter

1403 calorizing
f calorisation
e calorización
i calorizzazione
d Kalorisieren; Kalorisation

1404 camphane
f camphane
e canfano
i canfano
d Camphan

1405 camphene
f camphène
e canfeno
i canfene
d Camphen

1406 camphor
f camphre
e alcanfor
i canfora
d Kampfer

1407 camphoric acid
f acide camphorique
e ácido canfórico
i acido canforico
d Kampfersäure

1408 camphor oil
f huile de camphrées
e aceite de alcanfor
i olio di canfora
d Kampferöl

1409 campylite
f campylite
e campilita
i campilite
d Kampylit

1410 can
f bac
e recipiente
i bicchiere
d Becher

1411 Canada balsam
f baume du Canada
e bálsamo del Canadá
i balsamo del Canadà
d Kanadabalsam

1412 canary litharge
f jaune serin de litharge
e litargirio amarillo
i litargirio giallo
d Kanariengelb

1413 candelilla wax
f cire de candelilla
e cera de candelilla
i cera di candelilla
d Candelillawachs

1414 cane sugar
f sucre de canne
e azúcar de caña
i zucchero di canna
d Rohrzucker

1415 cannabis
f cannabis; chanvre indien
e cáñamo
i canapa
d Haschisch; indischer Hanf

1416 cannel coal
f cannel-coal
e cannel-coal; carbón de llama larga
i carbone cannel
d Kännelkohle; Flammkohle; Fackelkohle

1417 capillary flow
f courant capillaire
e flujo capilar
i flusso capillare
d Kapillarströmung

1418 capillary theory of separation
f théorie de séparation capillaire
e teoría de separación capilar
i teoria di separazione capillare
d Kapillaritätstheorie der Gastrennung

1419 capillary viscometer
f viscosimètre capillaire
e viscosímetro capilar
i viscosimetro capillare
d Kapillarviskosimeter

1420 capric acid
f acide caprique
e ácido cáprico
i acido caprico
d Caprinsäure

1421 caproic acid
f acide caproïque
e ácido caproico
i acido caproico; acido capronico
d Capronsäure

1422 caprolactam
f caprolactame
e caprolactama
i caprolattame
d Caprolactam

1423 caprylic acid
f acide caprylique
e ácido caprílico
i acido caprilico
d Caprylsäure

1424 capsicum; Cayenne pepper
f capsicum
e pimienta de Cayena
i capsico
d Capsicum; spanischer Pfeffer

1425 captan
f captan
e captano
i captano
d Captan

1426 carat
f carat
e quilate
i carato
d Karat

1427 carbamic acid
f acide carbamique
e ácido carbámico
i acido carbamico
d Carbaminsäure

1428 carbamide; urea
f carbamide; urée
e carbamida; urea
i carbamide; urea
d Carbamid; Harnstoff

1429 carbamide peroxide
f peroxyde de carbamide
e peróxido de carbamida
i perossido di carbamide
d Carbamidperoxyd; Harnstoffwasserstoff-
   superoxyd

1430 carbamide phosphoric acid
f acide uréo-phosphorique
e ácido carbamida fosfórico
i carbamido acido fosforico
d Carbamidphosphorsäure

1431 carbamyl chloride
f chlorure de carbamyle
e cloruro de carbamilo
i cloruro di carbamile
d Carbamylchlorid

1432 carbanilide; diphenyl urea
f carbanilide; diphénylurée
e carbanilida; difenilurea
i carbanilide; difenilurea
d Carbanilid; Diphenylharnstoff

1433 carbasone
f carbasone
e carbasona
i carbasone
d Carbason

1434 carbazole
f carbazol
e carbazol
i carbazolo
d Carbazol

1435 carbene
f carbène
e carbeno
i carbene
d Carben

1436 carbinol
f carbinol
e carbinol

i carbinolo
d Carbinol

1437 carbocyclic compounds
f composés carbocycliques
e compuestos carbocíclicos
i composti carbociclici
d carbocyclische Verbindungen

1438 carbohydrase
f carbohydrase
e carbohidrasa
i carboidrasi
d Carbohydrase

1439 carbohydrates
f hydrates de carbone
e hidratos de carbono
i carboidrati
d Kohlenhydrate

* **carbolic acid** → 6254

1440 carbomycin
f carbomycine
e carbomicina
i carbomicina
d Carbomycin

1441 carbon
f carbone
e carbono
i carbonio
d Kohlenstoff

* **carbonado** → 1013

1442 carbonator
f carbonisateur
e carbonatador; recipiente para
   carbonación
i carbonatore
d Saturationsapparat für Kohlendioxyd

1443 carbon black
f noir de carbone
e negro de carbón
i nerofumo
d Ruß

1444 carbon block
f bloc de carbone
e bloque de carbón
i blocco di carbonio
d Kohlenstoffblock

**1445 carbon-consuming cell; carbon-
   combustion cell**
f pile au charbon

e pila de carbón
i pila a carbone
d Kohlenstoffelement

1446 carbon cycle
f cycle du carbone
e ciclo del carbono
i ciclo del carbonio
d Kohlenstoffzyklus

1447 carbon dioxide
f bioxyde de carbone
e dióxido de carbono
i biossido di carbonio
d Kohlendioxyd

1448 carbon disulphide
f bisulfure de carbone
e disulfuro de carbono
i bisolfuro di carbonio
d Schwefelkohlenstoff

1449 carbon hexachloride; hexachloroethane
f hexachloréthane
e hexacloruro de carbono; hexacloroetano
i esacloruro di carbonio; esacloroetano
d Hexachloräthan

* **carbon hydride** → 4173

1450 carbonic acid
f acide carbonique
e ácido carbónico
i acido carbonico
d Kohlensäure

1451 carbonium ion
f ion carbénium
e ión carbonio
i carbonio ione
d Carboniumion

1452 carbonization of coal
f carbonisation de la houille
e carbonización del carbón
i carbonizzazione del carbone
d Steinkohlendestillation

1453 carbon monoxide
f oxyde de carbone
e monóxido de carbón
i ossido di carbonio; monossido di carbonio
d Kohlenoxyd; Kohlenmonoxyd

1454 carbon oxychloride
f oxychlorure de carbone
e oxicloruro de carbono
i ossicloruro di carbonio
d Kohlenoxychlorid

1455 carbon tetrachloride
f tétrachlorure de carbone
e tetracloruro de carbono
i tetracloruro di carbonio
d Tetrachlorkohlenstoff

* **carbonyl chloride** → 6320

1456 carbonyl cyanide
f cyanure carbonylé
e cianuro de carbonilo
i cianuro di carbonile
d Carbonylcyanid

1457 carbonyl group
f groupe carbonyle
e grupo carbonilo
i gruppo carbonile
d Carbonylgruppe

1458 carbonyl iron powder
f poudre de fer carbonylé
e polvo de hierro carbonilo
i polvere di ferro carbonile
d Carbonyleisenpulver

1459 carboxylic
f carboxylique
e carboxílico
i carbossilico
d karbonsauer; Karboxyl-

1460 carboy
f ballon; dame-jeanne
e damajuana; garrafa
i damigiana
d Ballon; Korbflasche

1461 carboy filter
f filtre pour bonbonne
e filtro para garrafa
i filtro da damigiana
d Ballonfilter

1462 carboy tipper
f bascule à bonbonnes
e dispositivo inclinador de garrafas
i apparecchio per inclinare le damigiane
d Ballonkipper

1463 carburetted water gas
f gaz à l'eau carburé
e gas de agua carburado
i gas d'acqua carburato
d karburiertes Wassergas

1464 carburizer
f carbonisant
e producto cementador

i carburante
d Aufkohlungsmittel

1465 carcinogens
f carcinogènes
e carcinógenos
i carcinogeni
d Cancerogene; krebserzeugende
Substanzen

1466 cardamom oil
f essence de cardamome
e aceite de cardamomo
i olio di cardamomo
d Kardamomöl

1467 caries
f carie
e caries
i carie
d Karies; Zahnfäule

1468 carminative
f carminatif
e carminativo
i carminativo
d Karminativum; Blähungsmittel

1469 carmine
f carmin
e carmín
i carminio
d Karminzinnober; Karminrot

1470 carnallite
f carnallite
e carnalita
i carnalite
d Karnallit

1471 carnauba wax; Brazil wax
f cire de carnauba
e cera carnauba; cera del Brasil
i cera di carnauba
d Karnaubawachs

1472 carnotite
f carnotite
e carnotita
i carnotite
d Carnotit

1473 Carnot's reagent
f réactif de Carnot
e reactivo de Carnot
i reagente di Carnot
d Carnotsche Reagenz

1474 carotene
f carotène
e caroteno
i carotina
d Karotin

1475 carrageenan; Irish moss
f mousse perlée; mousse d'Irlande
e musgo de Irlanda
i muschio d'Irlanda
d irländisches Moos; Perlmoos; Karrageen

1476 Carter process
f procédé Carter
e procedimiento de Carter
i processo di Carter
d Cartersches Verfahren

1477 cartridge
f cartouche à absorption
e cartuche de absorción
i cartuccio ad assorbimento
d Absorptionspatrone

1478 carvacrol
f carvacrol
e carvacrol
i carvacrolo
d Carvacrol

1479 carvone
f carvone
e carvona
i carvone
d Carvon

1480 caryogamy
f caryogamie
e cariogamia
i cariogamia
d Karyogamie

1481 caryophyllenes
f caryophyllènes
e cariofilenos
i cariofilline
d Karyophyllene

1482 Casale ammonia process
f procédé Casale
e proceso Casale
i processo Casale
d Casale-Verfahren

1483 cascade
f cascade
e cascada
i cascata
d Kaskade

1484 cascade isotope separation plant
  f installation de séparation d'isotopes en
  cascade
  e equipo de separación de isótopos en
  cascada
  i impianto di separazione d'isotopi in
  cascata
  d Kaskadenisotopentrennungsanlage

1485 case *(of a balance)*
  f coffret
  e cabinete
  i custodia
  d Gehäuse

1486 case-hardening compounds
  f composés de cémentation
  e compuestos de cementación
  i composti per cementazione
  d Einsatzhärtungsverbindungen

1487 casein
  f caséine
  e caseína
  i caseina
  d Casein; Kasein

1488 casein glue
  f colle à base caséine
  e cola a base de caseína
  i colla alla caseina
  d Caseinleim

1489 caseinogen
  f caséinogène
  e caseinógeno
  i caseinogeno
  d Caseinogen

1490 casein plastic
  f matière plastique de caséine
  e plástico de caseína
  i plastica di caseina
  d Caseinkunststoff

1491 caseous; cheesy
  f caséeux
  e caseoso
  i caseoso
  d käsig

1492 cashew resin
  f résine acajou
  e resina de anacardo
  i resina di anacardo
  d Akajouharz

1493 casinghead gasoline
  f essence de gaz naturel; gasoline naturelle
  e gasolina de gas natural
  i gasolina naturale
  d Naturgasolin

1494 cask
  f tonneau; foudre; fût
  e barril; barril de transporte
  i fusto; barile per trasporto
  d Faß; Transportfaß

1495 Cassel's yellow; lead oxychloride
  f jaune de Cassel; oxychlorure de plomb
  e amarillo de Cassel; oxicloruro de plomo
  i giallo di Cassel
  d Kasslergelb

1496 cassia oil
  f essence de cannelle de Chine
  e aceite de casia
  i olio di cassia
  d Kassiaöl

1497 cassiterite; tinstone
  f cassitérite; mine d'étain
  e casiterita; piedra de estaño
  i cassiterite
  d Kassiterit; Zinkerz; Zinkstein

1498 cast film
  f feuille mince coulée
  e película colada
  i pellicola fusa
  d Gußfilm

1499 casting
  f coulée; moulage
  e moldeo; colada
  i getto; fusione
  d Gießen

1500 casting resin
  f résine de coulée
  e resina de colada
  i resina da fusione
  d Gußharz

1501 Castner cell
  f élément Castner
  e elemento Castner
  i pila di Castner
  d Castner-Zelle

1502 castor oil
  f huile de ricin
  e aceite de ricino; aceite de castor
  i olio di ricino
  d Rizinusöl

1503 castor oil plant; Ricinus communis
  f ricin
  e planta de ricino
  i ricino
  d Rizinus; Wunderbaum

1504 catabiosis
  f catabiose
  e catabiosis
  i catabiosi
  d Altersdegeneration der Zellen

1505 catabolism
  f catabolisme
  e catabolismo; desasimilación
  i catabolismo; disassimilazione
  d Katabolismus; Abbau

1506 catagenesis
  f catagénèse
  e catagénesis
  i catagenesi
  d Katagenese

1507 catalase
  f catalase
  e catalasa
  i catalasi
  d Katalase

1508 catalysis
  f catalyse
  e catálisis
  i catalisi
  d Katalyse

1509 catalyst; catalyzer
  f catalyseur
  e catalizador
  i catalizzatore
  d Katalysator

1510 catalytic bomb
  f bombe catalytique
  e bomba catalítica
  i bomba catalitica
  d Kontaktbombe

1511 catalytic cracking; cat-cracking
  f craquage catalytique
  e cracking catalítico; termofragmentación catalítica
  i cracking catalítico; piroscissione per catalisi
  d katalytisches Kracken

1512 catalytic poison
  f inhibiteur catalytique
  e inhibidor catalítico

  i veleno catalitico
  d katalytischer Inhibitor

1513 catalytic reactor
  f réacteur catalytique
  e reactor catalítico
  i reattore catalitico
  d katalytischer Reaktor

1514 catalytic reforming
  f "reforming" catalytique
  e modificación catalítica
  i "reforming" catalitico
  d katalytisches "Reforming"

  * **catalyzer → 1509**

1515 cataphoresis
  f cataphorèse
  e cataforesis
  i cataforesi
  d Kataphorese

1516 cataplasm
  f cataplasme
  e cataplasma
  i cataplasma
  d Kataplasma

  * **cat-cracking → 1511**

1517 catechol; catechin
  f pyrocatéchine
  e catecol; pirocatequina
  i pirocatechina
  d Brenzkatechin

1518 catenane
  f caténane
  e catenano
  i catenano
  d Kettenverbindung

1519 catenation
  f caténation
  e catenación
  i catenazione
  d Kettenbildung

1520 cathepsin
  f cathepsine
  e catepsina
  i catepsina
  d Kathepsin

1521 cathetometer
  f cathétomètre
  e catetómetro

i catetometro
d Kathetometer

1522 cathode layer
f couche cathodique; cathode liquide
e capa catódica; cátodo líquido
i strato catodico; catodo liquido
d kathodische Schicht

1523 cathodic inhibitors
f inhibiteurs cathodiques
e inhibidores catódicos
i inibitori catodici
d kathodische Inhibitoren

1524 cathodic pickling
f décapage cathodique
e decapado catódico
i decapaggio catodico
d kathodisches Beizen

1525 cathodic polarization
f polarisation cathodique
e polarización catódica
i polarizzazione catodica
d Kathodenpolarisation

1526 cathodic protection
f protection cathodique
e protección catódica
i protezione catodica
d kathodischer Schutz; galvanischer
　 Korrosionsschutz

1527 cathodic reaction
f réaction cathodique
e reacción catódica
i reazione catodica
d Kathodenreaktion

1528 catholyte
f catholyte
e catolito
i catolito
d Katholyt

1529 cation
f cation
e catión
i catione
d Kation

1530 cationic reagents
f réactifs cathioniques
e agentes catiónicos
i reagenti cationici
d kationische Reagenten

1531 cationic soap
f savon cationique
e jabón catiónico
i sapone cationico
d kationische Seife

1532 caulking compound
f composé de calfatage
e compuesto para calafatear
i composto da calafatare
d Kalfaterverbindung

1533 caustic
f caustique
e cáustico
i caustico
d ätzend

1534 caustic potash
f potasse caustique
e potasa cáustica
i potassa caustica
d Ätzkali

1535 caustic soda; sodium hydroxide
f soude caustique; hydroxyde de sodium
e sosa cáustica; hidróxido sódico
i soda caustica; idrossido di sodio
d Ätznatron

1536 caustic soda cell
f pile à la soude
e pila de sosa cáustica
i pila a idrossido sodico
d Alkalielement

1537 cavitation
f cavitation
e cavitación
i cavitazione
d Hohlraum

* **cavity of a blastula** → 1030

* **Cayenne pepper** → 1424

1538 cedarwood oil
f huile de bois de cèdre
e aceite de madera de cedro
i olio di legno di cedro
d Zedernholzöl

1539 celestine; celestite
f célestine
e celestina
i celestite
d Cölestin

1540 cell
 f cellule
 e célula
 i cellula
 d Zelle

1541 Cellarius vessel
 f récipient Cellarius
 e recipiente Cellarius
 i vaso di gres a forma di sella
 d Cellarius-Kühler

1542 cell body
 f corps cellulaire
 e cuerpo celular
 i corpo cellulare
 d Zellkörper

1543 cell conjugation
 f conjugaison cellulaire
 e conjugación celular
 i coniugazione cellulare
 d Zellenkonjugation

1544 cell constant
 f constante d'une cellule électrolytique
 e constante de una celda electrolítica
 i costante di una cella
 d Elektrolysezellenkonstante

1545 cell count
 f dénombrement cellulaire
 e número de elementos
 i conteggio della cella
 d Zellwert

1546 cell division
 f division cellulaire
 e división celular
 i divisione cellulare
 d Zellteilung

1547 cellophane
 f cellophane; pellicule cellulosique
 e celofano; celofana
 i cellofane
 d Cellophan; Zellglas

1548 cellular
 f cellulaire; alvéolaire
 e celular; alveolar
 i cellulare
 d zellulär; zellig; zellförmig

1549 cellular insulation
 f isolation cellulaire
 e aislante alveolar
 i isolamento cellulare
 d Zellisolierung

1550 celluloid
 f celluloïde
 e celuloide
 i celluloide
 d Celluloid

1551 cellulose
 f cellulose
 e celulosa
 i cellulosa
 d Cellulose; Zellstoff

1552 cellulose acetate
 f acétate de cellulose
 e acetato de celulosa
 i acetato di cellulosa
 d Celluloseacetat

1553 cellulose acetate butyrate
 f acétobutyrate de cellulose
 e acetato butirato de celulosa
 i butirrato acetato di cellulosa
 d Celluloseacetobutyrat

1554 cellulose acetate propionate
 f acétate propionate de cellulose
 e acetato propionato de celulosa
 i acetato propionato di cellulosa
 d Celluloseacetopropionat

1555 cellulose ester
 f ester de cellulose
 e éster de celulosa
 i estere di cellulosa
 d Celluloseester

1556 cellulose ether
 f éther de cellulose
 e éter de celulosa
 i etere di cellulosa
 d Celluloseäther

1557 cellulose nitrate
 f nitrocellulose
 e nitrato de celulosa
 i nitrato di cellulosa
 d Cellulosenitrat

1558 cellulose nitrate sheeting
 f feuille en nitrocellulose
 e laminado de nitrato de celulosa
 i foglio di nitrato di cellulosa
 d Cellulosenitratfolie

1559 cellulose plastics
 f matières plastiques à base de cellulose
 e plásticos de celulosa
 i resine termoplastiche
 d Cellulosekunststoffe

1560 cellulose propionate
 f propionate de cellulose
 e propionato de celulosa
 i propionato di cellulosa
 d Cellulosepropionat

1561 cellulose sponge
 f éponge de cellulose
 e esponja celulósica
 i spugna di cellulosa
 d Celluloseschwamm

1562 cellulose triacetate
 f triacétate de cellulose
 e triacetato de celulosa
 i triacetilcellulosa
 d Cellulosetriacetat

1563 cellulose xanthate
 f xanthate de cellulose
 e xantato de celulosa
 i xantato di cellulosa
 d Cellulosexanthogenat

1564 cellulosics
 f dérivés cellulosiques
 e derivados celulósicos
 i cellulosici
 d Celluloseharze

1565 cell voltage
 f tension de bain
 e tensión de baño
 i tensione del bagno
 d Badspannung

1566 cement
 f ciment
 e cemento
 i cemento
 d Kitt; Zement

1567 cementation
 f cémentation
 e cementación
 i cementazione
 d Zementierung; Einsatzhärtung

1568 cementite
 f cémentite
 e cementita
 i cementite
 d Zementit; Eisenkarbid

1569 cement kiln
 f four à ciment
 e horno de cemento
 i forno da cemento
 d Zementbrennofen

1570 centigrade
 f centigrade
 e centígrado
 i scala centigrada
 d Celsiusskala

1571 centipoise
 f centipoise
 e centipoise
 i centipoise
 d Centipoise

* **central body** → 1587

1572 central spindle
 f fuseau central
 e huso central
 i fuso centrale
 d Zentralspindel

1573 centre-zero instrument
 f appareil à zéro au centre
 e instrumento de cero al centro
 i strumento con zero al centro
 d Instrument mit Nullpunkt in der Mitte

1574 centrifugal blender
 f mélangeur centrifuge
 e mezcladora centrífuga
 i mescolatore centrifugatore
 d Zentrifugalmischer

1575 centrifugal clarifying
 f purification par centrifugation
 e clarificación por centrifugación
 i purificazione mediante centrifugazione
 d Klärung durch Zentrifugieren

1576 centrifugal effect
 f effet centrifuge
 e efecto centrífugo
 i effetto centrifugo
 d Zentrifugaleffekt

1577 centrifugal filter
 f filtre centrifuge
 e filtro centrífugo
 i filtro centrifugo
 d Zentrifugalfilter

1578 centrifugal pump
 f pompe centrifuge
 e bomba centrífuga
 i pompa centrifuga
 d Schleuderpumpe; Zentrifugalpumpe

1579 centrifugal sugar
 f sucre centrifuge
 e azúcar centrífugo

i zucchero di centrifuga
d Rohzucker aus den Zentrifugen;
Zentrifugalzucker

1580 centrifuge
f séparateur centrifuge
e separador centrífugo
i separatore centrifugo
d Zentrifuge

1581 centrifuge *v*
f centrifuger
e centrifugar
i centrifugare
d abscheiden; ausschleudern;
zentrifugieren

1582 centriole
f centriole
e centriolo
i centriolo
d ⁣⁣⁣⁣

1583 centrochromatin
f centrochromatine
e centrocromatina
i centrocromatina
d Centrochromatin

1584 centromere
f centromère
e centrómero
i centromero
d Centromer

1585 centromere shift
f déplacement du centromère
e desplazamiento del centrómero
i scivolamento del centromero
d Centromerdislokation

1586 centronucleus
f noyau à chromocentre
c centronúcleo
i centronucleo
d Centronukleus

1587 centrosome; central body
f centrosome; corpuscule central
e centrosoma; corpúsculo central
i centrosoma; corpuscolo centrale
d Centrosom; Zentralkörper

1588 cephaeline
f céphaéline
e cefalina
i cefalina
d Cephaelin

1589 cera alba
f cire blanche (raffinée)
e cera blanca
i cera alba
d weißes Wachs

1590 cera flava
f cire jaune (brute)
e cera amarilla
i cera flava
d gelbes Bienenwachs

1591 ceramics
f céramique
e cerámica; alfarería
i ceramica
d Keramik

1592 ceramoplastics
f céramoplastiques
e ceramoplásticos
i plastocramiche
d Keramikkunststoffe

1593 cerargyrite; chlorargyrite; horn silver
f cérargyrite; argent corné
e querargirita; cerargirita; clorargirita;
plata córnea
i cerargirite; clorargirite; argento corneo
d Zerargyrit; Silberhornerz; Hornsilber;
Hornerz; Silberspat; Chlorargyrit

1594 cereal
f céréale
e cereal
i cereale
d Getreide

1595 cerebrose
f cérébrose
e cerebrosa
i cerebrosi
d Gehirnzucker

1596 cerebroside
f cérébroside
e cerebrosido
i cerebrosido
d Zerebrosid

1597 ceresin wax
f cire de cérésine
e cera de ceresina
i ceresina
d Ceresinwachs

1598 ceric ammonium nitrate
f nitrate cérique d'ammonium
e nitrato cérico amónico

i nitrato di ammonio cerico
d Ceriammoniumnitrat

1599 ceric hydroxide
f hydroxyde cérique
e hidróxido cérico
i idrossido cerico
d Cerihydroxyd

1600 ceric oxide
f oxyde cérique
e óxido cérico
i ossido cerico
d Cerioxyd

1601 ceric sulphate
f sulfate cérique
e sulfato cérico
i solfato cerico
d Cerisulfat

1602 cerite
f cérite
e cerita
i cerite
d Cererz; Cerinstein; Zerit

1603 cerium
f cérium
e cerio
i cerio
d Zer; Cer

1604 cerium naphthenate
f naphténate de cérium
e naftenato de cerio
i naftenato di cerio
d Cernaphthenat

1605 cermet
f composé métal-céramique
e cerametal
i metallo ceramico
d Cermet; Keramik-Metallgemisch

1606 cerous hydroxide
f hydroxyde céreux
e hidróxido ceroso
i idrossido ceroso
d Cerohydroxyd

* **certation** → 247

1607 cerumen
f cérumen
e cerumen
i cerume
d Ohrschmalz

1608 cerussite
f cérusite
c cerusita
i cerussite
d Cerussit; Bleispat

* **cesium** → 1317

1609 cetalkonium chloride
f chlorure de cétalkonium
e cloruro de cetalkonio
i cloruro di cetalconio
d Cetalkoniumchlorid

1610 cetane number
f nombre cétane
e número de cetano
i numero di cetano
d Cetanwert; Cetanzahl

1611 cetin
f cétine
e cetina
i cetina
d Zetin; Walratfett

1612 cetyl alcohol
f alcool cétylique
e alcohol cetílico
i alcool cetilico
d Cetylalkohol

1613 chain
f chaîne
e cadena
i catena
d Kette

1614 chain reaction
f réaction en chaîne
e reacción en cadena
i reazione in catena
d Kettenreaktion

1615 chalcanthite
f chalcanthite
e calcantita
i calcantite
d Chalkanthit; Blauvitriol

1616 chalcocite; copper glance
f chalcocite; cuivre éclatant
e calcocita; calcosina
i calcosina; calcosite
d Chalkosin; Kupferglanz; Kupferglas; Graukupfererz

1617 chalcopyrite; copper pyrite
f chalcopyrite; pyrite de cuivre

e calcopirita; pirita de cobre
i calcopirite
d Chalkopyrit; Kupferpyrit; Kupferkies

1618 chalcostibite
f chalcostibite
e calcostibita
i calcostibite
d Kupferantimonglanz; Chalkostibit

1619 chalcotrichite
f chalcotrichite
e calcotriquita
i calcotrichite
d Chalkotrichit; Kupferblüte

1620 chalk
f craie
e tiza; yeso
i creta; gesso
d Kreide

1621 chalking
f farinage
e formación de una capa de tiza
i difetto superficiale avente apparenza di macchia di gesso
d Ausschwitzen eines kreideähnlichen Belages

* **chalybdite** → 7369

1622 chalybeate
f ferrugineux
e calibeado
i ferruginoso
d eisenartig

1623 chamber acid
f acide de chambre
e ácido de cámaras
i acido solforico (delle camere di piombo)
d Kammersäure

1624 chamber crystals
f cristaux de chambre
e cristales de cámaras
i cristalli di acido nitroso solforico
d Kammerkristalle

1625 chamber process
f procédé des chambres de plomb
e método de las cámaras de plomo
i processo delle camere di piombo
d Bleikammerverfahren

1626 chamomille
f camomille
e manzanilla

i camomilla
d Kamille

1627 chamosite
f chamoisite
e chamosita
i camosite
d Chamosit

1628 change-can mixer
f malaxeur à cuve mobile
e mezcladora de paletas con recipientes
i mescolatore a recipiente amovibile
d Mischer mit auswechselbarem Behälter

1629 channel black
f noir au tunnel
e negro de canal; negro de túnel
i nero da canale
d Kanalruß

1630 char
f matière carbonisée
e carbón animal, carbón de huesos
i carbone animale
d verkohltes Material

1631 charcoal
f charbon de bois; braise
e carbón vegetal
i carbone di legna
d Holzkohle

1632 chaser mill
f mélangeur
e mezclador para colores
i mescolatore per colori
d Farbmühle

1633 checking
f craquelure
e cuarteado; agrietamiento en ángulo recto
i retinamento
d Netzaderbildung

1634 cheesiness
f mou
e blandura
i lentezza
d Klebrigkeit

* **cheesy** → 1491

1635 chelate
f chélate
e quelato
i chelato
d Chelat

1636 chelating agent
f chélateur
e quelator
i agente chelante
d Chelator

1637 chelation
f chélation
e quelación
i chelazione
d Chelatbildung; Chelierung

1638 chemical age
f âge chimique
e edad química
i età chimica
d chemisches Alter

1639 chemical assay
f essai chimique
e ensayo químico
i assaggio chimico
d chemische Prüfung

1640 chemical balance
f balance de précision
e balanza química de precisión
i bilancia chimica
d chemische Waage; Präzisionswaage

1641 chemical binding effect
f effet de liaison chimique
e efecto de enlace químico
i effetto di legame chimico
d Einfluß der chemischen Bindung

1642 chemical bond
f liaison chimique
e enlace químico
i legame chimico
d chemische Bindung

1643 chemical colouring
f colorage chimique
e coloreado químico
i colorazione chimica
d chemisches Färben

1644 chemical combustion
f combustion chimique
e combustión química
i combustione chimica
d chemische Verbrennung

1645 chemical constitution
f constitution chimique
e constitución química
i costituzione chimica
d chemische Zusammensetzung

1646 chemical diffusion
f diffusion chimique
e difusión química
i diffusione chimica
d chemische Diffusion

1647 chemical dosemeter
f dosimètre chimique
e dosímetro químico
i dosimetro chimico
d chemischer Dosismesser

1648 chemical element
f élément chimique
e elemento químico
i elemento chimico
d chemisches Element

1649 chemical energy
f énergie chimique
e energía química
i energia chimica
d chemische Energie

1650 chemical engineering
f génie chimique; construction chimique
e ingeniería química
i ingegneria chimica
d chemische Ingenieurwissenschaft;
  chemische Technik

1651 chemical equation
f équation chimique
e ecuación química
i equazione chimica
d Verbindungsgleichung

1652 chemical equivalent
f équivalent chimique
e equivalente químico
i equivalente chimico
d chemisches Äquivalent

1653 chemical exchange
f échange chimique
e intercambio químico
i scambio chimico
d chemischer Austausch

1654 chemical exchange reaction
f réaction à échange chimique
e reacción de cambio químico
i reazione di scambio chimico
d chemische Austauschreaktion

1655 chemical foaming
f production chimique de mousse
e alveolación química

i formazione chimica di schiuma
d chemische Schaumstoffherstellung

1656 chemical formula
f formule chimique
e fórmula química
i formula chimica
d chemische Formel

1657 chemical impurity
f quantité d'élément non-actif
e cantidad de elemento no activo
i quantità d'elemento non attivo
d Begleitelementmenge

1658 chemical indicator
f indicateur chimique
e indicador químico
i tracciante chimico
d chemischer Indikator

1659 chemical inertness
t inertie chimique
e inercia química
i inerzia chimica
d chemische Trägheit

1660 chemical lead
f plomb chimique
e plomo químico
i piombo duro
d Blei für chemische Zwecke

1661 chemically modified
f modifié chimiquement
e modificado químicamente
i modificato chimicamente
d chemisch modifiziert

1662 chemical machining
f usinage chimique
e mecanizado químico
i lavorazione chimica
d chemisches Bearbeiten

1663 chemical passivation
f passivation chimique
e pasivación química
i passivazione chimica
d chemische Passivierung

1664 chemical pickling
f décapage chimique
e decapado químico
i decapaggio chimico
d chemisches Beizen

1665 chemical plant
f usine chimique

e fábrica química
i stabilimento chimico
d chemische Fabrik

1666 chemical plating
f revêtement chimique
e galvanización química
i placcatura chimica
d chemische Plattierung

1667 chemical process
f procédé chimique
e procedimiento químico
i processo chimico
d chemischer Prozess

1668 chemical processing
f traitement chimique
e elaboracion química
i lavorazione chimica
d chemische Bearbeitung

1669 chemical property
f propriété chimique
e propiedad química
i proprietà chimica
d chemische Eigenschaft

1670 chemical protection
f addition de matériaux protecteurs
e adición de materiales protectores
i addizione di materiali protettivi
d Schutzstoffzugabe

1671 chemical protector
f protecteur chimique
e substancia protectora química
i sostanza protettrice chimica
d chemischer Schutzstoff

1672 chemical purity
f pureté chimique
e pureza química
i purezza chimica
d chemische Reinheit

1673 chemical reaction
f réaction chimique
e reacción química
i reazione chimica
d chemische Reaktion

1674 chemical reprocessing
f traitement de régénération chimique
e reprocesamiento químico
i rigenerazione chimica
d chemische Aufarbeitung

1675 **chemical resistance**
  f résistance aux agents chimiques
  e resistencia a los productos químicos
  i resistenza chimica
  d Beständigkeit gegen Chemikalien

1676 **chemical separation**
  f séparation chimique
  e separación química
  i separazione chimica
  d chemische Trennung

1677 **chemical stability**
  f stabilité chimique
  e estabilidad química
  i stabilità chimica
  d chemische Stabilität

1678 **chemical stoneware**
  f grès résistant aux acides et aux bases
  e gres químico
  i gres resistente agli acidi
  d gegen chemischen Angriff widerstands-
    fähiges Steingut

1679 **chemical stripping**
  f dépouillage chimique
  e desgalvanización química
  i degalvanizzazione chimica
  d chemische Entplattierung

1680 **chemical waste**
  f déchets chimiques
  e desechos químico
  i rifiuti chimici
  d chemischer Abfall

1681 **chemical woodpulp**
  f cellulose chimique
  e pasta de madera química
  i pasta di legno chimica
  d chemische Cellulose

1682 **chemiluminescence**
  f chimioluminescence
  e quimioluminiscencia
  i chemiluminescenza
  d Chemilumineszenz

1683 **chemisorption**
  f adsorption chimique
  e quimiadsorción
  i chemiassorbimento
  d Chemisorption

1684 **chemoceptor**
  f chimie-récepteur
  e quimiorreceptor
  i chemiorecettore
  d Chemorezeptor

1685 **chemolysis**
  f chimiolyse
  e quimiólisis
  i chemiolisi
  d Chemolyse

1686 **chemonuclear**
  f chimionucléaire
  e quimionuclear
  i chimiconucleare
  d kernchemisch

1687 **chemoresistance**
  f chimiorésistance
  e quimiorresistencia
  i chemioresistenza
  d Chemoresistenz

1688 **chemosmosis**
  f chimiosmose
  e quimiósmosis
  i chemosmosi
  d Chemosmose

1689 **chemotherapy**
  f chimiothérapie
  e quimioterapia
  i chemioterapia
  d Chemotherapie

1690 **chemotropism**
  f chimiotropisme
  e quimiotaxis
  i chemiotassi
  d Chemotaxis

1691 **chemotype**
  f chémotype
  e quimiotipo
  i chemotipo
  d Chemotyp

1692 **chenopodium oil**
  f essence de chénopode
  e aceite de quenopodio
  i essenza di chenopodio
  d Chenopodiumöl; Wurmsamenöl

1693 **Chesney process**
  f procédé Chesney
  e método Chesney
  i processo Chesney
  d Chesney-Verfahren

1694 **chessylite; azurite**
  f chessylite; azurite

e chesilita; azurita
i azzurrite
d Chessylith; Azurit; Kupferlasur;
Bergblau; Kupferblau; Lasurit

1695 chiasma
f chiasma
e quiasma
i chiasma
d Chiasma

1696 chiasma interference
f interférence chiasmatique
e interferencia quiasmática
i interferenza chiasmatica
d Chiasmainterferenz

1697 chicle
f chicle
e chicle
i chicle
d Chicle

1698 chicory
f chicorée
e achicoria
i cicoria
d Zichorie

1699 Chilean mill
f moulin chilien
e molino chileno
i molazza
d Kollergang

* **Chile salpetre** → **7571**

1700 chill-back
f réfrigérant
e aditivo refrigerador
i refrigerante
d Kühlmittel

1701 chilling
f refroidissement brusque
e enfriamiento
i gelatura
d Kühlung

1702 chilling roll
f rouleau refroidisseur
e rodillo refrigerador
i rullo refrigerante
d Kühlwalze

* **china clay** → **4601**

* **China wood oil** → **8568**

1703 Chinese blue
f bleu de Chine; bleu de Prusse de qualité
e azul chino
i blu di Cina
d Chinesischblau; Preußischblau

1704 Chinese red
f rouge de Chine
e rojo chino
i rosso di Cina
d Zinnoberrot

1705 Chinese white
f blanc de Chine
e blanco chino
i biacca all'ossido di zinco
d Zinkweiß

1706 chitin
f chitine
e quitina
i chitina
d Chitin

1707 chitinous
f chitineux
e quitinoso
i chitinoso
d chitinös

1708 chloracetaldehyde
f chloracétaldéhyde
e cloracetaldehído
i aldeide cloroacetica
d Chloracetaldehyd

1709 chloral
f chloral
e cloral
i cloralio
d Chloral

1710 chloral formamide
f formamide de chloral
e cloral formamida
i formammide di cloralio
d Chloralformamid

1711 chloral hydrate
f hydrate de chloral
e hidrato de cloral
i idrato di cloralio
d Chloralhydrat

1712 chloramine
f chloramine
e cloramina
i cloramina
d Chloramin

1713 **chloramphenicol**
  f chloramphénicol
  e cloramfenicol
  i cloromicetina
  d Chloramphenicol

1714 **chloranil**
  f chloranile
  e cloranilo
  i cloranile
  d Chloranil

  * **chlorargyrite** → 1593

1715 **chlorates**
  f chlorates
  e cloratos
  i clorati
  d Chlorate

1716 **chlorazide**
  f chlorazide
  e clorazida
  i clorazide
  d Chlorazid

1717 **chlorbenside**
  f chlorbenside
  e clorbensida
  i clorbenside
  d Chlorbensid

1718 **chlorbutanol**
  f chlorbutanol
  e clorbutanol
  i clorbutanolo
  d Chlorbutanol

1719 **chlordane**
  f chlordane
  e clordán
  i clordano
  d Chlordan

1720 **chloric**
  f chlorique
  e clórico
  i clorico
  d chlorsauer

1721 **chloric acid**
  f acide chlorique
  e ácido clórico
  i acido clorico
  d Chlorsäure

  * **chloride of lime** → 1037

1722 **chlorides**
  f chlorures
  e cloruros
  i cloruri
  d Chloride

  * **chlorinated lime** → 1037

1723 **chlorinated polypropylene**
  f polypropylène chloré
  e polipropileno clorado
  i polipropilene clorurato
  d Chlorpolypropylen

1724 **chlorinated rubber**
  f caoutchouc chloré
  e caucho clorado
  i gomma clorurata
  d Chlorkautschuk; chlorierter Kautschuk

1725 **chlorination**
  f chloration; chloruration
  e cloración
  i clorurazione
  d Chlorierung; Verchlorung

1726 **chlorine**
  f chlore
  e cloro
  i cloro
  d Chlor

1727 **chlorine dioxide**
  f dioxyde de chlore
  e dióxido de cloro
  i biossido di cloro
  d Chlordioxyd

1728 **chlorine trifluoride**
  f trifluorure de chlore
  e trifluoruro de cloro
  i trifluoruro di cloro
  d Chlortrifluorid

1729 **chlorine water**
  f eau de chlore
  e agua de cloro
  i acqua di cloro
  d Chlorwasser

1730 **chlorisondamine chloride**
  f chlorure de chlorisondamine
  e cloruro de clorisondamina
  i cloruro di clorisondamina
  d Chlorisondaminchlorid

1731 **chlormerodrin**
  f chlormérodrine
  e clormerodrina

i clormerodrina
d Chlormerodrin

1732 chloroacetic acid
  f acide chloroacétique
  e ácido cloroacético
  i acido cloroacetico
  d Chloressigsäure; Monochloressigsäure

1733 chloroacetone
  f chloroacétone
  e cloroacetona
  i cloroacetone
  d Chloraceton

1734 chloroacetyl chloride
  f chlorure de chloroacétyle
  e cloruro de cloroacetilo
  i cloruro di cloroacetile
  d Chloracetylchlorid

1735 chloroazodin
  f chloroazodine
  e cloroazodina
  i cloroazodina
  d Chlorazodin

1736 chlorobenzene
  f chlorobenzène
  e clorobenceno
  i clorobenzene
  d Chlorbenzol

1737 chlorobutanol
  f chlorobutanol
  e clorobutanol
  i clorobutanolo
  d Acetonchloroform; Chlorobutanol

1738 chlorodifluoroacetic acid
  f acide difluorochloracétique
  e ácido clorodifluoroacetico
  i acido clorodifluoroacetico
  d Chlordifluoressigsäure

1739 chlorodifluoromethane
  f difluorochlorométhane
  e clorodifluometano
  i clorodifluorometano
  d Chlordifluormethan

  * **chloroethylene → 8750**

1740 chloroform
  f chloroforme
  e cloroformo
  i cloroformio
  d Chloroform

1741 chloroguanidine hydrochloride
  f chlorhydrate de chloroguanidine
  e hidrocloruro de cloroguanidina
  i idrocloruro di cloroguanidina
  d Chlorguanidinhydrochlorid

1742 chlorohydrin
  f chlorhydrine
  e clorhidrina
  i cloroidrina
  d Chlorhydrin

1743 chlorohydroquinone
  f chlorhydroquinone
  e clorohidroquinona
  i cloroidrochinone
  d Chlorhydrochinon

1744 chloromaleic anhydride
  f anhydride chloromaleïque
  e anhídrido cloromaleico
  i anidride cloromaleica
  d Chloromaleinsäureanhydrid

1745 chlorometer
  f chloromètre
  e clorómetro
  i clorometro
  d Chlormesser

1746 chloromethylphosphonic acid
  f acide chlorométhylphosphonique
  c ácido clorometilfosfónico
  i acido clorometilfosfonico
  d Chlormethylphosphonsäure

1747 chloromethylphosphonic dichloride
  f dichlorure chlorométhylphosphonique
  e dicloruro clorometilfosfónico
  i bicloruro clorometilfosfonico
  d Chlormethylphosphonsäuredichlorid

1748 chloronaphthalene oils
  f huiles de chloronaphtalène
  e aceites de cloronaftaleno
  i oli di clornaftalina
  d Chlornaphthalinöle

1749 chloronaphthalene waxes
  f cires de chloronaphtalène
  e ceras de cloronaftaleno
  i cere di clornaftalina
  d Chlornaphthalinwachse

1750 chloronitroacetophenone
  f chloronitroacétophénone
  e cloronitroacetofenona
  i cloronitroacetofenone
  d Chlornitroacetophenon

1751 chlorophenol
 f chlorophénol
 e clorofenol
 i clorofenolo
 d Chlorphenol

1752 chlorophenol red
 f rouge de chlorophénol
 e rojo de clorofenol
 i rosso di clorofenolo
 d Chlorphenolrot

1753 chlorophyll
 f chlorophylle
 e clorofila
 i clorofilla
 d Chlorophyll

1754 chloropicrin
 f chloropicrine
 e cloropicrina
 i cloropicrina
 d Chlorpicrin

1755 chloroplast
 f chloroplaste
 e cloroplastidio; cloroplasto
 i cloroplasto
 d Chloroplast

1756 chloroprene
 f chloroprène
 e cloropreno
 i cloroprene
 d Chloropren

1757 chloroprocaine hydrochloride
 f chlorhydrate de chloroprocaïne
 e hidrocloruro de cloroprocaína
 i idrocloruro di cloroprocaina
 d Chlorprocainhydrochlorid

1758 chloroquine
 f chloroquine
 e cloroquina
 i clorochina
 d Chloroquin

1759 chlorosulphonic acid
 f acide chlorosulfonique
 e ácido clorosulfónico
 i acido clorosolfonico
 d Chlorsulfonsäure

1760 chlorothen citrate
 f citrate de chlorothène
 e citrato de cloroteno
 i citrato di cloroteno
 d Chlorothencitrat

1761 chlorothiophenol
 f chlorothiophénol
 e clorotiofcnol
 i clorotiofenolo
 d Chlorthiophenol

1762 chlorothymol
 f chlorothymol
 e clorotimol
 i clorotimolo
 d Chlorthymol

1763 chlorotoluene
 f chlorotoluène
 e clorotolueno
 i clorotoluene
 d Chlortoluol

1764 chlorotrifluoroethylene
 f chlorotrifluoréthylène
 e clorotrifluoetileno
 i clorotrifluoroetilene
 d Chlortrifluoräthylen

1765 chlorotrifluoromethane
 f chlorotrifluorométhane
 e clorotrifluometano
 i clorotrifluorometano
 d Chlortrifluormethan

* chlorovinyl dichlorarsine → 4842

1766 chlorpheniramine maleate
 f maléate de chlorphéniramine
 e maleato de clorofeniramina
 i maleato di clorfeniramina
 d Chlorpheniraminmaleat

1767 chlorpromazine
 f chlorpromazine
 e cloropromacina
 i cloropromazina
 d Chlorpromazin

1768 chlorquinadol
 f chlorquinadol
 e clorquinadol
 i clorochinadolo
 d Chlorchinadol

1769 chlortetracycline
 f chlortétracycline
 e clorotetraciclina
 i clorotetraciclina
 d Chlortetracyclin

1770 choke-damp
 f gas étouffant
 e mofeta

i biossido di carbonio; grisú
d Nachschwaden

1771 cholagogue
f cholagogue
e colagogo
i colagogo
d Cholekinetikum

1772 cholepoiesis
f choléopoïèse
e colepoyesis
i colepoiesi
d Gallenproduktion

1773 choleretic
f cholérétique
e colerético
i coleretico
d choleretisch

1774 cholesterase
f cholestérase
e colesterasa
i colesterasi
d Cholesterase

1775 cholesterol; cholesterin
f cholestérol
e colesterol
i colesterolo
d Cholestrin; Gallenfett

1776 cholic acid
f acide cholique
e ácido cólico
i acido colico
d Gallensäure

1777 choline
f choline
e colina
i colina
d Cholin

1778 choline esterase
f cholinestérase
e colinesterasa
i colinesterasi
d Cholinesterase

1779 chondriosomal mantle
f manteau chondriosomal
e manto condriosomal
i mantello condriosomico
d Chondriosomenmantel

1780 chondriosome
f chondriosome

e condriosoma
i condriosomi
d Chondriosom

1781 chondroblast
f chondroblaste
e condroblasto
i condroblasto
d Chondroblast

1782 chondrocyte
f chondrocyte
e condrocito
i condrocita
d Knorpelzelle

1783 choroidea
f choroïde
e coroide
i coroide
d Aderhaut

1784 chroma; chromaticity
f chromaticité
e cromaticidad
i cromaticità
d Farbenskala

1785 chromaffin system
f système chromaffine
e sistema cromafin
i sistema cromaffine
d Chromaffin-System

1786 chromates
f chromates
e cromatos
i cromati
d Chromate; Chromsalze

* **chromaticity** → **1784**

1787 chromatid
f chromatide
e cromatidio
i cromatidio
d Chromatide

1788 chromatid break
f rupture chromatique
e ruptura cromatídica
i rottura cromatica
d Chromatidenbrücke

1789 chromatin
f chromatine
e cromatina
i cromatina
d Chromatin

1790 **chromatography**
  f chromatographie
  c cromatografía
  i cromatografia
  d Chromatographie

1791 **chromatoid body**
  f chromatoïde
  e cromatoide
  i cromatoide
  d Chromatoidkörper

1792 **chromatometer**
  f chromatomètre
  e cromatómetro
  i cromatometro
  d Farbenkreisel

1793 **chromatophore**
  f chromatophore
  e cromatoforo
  i cellula pigmentata
  d Chromatophor

1794 **chromatoptometer**
  f optomètre chromatique
  e cromatoptómetro
  i cromatometro ottico
  d Farbempfindlichkeitsmesser

1795 **chrome alum**
  f alum de chrome
  e alumbre crómico
  i allume di cromo
  d Chromalaun

1796 **chrome green**
  f vert de chrome
  e verde cromo
  i verde di cromo
  d Chromgrün

* **chrome iron ore** → 1813

1797 **chrome orange**
  f orangé de chrome
  e anaranjado de cromo
  i arancione di cromo
  d Chromorange

1798 **chrome red**
  f rouge de chrome
  e rojo de cromo
  i rosso di cromo
  d Chromrot

1799 **chrome scarlet; scarlet chrome**
  f écarlate de chrome
  e escarlata de cromo; cromo escarlato

  i scarlatto di cromo
  d Chromscharlachrot; Scharlachrot

1800 **chrome yellow; yellow chrome**
  f jaune de chrome
  e amarillo de cromo; cromo de amarillo
  i giallo di cromo
  d Chromgelb

1801 **chromic acetate**
  f acétate chromique
  e acetato crómico
  i acetato cromico
  d Chromacetat

1802 **chromic acid**
  f acide chromique
  e ácido crómico
  i acido cromico
  d Chromsäure

1803 **chromic acid anodizing**
  f électrolyse à l'acide chromique
  e anodización al ácido crómico
  i ossidazione anodica all'acido cromico
  d Chromsäureeloxierung

1804 **chromic acid mist**
  f brouillard d'acide chromique
  e neblina de ácido crómico
  i nebbia d'acido cromico
  d Chromsäurenebel

1805 **chromic chloride**
  f chlorure chromique
  e cloruro crómico
  i cloruro cromico
  d Chromchlorid

1806 **chromic fluoride**
  f fluorure chromique
  e fluoruro crómico
  i fluoruro cromico
  d Chromfluorid

1807 **chromic hydroxide**
  f hydroxyde de chrome
  e hidróxido crómico
  i idrossido cromico
  d Chromhydroxyd

1808 **chromic oxide**
  f sesquioxyde de chrome
  e óxido crómico
  i ossido cromico
  d Chromoxyd

1809 **chromic phosphate**
  f phosphate de chrome

e fosfato crómico
i fosfato cromico
d Chromphosphat

1810 chromic salts
f sels de chrome
e sales crómicas
i sali di cromo
d Chromisalze

1811 chromic sulphate
f sulfate chromique
e sulfato crómico
i solfato cromico
d Chromisulfat

1812 chromidium
f chromidie
e cromidio
i cromidio
d Chromidie

1813 chromite; chrome iron ore
f chromite
e cromita
i cromite
d Chromit; Chromerz

1814 chromium
f chrome
e cromo
i cromo
d Chrom

1815 chromium ammonium sulphate
f sulfate double d'ammonium et de chrome
e sulfato crómico-amónico
i solfato di cromo e ammonio
d Chromammoniumsulfat

1816 chromium naphthenate
f naphténate de chrome
e naftenato de cromo
i naftenato di cromo
d Chromnaphthenat

1817 chromium plating
f chromage
e cromado
i cromatura
d Verchromung

1818 chromium potassium sulphate
f alum de chrome
e sulfato crómico-potásico
i solfato di cromo e potassio
d Kaliumchromalaun

1819 chromocentre
f chromocentre; nœud chromatique
e cromocentro
i cromocentro
d Chromozentrum

1820 chromocyte
f chromocyte
e cromocito
i cromocito
d Chromocyte

1821 chromogenic
f chromogène
e cromogénico
i cromogenico
d chromogen; farbenerzeugend

1822 chromoisomerism; chromotropy
f chromoisomérie; chromotropie
e cromoisomerismo; cromotropía
i cromoisomerismo; cromotropia
d Chromisomerie; Chromotropie

1823 chromomere
f chromomère
e cromomero
i cromomero
d Chromomer

1824 chromophilic
f chromophile
e cromofílico
i cromofilico
d chromophil; leicht färbbar

1825 chromophobic
f chromophobe
e cromofobio
i cromofobo
d chromophob; farbfeindlich

1826 chromophores
f chromophores
e cromóforos
i cromofori
d Chromophore

1827 chromosome
f chromosome
e cromosoma
i cromosoma
d Chromosom

1828 chromosome mottling
f coloration segmentaire alternée
e coloración alternada de los segmentos cromosómicos
i colorazione alternativa dei segmenti

cromosomici
d Chromosomenmottling

1829 chromosome ring
f chromosome en anneau
e cromosoma anular
i cromosoma anulare; anello cromosomico
d Chromosomenring

1830 chromosomin
f chromosomine
e cromosomina
i cromosomina
d Chromosomin

* **chromotropy** → 1822

1831 chromous chloride
f chlorure chromeux
e cloruro cromoso
i cloruro cromoso
d Chromochlorid; Chromchlorür

1832 chromous salts
f sels chromeux
e sales cromosas
i sali cromosi
d Chromosalze

1833 chromyl chloride
f chlorure de chromyle
e cloruro de cromilo
i cloruro di cromile
d Chromylchlorid

1834 chrysarobin
f chrysarobine
e crisarrobina
i crisarobina
d Chrysarobinum

1835 chrysene
f chrysène
e criseno
i crisene
d Chrysin

1836 chrysocolla
f chrysocolle
e crisocola
i crisocolla
d Chrysokollerz; Kupfergrün;
Malachitkiesel

1837 chrysotile
f chrysotile
e crisótilo
i crisotile
d Chrysotil

1838 churning
f barattage
e batido; agitación
i agitazione della massa fusa
d Durchschütteln; Kneten

1839 chyle
f chyle
e quilo
i chilo
d Chylus; Milchsaft

1840 chylification
f formation du chyle
e quilificación
i chilificazione
d Chylusbildung

1841 chymosin; rennin; rennase
f rénine
e quimosina; renina
i chimosina; rennina
d Chymosin; Rennin; Labferment

1842 chymotrypsins
f chymotrypsines
e quimotripsinas
i chimotripsine
d Chymotrypsine

1843 cider
f cidre
e sidra
i sidro
d Apfelmost

1844 cinchona bark
f écorce de quinquina
e corteza de quina
i corteccia di china
d Chinarinde

1845 cinchonidine
f cinchonidine
e cinconidina
i cinconidina
d Cinchonidin

1846 cinchophen
f cinchophène
e cincófeno
i cincofeno
d Cinchophen

1847 cinerin
f cinérine
e cinerina
i cinerina
d Cinerin

1848 cinnabar
f cinabre
e cinabrio
i cinabro
d Zinnober

1849 cinnamic alcohol
f alcool cinnamique
e alcohol cinámico
i alcool cinnamico
d Zimtalkohol

1850 cinnamic aldehyde
f aldéhyde cinnamique
e aldehído cinámico
i aldeide cinnamica
d Zimtaldehyd

1851 cinnamon oil
f essence de cannelle
e aceite de canela
i olio di cannella
d Zimtol

1852 cinnamyl acetate
f acétate de cinnamyle
e acetato de cinamilo
i acetato di cinnamile
d Cinnamylacetat

1853 circulating pump
f pompe de circulation
e bomba de circulación
i pompa centrifuga di alimentazione
d Umwälzpumpe; Zirkulationspumpe

1854 cis-
Prefix denoting an isomer having certain
groups of atoms on the same side of the
plane.
f cis-
e cis-
i cis-
d Cis-

1855 cis-trans test
f test cis-trans
e prueba cis-trans
i test cis-trans
d Cis-Trans-Test

1856 citral
f citral
e citral
i citrale
d Citral

1857 citrates
f citrates

e citratos
i citrati
d Citrate

1858 citric acid
f acide citrique
e ácido cítrico
i acido citrico
d Citronensäure; Zitronensäure

1859 citronellal
f citronellal
e citronelal
i citronellale
d Citronellal

1860 citronella oil
f essence de citronelle
e esencia de citronela
i olio di cedrina
d Citronellöl; Citronellaöl

1861 civet
f civette
e civeto
i zibetto
d Zibet

1862 Claisen flask
f flacon de Claisen
e frasco de Claisen
i bottiglia di Claisen
d Claisensche Flasche; Claisenscher Kolben

1863 clamminess
f état visqueux
e estado viscoso
i pastosità
d Klebrigsein

1864 clarify v (a liquid)
f purifier; clarifier
e clarificar; purificar
i chiarificare; purificare
d klären; reinigen

1865 clathrate
f clathrate
e clatrato
i clatrato
d Clathrat

1866 Claude ammonia process
f procédé Claude
e proceso de Claude
i processo di Claude
d Claude-Verfahren

1867 **clay**
f argile
e arcilla
i argilla
d Ton

1868 **clear gas**
f gaz sans goudrons
e gas sin aditivos
i gas senza additivi
d Klargas

1869 **cleavage**
f clivage
e escisión
i scissione
d Spaltung

1870 **cleveite**
f clévéite
e cleveita
i cleveite
d Cleveit

1871 **clinker**
f mâchefer; scorie
e escoria
i scoria
d Schlacke

1872 **clone**
f clône
e clon
i clone
d Klon

1873 **clonicotonic**
f clonico-tonique
e clonicotónico
i clonico-tonico
d klonisch und tonisch

1874 **close contact glue**
f adhésif de contact
e adhesivo para unión por contacto entre superficies
i colla per superfici a stretto contatto
d Kontaktklebstoff

1875 **clotting**
f coagulation
e coagulación
i coagulazione
d Koagulierung

1876 **cloud chamber**
f chambre à nuages
e cámara de niebla
i camera di nebbia
d Nebelkammer

1877 **cloud point**
f point de nuage d'une huile
e punto de turbiedad
i temperatura di cristallizzazione
d Kristallisationsbeginn; Trübepunkt

1878 **clove oil**
f essence de girofle
e esencia de clavillos
i olio di garofano
d Nelkenöl

1879 **clump**
f agglutinat
e bloque
i blocco
d Masse

1880 **Clusius and Starke heavy water production process**
f procédé de production d'eau lourde de Clusius et Starke
e procedimiento de producción de agua pesada de Clusius y Starke
i processo di produzione d'acqua pesante di Clusius e Starke
d Schwerwasserherstellungsverfahren nach Clusius und Starke

1881 **Clusius column**
f colonne de Clusius
e columna de Clusius
i colonna di Clusius
d Clusiussches Trennrohr

* C.N. → 2073

1882 **C-neutrons**
f neutrons C
e neutrones C
i neutroni C
d C-Neutronen

1883 **coagulant**
f coagulant
e coagulante
i coagulante
d Koagulierungsmittel

1884 **coagulase**
f coagulase
e coagulasa
i coagulasi
d Koagulase

1885 coagulation
f coagulation
e coagulación
i coagulazione
d Koagulation; Gerinnung; Ausflockung

1886 Coahran process
f procédé Coahran
e método Coahran
i processo Coahran
d Coahran-Verfahren

1887 coal
f charbon; houille
e hulla
i carbone
d Kohle

1888 coal carbonization
f carbonisation de la houille
e carbonización de la hulla
i carbonizzazione del carbone
d Kohlenentgasung

1889 coalescence
f coalescence
e coalescencia
i aderenza
d Verwachsung

1890 coal gas
f gaz d'éclairage; gaz de houille
e gas de hulla
i gas di carbone
d Leuchtgas; Steinkohlengas

1891 coal gasification
f gazéification de la houille
e gasificación de la hulla
i gassificazione del carbone
d Steinkohlengaserzeugung

1892 coal tar
f goudron de houille
e alquitrán de hulla
i catrame di carbone
d Steinkohlenteer

    * **coal tar resins** → 2129

1893 coal water gas
f gaz double
e gas de agua y hulla
i gas d'acqua arricchito
d Kohlenwassergas

1894 coating
f revêtement
e recubrimiento
i rivestimento
d Überzug

1895 cobalt
f cobalt
e cobalto
i cobalto
d Kobalt; Cobalt

1896 cobalt bloom; erythrite
f fleur de cobalt; érythrite
e flor de cobalto; eritrita
i eritrite
d Kobaltblüte; Kobaltschlag; Erythrit

1897 cobalt blue; king's blue
f bleu de cobalt; bleu royal
e azul de cobalto; azul real
i blu di cobalto
d Kobaltblau; Königsblau, Smalte

1898 cobalt carbide
f carbure de cobalt
e carburo de cobalto
i carburo di cobalto
d Kobaltkarbid

1899 cobalt driers
f siccatifs au cobalt
e desecantes de cobalto
i essicanti di cobalto
d Kobalttrockenmittel

1900 cobalt glance
f cobaltine
e cobalto brillante
i cobaltite
d Kobaltglanz

1901 cobalt green; Rinmann's green; zinc green
f vert de cobalt; vert de Rinmann; vert de zinc
e verde de cobalto; verde de Rinmann; verde de zinc
i verde di cobalto; verde de Rinmann; verde di zinco
d Kobaltgrün; Zinkgrün

1902 cobaltic hydroxide
f hydroxyde de cobalt
e hidróxido cobáltico
i idrossido di cobalto
d Kobalthydroxyd

1903 cobaltic oxide
f sesquioxyde de cobalt
e óxido cobáltico

i ossido di cobalto
d Kobaltgrün

1904 cobaltite
f cobaltine
e cobaltita
i cobaltina
d Kobaltglanz; Cobaltin

1905 cobalto-cobaltic oxide
f oxyde cobalto-cobaltique
e óxido cobaltoso-cobáltico
i ossido cobalto-cobaltico
d Kobaltoxyduloxyd

1906 cobaltous acetate
f acétate cobalteux
e acetato cobaltoso
i acetato cobaltoso
d essigsaures Kobalt

1907 cobaltous ammonium sulphate
f sulfate d'ammonium cobalteux
e sulfato cobaltoso-amónico
i solfato d'ammonio cobaltoso
d Kobaltoammoniumsulfat

1908 cobaltous arsenate
f arséniate cobalteux
e arseniato cobaltoso
i arseniato cobaltoso
d Kobaltoarseniat

1909 cobaltous carbonate
f carbonate cobalteux
e carbonato cobaltoso
i carbonato cobaltoso
d Kobaltocarbonat

1910 cobaltous chloride
f chlorure cobalteux
e cloruro cobaltoso
i cloruro cobaltoso
d Kobaltchlorür

1911 cobaltous chromate
f chromate cobalteux
e cromato cobaltoso
i cromato cobaltoso
d Kobaltochromat

1912 cobaltous hydroxide
f hydroxyde cobalteux
e hidróxido cobaltoso
i idrossido cobaltoso
d Kobaltohydroxyd

1913 cobaltous linoleate
f linoléate cobalteux; linoléate de cobalt

e linoleato cobaltoso
i linoleato cobaltoso
d Kobaltolinoleat

1914 cobaltous naphthenate
f naphténate cobalteux
e naftenato cobaltoso
i naftenato cobaltoso
d Kobaltonaphthenat

1915 cobaltous nitrate
f nitrate cobalteux
e nitrato cobaltoso
i nitrato cobaltoso
d Kobaltonitrat

1916 cobaltous oleate
f oléate cobalteux
e oleato cobaltoso
i oleato cobaltoso
d Kobaltooleat

1917 cobaltous oxalate
f oxalate cobalteux
e oxalato cobaltoso
i ossalato cobaltoso
d Kobaltooxalat

1918 cobaltous oxide
f oxyde cobalteux
e óxido cobaltoso
i ossido cobaltoso
d Kobaltoxydul

1919 cobaltous phosphate
f phosphate cobalteux
e fosfato cobaltoso
i fosfato cobaltoso
d Kobaltophosphat

1920 cobaltous resinate
f résinate cobalteux
e resinato cobaltoso
i resinato cobaltoso
d Kobaltresinat

1921 cobaltous sulphate
f sulfate cobalteux
e sulfato cobaltoso
i solfato cobaltoso
d Kobaltsulfat

1922 cobaltous tungstate
f tungstate cobalteux
e tungstato de cobalto
i tungstato cobaltoso
d Kobaltowolframat

1923 cobalt potassium nitrite
 f nitrite de cobalt et de potassium
 e nitrito de cobalto y potasio
 i nitrito di cobalto e potassio
 d Kaliumkobaltnitrit

1924 cobalt tetracarbonyl
 f cobalt tétracarbonylé
 e cobalto tetracarbonilo
 i tetracarbonile di cobalto
 d Kobalttetracarbonyl

1925 cobalt trifluoride
 f trifluorure de cobalt
 e trifluoruro de cobalto
 i trifluoruro di cobalto
 d Kobalttrifluorid

1926 cobalt ultramarine
 f bleu d'azur; bleu de cobalt
 e ultramar de cobalto
 i oltremare di cobalto
 d Kobaltultramarin

1927 cobalt violet
 f violet de cobalt
 e violeta de cobalto
 i violetto di cobalto
 d Kobaltviolett

1928 cobwebbing
 f rayonnage
 e formación de filamentos
 i formazione di filamenti
 d Netzbildung

1929 coca
 f coca
 e coca
 i coca
 d Coca

1930 cocaine
 f cocaïne
 e cocaina
 i cocaina
 d Cocain; Kokain

1931 cocculus
 f cocculus
 e coca de Levante
 i cocculus
 d Kokkelskorn

1932 cochineal
 f cochenille
 e cochinilla
 i cocciniglia
 d Koschenillefarbstoff

1933 coconut acid
 f acide de noix de coco
 e ácido de nuez de coco
 i acido di cocco
 d Kokossäure

1934 coconut oil
 f huile de coco
 e aceite de coco
 i olio di cocco
 d Kokosnußöl

1935 codeine
 f codéine
 e codeína
 i codeina
 d Codein; Kodein

1936 cod liver oil
 f huile de foie de morue
 e aceite de hígado de bacalao
 i olio di fegato di merluzzo
 d Dorschlebertran

1937 coefficient of electrolytic dissociation
 f constante de dissociation électrolytique
 e constante de disociación electrolitica
 i costante di dissociazione elettrolitica
 d Dissoziationskonstante

1938 coenocyte
 f coenocyte
 e cenocito
 i cenocito
 d Coenocyte

1939 coenogamete
 f coenogamète
 e cenogameto
 i cenogamete
 d Coenogamet

1940 coenzyme
 f coenzyme
 e coenzima
 i coenzima
 d Coenzym; Coferment

1941 coffinite
 f coffinite
 e cofinita
 i coffinite
 d Coffinit

1942 cohesion
 f cohésion
 e cohesión
 i coesione
 d Kohäsion

1943 coke
f coke
e coque; cok
i coke
d Koks

1944 coke breaker
f concasseur de coke
e trituradora de coque
i frantumatrice per coke
d Koksbrechmaschine

1945 coke breeze
f petit coke
e menudos de coque
i impasto di cemento e scorie di coke
d Gruskoks

1946 coke grading plant
f installation de triage de coke
e instalación de clasificación de coque
i impianto classifica del coke
d Kokssortieranlage

1947 coke mill
f broyeur de coke
e molino para pulverizar coque
i molino per polverizzazione di coke
d Koksmühle

1948 coke oven
f four à coke
e horno de coque
i forno a coke
d Koksofen; Kokereiofen

1949 coking
f cokéfaction
e coquificación
i cokizzazione
d Verkokung

1950 colature
f colature
e coladura
i colatura
d Kolatur

1951 Colburn process
f procédé Colburn
e método Colburn
i processo Colburn
d Colburn-Verfahren

1952 colchicine
f colchicine
e colquicina
i colchicina
d Kolchizin

1953 colcothar; iron oxide
f colcotar; oxyde de fer
e colcótar; óxido férrico
i ossido di ferro
d Kolkothar; Eisenoxyd

1954 cold cure
f vulcanisation à froid
e vulcanización en frío
i vulcanizzazione a freddo
d Kaltvulkanisation

1955 cold-cutting
f mélange à froid
e mezcla de ingredientes en frío
i mescolatura a freddo
d Kaltmischen

1956 cold flow
f fluage à froid
e escurrimiento en frío
i scorrimento a freddo
d kalter Fluß

1957 cold polymerization
f polymérisation à froid
e polimerización en frío
i polimerizzazione a freddo
d Kaltpolymerisation

1958 cold rubber
f caoutchouc froid
e goma fría
i gomma fredda
d Tieftemperaturkautschuk

1959 cold-setting adhesive; cold-setting glue
f colle durcissable à froid
e adhesivo de fraquado en frío
i adesivo a freddo
d kaltabbindender Klebstoff

1960 colemanite
f colémanite
e colemanita
i colemanite
d Colemanit

1961 colicin
f colicine
e colicina
i colicina
d Colicin

1962 collagen
f collagène
e colágeno
i collageno
d Kollagen

1963 colliquation
f dégénérescence vacuolaire
e degeneración vacuolar
i degenerazione vacuolare
d vakuölare Kolliquation

1964 colliquative
f colliquatif
e colicuativo
i colliquativo
d kolliquativ

1965 collochemistry
f chimie des colloïdes
e coloidoquímica
i chimica dei colloidi
d Kolloidchemie

1966 collodion
f collodion
e colodión
i collodio
d Collodium; Kollodium

1967 colloid
f colloïde
e coloide
i colloide
d Kolloid

1968 colloidal electrolyte
f électrolyte colloïdal
e electrólito coloidal
i elettrolito colloidale
d Mizelle

1969 colloidal particles
f particules colloïdales
e partículas coloidales
i particelle colloidali
d kolloidale Teilchen

1970 colloidal silver
f argent colloïdal
e plata coloidal
i argento colloidale
d Kolloidsilber

1971 colloidal solution
f solution colloïdale; dissolution colloïdale
e solución coloidal
i soluzione colloidale
d kolloidale Lösung

1972 colloidal state
f état colloïdal
e estado coloidal
i stato colloidale
d Kolloidalzustand

1973 colloid equivalent
f équivalent colloïdal
e equivalente coloidal
i equivalente colloidale
d Kolloidäquivalent

1974 colloid mill
f broyeur pour colloïdes
e molino para coloides
i molino per colloidi
d Kolloidmühle

1975 colloid suspension
f suspension colloïdale
e suspensión coloidea
i sospensione colloidale
d Kolloidsuspension

1976 collutory
f collutoire
e colutorio
i colluttorio
d Mundwasser

1977 collyrium
f collyre
e colirio
i collirio
d Augenwasser

1978 Cologne spirits
f eaux de Cologne
e alcohol de Colonia
i acqua di Colonia
d Feinspirit

* **colonial spirit** → 5287

1979 colony
f colonie
e colonia
i colonia
d Kolonie

* **colophony** → 7124

1980 colorimetric method
f méthode colorimétrique
e método colorimétrico
i metodo colorimetrico
d kolorimetrisches Verfahren

1981 colorimetry
f colorimétrie
e colorimetría
i colorimetria
d Farbmessung; Kolorimetrie

1982 colour comparator
 f comparateur de couleurs
 e comparador de colores
 i comparatore di colori
 d Farbvergleicher

1983 colour fastness
 f solidité de la couleur
 e firmeza de colorante
 i resistenza di colore
 d Farbechtheit

1984 colour filter
 f filtre des couleurs
 e filtro de color
 i filtro colorato
 d Farbfilter

1985 colour migration
 f migration de couleur
 e migración de color
 i cambiamento di colore
 d Farbwanderung

1986 colours in oil
 f pigments en huile
 e pigmentos en aceite
 i pigmenti in olio
 d in Öl abgeriebene Pigmente

1987 colour stability
 f stabilité de couleur
 e estabilidad de los colores
 i stabilità dei colori
 d Farbbeständigkeit

1988 colour temperature
 f température de couleur
 e temperatura del color
 i temperatura del colore
 d Farbtemperatur

1989 columbite
 f columbite
 e columbita
 i colombite
 d Columbit

 * **columbium** → 5733

1990 column; tower
 f colonne; tour
 e columna; torre
 i colonna; torre
 d Kolonne; Säule; Turm

1991 column *(of a balance)*
 f colonne
 e columna

 i montante
 d Waagsäule

1992 colza oil
 f huile de colza
 e aceite de semilla de colza
 i olio di colza
 d Klettenöl

1993 combining ability
 f aptitude à la combinaison
 e aptitud por la combinación
 i attitudine combinatoria
 d Kombinationseignung

1994 combustible
 f combustible
 e combustible
 i combustibile
 d brennbar

1995 combustion boat
 f godet à fusion; nacelle à fusion
 e navecilla de fusión
 i navicella di fusione
 d Einsetzer; Glühschiffchen

1996 combustion index
 f indice de combustion
 e índice de combustión
 i indice di combustione
 d Verbrennungsindex; Verbrennungszahl

1997 combustion tube furnace
 f fourneau à tube à fusion
 e horno de tubo de combustión
 i fornace a tubo di fusione
 d Verbrennungsofen

1998 combustion tubing
 f tube à fusion
 e tubo de fusión
 i tubo di fusione
 d Einschmelzröhre

1999 commercial tank
 f cuve commerciale
 e cuba comercial
 i cella industriale
 d Produktionsbad

2000 comminutor
 f broyeur
 e triturador
 i sminuzzatore
 d Brecher; Zerkleinerungsmaschine

2001 common lime
 f chaux vive

e cal viva
i calce viva
d Kalk

2002　compact
f comprimé; pastille
e comprimido; pieza aglomerada
i compressa; pastiglia
d Pastille; Preßkörper; Preßling

2003　compatibility
f compatibilité
e compatibilidad
i compatibilità
d Verträglichkeit

2004　compatible
f compatible
e compatible
i compatibile
d verträglich

2005　compensating chiasma
f chiasma de compensation
e quiasma de compensación
i chiasma di compensazione
d kompensierendes Chiasma

2006　competent donor bacterium cell
f bactérie donatrice compétente
e bacteria donante competente
i batteria donatrice competente; cellula donatrice competente
d kompetentes Donor-Bakterium

2007　complementary chiasma
f chiasma complémentaire
e quiasma complementario
i chiasma complementare
d komplementäres Chiasma

2008　complex compound
f composé complexé
e compuesto complejo
i composto complesso
d Komplexverbindung

2009　complexing agent
f agent complexant
e agente secuestrante
i agente sequestrante
d Abscheidemittel

2010　complex ion
f ion complexe
e ión complejo
i ione complesso
d komplexes Ion

2011　composite plate
f dépôt composite
e depósito compuesto
i deposito composto
d Mehrschichtenüberzug; Verbundüberzug

2012　compound
f composé
e compuesto
i composto
d Verbindung; Zusammensetzung

2013　compressible flow
f débit compressible
e flujo compresible; flujo comprimible
i flusso comprimibile
d kompressible Strömung

2014　compression ratio
f taux de compression
e relación de compresión
i rapporto di compressione
d Kompressionsverhältnis; Verdichtungszahl

2015　compressor
f compresseur
e compresor
i compressore
d Kompressor; Verdichter

2016　concentrate
f concentré
e concentrado
i concentrato
d Konzentrat

2017　concentration
f concentration; enrichissement
e concentración
i concentrazione
d Konzentration; Anreicherung

2018　concentration cell
f pile de concentration
e pila de concentración
i pila di concentrazione
d Konzentrationselement

2019　concentration control
f contrôle de la concentration
e control de la concentración
i controllo della concentrazione
d Konzentrationskontrolle

2020　concentration overvoltage
f surtension de concentration
e sobretensión de concentración

i sovratensione di concentrazione
d konzentrationsbedingte Überspannung

2021 concentration plant; concentrator
f installation de concentration
e instalación de concentración
i impianto di ricupero
d Konzentrationsanlage

2022 conchoidal
f conchoïdal
e concoidal
i concoidale
d schneckenlinienförmig

2023 concrete
f béton
e hormigón
i calcestruzzo
d Beton

2024 concurrent flow
f écoulement parallèle
e flujo paralelo
i flusso parallelo
d Gleichstrom

2025 condensate
f condensat; produit
e líquido condensado
i condensa; condensato
d Kondensat

2026 condensation
f condensation
e condensación
i condensazione
d Kondensation

2027 condensation agent; condensation catalyst
f agent de condensation
e agente de condensación
i agente di condensazione
d Kondensationsmittel

2028 condensation column
f colonne de condensation
e columna de condensación
i colonna di condensazione
d Kondensationssäule

2029 condensation polymerization
f polymérisation de condensation
e polimerización de condensación
i polimerizzazione per condensazione
d Kondensationspolymerisation

2030 condensation resins
f résines de condensation
e resinas de condensación
i resine di condensazione
d Kondensationsharze

2031 condenser
Apparatus to convert a substance from the gaseous to the liquid state.
f condensateur
e condensador
i condensatore
d Kondensator; Kondensor

2032 condenser
Lens or system of lenses used to concentrate light on the object.
f lentille convergente
e lente colectora; lente condensadora
i condensatore
d Sammellinse

2033 conductimetric analysis
f analyse conductométrique
e análisis conductométrico
i analisi conduttometrica
d konduktometrische Analyse

2034 conducting salts
f sels conducteurs
e sales conductoras
i sali conduttori
d Leitsalze

2035 conductive rubber
f caoutchouc conducteur de l'électricité
e goma conductora de electricidad
i gomma conduttiva di elettricità
d elektrisch leitender Gummi

2036 conformation; configuration
f configuration; conformation
e configuración; conformación
i conformazione; configurazione
d Gestaltung; Struktur; Zusammensetzung

2037 conforming anode
f anode préformée
e ánodo preformado
i anodo preformato
d umschließende Anode

2038 congenital
f congénital
e congénito
i congenito
d konnatal

2039 conglomerate
f conglomérat
e conglomerado
i conglomerato
d Konglomerat; Trümmergestein

2040 Congo red
f rouge Congo
e rojo Congo
i rosso di Congo
d Kongorot

2041 congression
f congression
e congresión
i congressione; adunanza
d Kongression

2042 coniferin
f coniférine
e coniferina
i coniferina
d Coniferin

2043 coniine
t conèine
e coniína; conina
i cicutina
d Coniin; Koniin

2044 conjugated double bonds
f liaisons doubles conjuguées
e enlaces dobles conjugados
i doppi legami coniugati
d konjugierte Doppelverbindungen

2045 conjugate layers
f couchers conjuguées
e capas conjugadas
i strati coniugati
d konjugierte Schichten

2046 connate salt
f sel connè
e sal connata; sal singenética
i sale singenetico
d fossiles Salz

2047 connate water
f eau connée
e agua connata; agua singenética
i acqua singenetica
d fossiles Wasser

2048 connecting cock
f robinet de jonction
e grifo de unión
i robinetto di congiunzione
d Verbindungshahn

2049 consistency
f consistance
e consistencia
i consistenza
d Konsistenz; Dickflüssigkeit

2050 consistometer
f consistomètre
e consistómetro
i consistometro
d Konsistenzmesser

2051 constant of electrolytic dissociation
f constante de dissociation électrolytique
e constante de disociación electrolítica
i costante di dissociazione elettrolitica
d Dissoziationskonstante

2052 constant rate drying period
f tempo de séchage à vitesse constant
e tiempo de secado con velocidad constante
i tempo di disseccamento a velocità
   constante
d Zeit der konstanten Trocknungs-
   geschwindigkeit

* **constant white → 1023**

2053 constitution diagram
f diagramme de phases
e diagrama de fases
i diagramma di fasi
d Zustandsdiagramm

2054 contact adhesive
f colle de contact
e adhesivo de contacto
i adesivo a contatto
d Haftkleber

2055 contact plating
f dépôt par contact
e depósito por contacto
i deposito per contatto
d Kontaktplattierung

2056 contact process
f procédé de contact
e proceso de contacto
i processo per contatto
d Kontaktverfahren

2057 contact resins
f résines pour le procédé à basse pression
e resinas de contacto
i resine di contatto
d Harze für das Niederdruckverfahren

2058 container transport
f transport par citerne amovible
e transporte por cisterna
i trasporto a mezzo casse mobili
d Beförderung in Spezialbehältern

2059 contamination
f contamination
e contaminación
i infezione
d Kontamination

2060 continuous diffusion
f diffusion continue
e difusión continua
i diffusione continua
d kontinuierliche Entzuckerung;
kontinuierliche Diffusion

2061 continuous distillation
f distillation continue
e destilación continua
i distillazione continua
d kontinuierliche Destillation

2062 continuous phase
f phase continue
e fase continua
i fase continua; mezzo disperdente
d kontinuierliche Phase;
Dispersionsmedium

2063 continuous polymerization
f polymérisation en continu
e polimerización continua
i polimerizzazione continua
d kontinuierliche Polymerisation

2064 continuous process
f procédé continu
e procedimiento continuo
i processo continuo
d kontinuierliches Verfahren

2065 convection current
f courant de convection
e corriente de convección
i corrente di convezione
d Konvektionsstrom

2066 conversion coating
f revêtement par conversion
e revestimiento por conversión
i rivestimento per conversione
d Passivierung

2067 convulsant
f convulsivant
e convulsante

i convulsivante
d Krampfgift

2068 convulsion
f convulsion
e convulsión
i convulsione
d Konvulsion; Krampf

2069 coolant
f réfrigérant; liquide de refroidissement
e refrigerante
i refrigerante
d Kühlmittel

2070 cooling coil
f serpentin de refroidissement
e tubería espiral para la refrigeración
i serpentino di refrigerazione
d Kühlschlange

2071 cooling tower
f tour de réfrigération
e torre de refrigeración
i torre refrigerante
d Kühlturm

2072 co-ordination compound
f composé de coordination
e compuesto de coordinación
i composto di coordinazione
d Koordinationsverbindung

2073 co-ordination number; C.N.
f indice de coordination
e índice de coordinación
i numero di coordinazione
d Koordinationszahl

2074 copaiba oil
f essence de copahu
e aceite de Copaifera officinalis; aceite de copaiba
i olio di capaiba
d Kopaivaöl

2075 copal
f copal
e goma copal
i coppale
d Kopal

2076 copolycondensation
f copolycondensation
e copolicondensación
i copolicondensazione
d Copolykondensation

2077 copolymer
 f copolymère
 e copolímero
 i copolimero
 d Copolymer; Mischpolymerisat

2078 copolymerization
 f copolymérisation
 e copolimerización
 i copolimerizzazione
 d Copolymerisation; Mischpolymerisation

**2079 copolymers of vinyl methyl ether and
    maleic anhydride**
 f copolymères de vinylméthyléther et
    d'anhydride maléique
 e copolímeros del vinilmetiléter del
    anhídrido maleico
 i copolimeri di vinilmetiletere ed anidrido
    maleica
 d Mischpolymerisat aus Polyvinylmethyl
    äther und Maleinsäureanhydrid

2080 copper
    Element.
 f cuivre
 e cobre
 i rame
 d Kupfer

2081 copper
    Open vessel liquids in brewing.
 f chaudière à moût
 e caldera de cocimiento; vasija para cocer
 i caldaia per mosto
 d Braukessel

2082 copper acetate
 f acétate de cuivre
 e acetato de cobre
 i acetato di rame
 d Kupferacetat

2083 copper acetoarsenite
 f acéto arsénite de cuivre
 e acetoarsenito de cobre
 i acetoarsenito di rame
 d Kupferacetoarsenit

2084 copper arsenate
 f arséniate de cuivre
 e arseniato de cobre
 i arseniato di rame
 d Kupferarsenat

 * **copperas → 3439**

2085 copper-beryllium alloys
 f alliages de cuivre et de béryllium

 e aleaciones de cobre-berilio
 i leghe di berillio e rame
 d Kupfer-Berylliumlegierungen

2086 copper blue
 f bleu de cuivre
 e caparrosa azul
 i blu azzurrite
 d Kupferblau

2087 copper carbonate
 f carbonate de cuivre
 e carbonato de cobre
 i carbonato di rame
 d Kupfercarbonat

2088 copper chloride
 f chlorure de cuivre
 e cloruro de cobre
 i cloruro di rame
 d Kupferchlorid

2089 copper chromate
 f chromate de cuivre
 e cromato de cobre
 i cromato di rame
 d Kupferchromat

2090 copper cyanide
 f cyanure de cuivre
 e cianuro de cobre
 i cianuro di rame
 d Kupfercyanid

2091 copper ferrocyanide
 f ferrocyanure de cuivre
 e ferrocianuro de cobre
 i ferrocianuro di rame
 d Kupfereisencyanid

2092 copper fluosilicate
 f fluosilicate de cuivre
 e fluosilicato de cobre
 i fluosilicato di rame
 d Kupferfluosilikat

 * **copper glance → 1616**

2093 copper glycinate
 f glycinate de cuivre
 e glicinato de cobre
 i glicinato di rame
 d Kupferglycinat

2094 copper hydroxide
 f hydroxyde de cuivre
 e hidróxido de cobre
 i idrossido di rame
 d Kupferhydroxyd

2095 copper lactate
f lactate de cuivre
e lactato de cobre
i lattato di rame
d Kupferlactat

2096 copper metaborate
f métaborate de cuivre
e metaborato de cobre
i metaborato di rame
d Kupfermetaborat

2097 copper naphthenate
f naphténate de cuivre
e naftenato de cobre
i naftenato di rame
d Kupfernaphthenat

2098 copper nitrate
f nitrate de cuivre
e nitrato de cobre
i nitrato di rame
d Kupfernitrat; salpetersaures Kupferoxyd

2099 copper number
f coefficient de cuivre
e coeficiente de cobre
i indice del rame
d Kupferzahl

2100 copper oleate
f oléate de cuivre
e oleato de cobre
i oleato di rame
d Kupferoleat

2101 copper oxide
f oxyde cuivrique
e óxido de cobre
i ossido di rame
d Kupferoxyd

2102 copper oxychloride
f oxychlorure de cuivre
e oxicloruro de cobre
i ossicloruro di rame
d Kupferoxychlorid

2103 copper plating
f cuivrage
e revestimiento electrolítico de cobre;
  cuprogalvanoplastia
i ramatura
d Verkupferung

* **copper pyrite → 1617**

2104 copper resinate
f résinate de cuivre

e resinato de cobre
i resinato di rame
d Kupferresinat; harzsaures Kupfer

2105 copper ricinoleate
f ricinoléate de cuivre
e ricinoleato de cobre
i ricinoleato di rame
d Kupferricinoleat

2106 copper soaps
f savons de cuivre
e jabones de cobre
i saponi di rame
d Kupferseifen

2107 copper sulphate
f sulfate de cuivre
e sulfato de cobre
i solfato di rame
d Kupfersulfat

* **copper sulphate, tribasic ~ → 8394**

2108 copra
f copra
e copra
i copra
d Kopra

2109 coprecipitation
f coprécipitation
e coprecipitación
i coprecipitazione
d Kopräzipitation

2110 coprolite
f coprolite
e coprolito
i coprolite
d Koprolith

2111 cordite
f cordite
e cordita
i cordite
d Kordit

2112 cork
f liège
e corcho
i sughero
d Kork

2113 corpuscle
f corpuscule
e corpúsculo
i corpuscolo
d Körperchen

2114 correlated phenotypic variation
 f variation phénotypique correspondante
 e variación fenotípica correspondiente
 i variazione fenotipica corrispondente
 d korrelierte Variation phonotypischer
 Merkmale

2115 corroding lead
 f plomb pur
 e plomo comercial purísimo
 i piombo puro
 d Korrosionsblei

2116 corroding pot
 f cuve de corrosion
 e vasija de corrosión
 i crogiuolo per corrosione
 d Tontopf

2117 corrosion
 f corrosion
 e corrosión
 i corrosione
 d Korrosion

2118 corrosive
 f corrodant
 e corrosivo
 i corrosivo
 d korrodierend; korrosiv; ätzend

2119 corrosive sublimate
 f sublimé corrosif
 e sublimado corrosivo
 i sublimato corrosivo
 d Ätzsublimat

2120 corticosterone
 f corticostérone
 e corticosterona
 i corticosterone
 d Kortikosteron

2121 cortisone
 t cortisone
 e cortisona
 i cortisone
 d Cortison; Kortison

2122 corundum
 f corindon
 e corindón
 i corindone
 d Korund

2123 cosette
 f cossette
 e rebanada de remolacha en forma de V

 i fettina di barbabietola
 d Zuckerrübenschnitzel

2124 cotton oil
 f huile de coton
 e aceite para algodón
 i olio per cotone
 d Cottonöl; Baumwollöl

2125 cottonseed oil
 f huile de coton
 e aceite de semillas de algodón
 i olio di semi di cotone
 d Baumwollsamenöl

2126 Cottrell precipitator
 f cottrell
 e precipitador dc Cottrell
 i precipitatore di Cottrell
 d Cottrellgasreiniger

2127 coumarin
 f coumarine
 e cumarina
 i cumarina
 d Cumarin

2128 coumarone
 f coumarone
 e cumarona
 i cumarone
 d Cumaron

**2129 coumarone-indene resins; coal tar
 resins; thermoplastic polymers**
 f résines coumarone-indènes
 e resinas de cumarona e indeno
 i resine cumeroindeniche
 d Cumaron-Indenharze

2130 coumarone resins
 f résines de coumarone
 e resinas de cumarona
 i resine cumeroniche
 d Cumaronharze

2131 countercurrent distillation
 f distillation à contre-courant
 e destilación por contracorriente
 i distillazione a controcorrente
 d Gegenstromdestillation

2132 countercurrent extraction process
 f procédé à contre-courants
 e proceso de contracorriente
 i processo di controcorrente
 d Gegenstromverfahren

2133 **countercurrent flow; counterflow**
 f contre-courant
 e contracorriente
 i controcorrente
 d Gegenstrom

2134 **countercurrent packed column**
 f colonne garnie à contre-courant
 e columna a testado de contracorriente
 i colonna a riempimento ed a controcorrente
 d Gegenstromfüllkörperkolonne

2135 **countercurrent pulsed column**
 f colonne à pulsations et à contre-courant
 e columna de pulsaciones y de contracorriente
 i colonna a pulsazioni ed a controcorrente
 d pulsierte Gegenstromkolonne

2136 **countercurrent rotary column**
 f colonne à rotation et à contre-courant
 e columna de rotación y de contracorriente
 i colonna a rotazione ed a controcorrente
 d Gegenstromdrehungskolonne

 * **counterflow** → 2133

2137 **coupling**
 f copulation
 e copulación
 i accoppiamento
 d Verbindung

2138 **covalence**
 f covalence
 e covalencia
 i covalenza
 d Kovalenz

2139 **covalent bond**
 f liaison covalente
 e enlace covalente
 i legame covalente
 d kovalente Bindung

2140 **covalent compound**
 f composé covalent
 e compuesto covalente
 i legame covalente
 d kovalente Verbindung

2141 **covellite**
 f covelline
 e covelita; covelina
 i covellite
 d Covellit; Kupferindigo

2142 **cover glass**
 f couvre-object
 e cubreobjetos
 i vetro protettivo
 d Deckgläschen

2143 **covering power**
 f opacité
 e poder cubriente
 i potere coprente
 d Deckkraft

2144 **cracking**
 Breaking of a carbon/carbon bond by heat and, usually, a catalyst.
 f cracking; craquage
 e cracking; termofragmentación catalítica
 i "cracking"; processo di crachizzazione
 d Cracken

2145 **cracking**
 Occurrence of small fissures in a coating.
 f craquelage; formation de craquelures
 e agrietamiento
 i fessurazione
 d Rißbildung

2146 **cracking process**
 f procédé par craquage
 e proceso de cracking; proceso de termofraccionamiento catalítico
 i piroscissione
 d Krackverfahren; Crackverfahren

2147 **cratometer**
 f cratomètre
 e instrumento medidor de amplificación; cratómetro
 i cratometro
 d Kratometer

2148 **crawling**
 Creep movement of sheet lead used as a tank-lining or the like.
 f fluage
 e reptación
 i scorrimento
 d Kriechen

2149 **crawling**
 Formation of wrinkles in paint before drying is complete.
 f plissage
 e arrugamiento
 i repellenza
 d Kriechen frisch aufgetragener Lacke

2150 **crazing**
 f craquelure

   e agrietamiento
   i screpolatura
   d Haarrisse; Rißbildung

2151 cream *v*
   f crémer
   e desnatar
   i scremare
   d aufrahmen

2152 cream
   f crème
   e crema
   i crema
   d Krem

2153 creaming
   f écrémage
   e desnate
   i scrematura
   d Aufrahmen

2154 cream of tartar
   f crème de tartre
   e cremor tártaro
   i cremor tartaro
   d weinsaures Kalium; Weinstein

2155 creep
   Formation of crystals.
   f formation de cristaux
   e formación de cristales reptantes encima
   de superficie de un líquido
   i formazione di cristalli
   d Kristallbildung

2156 creep
   Formation of wrinkles in a paint coating.
   f plissage
   e arrugamiento
   i raggrinzatura
   d Kriechen frisch aufgetragener Anstriche;
   Runzelbildung

2157 creeping of salts
   f grimpement des sels
   e viscofluencia de sales
   i salita dei sali
   d Kriechen der Salze

2158 creep limit
   f limite de fluage
   e límite de termodeformación
   i limite di scorrimento
   d Kriechfestigkeit

2159 creep test
   f essai de fluage
   e prueba al estiramiento

   i prova di scorrimento
   d Dauerstandversuch

2160 crenocyte
   f crénocyte
   e crenocito
   i crenocita
   d Crenocyt

2161 creosote
   f créosote
   e creosota
   i creosoto
   d Kreosot

2162 crepe rubber
   f caoutchouc-crêpe
   e goma crepe
   i gomma crepe
   d Kreppgummi

2163 cresol
   f crésol
   e cresol
   i cresolo
   d Kresol; Cresol

2164 cresol resin
   f résine crésolique
   e resina de cresol
   i resina cresolica
   d Kresolharz

2165 cresotinic acid
   f acide crésotique
   e ácido cresotínico
   i acido cresotinico
   d Kresotinsäure

2166 cresyl diphenyl phosphate
   f phosphate de crésyle et diphényle
   e fosfato de cresilo y difenilo
   i cresildifenilfosfato
   d Cresyldiphenylphosphat

2167 cresylic acid
   f acide crésylique
   e ácido cresílico
   i acido cresilico
   d Kresolsäure; Cresylsäure

2168 crimson antimony
   f antimoine cramoisi
   e bermellón de antimonio
   i vermiglio d'antimonio
   d Antimonrot

2169 crimson lake; florentine lake
   f vernis carmoisi; laque florentine; laque

carminée
e barniz de bermellón
i vernice cremisi; vernice fiorentina
d Karmesinlack

2170 critical point
f point critique
e punto crítico
i punto critico
d kritischer Punkt

2171 critical temperature
f température critique
e temperatura crítica
i temperatura critica
d kritische Temperatur

2172 critical velocity
f vitesse critique
e velocidad crítica
i velocità critica
d kritische Geschwindigkeit

2173 crizzling
f craquelage
e agrietamiento superficial
i incrinatura
d Rißbildung

* **crocidolite** → 1065

2174 crock
f cruche
e olla de barro; cazuela
i vaso di ceramica
d Steingutgefäß

2175 crocoite; crocoisite
f crokoïte
e crocoita; plomo rojo
i crocoite
d Krokoit; roter Bleispat

2176 crookesite
f crookésite
e croquesita
i crookesite
d Crookesit

2177 crossed Nicols
f nicols croisés
e nicoles cruzados
i nicols incrociati
d gekreuzte Nicolsche Prismen

2178 cross linking
f réticulation; liaison transversale
e reticulación

i legame trasversale
d Vernetzung

2179 cross-over chiasmata
f chiasmas de fusion
e quiasmas de fusión
i chiasmi di fusione
d Cross-over-Chiasmata

2180 crotamiton
f crotamitone
e crotamitón
i crotamitone
d Crotamiton

2181 crotonaldehyde
f crotonaldéhyde
e crotonaldehido
i crotonaldeide
d Crotonaldehyd

2182 crotonic acid
f acide crotonique
e ácido crotónico
i acido crotonico
d Crotonsäure

2183 crown glass
f crown; verre à boudines
e vidrio crown
i vetro corona
d Solinglas; Kronglas

2184 crows-footing
f rayonnage
e patas de gallo
i pieghe a zampe di galline
d Krähenfußbildung

2185 crucible
f creuset de fusion
e crisol para fusión
i crogiuolo di fusione
d Schmelztiegel

2186 crucible graphite
f graphite à creuset
e grafito para crisoles
i grafita per crogiuoli
d schmiedbarer Graphit

2187 crucible tongs
f pince à creuset
e tenazas para crisoles
i pinza per crogiuolo di fusione
d Schmelztiegelzange

2188 crude oil
f huile brute

e aceite bruto
i petrolio greggio
d Erdöl; Rohöl

2189 crumbs
f bavures
e pizcas; menudos
i grumi
d Schnitzel

2190 crushing rolls
f rouleaux concasseurs
e cilindros trituradores
i frantoio a rulli
d Zerkleinerungswalzen

2191 crutcher
f mélangeur
e mezclador
i mescolatore
d Seifenmischer

2192 cryogen
f cryogène; réfrigerant
e criogeno; mezcla refrigerante
i criogeno
d Kryogen; Kühlmittel

2193 cryolite; Greenland spar
f cryolite
e criolita; espato de Groenlandia
i criolite; spato di Groenlandia
d Kryolith; Eisstein; Grönlandspat

2194 cryoscopic method
f méthode cryoscopique
e método crioscópico
i metodo crioscopico
d Gefrierpunktmeßmethode

2195 cryoscopy
f cryoscopie
e crioscopia
i crioscopia
d Gefrierpunktlehre

2196 cryptic contamination
f contamination cryptique; contamination latente
e contaminacion genética críptica
i modificazione genetica criptica
d latente Verunreinigung

2197 cryptocrystalline
f cryptocristallin
e criptocristalino
i criptocristallino
d kryptokristallin; mikrokristallin

2198 cryptomere
f cryptomère
e criptomero
i criptomero
d Kryptomere

2199 cryptomerism
f cryptomérie
e criptomerismo
i criptomerismo
d Kryptomerie

2200 cryptometer
f cryptomètre
e criptómetro
i criptometro
d Deckkraftmesser

2201 crystal
f cristal
e cristal
i cristallo
d Kristall

2202 crystal boundaries
f faces de contact des cristaux
e superficies de contacto de los cristales
i superficie di contatto dei cristalli
d Kristallgrenzlinien

2203 crystal face
f facette
e superficie del cristal
i faccia
d Kristallfläche

2204 crystal lattice
f réseau cristallin; réseau spatial
e retículo cristalino; red cristalina
i reticolo cristallino
d Kristallgitter

2205 crystallinity
f cristallinite
e cristalinidad
i cristallinità
d Kristallinität

2206 crystallite
f cristallite
e cristalito
i cristallino
d Kristallit

2207 crystallization
f cristallisation
e cristalización
i cristallizzazione
d Kristallisation

2208 crystallizing finish
f fini cristallisant
e acabado cristalino
i finitura cristallizzante
d kristalliner Anstrich

2209 crystallizing pond
f bassin de cristallisation
e balsa de cristalización
i serbatoio di cristallizzazione
d Kristallisierschale

2210 crystallogram
f cristallogramme
e cristalograma
i cristallogramma
d Kristallogramm

2211 crystalloluminescence
f cristalloluminescence
e cristaloluminiscencia
i cristalloluminescenza
d Kristallumineszenz

2212 crystal spots
f taches de cristaux
e manchas cristalizadas
i macchie saline
d Schwefelflecken; Einschlüsse

2213 crystal varnish
f vernis cristallisé
e barniz cristalizado
i vernice cristallizzante
d Kristallack

2214 cudbear
f orseille
e polvo de púrpura
i polvere porporina
d Cudbear; Orseillefarbe

2215 culicide
f culicicide
e culicida
i culicida
d Mückenmittel

2216 cullet
f calcin
e vidrio de desecho
i rottame di vetro
d Glasscherbe

2217 culm
f poussière d'anthracite
e polvo de antracita
i antracite in polvere
d Grus

2218 culture dish
f plaque de culture
e placa de cultivo
i piastra da coltura
d Kulturplatte

2219 cumene
f cumène
e cumeno
i cumene
d Cumol

2220 cumene hydroperoxide
f hydroperoxyde de cumène
e hidroperóxido de cumeno
i cumene idroperossido
d Cumolhydroperoxyd

2221 cumin oil
f essence de cumin
e aceite de comino
i olio de comino
d Cuminöl

2222 cumyl phenol
f phénol de cumyle
e cumilfenol
i cumilfenolo
d Cumylphenol

2223 cupel
f coupelle
e copela
i coppella
d Kupelle

2224 cupel furnace
f four de coupellation
e horno de copelar
i forno di coppellazione
d Kupolofen; Kapellenofen

2225 cupellation
f coupellation
e copelación
i coppellazione
d Kupellation

2226 cup flow
f fluidité au gobelet
e prueba de plasticidad
i indice di plasticità
d Prüfbecherfluß

2227 cuprammonia
f oxyde de cuivre ammoniacal
e cuproamoníaco
i cuprammoniaca
d Kupferoxydammoniak

2228 cuprammonium process
  f procédé à l'oxyde de cuivre ammoniacal
  e método cuproamoniacal
  i processo all'ammoniuro di rame o al
     cuprammonio
  d Cuprammoniumverfahren

2229 cuprammonium rayon
  f rayonne cupro-ammoniacale
  e rayón cuproamoniacal
  i rayon cuproammoniacale
  d Kupferkunstseide

2230 cupric
  f cuivrique
  e cúprico
  i ramico
  d Kupfer-; Cupri-

2231 cupric oxide
  f oxyde cuivrique
  e óxido de cobre
  i ossido di rame
  d Kupferoxyd

2232 cuprite
  f cuprite
  e cuprita
  i cuprite
  d Cuprit; Kupferblüte

2233 cuprous
  f cuivreux
  e cuproso
  i rameoso
  d Kupfer-; Cupro-

2234 cuprous chloride
  f chlorure cuivreux
  e cloruro cuproso
  i cloruro rameoso
  d Kupferchlorür; Cuprochlorid

2235 cuprous cyanide
  f cyanure cuivreux
  e cianuro cuproso
  i cianuro rameoso
  d Kupfercyanür

2236 cuprous iodide
  f iodure cuivreux
  e yoduro cuproso
  i ioduro rameoso
  d Kupferjodür

2237 cuprous sulphide
  f sulfure cuivreux
  e sulfuro cuproso
  i solfuro rameoso
  d Kupfersulfür

2238 curare
  f curare
  e curare
  i curaro
  d Curare; Pfeilgift

2239 curarine
  f curarine
  e curarina
  i curarina
  d Curarin

2240 curarization
  f curarisation
  e curarización
  i curarizzazione
  d Kurarisierung

2241 curcumin
  f curcumine
  e curcumina
  i giallo di curcuma
  d Kurkumin

2242 curdle *v*
  f cailler
  e cuajar
  i coagulare
  d gerinnen

2243 cure
  f déshydratant
  e agente deshidratante
  i essicante
  d Entwässerungsmittel; Trockenmittel

2244 Curie
  f curie
  e curie
  i curie
  d Curie

2245 curing *(leather)*
  f séchage
  e curación
  i essiccazione
  d Trocknen

2246 curing *(plastics)*
  f prise
  e endurecimiento
  i indurimento
  d Härten

2247 curium
  f curium

e curio
i curio
d Curium

2248 current efficiency
f rendement en courant
e rendimiento de corriente
i rendimento di corrente
d Stromausbeute

2249 currying
f tannage
e tratamiento con substancias grasas
i concia
d Gerben

2250 cut-back
f "cut-back"
e betún reconstituido
i bitume ricostituito
d Verschnittbitumen

2251 cutting down
f aplanissage
e alisado
i sgrassatura
d Grobschleifen

2252 cutting oils
f huiles de coupe
e aceites para cortar
i oli impiegati per il taglio di metalli
d Schneideöle

2253 cyanamide
f cyanamide
e cianamida
i cianamide
d Cyanamid

2254 cyanamide process
f préparation de cyanamide
e proceso de la cianamida
i processo di preparazione della cianamide
d Cyanamidverfahren

* **cyanide, total ~ → 8347**

2255 cyaniding
f cyanuration
e cianuración
i cianurazione
d Cyanhärtung

2256 cyanines
f cyanines
e cianinas
i cianine
d Cyanine

2257 cyanite
f cyanite
e cianita
i cianite
d Cyanit

2258 cyanoacetamide
f cyanoacétamide
e cianoacetamida
i cianoacetammide
d Cyanacetamid

2259 cyanoacetic acid
f acide cyanoacétique
e ácido cianacético
i acido cianoacetico
d Cyanessigsäure

2260 cyanoethyl acrylate
f acrylate de cyanoéthyle
e acrilato de cianoetilo
i acrilato cianoetilico
d Cyanäthylacrylat

2261 cyanoethylated cotton
f coton cyanoéthylé
e algodón cianoetilado
i cotone cianoetilato
d cyanäthylierte Baumwolle

2262 cyanoethylation
f cyanoéthylation
e cianoetilación
i cianoetilazione
d Cyanäthylierung

2263 cyanogen
f cyanogène
e cianógeno
i cianogeno
d Cyangas; Cyan

2264 cyanogen bromide
f bromure de cyanogène
e bromuro de cianógeno
i bromuro di cianogeno
d Cyanbromid

2265 cyanogen chloride
f chlorure de cyanogène
e cloruro de cianógeno
i cloruro di cianogeno
d Cyanchlorid; Chlorcyan

2266 cyano-(methylmercuri)-guanidine
f cyano-(méthylmercuri)-guanidine
e ciano-(metilmercuri)-guanidina
i ciano-(metilmercurio)-guanidina
d Cyan-(methylmerkuri)-guanidin

2267 cyanotype; blueprinting
  f cyanotypie; ferrotypie; tirage en bleu
  e cianotipia
  i cianografia
  d Cyanotypie; Blaudruck

2268 cyanuric acid
  f acide cyanurique
  e ácido cianúrico
  i acido cianurico
  d Cyanursäure

  * **cyclanes** → **2288**

2269 cyclethrin
  f cycléthrine
  e cicletrina
  i cicletrina
  d Cyclethrin

2270 cyclic
  f cyclique
  e ciclico, periódico
  i ciclico
  d zyklisch

2271 cyclic compounds
  f composés cycliques
  e compuestos cíclicos
  i composti ciclici
  d zyklische Verbindungen; cyclische
     Verbindungen

2272 cycling
  f cyclage
  e ciclado; ciclaje
  i ciclizzazione
  d Kreisprozeß

2273 cyclizine hydrochloride
  f hydrochlorure de cyclizine
  e hidrocloruro de ciclicina
  i idrocloruro di ciclicina
  d Cyclizinhydrochlorid

2274 cyclo-
     Prefix denoting the presence of a closed
     carbon chain or carbon ring structure.
  f cyclo-
  e ciclo-
  i ciclo-
  d Cyclo-

2275 cyclobarbital
  f cyclobarbital
  e ciclobarbital
  i ciclobarbital
  d Cyclobarbital

2276 cyclocoumarol
  f cyclocoumarol
  e ciclocumarol
  i ciclocumarolo
  d Cyclocumarol

2277 cyclohexane
  f cyclohexane
  e ciclohexano
  i cicloesano
  d Cyclohexan

2278 cyclohexanedimethanol
  f cyclohexane-diméthanol
  e ciclohexandimetanol
  i cicloesanodimetanolo
  d Cyclohexandimethanol

2279 cyclohexanol
  f cyclohexanol
  e ciclohexanol
  i cicloesanolo
  d Adrónal; Hexalin

2280 cyclohexanol acetate
  t acétate de cyclohexanol
  e acetato de ciclohexanol
  i acetato di cicloesanolo
  d Cyclohexanolacetat

2281 cyclohexanone
  f cyclohexanone
  e ciclohexanona
  i cicloesanone
  d Cyclohexanon

2282 cyclohexene
  f cyclohexène
  e ciclohexeno
  i cicloesene
  d Cyclohexen

2283 cyclohexylamine
  f cyclohexylamine
  e ciclohexilamina
  i cicloesilammina
  d Cyclohexylamin

2284 cyclohexyl methacrylate
  f cyclohexylméthacrylate
  e metacrilato de ciclohexilo
  i cicloesilemetacrilato
  d Cyclohexylmetacrylsäureester;
     Cyclohexylmetacrylat

2285 cyclohexyl phenol
  f cyclohexylphénol
  e ciclohexilfenol

i cicloesilfenolo
d Cyclohexylphenol

2286 cyclohexyl stearate
f stéarate de cyclohexyle
e estearato de ciclohexilo
i stearato cicloesilico
d Cyclohexylstearat

2287 cyclone
f cyclone
e separador ciclónico
i separatore a ciclone
d Zyklonenscheider

2288 cycloparaffins; cyclanes
f cycloparaffines; cyclanes
e cicloparafinas; ciclanos
i cicloparaffine; ciclani
d Cycloparaffine

2289 cyclopentane
f cyclopentane
e ciclopentano
i ciclopentano
d Cyclopentan

2290 cyclopentanol
f cyclopentanol
e ciclopentanol
i ciclopentanolo
d Cyclopentanol

2291 cyclopentene
f cyclopentène
e ciclopenteno
i ciclopentina
d Cyclopenten

2292 cyclopentolate hydrochloride
f hydrochlorure de cyclopentolate
e hidrocloruro de ciclopentolato
i idrocloruro di ciclopentolato
d Cyclopentolathydrochlorid

2293 cyclopentyl bromide
f bromure cyclopentyle
e bromuro de ciclopentilo
i bromuro di ciclopentile
d Cyclopentylbromid

2294 cyclopropane
f cyclopropane
e ciclopropano
i ciclopropano
d Cyclopropan

2295 cyclosiloxane
f cyclosiloxane

e ciclosiloxano
i ciclosilossano
d Cyclosiloxan

2296 cycrimine hydrochloride
f hydrochlorure de cycrimine
e hidrocloruro de cicrimina
i idrocloruro di cicrimina
d Cycriminhydrochlorid

2297 cymene
f cymène
e cimeno
i cimene
d Cymol

2298 cystamine
f cystéamine
e quistamina
i cistamina
d Cystamin

2299 cystine
f cystine
e cistina
i cistina
d Cystin

2300 cytaster
f cytaster
e citáster
i citaster
d Cytaster

2301 cyte
f cyte; cellule
e cito
i cito
d Zelle

2302 cytoblast
f cytoblaste
e citoblasto
i citoblasto
d Cytoblast; Zellkern

2303 cytochromes
f cytochromes
e citocromos
i citocromi
d Cytochrome; Zellfarbstoffe

2304 cytocinesis
f cytocinèse
e citocinesis
i citocinesi
d Cytocinesis

2305 cytolysin
  f cytolysine
  e citolisina
  i citolisina
  d Cytolysin

2306 cytolysis
  f cytolyse
  e citólisis
  i citolisi
  d Zellzerfall

2307 cytome
  f cytome
  e citoma
  i citoma
  d Cytom

2308 cytopenia
  f cytopenie
  e citopenia
  i citopenia
  d Zellmangel

2309 cytoplasma
  f cytoplasme
  e citoplasma
  i citoplasma
  d Cytoplasma

2310 cytotoxic
  f cytotoxique
  e citotóxico
  i citotossico
  d cytotoxisch

D

2311 Dalton's law
f loi de Dalton
e ley de Dalton
i legge delle pressioni parziali
d Daltonsches Gesetz

2312 dammar
f dammar
e damara
i dammara
d Dammar

* **damp** → 3482

2313 dancer roll
f rouleau fou
e rodillo flotante
i ballerino
d Losrolle

2314 dandelion rubber
f caoutchouc de pissenlit
e goma de diente de león
i gomma da dandelion
d Dandelionkautschuk

2315 dark ground illumination
f éclairage sur fond noir
e iluminación de fondo oscuro
i illuminazione in campo oscuro
d Dunkelfeldbeleuchtung

2316 dative bond
f liaison semi-polaire
e enlace de coordinación
i legame semipolare
d semipolare Bindung

2317 daughter cell
f cellule fille
e célula hija
i cellula figlia
d Tochterzelle

2318 daughter chromatid
f chromatide fille
e cromatidio hijo
i cromatidio figlio
d Tochterchromatide

2319 daughter chromosome
f chromosome fil
e cromosoma hijo
i cromosoma figlio
d Tochterchromosom

2320 DDT; dichloro-diphenyl-trichloroethane
f DDT
e DDT
i DDT
d DDT

2321 Deacon's process
f procédé Deacon
e procedimiento Deacon
i processo di Deacon
d Deaconsches Verfahren

2322 dead-milled rubber
f caoutchouc malaxé à mort
e caucho masticado a muerte
i gomma masticata a morte
d tot mastizierter Kautschuk

2323 dead milling
f mastication à mort
e masticación a muerte
i masticazione a snervamento
d Totwalzen

2324 dead roasting
f grillage à mort
e tostación total; calcinación completa
i arrostimento totale
d Totbrennen

2325 dead-weight pressure gauge
f balance manométrique; manomètre à poids mort
e balanza manométrica
i bilancia manometrica; manometro a stantuffo
d manometrische Waage; Kolbenmanometer

2326 de-aeration
f désaération
e desaireación
i deaerazione
d Entlüftung

2327 de-aerator
f désaérateur
e desaireador
i disaeratore
d Luftabscheider

2328 de-airing
f évacuation de l'air
e desaireación
i disaerazione
d Entlüften

2329 de-airing machine
f désaérateur

e máquina de desaireación
i disaeratore
d Entlüfter

2330 Dean and Stark apparatus
f appareil de Dean et Stark
e aparato de Dean y Stark
i apparecchio di Dean e Stark
d Dean-und-Stark-Apparat

2331 dearsenicator
f désarsénicateur
e desarsenicador
i impianto per l'eliminazione di arsenico
d Arsenextraktor

2332 decaborane
f décaborane
c decaborano
i idruro di boro
d Decaboran

2333 decahydronaphthalene
f décahydronaphtalène
e decahidronaftaleno
i decaidronaftalene
d Decahydronaphthalin; Decalin

2334 decalcification
f décalcification
e decalcificación
i sottrazione del calcio; decalcificazione
d Entkalkung; Kalkentziehung

2335 decalescence
f décalescence
e decalescencia
i decalescenza
d Dekaleszenz

2336 decant *v*
f décanter
e decantar
i decantare
d dekantieren

2337 decantation
f décantation
e decantación
i decantazione
d Dekantieren

2338 decanter
f décanteur
e decantador
i decantatore
d Abklärgefäß; Dekantiergefäß

2339 decarburization
f décarbonisation
e descarburación; descarbonización
i decarburazione
d Entkohlen; Kohlenstoffentziehung

2340 decay
f déchéance
e decaimiento; caída
i disintegrazione; declino
d Verfall; Zerstörung

2341 deciduous
f caduc
e deciduo; caduco
i caduco; transitorio
d hinfällig; vergänglich

2342 deck *(of screen)*
f surface criblante
e superficie cribante
i superficie setacciante
d Siebblech

2343 de-cobalter
f appareil à retirer le cobalt
e aparato para retirar catalizador de cobalto
i impianto per l'eliminazione del catalizzatore di cobalto
d Kobaltextraktor

2344 decoction
f décoction
e decocción
i decotto
d Absud

* **decoloration** → 1035

2345 decomposition
f décomposition
c descomposición
i decomposizione
d Spaltung; Zersetzung

2346 decomposition voltage
f tension de dissociation
e voltaje de descomposición
i tensione di decomposizione
d Zersetzungsspannung

2347 decontamination
f décontamination
e descontaminación
i decontaminazione
d Entseuchung

2348 decorticator
f décortiqueuse
e descortezadora mecánica
i decorticatrice
d Schälmaschine; Entholzungsmaschine

2349 decyl mercaptan
f décylmercaptan
e decilmercaptano
i decilmercaptano
d Decylmercaptan

2350 decyl-octyl methacrylate
f méthacrylate décyle-octylique
e metacrilato decílico-octílico
i decilottilmetacrilato
d Decyloctylmethacrylate

2351 dedusting agent
f agent de dépoussiérage
e agente eliminador de polvo
i agente antipolvere
d Entstäubungsmittel

2352 deep flow
f écoulement profond
e largo camino de flujo
i percorso lungo
d langer Fließweg

2353 defecation
f défécation; clarification
e defecación
i defecazione
d Klären

2354 defense power
f forces défensives
e fuerza de defenza
i potere difensivo
d Abwehrkraft

2355 defibrillation
f défibrillation
e defibrilación
i defibrillazione
d Defibrillation

2356 definitive nucleus
f noyau définitif
e núcleo definitivo
i nucleo definitivo
d definitiver Zellkern

2357 deflagrating spoon
f cuillère de déflagration
e cucharilla de deflagración
i cucchiaio per deflagrazione
d Deflagrierlöffel

2358 deflagration
f déflagration
e deflagración
i deflagrazione
d Deflagration; Verpuffung

2359 deflocculation
f défloculation
e defloculación
i deflocculazione
d Entflockung

2360 DeFlorez process
f procédé DeFlorez
e proceso DeFlorez
i processo di DeFlorez
d DeFlorez-Verfahren

2361 defoamer
f agent antimoussant
e despumador
i antischiuma
d Antischaummittel

2362 degas *v*; **degasify** *v*
f dégazer
e desgasificar
i degassificare
d entgasen

2363 degasser
f dégazeur
e desgasificador
i degassificatore
d Entgaser

2364 degassing
f désaération
e desaéración
i degassificazione
d Entgasen

2365 degeneration
f dégénérescence
e degeneración
i degenerazione; decadenza
d Entartung

2366 degradation
f dégradation
e degradación; disgregación
i degradazione
d Abbau; Degradation

2367 degreasing
f dégraissage
e desengrase
i sgrassatura
d Entfettung

2368 degreasing tank
 f citerne de dégraissage
 e baño desengrasante; depósito
    desengrasante
 i recipiente per sgrassatura
 d Entfettungstank

2369 degree of dissociation
 f degré de dissociation
 e grado de disociación; grado de
    desdoblamiento
 i grado di dissociazione
 d Dissoziationsgrad

2370 degree of fluidity
 f degré de fluidité
 e grado de fluidez
 i grado di fluidità
 d Weichheitsgrad

2371 degree of grinding
 f degré de mouture
 e grado de molienda
 i grado di macinazione
 d Mahlgrad

2372 degree of ionization
 f degré d'ionisation
 e grado de ionización
 i grado di ionizzazione
 d Ionisationsgrad

2373 degree of polymerization
 f degré de polymérisation
 e grado de polimerización
 i grado di polimerizzazione
 d Polymerisationsgrad

2374 degree of superheat
 f degré de surchauffe
 e grado de sobrecalentamiento
 i grado di surriscaldamento
 d Überhitzungsgrad

2375 degrees of freedom
 f degrés de liberté
 e grados de libertad
 i gradi di libertà
 d Freiheitsgrade

2376 dehiscence
 f déhiscence
 e dehiscencia
 i deiscenza; fessura
 d Aufschlitzen

2377 dehydrate *v*
 f déshydrater
 e deshidratar

 i disidratare
 d entwässern; trocknen

2378 dehydration
 f déshydratation; anhydrisation
 e deshidratación
 i disidratazione
 d Dehydrierung; Entwässerung

2379 dehydrator
 f déshydratateur
 e deshidratador
 i disidratatore
 d Trockenapparat

2380 dehydroacetic acid
 f acide déhydracétique
 e ácido dehidroacético
 i acido deidroacetico
 d Dehydroessigsäure

2381 dehydrogenase
 f déhydrogénase
 e deshidrogenasa
 i deidrogenasi
 d Dehydrogenase

2382 dehydrogenation
 f déshydrogénation
 e deshidrogenación
 i deidrogenazione
 d Dehydrierung

2383 dehydrothiotoluidine
 f déhydrothiotoluidine
 e dehidrotiotoluidina
 i deidrotiotoluidina
 d Dehydrothiotoluidin

2384 deionization
 f désionisation
 e desionización
 i deionizzazione
 d Entionisierung

2385 Delf; Delft
 f faïence de Delft
 e porcelana de Holanda
 i porcellana olandese
 d Delfter Porzellan

2386 deliquescence
 f déliquescence
 e delicuescencia
 i deliquescenza
 d Zerfließen

2387 delivery
    Amount of fluid discharged by a pump

under given conditions.
f débit
e entrega
i erogazione
d Förderleistung

2388 delivery
Piping attached to the discharge opening
of a pump.
f tuyau de décharge
e tubería de entrega
i tubo di mandata
d Ableitung

2389 delphinine; ajacine
f delphinine; ajacine
e delfinina; ajacina
i delfinina; aiacina
d Delphinin

2390 delta metal
f métal delta
e metal delta
i metallo delta
d Deltametall

2391 delustrant
f agent de délustrage
e delustrante
i agente antilucido
d Mattierungsmittel

2392 demilune bodies
f corps semi-lunaires
e cuerpos semilunares
i corpi semilunari
d Halbmondkörper

2393 demulsification number
f indice de démulsification
e índice de desmulsificación
i indice di demulsificazione
d Entemulgierungszahl

2394 demulsifier
f agent démulsionnant
e agente desmulsificador
i agente demulsificante
d Mittel zum Brechen von Emulsionen

2395 denaturant
f dénaturant
e desnaturalizante
i denaturante
d Denaturierungsmittel

2396 denatured alcohol
f alcool dénaturé
e alcohol desnaturalizado; alcohol de

quemar
i alcool denaturato
d denaturierter Alkohol

2397 denaturing
f dénaturation
e desnaturalización
i denaturazione
d Denaturierung

2398 Denora cell
f pile Denora
e elemento Denora
i pila Denora
d Denora-Element

2399 density
f densité
e densidad
i densità
d Dichte

2400 density bottle
f bouteille densimétrique; flacon à densité
e frasco para determinación de densidades
i boccetta densimetrica
d Flasche zur Dichtebestimmung

2401 dentifrice
f dentifrice
e dentifricio
i dentifricio
d Zahnputzmittel

2402 denuder
f décomposeur
e descompositor; separador
i denudatore
d Amalgamzersetzer; Zersetzungszelle

2403 deodorant; deodorizer
f désodorisant
e desodorante; desodorizante;
desodorizador
i deodorante
d Desodoriermittel

2404 deoxidation
f désoxydation; désoxygénation
e desoxidación
i disossidazione
d Desoxydation

2405 deoxidizer
f désoxydant
e desoxidante
i riducente; disossidante
d Desoxydationsmittel

2406 deoxy-; desoxy-
Prefix denoting a compound in which
hydroxyl has been replaced by hydrogen.
f désoxy-
e desoxi-
i deossi-
d Desoxy-

2407 deoxycorticosterone
f désoxycorticostérone
e desoxicorticosterona
i deossicorticosterone
d Desoxycorticosteron

2408 deoxyribonucleic acid; DNA
f acide désoxyribonucléique
e ácido desoxirribonucleico
i acido deossiribonucleico
d Desoxyribonucleinsäure

2409 depassivation
f dépassivation
e desoasivación
i depassivazione
d Entpassivierung

2410 dephlegmator
f déphlegmateur
e desflegmador
i deflemmatore; colonna di frazionamento
d Dephlegmator

2411 dephosphorization
f déphosphoration
e desfosforación
i defosforizzazione
d Entphosphorung

2412 depigmentation
f dépigmentation
e despigmentación
i depigmentazione; vitiligine
d Pigmentverlust

2413 depilatory
f épilatoire; dépilatoire
e depilatorio
i depilatorio
d Haarentfernungsmittel

2414 depleted water
f eau appauvrie en eau lourde
e agua agotada
i acqua impoverita
d abgereichertes Wasser

2415 depletion
f appauvrissement
e agotamiento

i impoverimento
d Abreicherung

2416 depolarization
f dépolarisation
e despolarización
i depolarizzazione
d Depolarisation

2417 depolarizer
f dépolarisant
e despolarizador
i depolarizzante
d Depolarisator

2418 depolarizing mix
f mélange dépolarisant
e mezcla despolarizadora
i miscela depolarizzante
d Depolarisationsgemisch

2419 depolymerization
f dépolymérisation
e despolimerización
i depolimerizzazione
d Entpolymerisation

2420 depositing-out tank
f cuve de dépôt total
e cuba de depósito total
i cella di esaurimento
d Entmetallisierungsbad

2421 depressant
f sédatif
e sedante
i riduttore; calmante
d Beruhigungsmittel

2422 depropanizer
f dépropanisateur
e despropanizador
i impianto per l'eliminazione del propano
d Entpropanisierungsvorrichtung

2423 derby red
f rouge de chrome
e rojo chino; pigmento de cromato de
   plomo
i rosso cinese
d Chromrot

2424 derivative
f dérivé
e derivado
i derivato
d Derivat; Abkömmling

2425 dermoblast
f dermoblaste
e dermoblasto
i dermoblasto
d Hautblatt des Mesoderms

2426 desalting
f dessalure
e desalificación; desalación
i desalificazione
d Entsalzen

2427 desaminase
f désaminase
e desaminasa
i desaminasi
d Desaminase

2428 deserpidine
f déserpidine
e deserpidina
i deserpidina
d Deserpidin

2429 desiccant
f agent dessiccateur; dessiccatif
e desecador; desecante
i agente essiccatore; essicante
d Trockenmittel

2430 desiccator
f exsiccateur; dessiccateur
e desecador
i essicatore; apparecchio essicatore
d Exsiccator; Trockengefäß

2431 desmosome
f desmosome
e desmosoma
i ponte intercellulare
d Desmosom

2432 desorption
f désorption
e desorción
i dessorbimento
d Desorption

* desoxy- → **2406**

2433 despumation
f despumation
e despumación
i schiumatura
d Schaumentfernung

2434 destabilize v
f déstabiliser
e desestabilizar

i deassorbimento
d destabilisieren

2435 destructive distillation
f pyrogénation
e destilación destructiva
i distillazione distruttiva
d trockene Destillation

2436 destructor
f incinérateur
e horno incinerador de basuras
i impianto per incenerire rifiuti
d Abfallverbrennungsanlage;
   Abfallverbrennungsofen

2437 desuperheater
f désurchauffeur
e desrecalentador
i desurriscaldatore
d Heißdampfkühler

2438 desynapsis
f désynapsis
e desinapsis
i desinapsi
d Desynapsis

2439 detackifier
f antiadhésif
e antiadhesivo
i antiadesivo
d Antiklebemittel

2440 detention time
f temps de repos
e tiempo de detención; tiempo de reposo
i periodo della sedimentazione
d Stehzeit

2441 detergent
f détergent
e detergente
i detergente
d Reinigungsmittel

2442 detonation
f détonation
e detonación
i detonazione
d Detonation

2443 detonator
f amorce; détonateur
e detonador; fulminante
i detonatore
d Sprengkapsel

2444 detreading chips
  f déchets de décorticage
  e residuos de decortización
  i cascami di scorticatura di battistrada
  d Gummischälabfälle

2445 deuterium
  f deutérium
  e deuterio
  i deuterio
  d Deuterium

2446 deuteron
  f deutéron; deuton
  e deuterón
  i deuterio
  d Deuteron

2447 developed dye
  f colorant appliqué en réaction
  e colorante de desarrollo
  i colorante sviluppato
  d Entwicklungsfarbstoff

2448 developer
  f révélateur
  e revelador
  i rivelatore; sviluppatore
  d Entwickler

2449 devitrification
  f dévitrification
  e desvitrificación
  i devitrificazione
  d Versteinung; Entglasung

2450 devulcanization
  f dévulcanisation
  e desvulcanización
  i devulcanizzazione
  d Entvulkanisieren

2451 Dewar flask
  f vase Dewar
  e frasco Dewar
  i vaso di Dewar
  d Dewar-Gefäß

2452 dewatering
  f déhydratation
  e deshidratación; desagüe
  i drenaggio
  d Entwässerung

2453 dew point
  f point de rosée
  e punto de rocío
  i punto di rugiada
  d Taupunkt

2454 dextrin
  f dextrine
  e dextrina
  i destrina
  d Dextrin

2455 dextro-rotatory
  f dextrogyre
  e dextrorrotatorio; dextrorso
  i destro rotatorio
  d rechtsdrehend

2456 dextrose
  f dextrose
  e dextrosa
  i destrosio
  d Dextrose; Traubenzucker

2457 di-; bi-
      Prefix denoting twice or two
  f di-; bi-
  e di-; bi-
  i di-; bi-
  d Di-; Bi-

2458 diabase
  f diabase
  e diabasa
  i diabase
  d Diabas; Grünstein

2459 diacetin
  f diacétine
  e diacetina
  i diacetina
  d Diacetin

2460 diacetone alcohol
  f diacétone-alcool
  e diacetonalcohol
  i alcool diacetonico
  d Diacetonalkohol

2461 diacetyl
  f diacétyle
  e diacetilo
  i diacetile
  d Diacetyl

2462 diacetylmorphine
  f diacétylmorphine
  e diacetilmorfina
  i diacetilmorfina
  d Diacetylmorphin

2463 diadelphous
  f diadelphe
  e diadelfo

   i diadelfo
   d diadelphisch

2464 diakinesis
   f diacinèse
   e diacínesis
   i diacinesi
   d Diakinese

2465 diakinetic
   f diacinétique
   e diacinético
   i diacinetico
   d diakinetisch

2466 dialdehydes
   f dialdéhydes
   e dialdehídos
   i dialdeidi
   d Dialdehyde

2467 diallylbarbituric acid
   f acide diallylbarbiturique
   e ácido dialilbarbitúrico
   i acido diallilbarbiturico
   d Diallylbarbitursäure

2468 diallyl isophthalate
   f isophtalate de diallyle
   e isoftalato de dialilo
   i diallilisoftalato
   d Diallylisophthalat

2469 diallyl maleate
   f maléate de diallyle
   e maleato de dialilo
   i diallilmaleato
   d Diallylmaleat

2470 diallyl phthalate
   f phtalate de diallyle
   e ftalato de dialilo
   i diallilftalato
   d Diallylphthalat

2471 dialysable
   f dialysable
   e dializable
   i dializzabile
   d dialysierbar

2472 dialysis
   f dialyse
   e diálisis
   i dialisi
   d Dialyse

2473 diaminophenol hydrochloride
   f chlorhydrate de diaminophénol

   e hidrocloruro de diaminofenol
   i idrocloruro di diaminofenolo
   d Diaminophenolhydrochlorid

2474 diamond
   f diamant
   e diamante
   i diamante
   d Diamant

2475 diamylphenol
   f diamylphénol
   e diamilfenol
   i diamilfenolo
   d Diamylphenol

2476 diamyl phthalate
   f phtalate de diamyle
   e ftalato de diamilo
   i diamilftalato
   d Diamylphthalat

2477 diamyl sulphide
   f sulfure de diamyle
   e sulfuro de diamilo
   i solfuro diamilico
   d Diamylsulfid

2478 dianisidine
   f dianisidine
   e dianisidina
   i dianisidina
   d Dianisidin

2479 diaphanometer
   f diaphanomètre
   e diafanómetro
   i diafanometro
   d Diaphanometer; Lichtdurchlässigkeits-
      messer

2480 diaphragm
   f diaphragme; membrane
   e diafragma; membrana
   i diaframma; membrana
   d Diaphragma; Membran

2481 diaphragm cell
   f pile à diaphragme
   e pila de diafragma
   i diaframma elettrolitico
   d Diaphragmaelement

2482 diaphragm pump
   f pompe à diaphragme; pompe à
      membrane
   e bomba de diafragma; bomba de
      membrana

i pompa a membrana
d Membranpumpe

2483 diaphragm-type pressure gauge
f manomètre à membrane
e manómetro de membrana
i manometro a membrana
d Membranmanometer

2484 diaspore
f diaspore
e diásporo
i diasporo
d Diaspor

2485 diastase
f diastase
e diastasa
i diastase
d Diastase

2486 diastatic power
f pouvoir diastatique
e poder diastático
i potere diastatico
d diastatische Kraft

2487 diastem
f diastème
e diastema
i diastema
d Diastema

2488 diathermanous
f diathermique; diathermane
e diatermano
i diatermano
d wärmedurchlässig

2489 diatomite; kieselguhr
f diatomite; kieselgur; terre d'infusoires
e diatomita; kieselgur; tierra de infusorios
i diatomite; farina fossile; kieselgur; terra d'infusori
d Diatomit; Kieselgur; Diatomeenerde; Infusorienerde

2490 diatomite filter
f filtre à diatomite
e filtro de diatomita
i filtro a diatomite
d Diatomitfilter

2491 diazo compounds
f diazoïques
e compuestos diazo
i composti diazici
d Diazoverbindungen

2492 diazomethane
f diazométhane
e diazometano
i diazometano
d Diazomethan

2493 diazonium salts
f sels de diazonium
e sales de diazonio
i sali diazonici
d Diazoniumsalze

2494 diazotization
f diazotation
e diazotización
i diazotizzazione
d Diazotierung

2495 dibasic acids
f acides dibasiques
e ácidos dibásicos
i acidi bibasici
d zweibasische Säuren

2496 dibasic ammonium phosphate
f phosphate d'ammonium dibasique
e fosfato amónico bibásico
i fosfato di ammonio bibasico
d Ammoniumdiphosphat

2497 dibasic calcium phosphate
f phosphate bicalcique
e fosfato cálcico dibásico
i fosfato di calcio bibasico
d Dicalciumphosphat

2498 dibasic lead phosphate
f phosphate de plomb dibasique
e fosfato dibásico de plomo
i fosfato bibasico di piombo
d sekundäres Bleiphosphat

2499 dibasic magnesium phosphate
f phosphate de magnésium dibasique
e fosfato dibásico magnésico
i fosfato bibasico di magnesio
d Dimagnesiumphosphat

2500 dibasic potassium phosphate
f phosphate bipotassique
e fosfato potásico dibásico
i fosfato potassico bibasico
d Dikaliumphosphat

2501 dibasic sodium phosphate
f phosphate dibasique de sodium
e fosfato sódico dibásico
i fosfato bibasico di sodio
d Dinatriumphosphat

2502 dibenzanthrone
  f dibenzanthrone
  e dibenzantrona
  i dibenzantrone
  d Dibenzanthron

2503 dibenzyl ether
  f éther dibenzylique
  e éter dibencílico
  i etere dibenzilico
  d Dibenzyläther

2504 dibenzyl group
  f groupe dibenzylique
  e grupo dibencílico
  i gruppo dibenzile
  d Dibenzylgruppe

2505 dibenzyl sebacate
  f dibenzylsébacate
  e sebacato de dibencilo
  i dibenzilsebacato
  d Dibenzylsebacat

  * **diborane** → 1125

2506 dibromodiethyl sulphide
  f sulfure de dibromodiéthyle
  e sulfuro de dibromodietilo
  i solfuro dibromodietilico
  d Dibromdiäthylsulfid

2507 dibromodifluoromethane
  f dibromodifluorométhane
  e dibromodifluometano
  i dibromodifluorometano
  d Dibromdifluormethan

2508 dibromomalonic acid
  f acide dibromomalonique
  e ácido dibromomalónico
  i acido dibromomalonico
  d Dibrommalonsäure

2509 dibromopropanol
  f dibromopropanol
  e dibromopropanol
  i dibromopropanolo
  d Dibrompropanol

2510 dibucaine
  f dibucaïne
  e dibucaína
  i dibucaina
  d Dibucain

2511 dibutoline sulphate
  f sulfate de dibutoline
  e sulfato de dibutolina

  i solfato di dibutolina
  d Dibutolinsulfat

2512 dibutoxyethyl adipate
  f adipate dibutoxyéthylique
  e adipato de dibutoxietilo
  i dibutossietiladipato
  d Dibutoxyäthyladipat

2513 dibutoxyethyl phthalate
  f phtalate dibutoxyéthylique
  e ftalato de dibutoxietilo
  i dibutossietilftalato
  d Dibutoxyäthylphthalat

2514 dibutoxytetraglycol
  f dibutoxytétraglycol
  e dibutoxitetraglicol
  i dibutossitetraglicolo
  d Dibutoxytetraglykol

2515 dibutylbutyl phosphonate
  f phosphonate de dibutylbutyle
  e fosfonato de dibutilbutilo
  i dibutilbutilfosfonato
  d Dibutylbutylphosphonat

2516 dibutyl fumarate
  f fumarate de dibutyle
  e fumarato de dibutilo
  i dibutilfumarato
  d Dibutylfumarat

2517 dibutyl maleate
  f maléate de dibutyle
  e maleato de dibutilo
  i dibutilmaleato
  d Dibutylmaleat

2518 di-tert-butylmetacresol
  f di-tert-butylmétacrésol
  e di-ter-butilmetacresol
  i bi-ter-butilemetacresolo
  d Di-tert-butylmetakresol

2519 dibutyl oxalate
  f oxalate de dibutyle
  e oxalato de dibutilo
  i ossalato dibutilico
  d Dibutyloxalat

2520 di-tert-butyl peroxide
  f peroxyde de di-tert-butyle
  e peróxido de di-ter-butilo
  i perossido bi-ter-butilico
  d Di-tert-butylperoxyd

2521 dibutylphthalate
  f phtalate de dibutyle

e ftalato de dibutilo
i dibutilftalato
d Dibutylphthalat

2522 dibutyl sebacate
f sébacate de dibutyle
e sebacato de dibutilo
i sebacato dibutilico
d Dibutylsebacat

2523 dibutyl tartrate
f tartrate de dibutyle
e tartrato de dibutilo
i dibutiltartrato
d Dibutyltartrat; Weinsäuredibutylester

2524 dibutylthiourea
f dibutylthiourée
e dibutiltiourea
i dibutiltiourea
d Dibutylthioharnstoff

2525 dibutyltin diacetate
f diacétate de dibutylétain
e diacetato de dibutilestaño
i diacetato di dibutilstagno
d Dibutylzinndiacetat

2526 dibutyltin dilaureate
f dilaurate de dibutylétain
e dilaurato de dibutilestaño
i dilaurato di dibutilstagno
d Dibutylzinndilaurat

2527 dibutyltin maleate
f maléate de dibutylétain
e maleato de dibutilestaño
i maleato di dibutilstagno
d Dibutylzinnmaleat

2528 dibutyltin oxide
f oxyde de dibutylétain
e óxido de dibutilestaño
i ossido di dibutilstagno
d Dibutylzinnoxyd

2529 dibutyltin sulphide
f sulfure de dibutylétain
e sulfuro de dibutilestaño
i solfuro di dibutilstagno
d Dibutylzinnsulfid

2530 dicapryl adipate
f adipate dicaprylique
e adipato de dicaprilo
i dicapriladipato
d Dicapryladipat

2531 dicapryl phthalate
f phtalate dicaprylique
e ftalato de dicaprilo
i dicaprilftalato
d Dicaprylphthalat

2532 dicapryl sebacate
f sébacate dicaprylique
e sebacato de dicaprilo
i dicaprilsebacato
d Dicaprylsebacat

2533 dicetyl ether
f éther dicétylique
e éter dicetílico; dicetiléter
i etere dicetilico
d Dicetyläther

2534 dichlone
f dichlone
e diclona
i diclone
d Dichlon

2535 dichlorobenzene
f dichlorobenzène
e diclorobenceno
i diclorobenzene
d Dichlorbenzol

2536 dichlorodiethyl sulphide
f sulfure d'éthyle dichloré; gaz moutarde; ypérite
e sulfuro de diclorodietilo
i solfuro di diclorodietile
d Dichlordiäthylsulfid; Senfgas

2537 dichlorodifluoromethane
f dichlorodifluorométhane
e diclorodifluometano
i diclorodifluorometano
d Dichlordifluormethan

* dichloro-diphenyl-trichloroethane
→ 2320

2538 dichloroethyl ether
f éther dichloroéthylique
e éter dicloroetílico
i etere di dicloroetile
d Dichloräthyläther

2539 dichloroethyl formal
f dichloroéthylformal
e dicloroetilformal
i dicloroetilformal
d Dichloräthylformal

2540 dichlorofluoromethane
f dichlorofluorométhane
e diclorofluometano
i diclorofluorometano
d Dichlorfluormethan

2541 dichloroisopropyl ether
f éther dichloroisopropylique
e éter dicloroisopropílico
i etere dicloroisopropilico
d Dichlorisopropyläther

2542 dichloromethane
f dichlorométhane; chlorure de méthylène
e diclorometano
i diclorometano
d Dichlormethan; Methylenchlorid

2543 dichloropentane
f dichloropentane
e dicloropentano
i dicloropentano
d Dichlorpentan

2544 dichlorophene
f dichlorophène
e diclorofeno
i diclorofene
d Dichlorophen

2545 dichlorophenyltrichlorosilane
f dichlorophényltrichlorosilane
e diclorofeniltriclorosilano
i diclorofeniltriclorosilano
d Dichlorphenyltrichlorsilan

2546 dichlor phenamide
f dichlorphénamide
e diclorfenamida
i diclorfenammide
d Dichlorphenamid

2547 dichromate cell
f pile au dichromate
e pila de bicromato
i pila al bicromato
d Dichromatelement

2548 dictyate
f stade dictyotique
e estado dictiótico
i stadio dictiottivo
d retikuläres Stadium (der Chromosomen)

2549 dicumyl peroxide
f peroxyde de dicumyle
e peróxido de dicumilo
i perossido di dicumile
d Dicumylperoxyd

2550 dicyandiamide
f dicyandiamide
e diciandiamida
i diciandiammide
d Dicyandiamid

2551 dicyclohexylamine
f dicyclohexylamine
e diciclohexilamina
i dicicloesilammina
d Dicyclohexylamin

2552 dicyclohexyl phthalate
f phtalate de dicyclohéxyle
e ftalato de diciclohexilo
i ftalato di dicicloesile
d Dicyclohexylphthalat

2553 dicyclomine hydrochloride
f chlorhydrate de dicyclomine
e hidrocloruro de diciclomina
i idrocloruro di diciclomina
d Dicyclominhydrochlorid

2554 dicyclopentadiene
f dicyclopentadiène
e diciclopentadieno
i diciclopentadiene
d Dicyclopentadien

2555 dicyclopentadiene dioxide
f bioxyde de dicyclopentadiène
e dióxido de diciclopentadieno
i diciclopentadiene biossido
d Dicyclopentadiendioxyd

2556 didecyl adipate
f adipate de didécyle
e adipato de didecilo
i dideciladipato
d Didecyladipat

2557 didecyl ether
f éther didécylique
e éter didecílico; dideciléter
i etere didecilico
d Didecyläther

2558 didecyl phthalate
f phtalate de didécyle
e ftalato de didecilo
i didecilftalato
d Didecylphthalat

2559 didymium
f didymium
e didimio
i didimio
d Didym

2560 dieldrin
f dieldrine
e dieldrina
i dieldrina
d Dieldrin

2561 dielectric porcelain
f porcelaine diélectrique
e porcelana dieléctrica
i porcellana dielettrica
d dielektrisches Porzellan

2562 Diels-Alder reaction
f réaction de Diels-Alder
e reacción de Diels-Alder
i reazione Diels-Alder
d Diels-Alder-Reaktion

2563 diene
f diène
e dieno
i dieno
d Dien

2564 diene synthesis
f synthèse de Diels-Alder
e síntesis de dienos
i sintesi Diels-Alder
d Diels-Alder-Synthese

2565 diene value
f valeur diène
e valor en dienos
i valore di non saturazione
d Dienzahl

2566 dienoestrol
f diénestrol
e dienestrol
i dienoestrol
d Dienöstrol

2567 diesel oil
f huile lourde; mazout; fuel-oil
e aceite diesel
i olio pesante; olio diesel
d Dieselöl; Schweröl

2568 dietetics
f diététique
e dietética
i dietetica
d Diätetik

2569 diethanolamine
f diéthanolamine
e dietanolamina
i dietanolammina
d Diäthanolamin

2570 diethyl adipate
f adipate de diéthyle
e adipato de dietilo
i dietiladipato
d Diäthyladipat

2571 diethylaluminium chloride
f chlorure de diéthylaluminium
e cloruro de dietilaluminio
i cloruro di dietilalluminio
d Diäthylaluminiumchlorid

2572 diethylamine
f diéthylamine
e dietilamina
i dietilammina
d Diäthylamin

2573 diethylaminoethanol
f diéthylaminoéthanol
e dietilaminoetanol
i dietilamminoetanol
d Diäthylaminoäthanol

2574 diethylbenzene
f diéthylbenzene
e dietilbenceno
i dietilbenzene
d Diäthylbenzol

*   **diethylbromoacetamine** → 68

2575 diethylcarbamazine citrate
f citrate de diéthylcarbamazine
e citrato de dietilcarbamacina
i citrato di dietilcarbamacina
d Diäthylcarbamazincitrat

2576 diethyl carbonate
f carbonate de diéthyle
e carbonato de dietilo
i dietilcarbonato
d Diäthylcarbonat; Kohlensäurediäthylester

2577 diethyldiphenylurea
f diéthyldiphénylurée
e dietildifenilurea
i dietildifenilurea
d Diäthyldiphenylharnstoff

2578 diethyldithiocarbamic acid
f acide diéthyldithiocarbamique
e ácido dietilditiocarbámico
i acido dietilditiocarbamico
d Diäthyldithiocarbaminsäure

2579 diethylene glycol
f diéthylèneglycol; alcool diéthylénique
e dietilenglicol

i glicoldietilenico
d Diäthylenglykol

2580 diethylene glycol dibutyl ether
f éther dibutylique du diéthylèneglycol
e éter dibutílico del dietilenglicol
i etere dibutilglicoldietilenico
d Diäthylenglykoldibutyläther

2581 diethylene glycol diethyl ether
f éther diéthylique du diéthylèneglycol
e éter dietílico del dietilenglicol
i etere dietilglicoldietilenico
d Diäthylenglykoldiäthyläther

2582 diethylene glycol dimethyl ether
f éther diméthylique du diéthylèneglycol
e dimetiléter del dietilenglicol
i etere dimetilglicoldietilenico
d Diäthylenglykoldimethyläther

2583 diethylene glycol dinitrate
f dinitrate de diéthylèneglycol
e dinitrato de dietilenglicol
i dinitrato glicoldietilenico
d Diäthylenglykoldinitrat

2584 diethylene glycol monobutyl ether
f éther monobutylique de diéthylèneglycol
e éter monobutílico de dietilenglicol
i etere monobutilico del glicoldietilenico
d Diäthylenglykolmonobutyläther

2585 diethylene glycol monobutyl ether acetate
f acétate de l'éther monobutylique du diéthylèneglycol
e acetato del éter monobutílico del dietilenglicol
i acetato dell'etere monobutilglicoldietilenico
d Diäthylenglykolmonobutylätheracetat

2586 diethylene glycol monoethyl ether
f éther monoéthylique du diéthylèneglycol
e éter monoetílico del dietilenglicol
i etere monoetilglicoldietilenico
d Diäthylenglykolmonoäthyläther

2587 diethylene glycol monoethyl ether acetate
f acétate de l'éther monoéthylique du diéthylèneglycol
e acetato del éter monoetílico del dietilenglicol
i acetato dell'etere monoetilglicoldietilenico
d Diäthylenglykolmonoäthylätheracetat

2588 diethylene glycol monomethyl ethyl ether
f éther monométhylique du diéthylèneglycol
e éter monometílico del dietilenglicol
i etere monometilico glicoldietilenico
d Diäthylenglykolmonomethyläther

2589 diethylenetriamine
f diéthylènetriamine
e dietilentriamina
i dietilenetriammina
d Diäthylentriamin

* **diethyl ether** → 3128

2590 diethyl ethylphosphonate
f éthylphosphonate de diéthyle
e dietiletilfosfonato; etilfosfonato de dietilo
i etilfosfonato dietilico
d Diäthyläthylphosphonat

2591 diethyl maleate
f maléate de diéthyle
e maleato de dietilo
i maleato dietilico
d Maleinsäurediäthylester

2592 diethylmalonate
f malonate de diéthyle
e malonato de dietilo
i dietilmalonato
d Malonsäurediäthylester

* **diethyl-malonylurea** → 783

2593 diethyl phosphite
f phosphite de diéthyle
e fosfito de dietilo
i dietilfosfito
d Diäthylphosphit

2594 diethyl phthalate
f phtalate de diéthyle
e ftalato de dietilo
i dietilftalato
d Diäthylphthalat

2595 diethylstilboestrol
f diéthylstilbestrol
e dietilestilbestrol
i dietilstilbestrol
d Diäthylstilböstrol

2596 diethylstilboestrol dipropionate
f dipropionate de diéthylstilbestrol
e dipropionato de dietilestilbestrol
i dipropionato di dietilstilbestrol
d Diäthylstilböstroldipropionat

2597 diethyl succinate
 f succinate de diéthyle
 e succinato de dietilo
 i succinato dietilico
 d Diäthylsuccinat

2598 diethyl sulphate
 f sulfate de diéthyle
 e sulfato de dietilo
 i solfato dietilico
 d Diäthylsulfat

2599 diethyl tartrate
 f tartrate de diéthyle
 e tartrato de dietilo
 i tartrato dietilico
 d Diäthyltartrat

   diethylzinc → 8932

2600 diffluent
 f à flux divergent
 e difluente
 i diffluente
 d auseinanderfließend

2601 diffraction grating
 f réseau à diffraction
 e red de difracción
 i reticolo di diffrazione
 d Beugungsgitter

2602 diffusate
 f diffusat
 e difusado
 i diffusato
 d Diffusat

2603 diffusion
 f diffusion
 e difusión
 i diffusione
 d Diffusion

2604 diffusion coefficient
 f coefficient de diffusion
 e coeficiente de difusión
 i coefficiente di diffusione
 d Diffusionskoeffizient

2605 diffusion layer
 f couche de diffusion
 e capa de difusión
 i strato di diffusione
 d Diffusionsschicht

2606 digametic; heterogametic
 f digamétique; hétérogamétique
 e digamético; heterogamético

 i digametico; eterogametico
 d digametisch; heterogametisch

2607 digamety
 f digamétie
 e digametia
 i digametia
 d Digametie

2608 digitalin
 f digitaline
 e digitalina
 i digitalina
 d Digitalin

2609 digitalis
 f digitale
 e digital
 i digitale
 d Digitalis

2610 digitoxin
 f digitoxine
 e digitoxina
 i digitossina
 d Digitoxin

2611 diglycol carbamate
 f carbamate diglycolique
 e carbamato de diglicol
 i diglicolcarbammato
 d Diglykolcarbamat

2612 diglycol chloroformate
 f chloroformiate de diglycol
 e cloroformiato de diglicol
 i cloroformiato di diglicol
 d Diglykolchloroformiat

2613 diglycolic acid
 f acide diglycolique
 e ácido diglicólico
 i acido diglicolico
 d Diglykolsäure

2614 diglycol laurate
 f laurate de diglycol
 e laurato de diglicol
 i laurato di diglicol
 d Diglykollaurat

2615 diglycol monostearate
 f monostéarate de diglycol
 e monoestearato de diglicol
 i monostearato di diglicol
 d Diglykolmonostearat

2616 diglycol oleate
 f oléate de diglycol

e oleato de diglicol
i diglicololeato
d Diglykololeat

2617 diglycol ricinoleate
f ricinoléate de diglycol
e ricinoleato de diglicol
i diglicolricinoleato
d Diglykolricinoleat

2618 diglycol stearate
f stéarate de diglycol
e estearato de diglicol
i diglicolstearato
d Diglykolstearat

2619 dihexyl phthalate
f phtalate de dihexyle
e ftalato de dihexilo
i diesilftalato
d Dihexylphthalat

2620 dihexyl sebacate
f sébacate dihexylique
e sebacato de dihexilo
i diesilsebacato
d Dihexylsebacat

2621 dihydroabietyl alcohol
f alcool dihydroabiétylique
e alcohol dihidroabietílico
i alcool diidroabietilico
d Dihydroabietylalkohol

2622 dihydrocholesterol
f dihydrocholestérol
e dihidrocolesterol
i diidrocolesterolo
d Dihydrocholesterol

2623 dihydrodiethyl phthalate
f phtalate dihydrodiéthylique
e ftalato de dihidrodietilo
i diidrodietilftalato
d Dihydrodiäthylphthalat

2624 dihydromorphinone hydrochloride
f chlorhydrate de dihydromorphinone
e hidrocloruro de dihidromorfinona
i idrocloruro di diidromorfinone
d Dihydromorphinonhydrochlorid

2625 dihydrostreptomycin
f dihydrostreptomycine
e dihidroestreptomicina
i diidrostreptomicina
d Dihydrostreptomycin

2626 dihydroxyacetone
f dihydroxyacétone
e dihidroxiacetona
i diidrossiacetone
d Dihydroxyaceton

2627 dihydroxyaluminium aminoacetate
f aminoacétate de dihydroxyaluminium
e aminoacetato dihidroxialumínico
i amminoacetato di diidrossialluminio
d Dihydroxyaluminiumaminoacetat

2628 dihydroxydiphenyl sulphone
f dihydroxydiphénylsulfone
e dihidroxidifenilsulfona
i diidrossidifenilsolfone
d Dihydroxydiphenylsulfon

**2629 diiodohydroxyquinoline;
diiodohydroxyquin**
f diiodohydroxyquinoline
e diyodohidroxiquinolina
i diiodoidrossichinolina
d Dijodhydroxychinolin

2630 diiodothyronine
f diiodothyronine
e diyodotironina
i diiodotironina
d Dijodthyronin

2631 diisobutyl adipate
f adipate diisobutylique
e adipato de diisobutilo
i diisobutiladipato
d Diisobutyladipat

2632 diisobutyl aluminium chloride
f chlorure de diisobutylaluminium
e cloruro de diisobutilaluminio
i cloruro di disobutilalluminio
d Diisobutylaluminiumchlorid

2633 diisobutylamine
f diisobutylamine
e diisobutilamina
i diisobutilammina
d Diisobutylamin

2634 diisobutyl azelate
f azélate de diisobutyle
e azelato de diisobutilo
i diisobutilazelato
d Diisobutylazelat

2635 diisobutylene
f diisobutylène
e diisobutileno

i diisobutilene
d Diisobutylen

2636 diisobutyl ketone
f diisobutylcétone
e diisobutilcetona
i diisobutilchetone
d Diisobutylketon

2637 diisobutyl phthalate
f diisobutylphtalate
e ftalato de diisobutilo
i diisobutilftalato
d Diisobutylphthalat

2638 diisocyanates
f diisocyanates
e diisocianatos
i diisocianati
d Diisocyanate

2639 diisodecyl adipate
f adipate diisodécylique
e adipato de diisodecilo
i diisodeciladipato
d Diisodecyladipat

2640 diisodecyl phthalate
f phtalate diisodécylique
e ftalato de diisodecilo
i diisodecilftalato
d Diisodecylphthalat

2641 diisooctyl adipate
f adipate diisooctylique
e adipato de diisooctilo
i diisoottiladipato
d Diisooctyladipat

2642 diisooctyl azelate
f azélate de diisooctyle
e azelato de diisooctilo
i diisoottilazelato
d Diisooctylazelat

2643 diisooctyl phthalate
f phtalate diisooctylique
e ftalato de diisooctilo
i diisoottilftalato
d Diisooctylphthalat

2644 diisooctyl sebacate
f sébacate diisooctylique
e sebacato de diisooctilo
i diisoottilsebacato
d Diisooctylsebacat

2645 diisopropanolamine
f diisopropanolamine

e diisopropanolamina
i diisopropanolammina
d Diisopropanolamine

2646 diisopropyl benzene
f diisopropylbenzène
e diisopropilbenceno
i diisopropilbenzene
d Diisopropylbenzol

2647 diisopropyl carbinol
f diisopropylcarbinol
e diisopropilcarbinol
i diisopropilcarbinolo
d Diisopropylcarbinol

2648 diisopropyl cresol
f diisopropylcrésol
e diisopropilcresol
i diisopropilcresolo
d Diisopropylcresol

2649 diisopropyl dixanthogen
f dixanthogène de diisopropyle
e dixantogenato de diisopropilo
i dixantogenato isopropilico
d Diisopropyldixanthogen

2650 diisopropylene glycol salicylate
f salicylate de diisopropylèneglycol
e salicilato de diisopropilenglicol
i salicilato di diisopropilenglicol
d Diisopropylenglykolsalicylat

2651 diisopropyl fluophosphate
f fluophosphate de diisopropyle
e fluofosfato di diisopropilo
i fluofosfato di diisopropile
d Diisopropylfluophosphat

2652 diisopropylthiourea
f diisopropylthiourée
e diisopropiltiourea
i diisopropiltiourea
d Diisopropylthioharnstoff

2653 dikaryon
f dicaryon
e dicarión
i dicarion
d Dikaryon

2654 diketene
f dicétène
e dicetena
i dichetene
d Diketen

2655 diketones
f dicétones
e dicetonas
i dichetoni
d Diketone

2656 dilauryl ether
f éther dilaurylique
e éter dilaurílico
i etere dilaurilico
d Dilauryläther

2657 dilauryl sulphide
f sulfure de dilauryle
e sulfuro de dilaurilo
i solfuro di dilaurile
d Dilaurylsulfid

2658 dilauryl thiodipropionate
f thiodipropionate de dilauryle
e tiodipropionato de dilaurilo
i tiodipropionato di dilaurile
d Dilaurylthiodipropionat

2659 dilinoleic acid
f acide dilinoléique
e ácido dilinoleico
i acido dilinoleico
d Dilinoleinsäure

2660 dill
f essence d'aneth
e eneldo
i oneto
d Dillöl

2661 diluent *(adj.)*
f diluent
e diluente
i diluente
d verdünnend

2662 diluent *(noun)*; **diluting agent; thinner**
f diluant
e diluyente
i diluente; riempitivo
d Streckmittel; Verdünnungsmittel

2663 dilute *v*
f diluer
e diluir
i diluire
d verdünnen

2664 dilute solution
f solution dilué
e solución diluida
i soluzione diluita
d verdünnte Lösung

* **diluting agent** → **2662**

2665 dilution
f dilution
e dilución
i diluizione
d Verdünnung

2666 dimenhydrinate
f dimenhydrinate
e dimenhidrinato
i dimenidrinato
d Dimenhydrinat

2667 dimer
f dimère
e dímero
i dimero
d Dimer

2668 dimer acid
f acide dimérisé
e ácido dímero
i acido dimerico
d Dimersäure

2669 dimeric
f dimérique
e dimérico
i dimerico
d dimer

2670 dimery
f dimérie
e dimeria
i dimeria
d Dimerie

2671 dimethicone
f diméthicone
e dimeticona
i dimeticone
d Dimethicon

2672 dimethoxyethyl adipate
f adipate diméthoxyéthylique
e adipato de dimetoxietilo
i dimetossietiladipato
d Dimethoxyäthyladipat

2673 dimethoxytetraglycol
f diméthoxytétraglycol
e dimetoxitetraglicol
i dimetossitetraglicol
d Dimethoxytetraglykol

2674 dimethylacetal
f diméthylacétal
e dimetilacetal

i dimetilacetal
d Dimethylacetal

2675 dimethylacetamide
f acétamide diméthylique
e dimetilacetamida
i dimetilacetammide
d Dimethylacetamid

2676 dimethylamine
f diméthylamine
e dimetilamina
i dimetilammina
d Dimethylamin

2677 dimethylaminoazobenzene
f diméthylaminoazobenzène
e dimetilaminoazobenceno
i dimetilaminoazobenzene
d Dimethylaminoazobenzol

2678 dimethylaminopropylamine
f diméthylaminopropylamino
e dimetilaminopropilamina
i dimetilamminopropilammina
d Dimethylaminopropylamin

2679 dimethylaniline
f diméthylaniline
e dimetilanilina
i dimetilanilina
d Dimethylanilin

2680 dimethyl anthranilate
f anthranilate de diméthyle
e antranilato de dimetilo
i antranilato dimetilico
d Dimethylanthranilat

* **dimethylbenzene** → 8882

2681 dimethylbenzylcarbinol
f diméthylbenzylcarbinol
e dimetilbencilcarbinol
i dimetilbenzilcarbinolo
d Dimethylbenzylcarbinol

2682 dimethyl carbate
f carbate de diméthyle
e carbato de dimetilo
i carbato di dimetile
d Dimethylcarbat

2683 dimethylchloroacetal
f chloroacétal de diméthyle
e cloroacetal dimetílico
i cloroacetale di dimetile
d Dimethylchloracetal

2684 dimethylcyclohexane
f diméthylcyclohexane
e dimetilciclohexano
i dimetilcicloesano
d Dimethylcyclohexan

2685 dimethyldichlorosilane
f diméthyldichlorosilane
e dimetildiclorosilano
i dimetildiclorosilano
d Dimethyldichlorsilan

2686 dimethyldiphenylurea
f diméthyldiphénylurée
e dimetildifenilurea
i dimetildifenilurea
d Dimethyldiphenylharnstoff

2687 dimethyl ether
f éther diméthyllque
e éter dimetílico
i etere dimetilico
d Dimethyläther

2688 dimethyl glycol phthalate
f phtalate de diméthylglycol
e ftalato de dimetilglicol
i dimetilglicoleftalato
d Dimethylglykolphthalat

2689 dimethylglyoxime
f diméthylglyoxime
e dimetilglioxima
i dimetilgliossima
d Dimethylglyoxim

2690 dimethylhexanediol
f diméthylhexanediol
e dimetilhexanodiol
i dimetilesanediol
d Dimethylhexandiol

2691 dimethyl hexynol
f diméthylhexynol
e dimetilhexinol
i dimetilessinol
d Dimethylhexynol

2692 dimethylhydantoin
f diméthylhydantoïne
e dimetilhidantoína
i dimetilidantoina
d Dimethylhydantoin

2693 dimethylhydantoin resins
f résines de diméthylhydantoïne
e resinas de dimetilhidantoína
i resine dimetilidantoiniche
d Dimethylhydantoinharze

2694 dimethylisobutylcarbinyl phthalate
f phtalate de diméthylisobutylcarbinol
e ftalato de dimetilisobutilcarbinol
i dimetilisobutilcarbinilftalato
d Dimethylisobutylcarbinylphthalat

2695 dimethyl isophthalate
f isophtalate de diméthyle
e isoftalato de dimetilo
i isoftalato di dimetile
d Dimethylisophthalat

2696 dimethylisopropanolamine
f diméthylisopropanolamine
e dimetilisopropanolamina
i dimetilisopropanolammina
d Dimethylisopropanolamin

2697 dimethyl itaconate
f itaconate de diméthyle
e itaconato de dimetilo
i itaconato di dimetile
d Dimethylitaconat

2698 dimethyloctanediol
f diméthyloctanediol
e dimetiloctanodiol
i dimetilottanediol
d Dimethyloctandiol

2699 dimethyl octanol
f diméthyloctanol
e dimetiloctanol
i dimetiloctanol
d Dimethyloctanol

2700 dimethyloctynediol
f diméthyloctynediol
e dimetiloctinodiol
i dimetilottinediol
d Dimethyloctyndiol

2701 dimethylol ethylene urea
f diméthyloléthylèneurée
e dimetiloletilenurea
i dimetiloletilenurea
d Dimethyloläthylenharnstoff

2702 dimethylol urea
f diméthylolurée
e dimetilolurea
i dimetilolurea
d Dimethylolharnstoff

2703 dimethyl phosphite
f phosphite de diméthyle
e fosfito de dimetilo
i fosfito di dimetile
d Dimethylphosphit

2704 dimethyl phthalate
f phtalate de diméthyle
e ftalato de dimetilo
i dimetilftalato
d Dimethylphthalat

2705 dimethylpiperazine; lupetazine
f diméthylpipérazine
e dimetilpiperacina
i dimetilpiperazina
d Dimethylpiperazin

2706 dimethyl sebacate
f sébacate de diméthyle
e sebacato de dimetilo
i dimetilsebacato
d Dimethylsebacat

2707 dimethyl sulphate
f sulfate de diméthyle
e sulfato de dimetilo
i solfato di dimetile
d Dimethylsulfat

2708 dimethyl sulphide
f sulfure de diméthyle
e sulfuro de dimetilo
i solfuro di dimetile
d Dimethylsulfid

2709 dimethyl sulphoxide
f oxysulfure de diméthyle
e sulfóxido de dimetilo
i dimetilsolfossido
d Dimethylsulfoxyd

2710 dimethyl terephthalate
f téréphtalate de diméthyle
e tereftalato de dimetilo
i dimetiltereftalato
d Dimethylterephthalat

2711 dimethyl tubocurarine chloride
f chlorure de diméthyltubocurarine
e cloruro de dimetiltubocurarina
i cloruro dimetilico di tubocurarina
d Dimethyltubocurarinchlorid

2712 dimorphic
f dimorphe
e dimorfo
i dimorfo
d dimorph

2713 dimorphism
f dimorphisme
e dimorfismo
i dimorfismo
d Dimorphismus

2714 dimyristyl ether
f éther dimyristylique
e éter dimiristílico
i etere dimiristilico
d Dimyristyläther

2715 dimyristyl thioether
f thioéther dimyristylique
e tioéter dimiristílico
i tioetere di dimiristile
d Dimyristylthioäther

2716 dineric
f dinère
e dinérico
i dinerico
d dinerisch

2717 dinitrobenzene
f dinitrobenzène
e dinitrobenceno
i dinitrobenzene
d Dinitrobenzol

2718 dinitronaphthalene
f dinitronaphtalène
e dinitronaftaleno
i dinitronaftalene
d Dinitronaphthalin

2719 dinitrophenol
f dinitrophénol
e dinitrofenol
i dinitrofenolo
d Dinitrophenol

2720 dinonyl adipate
f adipate dinonylique
e adipato de dinonilo
i dinoniladipato
d Dinonyladipat

2721 dinonyl phenol
f dinonylphénol
e dinonilfenol
i dinonilfenolo
d Dinonylphenol

2722 dinonyl phthalate
f phtalate dinonylique
e ftalato de dinonilo
i dinonilftalato
d Dinonylphthalat

2723 dintrotoluene
f dinitrotoluène
e dinitrotolueno
i dinitrotoluene
d Dinitrotoluol

2724 dioctyl ether
f éther dioctylique
e éter dioctílico
i etere di diottile
d Dioctyläther

2725 dioctyl fumarate
f fumarate de dioctyle
e fumarato de dioctilo
i diottilfumarato
d Dioctylfumarat

2726 dioctyl isophthalate
f isophtalate de dioctyle
e isoftalato de dioctilo
i diottilisoftalato
d Dioctylisophthalat

2727 dioctyl phosphite
f phosphite de dioctyle
e fosfito de dioctilo
i fosfito di diottile
d Dioctylphosphit

2728 dioctyl phthalate
f phtalate de dioctyle
e ftalato de dioctilo
i ftalato diottilico
d Dioctylphthalat

2729 dioctyl sebacate
f sébacate de dioctyle
e sebacato de dioctilo
i diottilsebacato
d Dioctylsebacat

2730 diorite
f diorite
e diorita
i diorite
d Diorit; Grünstein

2731 dioxane
f dioxane
e dioxano
i diossano
d Dioxan

2732 dioxolane
f dioxolane
e dioxolano
i diossolano
d Dioxolan

* **dioxopurine** → 8873

2733 dip *v*
f tremper
e inmergir

   i immergere
   d tauchen

2734 dip
   f bain au trempé
   e baño por inmersión
   i bagno di tempra
   d Brenne; Beize

2735 dip coat *v*
   f enduire par trempage
   e revistir por inmersión
   i rivestire per immersione
   d durch Tauchen beschichten

2736 dip coating
   f revêtement par trempage
   e revestimiento por inmersión
   i rivestimento ad immersione
   d getauchter Überzug

2737 dipentaerythritol
   f dipentaérythritol
   e dipentaeritritol
   i dipentaeritritolo
   d Dipentaerythritol

2738 dipentene
   f dipentène
   e dipenteno
   i dipentene
   d Dipenten

2739 dipentene monoxide
   f monoxyde de dipentène
   e monóxido de dipenteno
   i monossido di dipentene
   d Dipentenmonoxyd

2740 diphemanil methylsulphate
   f méthylsulfate de diphémanile
   e metilsulfato de difemanilo
   i metilsolfato di difemanile
   d Diphemanilmethylsulfat

2741 diphenhydramine hydrochloride
   f chlorhydrate de diphénhydramine
   e hidrocloruro de difenhidramina
   i cloridrato di difenidrammina
   d Diphenhydraminhydrochlorid

2742 diphenic acid
   f acide diphénique
   e ácido difénico
   i acido difenico
   d Diphensäure

2743 diphenyl
   f biphényle; diphényle

   e difenilo
   i difenil
   d Diphenyl

2744 diphenylacetonitrile
   f diphénylacétonitrile
   e difenilacetonitrilo
   i difenilacetonitrile
   d Diphenylacetonitril

2745 diphenylamine
   f diphénylamine; phénylaniline
   e difenilamina
   i difenilammina
   d Diphenylamin

2746 diphenylbenzidine
   f diphénylbenzidine
   e difenilbencidina
   i difenilbenzidina
   d Diphenylbenzidin

2747 diphenylcarbazide
   f diphénylcarbazide
   e difenilcarbacida
   i difenilcarbazide
   d Diphenylcarbazid

2748 diphenyl carbonate
   f carbonate de diphényle
   e carbonato de difenilo
   i difenilcarbonato
   d Diphenylcarbonat

2749 diphenyl cresyl phosphate
   f phosphate de diphényle et crésyle
   e fosfato de difenilo y cresilo
   i difenilcresilfosfato
   d Diphenylkresylphosphat

2750 diphenyl decyl phosphite
   f phosphite de diphényldécyle
   e fosfito de difenildecilo
   i fosfito decildifenilico
   d Diphenyldecylphosphit

2751 diphenyldichlorosilane
   f diphényldichlorosilane
   e difenildiclorosilano
   i difenildiclorosilano
   d Diphenyldichlorsilan

2752 diphenylguanidine
   f diphénylguanidine
   e difenilguanidina
   i difenilguanidina
   d Diphenylguanidin

2753 diphenylguanidine phthalate
 f phtalate de diphénylguanidine
 e ftalato de difenilguanidina
 i ftalato di difenilguanidina
 d Diphenylguanidinphthalat

2754 diphenylhydantoin
 f diphénylhydantoïne
 e difenilhidantoína
 i difenilidantoina
 d Diphenylhydantoin

2755 diphenylmethane
 f diphénylméthane
 e difenilmetano
 i difenilmetano
 d Diphenylmethan

2756 diphenylmethane diisocyanate
 f diisocyanate de diphénylméthane
 e diisocianato de difenilmetano
 i difenilmetanodiisocianato
 d Diphenylmethandiisocyanat

2757 diphenylnaphthylenediamine
 f diphénylnaphtylènediamine
 e difenilnaftilendiamina
 i difenilnaftilenediammina
 d Diphenylnaphthylendiamin

2758 diphenyl octyl phosphate
 f phosphate de diphényle et d'octyle
 e fosfato de difenilo y octilo
 i difenilottilfosfato
 d Diphenyloctylphosphat

2759 diphenyl orthoxenyl phosphate
 f diphénylorthoxénylephosphate
 e fosfato de difenilo ortoxenilo
 i difenilortoxenilfosfato
 d Diphenylorthoxenylphosphat

2760 diphenyl oxide
 f oxyde de diphényle
 e óxido de difenilo
 i ossido di difenile
 d Diphenyloxyd

2761 diphenyl phthalate
 f phtalate de diphényle
 e ftalato de difenilo
 i difenilftalato
 d Diphenylphthalat

2762 diphenylpyraline hydrochloride
 f chlorhydrate de diphénylpyraline
 e hidrocloruro de difenilpiralina
 i idrocloruro di difenilpiralina
 d Diphenylpyralinhydrochlorid

2763 diphenylsilanediol
 f diphénylsilanediol
 e difenilsilanodiol
 i difenilsilanediol
 d Diphenylsilandiol

* **diphenylthiocarbazone** → 2826

* **diphenyl urea** → 1432

2764 diplobacillus
 f diplobacille
 e diplobacilo
 i diplobacillo
 d Diplobazillus

2765 diplochromosome
 f diplochromosome
 e diplocromosoma
 i diplocromosoma
 d Diplochromosom

2766 diploid
 f diploïde
 e diploide
 i diploide
 d Diploid

2767 diploidisation
 f diploïdisation
 e diploidización
 i diploidizzazione
 d Diploidisierung

2768 diploidy
 f diploïdie
 e diploidia
 i diploidismo
 d Diploidie

2769 diplont
 f diplonte
 e diplonte
 i diplonte
 d Diplont

2770 diplophase
 f diplophase
 e diplofase
 i diplofase
 d Diplophase

2771 diplosome
 f diplosome
 e diplosoma
 i diplosoma
 d doppeltes Zentralkörperchen

2772 dip moulding
f moulage par trempage
e moldeo por inmersión
i formatura ad immersione
d Tauchen

* **Dippel's oil** → **1111**

2773 dipping compound
f mélange de trempage
e mezcla por inmersión
i mescola per immersione
d Tauchmischung

2774 dipping varnish
f vernis au trempe
e barniz para inmersión
i vernice a immersione
d Tauchlack

2775 dip plating
f dépôt au trempé
e depósito por inmersión
i deposito per immersione
d Eintauchplattierung

2776 dipropylene glycol
f dipropylèneglycol
e dipropilenglicol
i glicolo dipropilenico
d Dipropylenglykol

2777 dipropylene glycol monomethyl ether
f éther monométhylique de dipropylèneglycol
e éter monometílico de dipropilenglicol
i etere monometilico di glicol dipropilenico
d Dipropylenglykolmonomethyläther

2778 dipropylene glycol monosalicylate
f monosalicylate de dipropylèneglycol
e monosalicilato de dipropilenglicol
i monosalicilato di glicol dipropilenico
d Dipropylenglykolmonosalicylat

2779 dipropyl ketone
f dipropylcétone
e dipropilcetona
i dipropilchetone
d Dipropylketon

2780 dipropyl phthalate
f phtalate de dipropyle
e ftalato de dipropilo
i dipropilftalato
d Dipropylphthalat

2781 dip tank
f cuve d'immersion
e cuba de inmersión
i vasca d'immersione
d Tauchbehälter

2782 dipyridylethyl sulphide
f sulfure dipyridyléthylique
e sulfuro de dipiridiletilo
i solfuro dipiridiletilico
d Dipyridyläthylsulfid

2783 direct-reading instrument
f instrument à lecture directe
e instrumento de lectura directa
i strumento a lettura diretta
d Instrument mit unmittelbarer Ablesung

2784 direct resistance furnace
f four à chauffage direct par résistance
e horno de calentamiento directo por
resistencia
i forno a riscaldamento diretto a resistenza
d direkter Widerstandsofen

2785 direct resistance heating
f chauffage direct par résistance
e calefacción directa por resistencia
i riscaldamento diretto a resistenza
d direkte Widerstandsheizung

2786 diresorcinol
f dirésorcinol
e diresorcina
i diresorcina
d Diresorcinol

2787 dirt content
f teneur en impuretés
e contenido en impurezas
i contenuto d'impurità
d Schmutzgehalt

2788 discharge cock
f robinet de purge
e grifo de descarga
i rubinetto di scarico
d Ablaßhahn

2789 discharge heat
f pression de refoulement
e presión de expulsión
i pressione d'uscita
d Druckhöhe

2790 discrete
f discrète
e separado
i separato; isolato
d abgesondert

2791 dish
 f cuvette; plat
 e cápsula; placa
 i capsula; piatto
 d Schüssel; Platte

2792 dished
 f en forme de soucoupe
 e cóncavo
 i concavo
 d schüsselförmig

2793 disinfectant
 f désinfectant
 e desinfectante
 i disinfettante
 d Desinfizierungsmittel

2794 disinfection
 f désinfection
 e desinfección
 i disinfezione
 d Desinfektion

2795 disinsectization
 f désinsectisation
 e desinsectización
 i disinfestazione
 d Entwesung; Desinsektion

2796 disintegrate *v*
 f désintégrer
 e desintegrar
 i disintegrare
 d zerkleinern

2797 disintegration
 f désagrégation
 e desintegración
 i disintegrazione
 d Zertrümmerung

2798 disintegrator
 f désintégrateur; broyeur
 e desintegrador; disgregador
 i disintegratore; rompitore
 d Zerkleinerungsmaschine

2799 disk mill
 f moulin à plateaux
 e molino de discos
 i molino a dischi
 d Stiftmühle

2800 disoxidation
 f désoxydation
 e desoxidación
 i desossidazione
 d Desoxydation

2801 disperse *v*
 f disperser
 e dispersar
 i disperdere
 d dispergieren

2802 dispersing agent
 f agent dispersant
 e agente dispersante
 i agente disperdente
 d Dispergiermittel

2803 dispersion
 f dispersion
 e dispersión
 i dispersione
 d Dispersion

2804 dispersion mixer
 f mélangeur à dispersion
 e amasador-dispersador
 i mescolatrice
 d Dispersionskneter

2805 dispersion stage
 f stade de dispersion
 e estado de disperción
 i stadio di dispersione
 d Zerstäubungsstadium

2806 dispersol colours
 f couleurs de dispersol
 e colores de dispersol
 i colori di dispersol
 d Dispersolfarbstoffe

2807 dissociation
 f dissociation
 e disociación
 i dissociazione
 d Dissoziation; Aufspaltung

2808 dissociation of gases
 f dissociation des gaz
 e disociación de gases
 i dissociazione dei gas
 d Dissoziation der Gase

2809 dissolved oxygen
 f oxygène dissous
 e oxígeno disuelto
 i ossigeno disciolto
 d gelöster Sauerstoff

2810 distal
 f distal
 e distal
 i distale
 d distal

2811 **distearyl ether**
 f éther distéarylique
 e éter diestearílico
 i etere distearilico
 d Distearyläther

2812 **distearyl thiodipropionate**
 f thiodipropionate de distéaryle
 e tiodipropionato de diestearilo
 i tiodipropionato distearilico
 d Distearylthiodipropionat

2813 **distemper**
 f badigeon; détrempe
 e pintura al temple
 i pittura a tempera
 d Temperafarbe; Wasserfarbe

2814 **distillate**
 f distillat
 e destilado
 i distillato
 d Destillat

2815 **distillate fuels** *(US)*
 f distillats
 e combustibles destilados
 i carburanti distillati
 d Destillatöle

2816 **distillation**
 f distillation
 e destilación
 i distillazione
 d Destillation

2817 **distillation flask**
 f ballon de distillation
 e frasco de destilación
 i matraccio da distillazione
 d Siedekolben

2818 **distillery**
 f distillerie
 e destilería
 i distilleria
 d Brennerei; Spiritusbrennerei

2819 **distilling column**
 f colonne de distillation
 e columna de destilación
 i colonna di distillazione
 d Destillationskolonne

2820 **distribution coefficient**
 f coefficient de distribution
 e coeficiente de distribución
 i coefficiente di distribuzione
 d Teilungskoeffizient; Verteilungskonstante

2821 **diterpenes**
 f diterpènes
 e diterpenos
 i diterpeni
 d Diterpene

2822 **ditetrahydrofurfuryl adipate**
 f adipate de ditétrahydrofurfuryle
 e adipato de ditetrahidrofurfurilo
 i ditetraidrofurfuriladipato
 d Ditetrahydrofurfuryladipat

2823 **dithiocarbamic acid**
 f acide dithiocarbamique
 e ácido ditiocarbámico
 i acido ditiocarbammico
 d Dithiocarbaminsäure

2824 **dithione**
 f dithione
 e dición
 i ditione
 d Dithion

2825 **dithiooxamide**
 f dithiooxamide
 e ditiooxamida
 i ditioossammide
 d Dithiooxamid

2826 **dithizone; diphenylthiocarbazone**
 f dithizone
 e ditizona
 i ditizone
 d Dithizon

2827 **ditridecyl phthalate**
 f ditridécylephtalate
 e ftalato de ditridecilo
 i ditridecilftalato
 d Ditridecylphthalat

2828 **ditridecyl thiodipropionate**
 f thiodipropionate de ditridécyle
 e tiodipropionato de ditridecilo
 i tiodipropionato di ditridecile
 d Ditridecylthiopropionat

2829 **divalent**
 f bivalent
 e bivalente
 i bivalente
 d zweiwertig

2830 **divinylbenzene**
 f divinylbenzène
 e divinilbenceno
 i divinilbenzene
 d Divinylbenzol

2831 divinyl sulphone
 f divinylsulfone
 e divinilsulfona
 i solfone di divinile
 d Divinylsulfon

* **DNA** → **2408**

2832 doctor *(papermaking)*
 f râcle; racloir
 e cuchillo rascador
 i coltello per spalmatura
 d Gegenrakel; Abstreichmesser

2833 doctor treatment
 f traitement au plombite
 e proceso para la desodorización de
  petróleos
 i raffinazione per eliminare cattivi odori
 d Plombitbehandlung

2834 dodecene
 f dodécène
 e dodeceno
 i dodecene
 d Dodecen

2835 dodecenylsuccinic acid
 f acide dodécénylsuccinique
 e ácido dodecenilsuccínico
 i acido dodecenilsuccinico
 d Dodecenylbernsteinsäure

2836 dodecenylsuccinic anhydride
 f anhydride dodécénylsuccinique
 e anhídrido dodecenilsuccínico
 i anidride dodecenilsuccinica
 d Dodecenylbernsteinsäureanhydrid

2837 dodecyl acetate
 f acétate de dodécyle
 e acetato de dodecilo
 i acetato di dodecile
 d Dodecylacetat

2838 dodecylbenzene
 f dodécylbenzène
 e dodecilbenceno
 i dodecilbenzene
 d Dodecylbenzol

2839 dodecylphenol
 f dodécylphénol
 e dodecilfenol
 i dodecilfenolo
 d Dodecylphenol

2840 dodecyltrimethylammonium chloride
 f chlorure de dodécyltriméthylammonium

 e cloruro de dodeciltrimetilamonio
 i cloruro di dodeciltrimetilammonio
 d Dodecyltrimethylammoniumchlorid

2841 dolomite; bitter spar
 f dolomie; dolomite
 e dolomia; magnesita
 i creta dolomitica
 d Bitterspat; Bitterkalkspat

2842 domiphen bromide
 f bromure de domiphène
 e bromuro de domifeno
 i bromuro di domifene
 d Domiphenbromid

2843 Donnan potential
 f potentiel de Donnan
 e potencial de Donnan
 i potenziale di Donnan
 d Donnanpotential

2844 donor cell
 f cellule donatrice
 e célula donante
 i cellula donatrice
 d Donor-Zelle

2845 dosage
 f dosage; posologie
 e dosificación; posología
 i dosaggio; posologia
 d Dosierung; Arzneigabe

2846 dosage compensation
 f compensation de dosage
 e compensación de dosis
 i compensazione di dosaggio
 d Dosiskompensation

2847 dose
 f dose
 e dosis
 i dose
 d Dosis

2848 dosemeter
 f dosimètre
 e dosímetro
 i dosimetro
 d Dosimeter; Dosismesser

2849 dosing feeder
 f dispositif de dosage
 e dispositivo de dosificación
 i dosatore
 d Dosiervorrichtung

2850 **double bond**
  f double liaison
  e ligadura doble
  i doppio legame
  d Doppelbindung

2851 **double-clad vessel**
  f vase à double chemise
  e vaso con doble revestimiento
  i vaso a doppio rivestimento
  d doppelt umkleideter Behälter

2852 **doublet; duplet**
  f doublet
  e doblete
  i doppietto
  d Dublett

2853 **doubling time**
  f temps de duplication
  e tiempo de duplicación
  i tempo di doppiatura
  d Verdopplungszeit

2854 **dough**
  f pâte
  e pasta
  i pasta
  d Teig

2855 **Downs process**
  f procédé Downs
  e proceso Downs
  i processo Downs
  d Downs-Verfahren

2856 **downstream**
  f en aval
  e de aguas abajo
  i a valle
  d abwärts

2857 **Dragendorff's solution**
  f solution de Dragendorff
  e solución de Dragendorff
  i soluzione di Dragendorff
  d Dragendorffsche Lösung

2858 **drag in**
  f solution adhérente
  e solución adherida
  i soluzione aderente
  d eingeschleppte Lösung

2859 **dragon's blood**
  f sang-dragon
  e sangre de drago
  i sangue di drago
  d Drachenblut

2860 **drag out**
  f solution entraînée
  e solución arrastrada
  i soluzione estratta
  d herausgeschleppte Lösung

2861 **drastic**
  f purgatif violent
  e purgante enérgico
  i purgativo catartico
  d Drasticum

2862 **draught gauge**
  f déprimomètre
  e deprimómetro; indicador de vacío
  i deprimometro
  d Zugmesser; Unterdruckmesser

2863 **draw** v *(a kiln)*
  f puiser
  e sacar
  i prelevare
  d schöpfen

2864 **draw-down**
  f réduction à l'étirage
  e estirado
  i riduzione di diametro per trafilatura
  d Ziehabnahme

2865 **dregs**
  f dépôt
  e depósito; sedimento
  i feccia
  d Geläger

2866 **drier; siccative**
  f desséchant; siccatif
  e secador; secante; secativo
  i essiccante; essicativo
  d Trockenmittel; Sikkativ

2867 **drip-proof**
  f à l'épreuve des éclaboussures
  e a prueba de goteo
  i protetto contro lo stillicidio
  d tropfdicht

2868 **drocarbil**
  f drocarbil
  e drocarbil
  i drocarbile
  d Drocarbil

2869 **droplet**
  f gouttelette
  e gotita
  i gocciolina
  d Tröpfchen

2870 **dropping bottle**
  f flacon compte-gouttes
  e frasco cuentagotas
  i sgocciolatore
  d Tropfglas

2871 **dropping funnel**
  f entonnoir compte-gouttes
  e embudo cuentagotas
  i imbuto contagoccie
  d Tropftrichter

2872 **dropwise condensation**
  f condensation en gouttelettes
  e condensación en forma de gotas
  i condensazione a goccie
  d tropfenweise Kondensation

2873 **drossing**
  f écrémage
  e extracción del cobre del plomo de obra
  i rimozione di ossidi o metalli non richiesti da metalli fusi
  d Schaumabheben

2874 **drossing kettle**
  f écrémeuse
  e vasija para extracción del cobre del plomo de obra
  i recipiente per la rimozione dei metalli non desiderati dai metalli fusi
  d Abschaumkessel

2875 **drug**
  f drogue; médicament
  e droga; medicamento
  i medicamento; droga; farmaco
  d Arznei; Droge

2876 **drum dryer**
  f séchoir à tambour
  e secador rotativo
  i essiccatore rotativo
  d Trommeltrockner

2877 **drum-type magnetic separator**
  f séparateur magnétique à tambour
  e separador magnético tipo tambor
  i separatore magnetico a tamburo
  d magnetischer Trommelscheider

2878 **dryability**
  f pouvoir de séchage
  e grado al que una substancia puede secarse
  i grado di seccabilità di una sostanza
  d Trockenvermögen

2879 **dry assay**
  f détermination à sec
  e análisis por vía seca
  i saggio per via secca
  d Trockenprobe

2880 **dry binder**
  f liant à sec
  e aglutinador a seco
  i legante a secco
  d Trockenbindemittel

2881 **dry cell**
  f pile sèche
  e pila seca
  i pila secco
  d Trockenelement

2882 **dry ice**
  f glace carbonique; carboglace
  e hielo seco
  i ghiaccio secco
  d Trockeneis; festes Kohlendioxyd

2883 **drying oil**
  f huile siccative; siccatif
  e aceite secante
  i olio essiccativo
  d trocknendes Öl

2884 **drying oven**
  f étuve
  e estufa
  i stufa d'essiccazione
  d Trockenkasten

2885 **drying tube**
  f tube sécheur
  e tubo secador
  i tubo essiccatore
  d Trockenrohr

2886 **dry process**
  f procédé par voie sèche
  e proceso por vía seca
  i processo a secco
  d Trockenverfahren

2887 **dry solids content**
  f teneur en matière solide totale
  e contenido en sustancias secas
  i contenuto totale di materiale solido
  d Trockenstoffgehalt

2888 **dry spinning**
  f filage au sec
  e hilado en seco; hilatura en seco
  i filatura a secco
  d Trockenspinnverfahren

2889 dry steam
f vapeur sèche
e vapor seco
i vapore secco
d trockener Dampf

2890 dual system coagulation
f coagulation double
e coagulación por sistema doble
i coagulazione a sistema doppio
d Doppelkoagulierung

2891 ductility
f ductilité
e ductilidad; maleabilidad
i duttibilità
d Dehnbarkeit

2892 Dühring's rule
f règle de Dühring
e regla de Dühring
i regola di Dühring
d Dühringsche Regel

2893 dulcin; sucrol
f dulcine
e dulcina
i dulcina
d Dulcin

2894 dull
f terne; mat
e mate; sin sonoridad
i poaco; ottuso
d glanzlos; trübe

2895 Dulong and Petit's law
f loi de Dulong et Petit
e ley de Dulong y Petit
i legge di Dulong e Petit
d Dulong-Petitsches Gesetz

2896 dumortierite
f dumortiérite
e dumortierita
i dumortierite
d Dumortierit

* **duplet** → 2852

2897 durain
f durain
e constituyente mate del carbón
i duraina
d Mattkohle

2898 durene
f durène
e dureno

i durene
d Durol

* **Dutch liquid** → 3187

2899 Dutch metal
f similor
e similor
i similoro
d Schaumgold

2900 Dutch process
f procédé hollandais
e método holandés
i metodo olandese
d holländisches Topfverfahren

2901 "duty"
f service
e servicio
i rendimento; servizio
d Betriebsleistung

2902 dyad
f diade
e diada
i diade
d Dyade

2903 dyclonine hydrochloride
f chlorhydrate de dyclonine
e hidrocloruro de diclonina
i cloridrato di diclonina
d Dycloninhydrochlorid

2904 dyes
f colorants; teintures
e colorantes
i coloranti
d Farbstoffe

2905 dynamic electrode potential
f tension dynamique d'une électrode
e tensión dinámica de un electrodo
i potenziale dinamico di elettrodo
d dynamisches Elektrodenpotential

* **dynamic isomerism** → 8054

2906 dynamite
f dynamite
e dinamita
i dinamite
d Dynamit

2907 dyphylline
f dyphylline
e difilina

    i difilina
    d Dyphyllin

2908 dypnone
    f dypnone
    e dipnona
    i dipnone
    d Dypnon

2909 dyscrasite
    f dyscrasite
    e discrasita
    i discrasite
    d Antimonsilber; Dyskrasit

2910 dysprosium
    f disprosium
    e disprosio
    i disprosio
    d Dysprosium

2911 dystectic mixture
    f mélange dystectique
    e mezcla distéctica
    i miscela distettica
    d dystektische Mischung

E

2912 earth colours
f couleurs terreuses
e pigmentos de tierras naturales
i terre coloranti
d Erdfarben

2913 earthenware
f faïence
e alfarería; loza de barro
i terraglia
d Steingut

* **earthy cobalt** → 683

* **easy flow** → 7634

2914 ebonite
f ébonite
e ebonita
i ebonite
d Ebonit

2915 ebullioscopy
f ébullioscopie
e ebulloscopia
i ebullioscopia
d Ebullioskopie

2916 ebullition
f ébullition
e ebullición
i ebollizione
d Sieden

2917 ecad
f écade
e écade
i ecade
d Ökade

* **eccrine** → 3306

2918 ecobiotic adaptation
f adaptation écobiotique
e adaptación ecoclimática
i adattamento ecoclimatico
d ökobiotische Anpassung

2919 ecocline
f écocline
e ecocline
i ecocline
d Ökoklin

2920 economizer
f économiseur

e economizador
i economizzatore
d Speisewasservorwärmer

2921 ecotype
f écotype
e ecotipo
i ecotipo
d Ökotyp

2922 ectoplasm
f ectoplasme
e ectoplasma
i ectoplasma
d Ektoplasma

2923 ectosome
f ectosome
e ectosoma
i ectosoma
d Ektosom

2924 ectosphere
f ectosphère
e ectosfera
i ectosfera
d Ektosphäre

2925 Edeleanu process
f procédé Edeleanu
e proceso de Edeleanu
i processo Edeleanu
d Edeleanu-Verfahren

2926 edge-jointing adhesive
f colle pour joints
e cola para unir los bordes
i adesivo per saldatura dei bordi
d Randverbindungsleim

2927 edge-jointing cement
f colle d'assemblage bord à bord
e adhesivo para bordes
i mastice per giunzioni
d Leim zum Verbinden von
  Sperrholzplatten

2928 edge runner
f meule verticale
e muela vertical
i molazza a ruote
d Läufer

2929 edge-runner mill
f broyeur à meules verticales
e molino de muelas verticales
i rompitore
d Kollergang

2930 effective collision cross-section
 f section efficace de choc
 e sección eficaz de choque
 i sezione efficace di urto
 d Wirkungsquerschnitt

2931 efficiency
 f rendement en quantité
 e rendimiento en cantidad
 i rendimento di quantità
 d Wirkungsgrad

2932 efflorescence; bloom
 f efflorescence
 e eflorescencia
 i efflorescenza
 d Ausblühung; Effloreszenz

2933 effuser
 f turbine
 e turbina
 i diffusore
 d Effusionsvorrichtung

2934 effusion
 f effusion
 e efusión
 i effusione
 d Erguß

2935 egg *v*
 f refouler
 e bombear por medio de una vasija de aire comprimido
 i pompare mediante serbatoio ad aria compressa
 d mit Druckluft fördern

2936 egg
 Pressure vessel to transfer acid from one container to another by compressed air.
 f monte-jus
 e vasija a presión para trasvase de ácidos
 i apparecchio per muovere liquidi mediante pressione d'aria
 d Druckbirne

2937 egg albumen
 f albumine du blanc d'œuf
 e albúmina de huevo
 i albume d'uovo
 d Eialbumin

2938 Eichhorn's hydrometer
 f aréopycnomètre d'Eichhorn
 e areopicnómetro de Eichhorn
 i areopicnometro secondo Eichhorn
 d Aräopyknometer nach Eichhorn

2939 eicosane
 f eicosane
 e eicosano
 i eicosan
 d Eikosan

2940 eikonogen
 f iconogène
 e eiconógeno
 i eiconogeno
 d Eikonogen

2941 einsteinium
 f einsteinium
 e einsteinio
 i einsteinio
 d Einsteinium

2942 Einstein's law of photochemical equivalence
 f loi d'Einstein sur l'équivalence photochimique
 e ley de Einstein de la equivalencia fotoquímica
 i legge di Einstein dell'equivalenza fotochimica
 d Einsteinsches Gesetz der photochemischen Äquivalenz

2943 elasticator
 f agent élastifiant
 e agente elastificante
 i agente elasticante
 d elastisch machendes Mittel

2944 elastic nylon
 f nylon élastique
 e nylon elástico
 i nylon elasticizzato
 d elastisches Nylon

2945 elastomer
 f élastomère
 e elastómero
 i elastomero
 d Elastomer

2946 elastomeric
 f élastomère
 e elastomérico
 i elastomero
 d elastomer

2947 elastometer
 f élastomètre
 e elastómetro
 i elastometro
 d Elastizitätsmesser

2948 elaterite
f élatérite; bitume élastique
e elaterita
i elaterite
d Elaterit

2949 electrical porcelain
f porcelaine isolante
e porcelana aislante
i porcellana da isolatori
d Porzellan für elektrische Isolierungen

2950 electric migration
f migration électrique
e emigración eléctrica
i migrazione elettrica
d elektrische Ionenwanderung

2951 electric salinometer
f salinomètre électrique
e salinómetro eléctrico
i salinometro elettrico
d elektrischer Salzgehaltmesser

2952 electric steam generator
f générateur de vapeur électrique
e generador de vapor eléctrico
i generatore di vapore elettrico
d Elektrodampfgenerator

2953 electro-analysis
f électroanalyse
e electroanálisis
i analisi elettrolitica
d elektrolytische Analyse

2954 electrochemical diffusion
f diffusion électrochimique
e difusión electroquímica
i diffusione elettrochimica
d elektrochemische Ionenwanderung

2955 electrochemical equivalent
f équivalent électrochimique
e equivalente electroquímico
i equivalente elettrochimico
d elektrochemisches Äquivalent

2956 electrochemical migration
f migration électrochimique
e emigración electroquímica
i migrazione elettrochimica
d elektrochemische Ionenwanderung

2957 electrochemical oxidation
f oxydation électrochimique
e oxidación electroquímica
i ossidazione elettrochimica
d elektrochemische Oxydation

2958 electrochemical passivation
f passivation électrochimique
e pasivación electroquímica
i passivazione elettrochimica
d elektrochemische Passivierung

2959 electrochemical reduction
f réduction électrochimique
e reducción electroquimica
i riduzione elettrochimica
d elektrochemische Reduktion

2960 electrochemical series
f série électrochimique
e serie electroquímica
i serie elettrochimica
d elektrochemische Spannungsreihe

2961 electrochemical valve
f soupape électrochimique
e válvula electroquímica
i raddrizzatore elettrolitico
d elektrochemisches Ventil

2962 electrochemistry
f électrochimie
e electroquímica
i elettrochimica
d Elektrochemie

2963 electrode active surface
f surface active d'une électrode
e superficie activa de un electrodo
i superficie attiva di un elettrodo
d wirksame Elektrodenfläche

2964 electrodecantation
f électrodécantation
e electrodecantación
i elettrodecantazione
d Elektrodekantieren

2965 electrode current density
f densité de courant
e densidad de corriente
i densità di corrente
d Stromdichte

2966 electro-deposition
f dépôts électrolytiques
e depósitos electrolíticos
i deposizione elettrolitica
d elektrolytische Abscheidung

2967 electrode reaction
f réaction d'électrode
e reacción de electrodo
i reazione all'elettrodo
d Elektrodenreaktion

2968 electrode salt bath
f four à bain de sel à électrodes
e horno de baño salino de electrodos
i forno a bagno salino a elettrodi
d Elektrodensalzbadofen

2969 electrode steam generator
f chaudière à vapeur à électrodes
e caldera de electrodos
i caldaia a elettrodi
d Elektrodenkessel

2970 electrodialysis purification process
f procédé de purification par électrodialyse
e método de purificación por electrodiálisis
i purificazione per elettrodialisi
d elektrodialysische Wasserreinigung

2971 electro-dissolution
f électrodissolution
e electrodisolución
i dissoluzione elettrolitica
d elektrolytische Auflösung

2972 electro-extraction
f extraction électrochimique
e extracción electroquímica
i estrazione elettrochimica
d elektrolytische Extraktion

2973 electro-forming
f électroformage
e electroformación
i galvanoplastica
d Galvanoplastik

2974 electrolysis
f électrolyse
e electrólisis
i elettrolisi
d Elektrolyse

2975 electrolyte
f électrolyte
e electrólito
i elettrolito
d Elektrolyt

2976 electrolytic cell
f cellule électrolytique
e celda electrolítica; cuba electrolítica
i cella elettrolitica
d elektrolytische Zelle

2977 electrolytic cleaning
f dégraissage électrolytique
e desengrase electrolítico
i sgrassaggio elettrolitico
d elektrolytische Reinigung

2978 electrolytic copper
f cuivre électrolytique
e cobre electrolítico
i rame elettrolitico
d Elektrolytkupfer

2979 electrolytic dissociation
f dissociation électrolytique
e disociación electrolítica
i dissociazione elettrolitica
d elektrolytische Dissoziation

2980 electrolytic dissociation constant
f constante de dissociation électrolytique
e constante de disociación electrolítica
i costante di dissociazione elettrolitica
d Dissoziationskonstante

2981 electrolytic parting
f séparation électrolytique
e separación electrolítica
i separazione elettrolitica
d elektrolytische Metallscheidung

2982 electrolytic pickling
f décapage électrolytique
e decapado electrolítico
i decapaggio elettrolitico
d elektrolytisches Beizen

2983 electrolytic polarization
f polarisation électrolytique
e polarización electrolítica
i polarizzazione elettrolitica
d elektrolytische Polarisation

2984 electrolytic rectifier
f redresseur électrolytique
e rectificador electrolítico
i raddrizzatore elettrolitico
d Elektrolytgleichrichter

2985 electrolytic stripping
f dépouillage électrolytique
e eliminación electrolítica
i degalvanizzazione elettrolitica
d elektrolytische Entplattierung

2986 electrolytic valve ratio
f rapport de soupape électrolytique
e relación de válvula electrolítica
i rapporto di raddrizzamento
d Sperrwirkung

2987 electrolyzer
f électrolyseur
e electrolizador
i cella elettrolitica
d Elektrolyseur

2988 electromagnetic pulley
f poulie électromagnétique
e polea electromagnética
i puleggia elettromagnetica
d elektromagnetische Scheibenrolle

2989 electrometallurgy
f électrométallurgie
e electrometalurgia
i elettrometallurgia
d Elektrometallurgie

2990 electromotive series
f série électrochimique
e serie electroquímica
i serie elettrochimica
d elektrochemische Spannungsreihe

2991 electron
f électron
e electrón
i elettrone
d Elektron

2992 electron shell
f couche électronique
e capa electrónica
i strato elettronico
d Elektronenschale

2993 electro-osmosis
f électroosmose
e electroosmosis
i elettroosmosi
d Elektroosmose

2994 electro-osmotic potential
f potentiel électroosmotique
e potencial electroosmótico
i potenziale elettroosmotico
d elektroosmotisches Potential

2995 electrophoresis
f électrophorèse
e electrofóresis
i elettroforesi
d Elektrophorese

2996 electrophoretic mobility
f mobilité électrophorétique
e mobilidad electroforética
i mobilità elettroforetica
d elektrophoretische Beweglichkeit

2997 electrophoretic potential
f potentiel électrophorétique
e potencial electroforético
i potenziale elettroforetico
d elektrophoretisches Potential

2998 electro-plating
f galvanoplastie
e galvanoplastia
i galvanoplastica
d Elektroplattierung; Galvanostegie

2999 electro-plating generator
f génératrice pour galvanoplastie
e generador para galvanoplastia
i generatore per galvanoplastica
d Plattierungsgenerator

3000 electro-polishing
f polissage électrolytique
e pulido electrolítico
i pulitura elettrolitica
d elektrolytisches Polieren; Elektropolieren

3001 electro-refining
f électroraffinage
e afinado electrolítico; electrorefinado
i elettroraffinazione
d elektrolytische Raffination

3002 electrospectrogram
f électrospectrogramme
e electroespectrograma
i elettrospettrogramma
d Elektrospektrogramm

3003 electrostatic precipitation
f précipitation électrostatique
e precipitación electrostática
i precipitazione elettrostatica
d elektrostatische Abscheidung

3004 electrostatic separator
f séparateur électrostatique
e separador electrostático
i separatore elettrostatico
d Vorrichtung zur elektrostatischen Trennung

3005 electrothermics
f électrothermie
e electrotermia
i elettrotermica
d Elektrowärmelehre

3006 electro-typing
f électrotypie
e electrotipia
i elettrotipia
d Elektrotypie

3007 electrovalve
f vanne à solénoïde
e válvula controlada por solenoide

i valvola a solenoide per liquidi
d Elektroventil

3008 electro-winning
f extraction par voie électrolytique
e extracción por vía electrolítica
i estrazione per via elettrolitica
d elektrolytische Metallgewinnung

3009 electrum
f électre
e electro
i lega natura di oro e argento
d Elektrum

3010 element
f élément
e elemento
i elemento
d Element

3011 elixir
f élixir
e elixir
i elisir
d Elixir

3012 elongation at break
f élongation de rupture; allongement de
rupture
e alargamiento de rotura
i allungamento di rottura
d Bruchstreckung

3013 elution
f élution
e elución
i eluzione
d Elution

3014 elutriation
f élutriation
c elutriación
i separazione per decantazione
d Auswaschung; Schlämmung

3015 emanation
f émanation
e emanación
i emanazione
d Emanation

* emanon → 6915

3016 embedding compound
f matière d'enrobage
e material de inclusión
i materiale per inclusioni
d Einbettmaterial

3017 embryology
f embryologie
e embriología
i embriologia
d Embryologie

3018 emerald green; Paris green
f vert émeraude
e verde esmeralda; verde de París
i verde smeraldo
d Smaragdgrün; Mitisgrün; Deckgrün

3019 emetic
f émétique
e emético
i emetico
d Brechmittel

3020 emollient
f émollient
e emolliente
i emolliente; lenitivo
d erweichend

3021 empirical formula
f formule empirique; formule brute
e fórmula empírica
i formula empirica
d empirische Formel

3022 emulsification
f émulsionnement
e emulsionamiento
i emulsionamento
d Emulgierung

3023 emulsifier
Substance which promotes the formation
of an emulsion.
f émulsionnant; émulseur
e agente emulsionante
i emulsionante
d Emulgator

3024 emulsifier
Mixing machinery to disperse one liquid
in another.
f émulgateur
e aparato para emulsionar
i apparecchio di emulsione
d Emulgiermaschine

3025 emulsion
f émulsion
e emulsión
i emulsione
d Emulsion

3026 emulsion adhesive
f colle en émulsion
e adhesivo de emulsión
i emulsione adesiva
d Emulsionskleber

3027 emulsion paint
f peinture à base d'émulsion; peinture-émulsion
e pintura de emulsión
i vernice ad emulsione
d Emulsionsfarbe

3028 emulsoid
f émulsoïde
e emulsoide
i emulsoide
d Emulsoid

3029 enamel
f émail
e esmalte
i smalto
d Glasur; Emaille

3030 enantiomorphous
f énantiomorphe
e enantiomorfo
i enantiomorfo
d enantiomorph

3031 enantiotropic
f énantiotropique
e enantiotrópico
i enantiotropico
d enantiotrop

3032 enargite
f énargite
e enargita
i enargite
d Enargit

3033 encapsulating compound
f matière d'enrobage
e material de inclusión
i materiale per inclusioni
d Einbettmaterial

3034 encaustic
f encaustique
e encáustico
i encaustico
d enkaustisch

3035 end cell
f élément de régulation
e elemento de regulación

i elemento di regolazione
d Endzelle

3036 endelionite
f endélionite
e endelionita
i endelionite
d Endelionit

3037 endo-
Prefix meaning "internal", "inner", "inward".
f endo-
e endo-
i endo-
d Endo-

3038 endocrine
f endocrine
e endocrino
i endocrino
d endokrin; innersekretorisch

3039 endocrine glands
f glandes endocrines
e glándulas de secreción interna
i ghiandole endocrine
d endokrine Drüsen

3040 endomitosis
f endomitose
e endomitosis
i endomitosi
d Endomitose

3041 endomixis
f endomixie
e endomixis
i endomissi
d Endomixis

3042 endoplasm
f endoplasme
e endoplasma
i endoplasma
d Endoplasma

3043 endopolyploidy
f endopolyploïdie
e endopoliploidia
i endopoliploidismo
d Endopolyploidie

3044 endosome
f endosome
e endosoma
i endosoma
d Endosom

3045 endothermic
f endothermique
e endotérmico
i endotermico
d endotherm(isch); wärmeverzehrend

3046 end point
f point d'achèvement; virage au titrage
e punto final; punto de valoración
i punto di viraggio
d Abtitrierungspunkt; Endpunkt

3047 endrophonium chloride
f chlorure d'endrophonium
e cloruro de endrofonio
i cloruro d'endrofonio
d Endrophoniumchlorid

3048 energic nucleus
i noyau quiescent
e núcleo enérgico
i nucleo energico
d Arbeitskern; Ruhekern

3010 energic stage
f phase métabolique
e estado enérgico
i stadio energico
d Ruhephase

3050 energy
f énergie
e energía
i energia
d Energie

3051 Engler distillation
f distillation Engler
e destilación Engler
i distillazione Engler
d Englersche Destillation

3052 Engler flask
f vallon d'Engler
e frasco de Engler
i matraccio d'Engler
d Englerscher Kolben

3053 English vermilion
f rouge d'Angleterre
e rojo inglés
i rosso inglese
d Englischrot

3054 enriched water
f eau enrichie
e agua enriquecida
i acqua arricchita
d angereichertes Wasser

3055 enriching section
f zone d'enrichissement
e sección de enriquecimiento
i colonna d'arricchimento
d Anreicherungszone

3056 enthalpy
f enthalpie
e calor específico; entalpia
i entalpia
d Enthalpie

3057 entrainer
f entraïneur
e agente arrastrante
i trasportatore
d Mitschleppmittel

3058 entrainment
f entraînement
e arrastre
i trascinamento
d Mitschleppen

3059 entropy
f entropie
e entropia
i entropia
d Entropie

3060 environment
f milieu; environnement
e medio ambiente
i ambiente
d Umwelt; Umgebung; Milieu

3061 environmental
f ambiant; influencé par le milieu
e ambiental; influenciado por el medio
i ambientale
d Umwelt-; Außen-

3062 environmental conditions
f conditions du milieu extérieur
e condiciones ambientales
i condizioni dell'ambiente
d Umweltbedingungen

3063 environmental correlation
f corrélation due au milieu
e correlación ambiental
i correlazione dovuta all'ambiente
d umweltbedingte Korrelation

3064 environmental variation
f variation influencée par le milieu
e variación ambiental
i variazione dovuta all'ambiente
d umweltbedingte Variation

3065 enzymatic
 f enzymique
 e enzimático
 i enzimico
 d enzymisch

3066 enzymatic coagulation
 f coagulation enzymatique
 e coagulación enzímica
 i coagulazione per enzimi
 d Enzymkoagulation

3067 enzymatic deproteinization
 f déprotéinisation par enzymes
 e desproteinización enzímica
 i deproteinizzazione per enzimi
 d Aufspalten von Eiweißstoffen durch
   Enzyme

3068 enzyme
 f enzyme
 e enzima
 i enzima
 d Enzym

3069 enzymology
 f enzymologie
 e enzimología
 i enzimologia
 d Fermentlehre

3070 eosin
 f éosine
 e eosina
 i eosina
 d Eosin

3071 eosinophil
 f éosinophile
 e eosinófilo
 i eosinofilo
 d Eosinophil

3072 ephemeral
 f éphémère
 e efémera
 i effimero
 d vorübergehend

3073 epichlorohydrin
 f épichlorhydrine
 e epiclorhidrina
 i epicloridrina
 d Epichlorhydrin

3074 epidote
 f épidote
 e epidota

 i epidoto
 d Epidot

3075 epigenetics
 f épigénétique
 e epigenética
 i epigenetica
 d Epigenetik

3076 epinephrine; adrenalin
 f épinéphrine
 e epinefrina
 i epinefrina
 d Adrenalin; Epinephrin

3077 epistasis
 f épistasie
 e epistasis
 i epistasia
 d Epistasis

3078 epithermal neutrons
 f neutrons épithermiques
 e neutrones epitérmicos
 i neutroni epitermici
 d epitherme Neutronen

3079 epoxy-
   Prefix denoting an oxygen atom bonded
   to two other atoms which are already
   united by other bonds.
 f époxy-
 e epoxi-
 i epossi-
 d Epoxy-

3080 epoxy resins
 f résines époxydes
 e resinas epóxido; resinas epoxídicas
 i resine epossiliche; resine ossidate
 d Epoxyharze

3081 epsomite
 f epsomite
 e epsomita
 i epsomite
 d Epsomit

**3082 Epsom salts; hydrated magnesium
   sulphate**
 f sels d'Epsom
 e sales de Epsom
 i sali d'Epsom
 d Epsomsalze; Bittersalze; schwefelsaures
   Magnesium

3083 equation of state
 f équation d'état
 e ecuación de estado

i equazione di stato
d Zustandsgleichung

3084 equatorial body
f cellule équatoriale
e cuerpo ecuatorial
i cellula equatoriale
d Äquatorialkörper

3085 equilibrium
f équilibre
e equilibrio
i equilibrio
d Gleichgewicht

3086 equilibrium constant
f constante d'équilibre
e constante de equilibrio
i costante d'equilibrio
d Gleichgewichtskonstante

3087 equilibrium electrode potential
f tension d'équilibre d'une électrode
e tensión de equilibrio de un electrodo
i potenziale d'equilibrio di elettrodo
d Gleichgewichtspotential einer Elektrode

3088 equilibrium enrichment factor
f facteur d'enrichissement équilibré
e factor de enriquecimiento equilibrado
i fattore di arricchimento all'equilibrio
d Gleichgewichtsanreicherungsfaktor

3089 equilibrium reaction potential
f tension d'équilibre d'une réaction
e tensión de equilibrio de una reacción
i tensione di equilibrio di una reazione
d statisches Gleichgewichtspotential der Reaktion

3090 equimolecular
f équimoléculaire
e equimolecular
i equimolecolare
d äquimolecular

3091 equivalent concentration
f concentration équivalente
e concentración equivalente
i concentrazione equivalente
d Äquivalentkonzentration

3092 equivalent conductance
f conductance équivalente
e conductancia equivalente
i conduttanza equivalente
d Äquivalentleitfähigkeit

3093 equivalent weight
f poids équivalent
e equivalente químico
i peso equivalente
d Äquivalenzgewicht

3094 erbium
f erbium
e erbio
i erbio
d Erbium

3095 Erdmann float
f flotteur d'Erdmann
e flotador de Erdmann
i galleggiante di Erdmann
d Erdmannscher Schwimmer

3096 ergosterol
f ergostérol; ergostérine
e ergosterol
i ergosterol
d Ergosterin

3097 Erlenmeyer flask
f ballon en verre
e matraz de vidrio
i matraccio conico
d Erlenmeyerkolben

3098 erubescite; bournite
f érubescite
e erubescita
i erubescite; bornite
d Bornit; Buntkupferkies

3099 erucamide
f érucamide
e erucamida
i erucammide
d Erucamid

3100 erucyl alcohol
f alcool érucylique
e alcohol erucílico
i alcool erucilico
d Erucylalkohol

3101 erythorbic acid
f acide érythorbique
e ácido eritórbico
i acido eritorbico
d Erythorbinsäure

3102 erythrene
f érythrène
e eritreno
i eritrene
d Erythren

* **erythrite** → 1896

3103 erythritol
f érythritol
e eritritol
i eritritolo
d Erythritol

3104 erythroblast
f érythroblaste
e eritroblasto
i eritroblasto
d Erythroblast

3105 erythrocyte
f érythrocyte
e eritrocito
i eritrocita
d Erythrozyt

3106 erythrodextrin
f érythrodextrine
e eritrodextrina
i eritrodestrina
d Erythrodextrin

3107 erythrogonium
f érythrogonie
e eritrogonio
i eritrogonio
d Erythrogonie

3108 erythrosine
f érythrosine
e eritrosina
i eritrosina
d Erythrosin

3109 esculin
f esculine
e esculina
i esculina
d Äskulin; Schillerstoff

3110 esparto
f sparte
e esparto
i sparto
d Faengras; Espartogras

3111 esparto wax
f cire de sparte
e cera de esparto
i cera di sparto
d Espartowachs

3112 essential oils
f huiles essentielles; huiles volatiles
e aceites esenciales; aceites volátiles
i oli essenziali
d ätherische Öle

3113 esterase
f estérase
e esterasa
i esterasi
d Esterase

3114 ester gums
f gommes esters
e gomas éster; ésteres de colofonia
i gomme esterificate
d Estergummi; Harzester

3115 esterification
f estérification
e esterificación
i esterificazione
d Veresterung

3116 esters
f esters
e ésteres
i esteri
d Ester

3117 estradiol; oestradiol
f oestradiol
e estradiol
i estradiolo
d Östradiol

3118 estragole
f estragol
e estragol
i estragolo
d Estragol

3119 estrin
f hormone menstrogène
e hormona folicular
i ormone estrogeno
d Follikelhormon

3120 estriol; oestriol
f oestriol
e estriol
i estriolo
d Östriol

3121 estrogen; oestrogen
f oestrogène
e estrógeno
i estrogen
d Östrogen

3122 estrogenic; oestrogenic
f oestrogène

e estrógeno
i estrogeno
d östrogen

3123 estrone; oestrone
f oestrone
e estrón
i estrone
d Östron

3124 ethane
f éthane
e etano
i etano
d Äthan

* **ethanol** → 3138

3125 ethanolamine
f éthanolamine
e etanolamina
i etanolammina
d Äthanolamin

* **ethene** → 3180

3126 ethenoid
f éthylénique
e etenoide
i etenoidi
d Äthenoid

3127 ethenoid plastics
f plastiques éthénoïdiques
e resinas etenoides
i materie plastiche etenoidi
d Äthenoidharze

3128 ether; diethyl ether
f éther
e éter
i etere
d Äther

3129 ether index
f indice d'éther
e índice de éter
i indice di etere
d Ätherzahl

3130 ethers
f éthers; éthers-oxydes
e éteres
i eteri
d Äther

3131 ethion
f éthion
e etión

i ethion
d Äthion

* **ethocaine hydrochloride** → 6715

3132 ethoxytriglycol
f éthoxytriglycol
e etoxitriglicol
i etossitriglicolo
d Äthoxytriglykol

3133 ethyl acetanilide
f éthylacétanilide
e etilacetanilida
i etil acetanilide
d Äthylacetanilid

3134 ethyl acetate
f acétate d'éthyle
e acetato de etilo
i etil acetato
d Äthylacetat

3135 ethyl acetoacetate
f acéto-acétate d'éthyle
e acetoacetato de etilo
i acetacetato di etile
d Äthylacetoacetat

3136 ethyl acetylene
f éthylacétylène
e etilacetileno
i etilacetilene
d Äthylacetylen

3137 ethyl acrylate
f acrylate d'éthyle
e acrilato de etilo
i acrilato di etile
d Äthylacrylat

3138 ethyl alcohol; ethanol, spirits of wine
f alcool éthylique, éthanol, esprit de vin
e alcohol etílico; etanol; espíritu de vino
i alcool etilico; etanolo
d Äthylalkohol; Äthanol; Weingeist

3139 ethyl aluminium chloride
f aluminium dichloréthylique
e dicloruro de etilo y aluminio
i dicloruro di alluminio-etile
d Aluminiumdichloräthylen

3140 ethyl aluminium sesquichloride
f aluminium sesquichloréthylique
e sesquicloruro de etilo y aluminio
i sesquicloruro di alluminio-etile
d Äthylaluminiumsesquichlorid

3141 **ethylamine**
  f éthylamine
  e etilamina
  i etilammina
  d Äthylamin

3142 **ethyl aminobenzoate**
  f aminobenzoate d'éthyle
  e aminobenzoato de etilo
  i amminobenzoato d'etile
  d Äthylaminobenzoat

3143 **ethyl amyl ketone**
  f éthylamylcétone
  e etilaminocetona
  i etilamilchetone
  d Äthylamylketon

3144 **ethyl anthranilate**
  f anthranilate d'éthyle
  e etilantranilato
  i etilantranilato
  d Äthylanthranilat

3145 **ethylarsenious oxide**
  f oxyde éthylarsénieux
  e óxido etilarsenioso
  i ossido etilarsenioso
  d Äthylarsentrioxyd

3146 **ethylbenzene**
  f benzène éthylé
  e etilbenceno
  i etilbenzene
  d Äthylbenzol

3147 **ethyl benzoacetate**
  f benzoacétate d'éthyle
  e benzoacetato de etilo
  i benzoacetato d'etile
  d Äthylbenzoacetat

3148 **ethyl benzoate**
  f benzoate d'éthyle
  e benzoato de etilo
  i benzoato di etile
  d Äthylbenzoat

3149 **ethyl benzoylbenzoate**
  f benzoylbenzoate d'éthyle
  e benzoilbenzoato de etilo
  i etilbenzoilbenzoato
  d Äthylbenzoylbenzoat

3150 **ethylbenzylaniline**
  f éthylbenzylaniline
  e etilbencilanilina
  i etilbenzilanilina
  d Äthylbenzylanilin

3151 **ethylbenzyl chloride**
  f chlorure d'éthylbenzyle
  e cloruro de etilbencilo
  i cloruro di etilbenzile
  d Äthylbenzylchlorid

3152 **ethyl borate**
  f borate d'éthyle
  e borato de etilo
  i borato d'etile
  d Äthylborat

3153 **ethyl bromide**
  f bromure d'éthyle
  e bromuro de etilo
  i bromuro d'etile
  d Äthylbromid

3154 **ethylbutyl alcohol**
  f alcool éthylbutylique
  e alcohol etilbutílico
  i alcool etilbutilico
  d Äthylbutylalkohol

3155 **ethyl butyl carbonate**
  f éthylbutylcarbonate
  e etilbutilcarbonato
  i etilbutilcarbonato
  d Äthylbutylcarbonat

3156 **ethyl butyl ether**
  f éther éthylbutylique
  e éter etilbutílico
  i etere etilbutilico
  d Äthylbutyläther

3157 **ethyl butyl ketone**
  f éthylbutylcétone
  e etilbutilcetona
  i etilbutilchetone
  d Äthylbutylketon

3158 **ethylbutyl silicate**
  f silicate d'éthylbutyle
  e silicato etilbutílico
  i silicato etilbutilico
  d Äthylbutylsilikat

3159 **ethylbutyraldehyde**
  f éthylbutyraldéhyde
  e etilbutiraldehído
  i etilbutirraldeide
  d Äthylbutyraldehyd

3160 **ethyl butyrate**
  f butyrate d'éthyle
  e butirato de etilo
  i butirrato di etile
  d Äthylbutyrat

3161 ethylbutyric acid
 f acide éthylbutyrique
 e ácido etilbutírico
 i acido etilbutirrico
 d Äthylbuttersäure

3162 ethyl caffeate
 f cafféate d'éthyle
 e cafeato de etilo
 i caffeato d'etile
 d koffeinsaurer Ethylester

3163 ethyl caprate
 f caprate d'éthyle
 e caprato de etilo
 i caprato d'etile
 d Äthylcaprat

3164 ethyl caproate
 f caproate d'éthyle
 e caproato de etilo
 i caproato d'etile
 d Äthylcaproat

3165 ethyl caprylate
 f caprylate d'éthyle
 e caprilato de etilo
 i caprilato d'etile
 d Äthylcaprylat

3166 ethyl carbamate
 f carbamate éthylé
 e carbamato de etilo
 i carbammato d'etile
 d Äthylcarbamat

3167 ethyl cellulose
 f éthylcellulose
 e etilcelulosa
 i etil cellulosa
 d Äthylcellulose

3168 ethyl chloride
 f chlorure d'éthyle
 e cloruro de etilo
 i cloruro d'etile
 d Äthylchlorid; Chloräthan

3169 ethyl chloroacetate
 f acétate de chlore éthylé
 e cloroacetato de etilo
 i cloroacetato d'etile
 d Äthylchloracetat

3170 ethyl chlorocarbonate
 f chlorocarbonate d'éthyle
 e clorocarbonato de etilo
 i clorocarbonato d'etile
 d Äthylchlorcarbonat

3171 ethyl chlorosulphonate
 f éthylchlorosulfonate
 e etilclorosulfonato
 i etilclorosolfonato
 d Äthylchlorsulfonat

3172 ethyl cinnamate
 f cinnamate d'éthyle
 e cinamato de etilo
 i cinnamato d'etile
 d Äthylcinnamat

3173 ethyl cocoinate
 f cocoïnate d'éthyle
 e cocoinato de etilo
 i cocoinato d'etile
 d Äthylcocoinat

3174 ethyl crotonate
 f crotonate d'éthyle
 e crotonato de etilo
 i crotonato d'etile
 d Äthylcrotonat

3175 ethyl cyanide
 f cyanure d'éthyle
 e cianuro de etilo
 i cianuro d'etile
 d Äthylcyanid

3176 ethyl cyanoacetate
 f acétate cyano-éthylique; cyanoacétate
 d'éthyle
 e cianoacetato de etilo
 i cianoacetato d'etile
 d Äthylcyanacetat

3177 ethylcyclohexane
 f éthylcyclohexane
 e etilciclohexano
 i etilcicloesano
 d Äthylcyclohexan

3178 ethyldichlorosilane
 f ethyldichlorosilane
 e etildiclorosilano
 i etildiclorosilano
 d Äthyldichlorsilan

3179 ethyldiethanolamine
 f éthyldiéthanolamine
 e etildietanolamina
 i etildietanolammina
 d Äthyldiäthanolamin

3180 ethylene; ethene; olefiant gas
 f éthylène
 e etileno

i etilene
d Äthylen

3181 ethylene carbonate
f carbonate d'éthylène
e carbonato de etileno
i carbonato d'etilene
d Äthylencarbonat

3182 ethylene chlorohydrin
f chlorhydrine d'éthylène
e clorhidrina etilénica
i cloridrina d'etilene
d Äthylenchlorhydrin

3183 ethylene cyanide
f cyanure d'éthylène
e cianuro de etileno
i cianuro d'etilene
d Äthylencyanid

3184 ethylene cyanohydrin
f cyanohydrine d'éthylène
e cianohidrina de etileno
i cianoidrina d'etilene
d Äthylencyanhydrin

3185 ethylenediamine
f éthylènediamine
e etilendiamina
i etilendiammina
d Äthylendiamin

3186 ethylene dibromide
f dibromure d'éthylène
e dibromuro de etileno
i bibromuro d'etilene
d Äthylendibromid

3187 ethylene dichloride; Dutch liquid
f dichlorure d'éthylène; liqueur des
Hollandais
e dicloruro de etileno; líquido de Holanda
i bicloruro di etilene; liquido olandese
d Dichloräthylen; Äthylendichlorid;
holländische Flüssigkeit

3188 ethylene glycol
f éthylèneglycol
e etilenglicol
i glicoletilenico
d Äthylenglykol

3189 ethylene glycol diacetate
f diacétate d'éthylèneglycol
e diacetato de etilenglicol
i diacetato glicoletilenico
d Äthylenglykoldiacetat

3190 ethylene glycol dibutyl ether
f éther dibutylique du éthylèneglycol
e éter dibutílico del etilenglicol
i etere dibutilglicoletilenico
d Äthylenglykoldibutyläther

3191 ethylene glycol dibutyrate
f dibutyrate d'éthylèneglycol
e dibutirato de etilenglicol
i bibutirrato glicoletilenico
d Äthylenglykoldibutyrat

3192 ethylene glycol diethyl ether
f éther diéthylique du éthylèneglycol
e éter dietílico del etilenglicol
i etere glicoldietiletilenico
d Äthylenglykoldiäthyläther

3193 ethylene glycol diformate
f diformiate d'éthylèneglycol
e diformiato de etilenglicol
i diformiato glicoletilenico
d Äthylenglykoldiformiat

3194 ethylene glycol dimethyl ether
f éther diméthylique du éthylèneglycol
e éter dimetílico del etilenglicol
i etere dimetilglicoletilenico
d Äthylenglykoldimethyläther

3195 ethylene glycol dipropionate
f dipropionate d'éthylèneglycol
e dipropionato de etilenglicol
i etileneglicoldipropionato
d Äthylenglykoldipropionat

3196 ethylene glycol monoacetate
f monoacétate d'éthylèneglycol
e monoacetato de etilenglicol
i etileneglicolmonoacetato
d Äthylenglykolmonoacetat

3197 ethylene glycol monobenzyl ether
f éther monobenzylique d'éthylèneglycol
e éter monobencílico de etilenglicol
i etere monobenzilico del glicoletilenico
d Äthylenglykolmonobenzyläther

3198 ethylene glycol monobutyl ether
f monobutyléther d'éthylèneglycol
e monobutiléter de etilenglicol
i etere monobutilico del glicoletilenico
d Äthylenglykolmonobutyläther

3199 ethylene glycol monobutyl ether acetate
f monobutyléther-acétate d'éthylèneglycol
e monobutiléter del acetato de etilenglicol
i acetato dell'etere monobutilico del glicol-

etilenico
d Äthylenglykolmonobutylätheracetat

3200 ethylene glycol monobutyl ether laurate
f monobutyléther-laurate d'éthylèneglycol
e monobutiléter del laurato de etilenglycol
i laurato dell'etere monobutilico del glicoletilenico
d Äthylenglykolmonobutylätherlaurat

3201 ethylene glycol monobutyl ether oleate
f monobutyléther-oléate d'éthylèneglycol
e monobutiléter del oleato de etilenglicol
i oleato dell'etere monobutilico del glicoletilenico
d Äthylenglykolmonobutylätheroleat

3202 ethylene glycol monoethyl ether
f monoéthyléther d'éthylèneglycol
e monoetiléter de etilenglicol
i etere monoetilico del glicoletilenico
d Äthylenglykolmonoäthyläther

3203 ethylene glycol monoethyl ether acetate
f monoéthyléther acétate d'éthylèneglycol
e monoetiléter del acetato de etilenglicol
i acetato dell'etere monoetilico del glicoletilenico
d Äthylenglykolmonoäthylätheracetat

3204 ethylene glycol monoethyl ether laurate
f laurate de monoéthyléther d'éthylèneglycol
e monoetiléter del laurato de etilenglicol
i etileneglicolmonoetileterelaurato
d laurinsaurer Äthylenglykolmonoäthyläther

**3205 ethylene glycol monoethyl ether
ricinoleate**
f monoéthyléther-ricinoléate d'éthylèneglycol
e monoetiléter del ricinoleato de
etilenglicol
i ricinoleato dell'etere monoetilico del
glicoletilenico
d Äthylenglykolmonoäthylätherricinoleat

3206 ethylene glycol monohexyl ether
f éther monohexylique du éthylèneglycol
e éter monohexílico del etilenglicol
i etere monoesilicoglicoletilenico
d Äthylenglykolmonohexyläther

3207 ethylene glycol monomethyl ether
f monométhyléther d'éthylèneglycol
e monometiléter de etilenglicol
i etere monometilico del glicoletilenico
d Äthylenglykolmonomethyläther

**3208 ethylene glycol monomethyl ether
acetate**
f monométhyléther-acétate d'éthylèneglycol
e monometiléter del acetato de etilenglicol
i acetato dell'etere monometilico del glicoletilenico
d Äthylenglykolmonomethylätheracetat

**3209 ethylene glycol monomethyl ether acetyl
ricinoleate**
f monométhyléther-acétylricinoléate
d'éthylèneglycol
e monometiléter del acetilricinoleato de
etilenglicol
i acetilricinoleato dell'etere monometilico
del glicoletilenico
d Äthylenglykolmonomethylätheracetyl-
ricinoleat

**3210 ethylene glycol monomethyl ether
ricinoleate**
f monométhyléther-ricinoléate d'éthylèneglycol
e monometiléter del ricinoleato de
etilenglicol
i ricinoleato dell'etere monometilico del
glicoletilenico
d Äthylenglykolmonomethylätherricinoleat

**3211 ethylene glycol monomethyl ether
stearate**
f monométhyléther-stéarate d'éthylèneglycol
e monometiléter del estearato de
etilenglicol
i stearato dell'etere monometilico del glicoletilenico
d Äthylenglykolmonomethylätherstearat

3212 ethylene glycol monooctyl ether
f monooctyléther d'éthylèneglycol
e éter monooctílico del etilenglicol
i etere monoottilico del glicoletilenico
d Äthylenglykolmonooctyläther

3213 ethylene glycol monophenyl ether
f monophényléther d'éthylèneglycol
e éter monofenílico del etilenglicol
i etere monofenilico del glicoletilenico
d Äthylenglykolmonophenyläther

3214 ethylene glycol monoricinoleate
f monoricinoléate d'éthylèneglycol
e monorricinoleato de etilenglicol
i monoricinoleato del glicoletilenico
d Äthylenglykolmonoricinoleat

3215 ethylene glycol silicate
f silicate d'éthylèneglycol
e silicato de etilenglicol
i silicato glicoletilenico
d Äthylenglykolsilikat

3216 ethyleneimine
f éthylénimine
e etilenimina
i etilenimmina
d Äthylenimin

3217 ethylene oxide
f oxyde d'éthylène
e óxido de etileno
i ossido di etilene
d Äthylenoxyd

3218 ethylene propylene rubber
f caoutchouc éthylènepropylénique
e goma etilenpropilénica
i gomma propilenetilenica
d Äthylenpropylenkautschuk

3219 ethylene thiourea
f éthylènethiourée
e etilentiourea
i etilentiourea
d Äthylenthioharnstoff

3220 ethylethanolamine
f éthyléthanolamine
e etiletanolamina
i etiletanolammina
d Äthyläthanolamin

3221 ethyl formate
f formiate d'éthyle
e formiato de etilo
i formiato di etile
d Äthylformiat

3222 ethyl glycol monohexyl ether
f éther monohexylique du glycol
e éter etilglicolmonohexilico
i etere etilglicolmonoesilico
d Äthylglykolmonohexyläther

3223 ethylhexoic acid
f acide éthylhexoïque
e ácido etilhexoico
i acido etilesoico
d Äthylhexonsäure

3224 ethylhexyl acetate
f acétate d'éthylhexyle
e acetato de etilhexilo
i acetato etilesilico
d Äthylhexylacetat

3225 ethylhexyl acrylate
f acrylate d'éthylhexyle
e acrilato de etilhexilo
i acrilato etilesilico
d Äthylhexylacrylat

3226 ethylhexyl alcohol
f alcool éthylhexylique
e alcohol etilhexílico
i alcool etilesilico
d Äthylhexylalkohol

3227 ethylhexylamine
f éthylhexylamine
e etilhexilamina
i etilesilammina
d Äthylhexylamin

3228 ethylhexyl bromide
f bromure d'éthylhexyle
e bromuro de etilhexilo
i bromuro etilesilico
d Äthylhexylbromid

3229 ethylhexyl chloride
f chlorure d'éthylhexyle
e cloruro de etilhexilo
i cloruro etilesilico
d Äthylhexylchlorid

3230 ethylhexyl octylphenyl phosphite
f phosphite d'éthylhexyloctylphényle
e fosfito de etilhexiloctilfenílo
i fosfito ottilfeniletilesilico
d Äthylhexyloctylphenylphosphit

3231 ethyl hydroxyisobutyrate
f hydroxyisobutyrate d'éthyle
e hidroxiisobutirato de etilo
i idrossisobutirrato d'etile
d Äthylhydroxyisobutyrat

3232 ethylidene acetobenzoate
f acétobenzoate d'éthylidène
e acetobenzoato de etilideno
i etilidenacetobenzoato
d Äthylidenacetobenzoat

3233 ethylidene aniline
f éthylidèneaniline
e etilidenanilina
i etilidenanilina
d Äthylidenanilin

3234 ethyl iodide
f iodure d'éthyle
e yoduro de etilo
i ioduro d'etile
d Äthyljodid

3235 **ethyl isobutyrate**
   f isobutyrate d'éthyle
   e isobutirato de etilo
   i isobutirrato d'etile
   d Äthylisobutyrat

3236 **ethyl isocyanate**
   f isocyanate d'éthyle
   e isocianato de etilo
   i isocianato d'etile
   d Äthylisocyanat

3237 **ethyl isovalerate**
   f isovalérate d'éthyle
   e isovalerato de etilo
   i isovalerato d'etile
   d Äthylisovalerianat

3238 **ethyl lactate**
   f lactate d'éthyle
   e lactato de etilo
   i lattato di etile
   d Äthyllactat

3239 **ethyl levulinate**
   f lévulinate d'éthyle
   e levulinato de etilo
   i levulinato d'etile
   d Äthyllävulinat

3240 **ethyl magnesium bromide**
   f bromure de magnésium et d'éthyle
   e bromuro de magnesio y etilo
   i bromuro di magnesio e di etile
   d Äthylmagnesiumbromid

3241 **ethyl magnesium chloride**
   f chlorure de magnésium et d'éthyle
   e cloruro de magnesio y etilo
   i cloruro di magnesio e di etile
   d Äthylmagnesiumchlorid

3242 **ethyl malonate**
   f malonate d'éthyle, ether malonique
   e malonato de etilo
   i malonato d'etile
   d Malonester; Äthylmalonat

3243 **ethylmercuric acetate**
   f acétate éthylmercurique
   e acetato etilmercúrico
   i acetato etilmercurico
   d Äthylmerkuriacetat

3244 **ethylmercuric chloride**
   f chlorure éthylmercurique
   e cloruro etilmercúrico
   i cloruro etilmercurico
   d Äthylmerkurichlorid

3245 **ethylmercuric phosphate**
   f phosphate éthylmercurique
   e fosfato etilmercúrico
   i fosfato etilmercurico
   d Äthylmerkuriphosphat

3246 **ethylmethacrylate**
   f méthacrylate d'éthyle
   e metacrilato de etilo
   i metacrilato di etile
   d Äthylmetacrylat

3247 **ethyl methyl phenyl glycidate**
   f éthylméthylphénylglycidate
   e etilmetilfenilglicidato
   i etilmetilfenilglicidato
   d Äthylmethylphenylglycidat

3248 **ethylmorphine hydrochloride**
   f hydrochlorure d'éthylmorphine
   e hidrocloruro de etilmorfina
   i cloridrato di etilmorfina
   d Äthylmorphinhydrochlorid

3249 **ethyl morpholine**
   f éthylmorpholine
   e etilmorfolina
   i etilmorfolina
   d Äthylmorpholin

3250 **ethyl nitrate**
   f nitrate d'éthyle
   e nitrato de etilo
   i nitrato d'etile
   d Äthylnitrat

3251 **ethyl nitrite**
   f nitrite d'éthyle
   e nitrito de etilo
   i nitrito d'etile
   d Äthylnitrit

3252 **ethyl oenanthate**
   f oenanthate d'éthyle
   e enantato de etilo
   i enantato d'etile
   d Äthylönanthat

3253 **ethyl oleate**
   f oléate d'éthyle
   e oleato de etilo
   i oleato di etile
   d Äthyloleat

3254 **ethyl oxalate**
   f oxalate d'éthyle
   e oxalato de etilo
   i ossalato di etile
   d Äthyloxalat

3255 ethyl pelargonate
 f pélargonate d'éthyle
 e pelargonato de etilo
 i pelargonato d'etile
 d Äthylpelargonat

3256 ethyl phenylacetate
 f acétate phénylique éthylé
 e fenilacetato de etilo
 i fenilacetato d'etilo
 d Äthylphenylacetat

3257 ethyl phenyl ethanolamine
 f éthanolamine d'éthylphényle
 e etilfeniletanolamina
 i etilfeniletanolammina
 d Äthylphenyläthanolamin

3258 ethylphosphoric acid
 f acide éthylphosphorique
 e ácido etilfosfórico
 i acido etilfosforico
 d Äthylphosphorsäure

3259 ethyl phthalyl ethyl glycollate
 f éthylphtalyléthylglycollate
 e etilftaliletilglicolato
 i etilftaliletilglicollato
 d Äthylphthalyläthylglycollat

3260 ethyl propionate
 f propionate d'éthyle
 e propionato de etilo
 i propionato di etile
 d Äthylpropionat

3261 ethyl propylacrolein
 f propylacroléine d'éthyle
 e propilacroleína de etilo
 i propilacroleina d'etile
 d Äthylpropylacrolein

3262 ethyl pyridylethylacrylate
 f pyridyléthylacrylate d'éthyle
 e piridiletilacrilato de etilo
 i etilpiridiletilacrilato
 d Äthylpyridyläthylacrylat

3263 ethyl salicylate
 f salicylate d'éthyle
 e salicilato de etilo
 i salicilato d'etile
 d Salicylsäureäthylester

3264 ethyl silicate
 f silicate d'éthyle
 e silicato de etilo
 i silicato d'etile
 d Äthylsilikat

3265 ethyl sodium oxalacetate
 f oxalacétate de sodium et d'éthyle
 e oxalacetato de sodio y etilo
 i ossalacetato di sodio e etile
 d Natriumäthyloxalacetat

3266 ethyl sulphide
 f sulfure d'éthyle
 e sulfuro de etilo
 i solfuro d'etile
 d Äthylsulfid

3267 ethylsulphuric acid
 f acide éthylsulfurique
 e ácido etilsulfúrico
 i acido etilsolforico
 d Äthylschwefelsäure

3268 ethylsulphurous acid
 f acide éthylsulfureux
 e ácido etilsulfuroso
 i acido etilsolforoso
 d Äthylschwefligsäure

3269 ethyl thioethanol
 f thioéthanol d'éthyle
 e etiltioetanol
 i tioetanol d'etile
 d Äthylthioäthanol

3270 ethyl toluenesulphonate
 f toluène-sulfonate éthylé
 e toluensulfonato de etilo
 i toluenesolfonato d'etile
 d Äthyltoluolsulfonat

3271 ethyltrichlorosilane
 f éthyltrichlorosilane
 e etiltriclorosilano
 i etiltriclorosilano
 d Äthyltrichlorsilan

3272 ethyl vanillin
 f vanilline d'éthyle
 e vainillina de etilo
 i vanillina d'etile
 d Äthylvanillin

3273 ethyl vinyl ether; EVE
 f éther éthylvinylique
 e viniletiléter
 i etere viniletilico
 d Äthylvinyläther

3274 eucaine lactate
 f lactate d'eucaïne
 e lactato de eucaína
 i lattato di eucaina
 d Eucainlactat; milchsaures Eucain

3275 eucalyptol
 f eucalyptol
 e eucaliptol
 i eucaliptol
 d Eucalyptol

3276 eucalyptus oil
 f huile d'eucalyptus
 e aceite de eucalipto
 i olio di eucalipto
 d Eucalyptusöl

3277 euchlorine
 f euchlorine
 e euclorina
 i euclorina
 d Euchlorin

3278 euchromatin
 f euchromatine
 e eucromatina
 i eucromatina
 d Euchromatin

3279 euchromosome; autosome
 f euchromosome; autosome
 e eucromosoma; autosoma
 i eucromosoma; autosoma
 d Euchromosom; Autosom

 * **eudiometer → 3312**

3280 eugenol
 f eugénol
 e eugenol
 i eugenol
 d Eugenol

3281 eugenol acetate
 f acétate d'eugénol
 e acetato de eugenol
 i acetato di eugenol
 d Eugenolacetat

3282 euploid
 f euploïde
 e euploide
 i euploide
 d euploid

3283 euploidy
 f euploïdie
 e euploidia
 i euploidismo
 d Euploidie

3284 europium
 f europium
 e europio

 i europio
 d Europium

3285 Euston process
 f procédé Euston
 e proceso Euston
 i processo Euston
 d Euston-Verfahren

3286 eutectic
 f mélange eutectique
 e eutéctico
 i eutettico
 d Eutektikum

3287 eutectic change
 f changement eutectique
 e cambio eutéctico
 i mutamento eutettico
 d eutektischer Wechsel

3288 eutectic point
 f point eutectique
 e punto eutéctico
 i punto eutettico
 d eutektischer Punkt

3289 eutropic series
 f série eutropique
 e serie eutrópica
 i serie eutropica
 d eutropische Reihe

3290 eutropy
 f eutropie
 e eutropía
 i eutropia
 d Eutropie

3291 euxenite
 f euxénite
 e euxenita
 i euxenita
 d Euxenit

3292 evaporating dish
 f vase d'évaporation en porcelaine
 e vaso de porcelana para evaporar
 i bacinella d'evaporazione di porcellana
 d Porzellanabdampfschale; Abdampfschale

3293 evaporation
 f évaporation
 e evaporación
 i evaporazione
 d Verdampfung

3294 evaporation pond
 f bassin d'évaporation

e estanque de evaporación
i vasca d'evaporazione; bacino d'evaporazione
d Abdampfbecken

3295 evaporator
f appareil à concentrer
e aparato de evaporación
i apparato per evaporare
d Eindampfer

* EVE → 3273

3296 examination
f examen
e examen
i esame
d Examen

3297 excess air factor; excess air ratio
f facteur d'excès d'air
e factor de exceso de aire
i fattore d'eccesso d'aria
d Luftüberschußfaktor

3298 exchange reaction
f réaction de substitution
e reacción de intercambio
i reazione di sostituzione
d Austauschreaktion

3299 excitoanabolic
f excito-anabolique
e excitoanabólico
i eccitoanabolico
d anabolismusfördernd

3300 excitocatabolic
f excito-catabolique
e excitocatabólico
i eccitocatabolico
d katabolischwirkend

3301 exhaust duct
f canal de désaération
e conducto de aspiración
i condotto di sfiato
d Entlüftungsleitung

3302 exhaust fan
f ventilateur par aspiration
e ventilador extractor; ventilador de succión
i aspiratore; estrattore
d Exhaustor; Sauger

3303 exhaust steam
f vapeur d'échappement
e vapor de escape

i vapore di scarico
d Abdampf

3304 exit dose
f dose de sortie
e dosis emergente
i dose emergente
d Austrittsdosis

3305 exo-
Prefix meaning "external".
f exo-
e exo-
i esso-
d Exo-

3306 exocrine; eccrine
f exocrine
e exocrina
i esocrino
d exokrin

3307 exoenzyme
f exoenzyme
e exoenzima
i esoenzima
d Exo-Ferment

3308 exogenic
f exogène
e exógeno
i esogeno
d exogen

3309 exotherm curve
f courbe exothermique
e gráfico exotérmico
i curva esotermica
d exotherme Kurve

3310 exothermic
f exothermique
e exotérmico
i essotermico
d exotherm

3311 expanding agent
f agent moussant
e agente expansor
i agente di espansione
d Porenbildner

3312 explosion burette; eudiometer
f burette à explosion; eudiomètre
e bureta de explosión; eudiómetro
i buretta d'esplosione; eudiometro
d Explosionsbürette; Eudiometer

3313 explosive
 f explosif
 e explosivo
 i esplosivo
 d Sprengstoff

3314 exponential death phase
 f phase de mort exponentielle
 e fase de morte exponencial
 i fase esponenziale di decesso
 d exponentielle Sterbephase

3315 exposed stem correction
 f correction de colonne émergente
 e corrección de columna emergente
 i correzione del filo emergente
 d Fadenkorrektur

3316 oxoiooant
 f dessiccatif
 e deser nolr
 i essicante
 d Austrocknungsmittel

3317 extend *v*
 f étendre
 e diluir un ingrediente
 i diluire
 d strecken

3318 extender
 f agent diluant
 e diluyente
 i agente diluente
 d Streckmittel; Verschnitt

3319 extension
 f extension
 e extensión
 i estensione
 d Dehnung

3320 oxterior finish
 f couche de finissage extérieur
 e barniz para exteriores
 i vernice per esterni
 d Außenanstrich

3321 external indicator
 f indicateur externe
 e indicador externo
 i indicatore esterno
 d externer Anzeiger

3322 external quenching
 f coupure externe
 e extinción externa
 i spegnimento esterno
 d externe Löschung

3323 extract
 f extrait
 e extracto
 i estratto
 d Extrakt

3324 extractant
 f solvant d'extraction
 e solvente de extracción
 i solvente di estrazione
 d Extraktionsmittel

3325 extraction
 f extraction
 e extracción
 i estrazione
 d Extraktion

3326 oxtraction apparatus
 f appareil d'extraction
 e aparato de extracción
 i apparato d uotrazione
 d Extraktionsapparat

3327 extraction funnel
 f entonnoir d'extracteur
 e embudo de extracción
 i imbuto d'estrazione
 d Extraktionstrichter; Auslaugtrichter

3328 extraction liquor
 f liqueur d'extraction
 e líquido de extracción
 i liquido lisciviante
 d Extraktionsflüssigkeit; Lauge

3329 extraction thimble
 f enveloppe d'extracteur
 e envoltura de aparato de extracción
 i inviluppo d'estrazione
 d Auslaughülse; Extraktionshülse

3330 extractive distillation
 f distillation par fractionnement
 e destilación extractiva
 i estrazione
 d Extraktivdestillation

3331 extra-nuclear process
 f processus extranucléaire
 e proceso extranuclear
 i processo estranucleare
 d Prozeß außerhalb des Kerns

3332 extra-nuclear structure
 f structure extranucléaire
 e estructura extranuclear
 i struttura estranucleare
 d Struktur außerhalb des Kerns

F

3333 face-centred cubic structure
f réseau cubique à faces centrées
e red cúbica de caras centradas
i reticolo cubico di faccie centrate
d kubisch-flächenzentriertes Gitter

3334 facet
f facette
e faceta
i faccetta
d Facette; kleine Fläche

3335 factice
f factice
e facticio
i fatturato
d Faktis

3336 factor
f facteur
e factor
i fattore
d Faktor

3337 factor map
f carte factorielle
e mapa factorial
i carta fattoriale
d Genkarte

3338 fadeometer
f fadéomètre
e fadeómetro
i misuratore di decolorazione
d Fadeometer; Lichtbeständigkeitsprüfer

3339 Fahraeus' phenomenon
f réaction de sédimentation
e fenómeno de Fahraeus
i reazione di sedimentazione delle emazie
d Blutkörperchensenkungsreaktion

3340 falling back (of fermentation)
f ralentissement
e disminución; descenso (de la fermentación)
i diminuzione (della fermentazione)
d Zurückgehen

3341 falling rate drying period
f période de séchage à vitesse décroissante
e período de disminución del régimen de secado
i periodo d'essicazione a tasso decrescente
d Periode mit abnehmender Trockengeschwindigkeit

3342 false body
f corps apparent
e disminución de la viscosidad
i falso corpo
d falscher Körpergehalt

3343 false grain *(sugar)*
f faux grains
e falta de núcleos para formación de cristales
i grana falsa
d Feinkorn

3344 famatinite
f famatinite
e famatinita
i famatinite
d Famatinit

3345 family *(of elements)*
f groupe
e grupo
i gruppo
d Gruppe

3346 family of curves
f famille de courbes
e familia de curvas
i famiglia di curve
d Kurvenschar

3347 faraday
f faraday
e faraday
i faraday
d Faraday

3348 Faraday's law of electrolysis
f loi de Faraday de l'électrolyse
e leye de Faraday sobre la electrólisis
i legge dell'elettrolisi di Faraday
d Faradaysches Gesetz

3349 farinaceous
f farineux
e farináceo
i farinaceo
d mehlig

3350 farnesol
f farnésol
e farnesol
i farnesol
d Farnesol

3351 fascicular
f fasciculé
e fasciculado

i fascicolare
d faszikular; büschelförmig

3352 fast
f grand teint
e estable; indeleble
i indelebile; stabile
d beständig; echt

3353 fast milling pigment
f pigments à dispersion rapide
e pigmentos de rápida dispersión
i pigmenti a dispersione rapida
d gut mischende Pigmente

3354 fastness
f résistance
e resistencia
i resistenza
d Festigkeit

3355 fat
f graisse
e grasa
i grasso
d Fett

3356 fat edge
f surépaisseur en bordure
e borde craso; exceso de pintura en el borde
i bordo con eccesso di pittura
d verdickter Rand

3357 fatigue
f fatigue
e fatiga
i fatica
d Ermüdung

3358 fatigue poison
f toxine de la fatigue
e toxina de fatiga
i miotossina della fatica
d Ermüdungsstoff

3359 fat turpentine; fat oil; oxidized turpentine
f térébenthine soufflée; térébenthine oxydée
e trementina grasa; trementina oxidada; trementina soplada
i trementina grassa; trementina ossidata
d geblasenes Terpentin

3360 fatty
f graisseux; gras
e gordo; graso

i grasso
d fettig

3361 fatty acid
f acide gras
e ácido graso
i acido grasso
d Fettsäure

3362 fatty acid pitch
f poix d'acides gras
e pez de ácidos grasos
i pece di acidi grassi; pece di acidi alifatici
d Fettsäurepech

3363 fatty alcohols
f alcools gras
e alcoholes grasos
i alcool grassi
d Fettalkohole

3364 fatty amines
f amines aliphatiques
e aminas alifáticas
i ammine alifatiche
d aliphatische Amine

3365 fatty esters
f esters d'acides gras
e ésteres grasos
i esteri grassi
d Fettester

3366 fatty nitriles
f nitriles aliphatiques
e nitrilos alifáticos
i nitrili alifatici
d Fettnitrile

3367 Fauser ammonia process
f procédé ammoniacal de Fauser
e proceso de Fauser al amoníaco
i processo Fauser all'ammoniaca
d Fausersche Ammoniaksynthese

3368 Fauser processes
f procédés Fauser
e procesos Fauser
i processi Fauser
d Fauser-Verfahren

3369 febricant
f pyrétogène
e pirógeno
i piretogeno
d fiebererzeugend

3370 febrifuge; febricide
f fébrifuge; antithermique

e febrífugo; febricida
i febbrifugo
d Fiebermittel

3371 fecula
f fécule
e fécula
i fecola
d Stärke

3372 feculent
f féculent
e fécula
i sedimentoso
d stärkemehlartig

3373 fecundation
f fécondation
e fecundación
i fecondazione; fertilizzazione
d Befruchtung

3374 feeding
f épaississement
e aumento de viscosidad
i impolmonimento
d Eindicken

3375 Fehling's solution
f liqueur de Fehling
e solución de Fehling
i soluzione di Fehling
d Fehlingsche Lösung

3376 feldspar
f spath adulaire; feldspath
e feldespato
i feldspato
d Feldspat

3377 fenchyl alcohol
f alcool fenchylique
e alcohol fenquílico
i alcool fencilico
d Fenchylalkohol

3378 fennel
f fenouil
e hinojo
i finocchio
d Fenchel

3379 ferbam
f ferbam
e ferbam
i ferbam
d Ferbam

3380 ferberite
f ferbérite
e ferberita
i ferberite
d Ferberit

3381 fergusonite
f fergusonite
e fergusonita
i fergusonite
d Fergusonit

3382 ferment
f ferment
e fermento
i fermento
d Ferment

3383 ferment *v*
f fermenter
e fermentar
i fermentare
d gären

3384 fermentation
f fermentation
e fermentación
i fermentazione
d Fermentation; Gärung

3385 fermentation-inhibiting
f freinant la fermentation
e que detiene la fermentación
i inibente la fermentazione
d gärungshemmend

3386 fermenter; fermentor
f cuve de fermentation
e cuba de fermentación
i tino di fermentazione
d Gärbottich

3387 fermenting cellar
f cave de fermentation
e bodega de fermentación
i cantina di fermentazione
d Gärkeller

3388 fermenting tank
f cuve de fermentation
e tanque de fermentación
i tank di fermentazione
d Gärtank

* **fermentor** → 3386

3389 fermium
f fermium
e fermio

i fermio
d Fermium

3390 fern
f fougère
e helecho
i felce
d Farn

3391 ferric acetylacetonate
f acétylacétonate ferrique
e acetilacetonato férrico
i acetilacetonato ferrico
d Eisenacetylacetonat

3392 ferric ammonium citrate
f citrate de fer ammoniacal
e citrato férrico-amónico
i citrato ferrico d'ammonio
d Ferriammoniumcitrat;
Eisenammoniumcitrat

3393 ferric ammonium oxalate
f oxalate de fer ammoniacal
e oxalato férrico-amónico
i ossalato ferrico d'ammonio
d Ferriammoniumoxalat

3394 ferric ammonium sulphate
f alun ferrique ammoniacal
e sulfato férrico-amónico
i solfato ferrico d'ammonio
d Ferriammoniumsulfat;
Ammoniumeisenalaun

3395 ferric arsenate
f arséniate ferrique
e arseniato férrico
i arseniato ferrico
d Ferriarsenat

3396 ferric cacodylate
f cacodylate ferrique
e cacodilato férrico
i cacodilato ferrico
d Ferricacodylat

3397 ferric chloride
f chlorure ferrique
e cloruro férrico
i cloruro ferrico
d Eisenchlorid; Ferrichlorid; Chloreisen

3398 ferric chromate
f chromate ferrique
e cromato férrico
i cromato ferrico
d Eisenchromat

3399 ferric ferrocyanide
f ferrocyanure ferrique
e ferrocianuro férrico
i ferrocianuro ferrico
d Ferriferrocyanid

3400 ferric fluoride
f fluorure ferrique
e fluoruro férrico
i fluoruro ferrico
d Eisenfluorid; Ferrifluorid

3401 ferric glycerophosphate
f glycérophosphate ferrique
e glicerofosfato férrico
i glicerofosfato ferrico
d Eisenglycerophosphat

3402 ferric hydroxide
f hydroxyde ferrique
e hidróxido férrico
i idrossido ferrico
d Ferrihydroxyd; Eisenhydroxyd

3403 ferric hypophosphite
f hypophosphite ferrique
e hipofosfito férrico
i ipofosfito ferrico
d Ferrihypophosphit; Eisenhypophosphit

3404 ferric naphthenate
f naphténate ferrique
e naftenato férrico
i naftenato ferrico
d Ferrinaphthenat; Eisennaphthenat

3405 ferric nitrate
f nitrate de fer
e nitrato férrico
i nitrato di ferro
d Ferrinitrat; Eisenoxydnitrat

3406 ferric oxalate
f oxalate ferrique
e oxalato férrico
i ossalato di ferro
d Ferrioxalat

3407 ferric oxide
f oxyde ferrique
e óxido férrico
i ossido ferrico
d Eisenoxyd; Ferrioxyd

3408 ferric phosphate
f phosphate ferrique
e fosfato férrico
i fosfato ferrico
d Ferriphosphat; Eisenphosphat

3409 ferric potassium citrate
 f citrate ferrique de potassium
 e citrato férrico-potásico
 i citrato ferrico di potassio
 d Ferrikaliumcitrat

3410 ferric potassium tartrate
 f tartrate ferrique de potassium
 e tartrato férrico-potásico
 i tartrato ferrico di potassio
 d Ferrikaliumtartrat

3411 ferric resinate
 f résinate ferrique
 e resinato férrico
 i resinato ferrico
 d Eisenresinat

3412 ferric sodium oxalate
 f oxalate ferrique de sodium
 e oxalato férrico-sódico
 i ossalato ferrico di sodio
 d Eisennatriumoxalat

3413 ferric stearate
 f stéarate de fer
 e estearato férrico
 i stearato ferrico
 d Eisenstearat

3414 ferric sulphate
 f sulfate ferrique
 e sulfato férrico
 i solfato ferrico
 d Ferrisulfat; schwefelsaures Eisenoxyd

3415 ferrite
 f ferrite
 e ferrita
 i ferrite
 d Ferrit

3416 ferrite yellows
 f jaunes de ferrite; oxydes de fer jaunes
 e amarillos de ferrita
 i gialli di ferrite; ossido di ferro
 d Eisengelb

3417 ferrocene
 f ferrocène
 e ferroceno
 i ferrocene
 d Ferrocen

3418 ferro-chrome; ferro-chromium
 f ferrochrôme
 e ferrocromo
 i ferrocromo
 d Ferrochrom

3419 ferrocyanide
 f ferrocyanure
 e ferrocianuro
 i ferrocianuro
 d Ferrocyanid

3420 ferro-manganese
 f ferromanganèse
 e ferromanganeso
 i ferromanganese
 d Ferromangan; Eisenmangan

3421 ferro-molybdenum
 f ferromolybdène
 e ferromolibdeno
 i ferromolibdeno
 d Ferromolybdän

3422 ferro-nickel
 f ferronickel
 e ferroníquel
 i ferronichel
 d Ferronickel

3423 ferro-silicon
 f ferrosilicium
 e ferrosilicio
 i ferrosilicio
 d Ferrosilizium; Siliziumeisen

3424 ferro-titanium
 f ferrotitane
 e ferrotitanio
 i ferrotitanio
 d Ferrotitan

3425 ferro-tungsten
 f ferrotungstène
 e ferrotungsteno
 i ferrotungsteno
 d Ferrowolfram

3426 ferrous acetate
 f acétate ferreux
 e acetato ferroso
 i acetato ferroso
 d Eisenacetat; Ferroacetat; essigsaures Eisenoxydul

3427 ferrous ammonium sulphate
 f sulfate ferreux ammoniacal
 e sulfato ferroso-amónico
 i solfato ferroso di ammonio
 d Ferroammonsulfat

3428 ferrous arsenate
 f arséniate de fer
 e arseniato ferroso

i arseniato ferroso
d Ferroarsenat

3429 ferrous chloride
f chlorure ferreux
e cloruro ferroso
i cloruro ferroso
d Ferrochlorid; Eisenchlorür

3430 ferrous fluoride
f fluorure ferreux
e fluoruro ferroso
i fluoruro ferroso
d Ferrofluorid

3431 ferrous fumarate
f fumarate ferreux
e fumarato ferroso
i fumarato ferroso
d Ferrofumarat

3432 ferrous gluconate
f gluconate ferreux
e gluconato ferroso
i gluconato ferroso
d Ferrogluconat

3433 ferrous iodide
f iodure ferreux
e yoduro ferroso
i ioduro ferroso
d Ferrojodid; Jodeisen

3434 ferrous lactate
f lactate ferreux
e lactato ferroso
i lattato ferroso
d Ferrolactat; milchsaures Eisen

3435 ferrous oxide
f oxyde ferreux
e óxido ferroso
i protossido di ferro; ossido ferroso
d Eisenoxydul

3436 ferrous phosphate
f phosphate ferreux
e fosfato ferroso
i fosfato ferroso
d Ferrophosphat

3437 ferrous phosphide
f phosphure ferreux
e fosfuro ferroso
i fosfuro ferroso
d Ferrophosphid

3438 ferrous quinine citrate
f citrate ferreux de quinine

e citrato ferroso-quinina
i citrato ferroso di chinino
d Ferrochinincitrat

3439 ferrous sulphate; copperas
f sulfate ferreux; couperose verte
e sulfato ferroso
i solfato ferroso; copperosa verde
d Ferrosulfat; schwefelsaures Eisenoxydul;
   Eisenvitriol

3440 ferrous sulphide
f sulfure ferreux
e sulfuro ferroso
i solfuro ferroso
d Eisensulfür; Ferrosulfid; Schwefeleisen

3441 ferrovanadium
f ferrovanadium
e ferrovanadio
i ferrovanadio
d Vanadiumeisen; Ferrovanadium

3442 ferroxyl indicator
f indicateur à ferroxyle
e indicador al ferroxilo
i indicatore al ferrossile
d Ferroxylindikator

3443 fertilizer
f engrais
e fertilizante
i concime; fertilizzante
d Düngemittel; Kunstdünger

3444 fertilizing capacity
f pouvoir fécondant
e poder fecundante
i potere fertilizzante
d Befruchtungsfähigkeit

3445 Fessler compound
f composé Fessler
e compuesto Fessler
i composto Fessler
d Fesslersche Verbindung

3446 festination
f festination
e festinación
i festinazione
d Festination

3447 fetid
f fétide
e fétido
i fetido
d stinkend

3448 **fetor**
  f mauvaise odeur
  e hedor
  i fetore
  d Gestank

3449 **fibril**
  f fibrille
  e fibrilla
  i fibrilla
  d Fibrille

3450 **fibrin**
  f fibrine
  e fibrina
  i fibrina
  d Fibrin

3451 **fibrinogen**
  f fibrinogène
  e fibrinógeno
  i fibrinogeno
  d Fibrinogen

3452 **fibrinolysin**
  f fibrinolysine
  e fibrinolisina
  i fibrinolisina
  d Fibrinolysin

3453 **fibroblast**
  f fibroblaste
  e célula conjuntiva
  i cellula congiuntiva
  d Bindegewebszelle

3454 **ficin**
  f ficine
  e ficina
  i ficina
  d Ficin

3455 **fictile**
  f céramique; plastique
  e plástico; figulino
  i fittile
  d formbar

3456 **field of view**
  f champ de la lunette
  e campo de la lente
  i campo visivo
  d Gesichtsfeld

3457 **filament**
  f filament
  e filamento
  i filamento
  d Filament

3458 **filaricide**
  f filaricide
  e filaricida
  i filaricida
  d Filarienmittel

3459 **filiform**
  f filiforme
  e filiforme
  i filiforme
  d fadenförmig

3460 **filler** *(paint)*
  f charge
  e mastique; aparejo
  i fondo; riempitivo
  d Füllmasse; Grundmasse

3461 **filler** *(plastics)*
  f mastic
  e masilla; mastique
  i fondo
  d Füllstoff

3462 **film**
  f pellicule; feuille
  e película
  i pellicola; film
  d Film

3463 **film building properties**
  f capacité de former une pellicule
  e capacidad de formación de una película
  i capacità di produrre una pellicola
  d Filmbildungseigenschaften

3464 **film glue**
  f colle en feuille
  e cola en hojas
  i colla in film; colla in fogli
  d Klebefilm; Klebefolie

3465 **filter**
  f filtre
  e filtro
  i filtro
  d Filter

3466 **filtering crucible**
  f creuset de filtration
  e crisol de filtración
  i crogiuolo di filtrazione
  d Filtriertiegel

3467 **filtering flask**
  f flacon à filtrer
  e frasco de filtrar
  i bottiglia per filtrazione
  d Filterflasche

3468 filter press
f presse à filtrer
e prensa para filtrar
i pressa da filtro
d Filterpresse

3469 filter pump
f pompe à filtrer
e bomba de filtración
i pompa filtrante
d Filterpumpe

3470 filtrable
f filtrable
e filtrable
i filtrabile
d filtrierbar

3471 filtrate
f produit filtré
e líquido filtrado
i prodotto filtrato
d Filtrat

3472 filtration
f filtration
e filtración
i filtratura
d Filtration; Filtrieren

3473 fimbriated
f frangé
e fimbriado
i frangiato
d fimbriatus

3474 fine v
f clarifier
e aclarar; clarificar
i chiarificare
d abklären

3475 fines
f fines
e granos finos
i materiali fini
d pulvriges Material

3476 finings
f colle de poisson
e cola de pescado
i colla di pesce
d Fischleim

3477 finished beer
f bière prête au débit
e cerveza lista para consumo
i birra pronta per il consumo
d fertiges Bier

3478 fire v
f brûler
e quemar
i bruciare
d verfeuern

3479 fire assay
f essai pyrognostique
e ensayo pirognóstico; copelación
i coppellazione
d Brandprobe

3480 fire-clay
f argile réfractaire
e arcilla refractaria
i argilla refrattaria
d feuerfester Ton

3481 fire-cracked
f craquelé (au feu)
e agrietado en el horno
i spaccato nella cottura
d gesprungen

3482 fire-damp; damp
f grisou
e grisú
i grisú
d Grubengas

3483 fire point
f point d'inflammation; point de feu
e punto de inflamación
i punto di infiammabilità
d Flammpunkt

3484 fire-proof
f résistant au feu
e resistente al fuego
i resistente al fuoco
d feuerfest

3485 firing
f cuisson
e cocido
i cottura
d Einbrennen

3486 firkin
f petit fût (demi)
e barrilito medio
i fusticino mezzo
d Fäßchen (halbes)

3487 first runnings
f têtes
e primeros productos de la destilación fraccionada
i prima frazione della distillazione;

prodotto di testa
d Vorlauf; Vorprodukt

3488 Fischer's reagent
f réactif de Fischer
e reactivo de Fischer
i reagente di Fischer
d Fischersches Reagens

3489 Fischer-Tropsch process
f procédé Fischer-Tropsch
e síntesis Fischer-Tropsch
i sintesi Fischer-Tropsch
d Fischer-Tropsch-Synthese

3490 fish-glue
f isinglass; colle de poisson
e cola de pez
i colla di pesce
d Fischleim

3491 Fittig's synthesis
f synthèse de Fittig
e síntesis de Fittig
i sintesi di Fittig
d Fittigsche Synthese

3492 fixatives
f fixateurs; agents fixatifs
e fijadores; mordentes
i fissatore
d Fixiermittel; Fixative

3493 fixed ammonia
f ammoniac combiné
e amoníaco combinado
i ammoniaca combinata
d gebundenes Ammoniak

3494 flake graphite
f graphite en flocons
e grafito lamelar; grafito cristalino
i grafite lamellare
d Flockengraphit

* **flake white** → 987

3495 flaking
f boursoufflure; écaillement
e exfoliación
i scagliatura
d Schuppenbildung; Abblättern

3496 flame test
f essai de coloration
e prueba a la llama
i analisi alla fiamma
d Flammenprobe

3497 flammability
f inflammabilité
e inflamabilidad
i infiammabilità
d Entflammbarkeit

3498 flash distillation
f distillation flash
e destilación en corriente de vapor
i distillazione equilibrata in corrente di vapore
d Gleichgewichtsdestillation

3499 flash dryer
f séchoir-éclair
e secador instantáneo
i seccatore rapido
d Schnelltrockner

3500 flash drying
f séchage instantané
e secado instantáneo
i essiccamento al lampo
d Schnelltrocknung

3501 flash evaporation
f évaporation instantanée
e evaporación instantánea
i evaporazione instantanea
d Schnellverdampfung

3502 flash-off
f séchage instantané
e vaporización instantánea
i vaporizzazione instantanea
d schnelles Trocknen

3503 flash plate; flash plating
f voile
e velo
i velo
d Schnellüberzug

3504 flash point
f point d'inflammabilité
e punto de inflamabilidad
i punto di infiammabilità
d Flammpunkt

3505 flask
f ballon; flacon
e frasco; matraz
i matraccio; pallone
d Flasche; Kolben

3506 flask clamp
f pince de support
e soporte de matraz

i morsetto per treppiede
d Ständerklemme

3507 flat v
f mater
e allanar
i spianare
d glätten

3508 flat bottomed flask
f ballon à fond plat
e matraz de fondo plano
i matraccio a fondo piatto
d Stehkolben

3509 flatting agents
f agents pour opacifier
e productos para opacificar
i agenti opacizzanti
d Mattierungsmittel

3510 flatting varnish
f vernis mat d'apprêt
e barniz mate para aparejos
i prima mano di vernice per farla essiccare
  senza lucido
d Grundierfarbe

3511 flat varnish
f vernis mat
e barniz opaco
i vernice opaca
d Mattlack

3512 flavanthrene
f flavanthrène
e flavantreno
i flavantrene
d Flavanthren

3513 flavine
f flavine
e flavina
i flavina
d Flavin

3514 flavone
f flavone
e flavona
i flavone
d Flavon

3515 flavoproteins
f flavoprotéines
e flavoproteínas
i flavoproteine
d Flavoproteine

3516 flint
f silex
e pedernal
i flint; selce
d Flintstein; Feuerstein

3517 flint glass
f flint
e flint-glass
i vetro flint
d Flintglas; Kieselglas

3518 flocculation
f floculation
e floculación
i flocculazione
d Ausflockung; Flockenbildung

3519 flocculent
f floconneux
e floculento
i fioccoso
d flocking

3520 flock v
f floquer
e aterciopelar; cubrir con algodón
i vellutare
d aufflocken

3521 flock
f flocon
e vorra; pelusilla
i cascami
d Ausschußwolle

3522 flocking
f flocage
e aterciopelado
i vellutazione
d Beflocken

3523 flock spraying
f floconnage
e rociadura de pelusilla
i spruzzatura di fiocco
d Flockieren

3524 flocs
f flocons; précipité
e copos; precipitado
i fiocchi
d Flocken

3525 flooding *(of distillation column)*
f noyage
e inundación
i inondazione
d Verwässerung

* **florentine lake** → 2169

3526 flotation
f flottation
e flotación
i flottazione
d Flotation; Schwimmverfahren

3527 flow
f fluidité; fluage
e capacidad de flujo
i fluidità
d Fließvermögen

3528 flow chart; flow sheet
f schéma de fabrication
e esquema de fabricación
i diagramma di lavorazione
d Fabrikationsschema

3529 flowers of sulphur
f fleurs de soufre
e flores de azufre
i fiori di zolfo
d Schwefelblüte

* **flow sheet** → 3528

3530 flow temperature
f température d'écoulement
e temperatura de fluidez
i temperatura di flusso
d Fließtemperatur

3531 flue gases
f fumées
e humos
i fumi
d Abgase; Rauchgase

3532 flue sweat
f bistre
e agua de condensación
i acqua di condensazione
d Schwitzwasser

3533 flue wall
f piédroit de chauffage
e pared de calefacción
i piedritto di riscaldamento
d Heizwand

3534 fluophosphoric acid
f acide fluophosphorique
e ácido fluofosfórico
i acido fluofosforico
d Fluorphosphorsäure

3535 fluoracetophenone
f fluoracétophénone
e fluoacetofenona
i fluoracetofenone
d Fluoracetophenon

3536 fluorescein
f fluorescéine
e fluoresceína
i fluoresceina
d Fluoreszein

3537 fluorescence
f fluorescence
e fluorescencia
i fluorescenza
d Fluoreszenz

3538 fluorescent pigments
f pigments fluorescents
e pigmentos fluorescentes
i pigmenti fluorescenti
d fluoreszierende Pigmente

3539 fluorinated ethylene propylene
f éthylène-propylène fluoré
e etileno y propileno fluorados
i etilene-propilene fluorinati
d Fluoräthylenpropylen

3540 fluorinated paraffin
f paraffine fluorinée
e parafina fluorada
i paraffina fluorata
d fluoriertes Paraffin

3541 fluorine
f fluor
e flúor
i fluoro
d Fluor

* **fluorite** → 3550

3542 fluoroaniline
f fluoraniline
e fluoanilina
i fluoroanilina
d Fluoranilin

3543 fluorobenzene
f fluorobenzène
e fluobenceno
i fluorobenzina
d Fluorbenzol

3544 fluoroboric acid
f acide fluoroborique
e ácido fluobórico

i acido fluoroborico
d Fluorborsäure

3545 fluorocarbon resins
f résines de fluorocarbone
e resinas de fluocarbono
i resine al fluorocarburo
d Fluorkohlenstoffharze

3546 fluorocarbons
f fluorocarbones
e fluocarbonos
i fluorocarburi
d Fluorcarbon; Fluorkohlenstoffe

3547 fluorochemicals
f composés fluorés
e productos químicos fluorados
i ▓▓▓▓▓▓▓ ▓▓▓▓▓▓ ▓▓▓▓▓▓▓▓▓
d Fluorchemikalien

3548 fluoroethylene
f fluoréthylène
e fluoretileno
i fluoroetilene
d Fluoräthylen

3549 fluorothene
f fluorothène
e fluoteno
i fluorotene
d Tetrafluoräthylen

3550 fluorspar; fluorite
f spath fluor; fluorite
e esparto flúor; fluorina
i fluorite
d Flußspat; Fluorit

3551 fluosilicic acid
f acide fluosilicique
e ácido fluosilícico
i acido fluosilicico
d Kieselfluorwasserstoffsäure

3552 fluosulphonic acid
f acide fluosulfonique
e ácido fluosulfónico
i acido fluosolfonico
d Fluorsulfonsäure

3553 flux
f fondant; flux
e fundente; fluidificante
i fondente
d Schmelzmittel

3554 fly ash
f cendres volantes
e cenizas volátiles
i ceneri volanti
d Flugasche

3555 foam glue
f colle battue en mousse
e cola de espuma
i colla a schiuma
d Schaumkleber

3556 foaming agent
f agent moussant
e agente espumador; agente de alveolación
i agente schiumogeno
d Schaummittel

3557 foaming process
f méthode de fabrication de mousse
e proceso de espumación
i procedimento d'espansione
d Schäumverfahren

3558 focal length
f distance focale
e distancia focal
i distanza focale
d Brennweite

3559 folic acid
f acide folique
e ácido fólico
i acido folico
d Folinsäure

3560 foots
f dépôts sédiments
e sedimentos
i morchia
d Bodensatz

3561 foot-valve
f clapet de pied
e válvula de pie; válvula de aspiración
i valvola di fondo
d Fußventil

3562 forced circulation
f circulation forcée
e circulación forzada
i circolazione forzata
d Zwangsumlauf

3563 forced draught
f tirage forcé
e tiro forzado
i tiraggio forzato
d Druckluftstrom

3564 forced drying temperature
f température de séchage accéléré
e temperatura de desecación acelerada
i temperatura di essicamento accelerato
d Temperatur bei
Beschleunigungstrocknung

3565 forceps
f pincette
e pinzas
i pinzetta
d Federzange; Pinzette

3566 forecooler
f préréfrigérant
e preenfriador
i preraffreddatore
d Vorkühler

3567 fore-runnings; foreshots
f produit de tête
e producto de cabeza
i prodotto di testa
d Kopfprodukt; Vorlauf

3568 fore vacuum
f vide préalable à basse pression
e vacío preliminar
i prevuoto
d niedriges Vorvakuum

* formal → 5312

3569 formaldehyde
f formaldéhyde
e aldehído fórmico
i formaldeide
d Formaldehyd

3570 formaldehyde aniline
f aniline de formaldéhyde
e formaldehído anilina
i anilina alla formaldeide
d Formaldehydanilin

3571 formalin
f formaline
e formalina
i formalina
d Formalin

3572 formalize v
f formoler
e formaldehidar; tratar con formaldehído
i trattare con formaldeide
d mit Formaldehyd härten

3573 formamide
f formamide; formiamide
e formamida
i formammide
d Formamid

3574 formic acid
f acide formique
e ácido fórmico
i acido formico
d Ameisensäure

3575 formicin
f formicine
e formicina
i formicina
d Formicin

3576 formula
f formule
e fórmula
i formula
d Formel

3577 formula weight
f poids de formule
e peso de fórmula
i pesi atomici complessivi degli elementi di
una formula
d Formelgewicht

3578 formyl fluoride
f fluorure de formyle
e fluoruro de formilo
i fluoruro di formile
d Formylfluorid

3579 fortify v
f fortifier
e fortificar
i fortificare
d verstärken

3580 foul electrolyte
f électrolyte impropre
e electrólito impuro
i elettrolito impuro
d verbrauchter Elektrolyt

3581 four-stage compressor
f compresseur à quatre étages
e compresor de cuatro escalones
i compressore quadrifase
d vierstufiger Kompressor

3582 four-way cross
f croisement double
e cruzamiento doble
i doppio incrocio
d doppelte Kreuzung

3583 **Fowler's solution**
 f liqueur de Fowler
 e solución de Fowler
 i soluzione di Fowler
 d Fowlersche Lösung

3584 **fractional crystallization**
 f cristallisation fractionnée
 e cristalización fraccionada
 i cristallizzazione frazionata
 d Umkristallisation

3585 **fractional distillation**
 f distillation fractionnée
 e destilación fraccionada
 i distillazione frazionata
 d fraktionierte Destillation

3586 **fractionation**
 f fractionnement
 e fraccionamiento
 i frazionamento
 d Fraktionieren

3587 **fragility**
 f fragilité
 e fragilidad
 i fragilità
 d Zerbrechlichkeit

3588 **fragmentation**
 f fragmentation
 e fragmentación
 i frammentazione
 d Fragmentation

3589 **francium**
 f francium
 e francio
 i francio
 d Francium

3590 **Frankfort black**
 f noir d'Allemagne
 e negro de Franckfort
 i nero di Francoforte
 d Frankfurterschwarz

3591 **franklinite**
 f franklinite
 e franklinita
 i franklinite
 d Franklinit; Zinkeisenerz

3592 **Frasch process**
 f procédé Frasch
 e proceso Frasch
 i processo Frasch
 d Frasch-Verfahren

3593 **free air dose**
 f dose dans l'air
 e dosis en aire
 i dose in aria
 d Luftdosis

3594 **free ammonia**
 f ammoniac libre
 e amoníaco libre
 i ammoniaca libera
 d freies Ammoniak

3595 **free cyanide**
 f cyanure libre
 e cianuro libre
 i cianuro libero
 d freies Cyanid

 * **free flowing** → 7634

3596 **free phenol**
 f phénol libre
 e fenol libre
 i fenolo libero
 d freies Phenol

3597 **free radical**
 f radical libre
 e radical libre
 i radicale libero
 d freies Radikal

3598 **free sulphur**
 f soufre libre
 e azufre libre
 i zolfo libero
 d freier Schwefel

3599 **freeze-thaw stable**
 f résistant au gel et à la rosée
 e resistente a la congelación y deshielo cíclicos
 i stabile al congelamento e disgelamento ripetuto
 d frost- und tauwasserbeständig

3600 **freezing point**
 f point de congélation
 e punto de congelación
 i punto di congelamento
 d Gefrierpunkt

3601 **freibergite**
 f freibergite
 e freibergita
 i freibergite
 d Freibergit; Silberfahlerz

 * **French blue** → 8601

3602 French chalk
f talc; craie de Meudon
e talco; yeso de sastres
i gesso francese
d Talkum

3603 French ochre
f ocre
e ocre
i ocra francese
d Ocker

3604 friability
f friabilité
e friabilidad
i fragilità
d Brüchigkeit

3605 friction head
f perte de charge par la friction
e pérdida de carga por rozamiento
i perdità di prevalenza per attrito
d Widerstandshöhe

3606 Friedel-Crafts reaction
f réaction Friedel-Crafts
e reacción de Friedel-Crafts
i reazione Friedel-Crafts
d Friedel-Craftsche-Reaktion

3607 frit
f fritte
e frita
i fritta
d Fritte

3608 frit v
f fritter
e fritar
i formare scorie
d fritten

3609 fritted glass filtering crucible
f creuset de filtrage à verre fritté
e crisol para filtrar de vidrio fritado
i crogiuolo di fusione a dischi di fritta
d Frittglasfiltertiegel

3610 frosting
f matage
e deslustración
i inzuccheramento
d Mattieren

3611 fructose
f fructose
e fructosa
i fruttosio
d Fructose; Fruchtzucker

3612 fuchsin
f fuchsine
e fucsina
i fucsina
d Fuchsin

3613 fuel oil
f mazout
e aceite pesado
i olio combustibile pesante
d Heizöl; Schweröl

3614 fugitive pigments
f pigments fugitifs
e pigmentos fugitivos
i pigmenti che sbiadiscono
d lichtunechte Pigmente

3615 fugitometer
f fugitomètre
e fugitómetro
i apparecchio per la misurazione dello
  sbiadimento
d Lichtechtheitsmesser

3616 full-automatic electroplating
f galvanoplastie automatique
e galvanoplastia automática
i galvanoplastica automatica
d vollautomatische Galvanisierung

3617 Fuller's cell
f élément de Fuller
e célula de Fuller
i cella di Fuller
d Fullersches Element

3618 Fuller's earth
f terre à foulon
e tierra decolorante
i terra per folloni
d Fullererde

3619 fulminates
f fulminates
e fulminatos
i fulminati
d Fulminate

3620 fumaric acid
f acide fumarique
e ácido fumárico
i acido fumarico
d Fumarsäure

3621 fumaryl chloride
f chlorure de fumaryle
e cloruro de fumarilo

i cloruro di fumarile
d Fumarylchlorid

3622 fume cupboard
f canal d'aspiration
e conducto de ventilación
i cappa d'aspirazione
d Abzug

3623 fumigants
f produits fumigatoires
e productos fumigantes
i fumiganti
d Räuchermittel

3624 fuming sulphuric acid; oleum
f acide sulfurique fumant; oléum
e ácido sulfúrico fumante
i acido solforico fumante; olio vetriolo
d rauchende Schwefelsäure; Oleum

3625 functional diploid
f diploïde fonctionnel
e diploide functional
i diploide funzionale
d funktionelle Diploide

3626 fundamental particle
f particule élémentaire
e partícula fundamental
i particella fondamentale
d Grundteilchen

3627 fungicidal paints
f peintures fungicides
e pinturas fungicidas
i vernici fungicide
d Farben zur Verhinderung von
   Schimmelbildung

3628 fungicides
f fungicides
e fungicidas
i fungicidi
d Fungizide; Schimmelvernichtungsmittel

3629 fungistats
f produits fungicides
e fungistatos
i preventivi fungicidi
d Fungistate

3630 funnel holder
f support de filtre
e soporte del filtro
i sostegno del filtro
d Filtergestell

3631 furan
f furanne
e furano
i furano
d Furan

3632 furan resins
f résines furanniques
e resinas de furano
i resine al furano
d Furanharze

3633 furfural
f furfural
e furfural
i furfurolo
d Furfural

3634 furfuryl acetate
f acétate de furfuryle
e acetato de furfurilo
i acetato di furfurilo
d Furfurylacetat

3635 furfuryl alcohol
f alcool furfurylique
e alcohol furfurílico
i alcool furfurilico
d Furfurylalkohol

3636 furnace black
f noir fourneau
e negro de horno
i nero di forno
d Ofenruß

3637 furnace charge
f charge d'un four
e carga
i carica
d Charge; Gut

3638 furylacrylic acid
f acide furylacrylique
e ácido furilacrílico
i acido furilacrilico
d Furylacrylsäure

3639 fusain
f fusain
e fuseno
i fusano
d Fusit; Faserkohle

3640 fused driers
f siccatifs fondus
e secantes fundidos
i essiccatori per refusione
d geschmolzene Trockenmittel

3641 fused electrolyte
 f électrolyte fondu
 e electrólito fundido
 i elettrolito fuso
 d Schmelze

3642 fused electrolyte cell
 f pile à électrolyte fondu
 e pila de electrólito fundido
 i pila a elettrolito fuso
 d Hochtemperaturelement

3643 fusel oil
 f huile de fusel
 e aceite de fusel
 i olio di flemma
 d Fuselöl

3644 fusible alloys
 f alliages fusibles
 e aleaciones fusibles
 i leghe fusibili
 d Schmelzlegierungen

3645 fusiform
 f fusiforme
 e fusiforme
 i fusiforme
 d spindelförmig

3646 fusion; melting
 f fusion
 e fusión
 i fusione
 d Schmelzung

3647 fusion nucleus
 f noyau de fusion
 e núcleo de fusión
 i nucleo di fusione
 d Fusionskern

3648 fusion process
 f procédé de fonte; procédé de fusion
 e proceso de fusión
 i processo di fusione
 d Schmelzverfahren

3649 fustic
 f fustet; bois jaune
 e fustete
 i fustetto
 d Fustikholz; Gelbholz

G

3650 gadolinite
f gadolinite
e gadolinita
i gadolinite
d Gadolinit

3651 gadolinium
f gadolinium
e gadolinio
i gadolinio
d Gadolinium

3652 gadolinium oxide
f oxyde de gadolinium
e óxido de gadolinio
i ossido di gadolinio
d Gadoliniierde

3653 galactose
f galactose
e galactosa
i galattosio
d Galaktose

3654 galena
f galène
e galena
i galena
d Galenit; Bleiglanz

3655 galipot; Bordeaux turpentine
f poix de Bourgogne; galipot
e galipodio; trementina de Burdeos
i galipot
d Galipot

3656 gall
f bile
e bilis
i bile
d Galle

3657 gallic acid
f acide gallique
e ácido gálico
i acido gallico
d Gallussäure

  * **galliolino → 5652**

3658 gallium
f gallium
e galio
i gallio
d Gallium

3659 gallocyanine
f gallocyanine
e galocianina
i gallocianina
d Gallocyanin

3660 galvanizing
f galvanisation
e galvanización; cincado
i galvanizzazione; zincatura
d Galvanisierung; Verzinkung

3661 gamboge
f gomme-gutte
e gomaguta
i gommagutta
d Gummigutt

3662 gamete
f gamète
e gameto
i gameto
d Gamet

3663 gametic
f gamétique
e gamético
i gametico
d gametisch

3664 gametic chromosome number
f nombre de chromosomes du gamète
e nombre de cromosomas gaméticos
i numero cromosomico del gameto
d gametische Chromosomenzahl

3665 gametocyte
f gamétocyte
e gametocito
i gametocito
d Gametocyte

3666 gametoid
f gametoïde
e gametoide
i gametoide
d Gametoid

3667 gamma brass
f laiton gamma
e latón gamma
i ottone gamma
d Gammamessing

3668 gamma compounds
f composés gamma
e compuestos gamma
i composti gamma
d Gammaverbindungen

3669 gamma iron
f fer gamma
e hierro gamma
i ferro gamma
d Gammaeisen; Austenit

3670 gamodeme
f gamodème
e gamodemo
i gamodemo
d Gamodeme

3671 gangliocyte
f cellule ganglionnaire
e gangliocito
i gangliocita
d Ganglienzelle

3672 ganister; gannister
f ganister
e ganíster
i ganisto
d Ganister

3673 gap-filling adhesive
f mastic bouche-pores
e adhesivo para unión de superficies
  separadas
i adesivo per riempire di soluzioni di
  continuità
d Fugenkitt

3674 Gardner-Holt viscosity tubes
f tubes Gardner-Holt de mesure de
  viscosité
e tubos de viscosidad Gardner-Holt
i tubi Gardner-Holt per misurare la
  viscosità
d Gardner-Holt-Viskositätsröhren

3675 garlic oil
f essence d'ail
e aceite de ajo
i olio essenziale di aglio
d Knoblauchöl

3676 garnet
f grenat
e granate
i granato
d Granat

3677 garnet lac
f laque grenat
e laca granate
i gommalacca granata
d roter Schellack

3678 garnierite
f garniérite
e garnierita
i garnierite
d Garnierit

3679 gas
f gaz
e gas
i gas
d Gas

3680 gas analyzer
f analyseur de gaz
e analizador de gas
i analizzatore di gas
d Gasanalysengerät

3681 gas balance
f balance à gaz
e balanza para gas
i bilancia per gas
d Gaswaage

3682 gas black
f noir de gaz
e negro de gas
i nero da gas
d Gasruß

3683 gas burette
f burette à gaz
e probeta de gas
i buretta a gas
d Gasmeßröhre; Gasbürette

3684 gas calorimeter
f calorimètre analyseur de gaz
e calorímetro analizador de gases
i calorimetro analizzatore di gas
d Gaskalorimeter

3685 gas cell
f pile à gaz
e pila de gas
i pila a gas
d Brennstoffelement

3686 gas chromatography
f chromatographie des gaz
e cromatografía a gas
i cromatografia a gas
d Gaschromatographie

3687 gas coke
f coke de gaz
e coque de gas
i coke di gas
d Gaskoks

3688 gas density meter
  f densimètre
  e densímetro
  i densimetro
  d Dichtenmesser

3689 gas discharge gauge
  f manomètre à décharge lumineuse
  e manómetro de descarga luminosa
  i manometro a scarica luminosa
  d Gasentladungsmanometer

3690 gas expansion thermometer
  f thermomètre à dilatation de gaz
  e termómetro de dilatación de gas
  i termometro a dilatazione di gas
  d Gasthermometer

3691 gas grooves
  f ondulations dues aux gaz
  e ondulaciones debidas a los gases
  i difetti da gas
  d Gasmarken

3692 gas hydrates
  f hydrates de gaz
  e hidratos de gas
  i gas idrati
  d Gashydrate

3693 gas lime
  f chaux de gaz
  e cal de gas
  i calce per la purificazione del gas
  d Gaskalk

3694 gas liquor
  f eau de gazomètre; eau ammoniacale
  e agua de gas; agua amoniacal
  i acqua ammoniacale
  d Gaswasser

3695 gas oil
  f gasoil; gasole
  e gasoil
  i gasolio
  d Gasöl

3696 gasoline
  f essence de pétrole
  e gasolina; esencia de petróleo
  i benzina
  d Benzin

3697 gas permeability
  f perméabilité au gas
  e permeabilidad al gas
  i permeabilità al gas
  d Gasdurchlässigkeit

3698 gas poisoning
  f asphyxie
  e intoxicación por respirar gases
  i intossicazione da gas
  d Vergasen

3699 gas-proof
  f à l'épreuve des gaz
  e a prueba de gas
  i a prova di gas
  d gasbeständig

3700 gas scrubber
  f laveur de gaz
  e lavador de gas; lavagases
  i depuratore del gas
  d Gaswäscher; Skrubber

3701 gassing
  f ébullition
  e desprendimiento de gases
  i ebollizione
  d Gasentwicklung

3702 gas thermometer
  f thermomètre à gaz
  e termómetro de gas
  i termometro a gas
  d Gasthermometer

3703 gas welding
  f soudure au gaz
  e soldadura con gas
  i saldatura per fusione
  d Gasschweißen

3704 gate paddle agitator
  f agitateur à treillis; agitateur à grille
  e mezcladora de paletas en forma de
     parrilla
  i mescolatore a pale in forma di griglia
  d Gitterrührer

3705 gauge
  f jauge; mesure
  e calibre
  i calibro
  d Eichmaß; Lehre

3706 gauge cock
  f robinet de jauge; robinet d'essai
  e grifo de ensayo
  i rubinetto di prova
  d Probierhahn

3707 Gay-Lussac acid
  f acide de Gay-Lussac
  e ácido de Gay-Lussac

i acido di Gay-Lussac
d Gay-Lussacsche Säure

3708 Gay-Lussac's law of volumes
f loi de Gay-Lussac
e ley de Gay-Lussac
i legge di Gay-Lussac
d Gay-Lussacsches Gesetz

3709 Gay-Lussac tower
f tour de Gay-Lussac
e torre de Gay-Lussac
i torre di Gay-Lussac
d Gay-Lussacscher Turm

3710 gaylussite
f gaylussite
e gaylusita
i gaylussite
d Gaylussit

3711 Geissler pump
f pompe Geissler
e bomba de Geissler
i pompa di Geissler
d Geissler Pumpe

3712 gel
f gel
e gel
i gel
d Gel

3713 gelatin
f gélatine
e gelatina
i gelatine
d Gelatine

3714 gelsemine
f gelsémine
e gelsemina
i gelsemina
d Gelsemin

3715 gel time
f temps de gélification
e tiempo de gelificación
i tempo di gelificazione
d Gelbildungszeit

3716 gene
f gène
e gen
i gene
d Gen

3717 gene chromatin
f chromatine du gène

e cromatina génica
i cromatina genica
d Genchromatin

3718 gene-cytoplasm isolation
f séparation du gène du cytoplasme
e aislamiento genocitoplásmico
i isolamento del gene dal citoplasma
d Gen-Cytoplasma-Isolation

3719 gene interaction
f interaction des gènes
e interacción de los genes
i interazione fra geni
d Wechselwirkung zwischen Genen

3720 general stain
f coloration générale
e coloración general
i colorazione generale
d gleichmäßige Färbung

3721 generative nucleus
f noyau génératif
e núcleo generativo
i nucleo generativo
d generativer Kern

3722 gene string
f chromatide; cordon de gènes
e filamento de los genes
i filamento dei geni
d Gen-"string"

3723 gene tagged chromosomes
f chromosomes couverts de gènes
e cromosomas cubiertos de genes
i cromosomi coperti di geni
d genetisch markierte Chromosomen

3724 genetic background
f milieu génétique
e ambiente genético
i ambiente genetico
d genotypisches Milieu

3725 genetic variation
f variation génotypique
e variación genotípica
i variazione genotipica
d genetische Variation

3726 genoblast
f génoblaste
e genoblasto
i genoblasto
d reife Geschlechtszelle

3727 genoid
 f génoïd
 e genoide
 i genoide
 d Genoid

3728 genome
 f génome; garniture chromosomique
 e genomio; genomia
 i genoma; genomio
 d Genom

3729 genome analysis
 f analyse génomique
 e análisis genómico
 i analisi genomica
 d Genomananalyse

3730 genosome
 f génosome
 e genosoma
 i genosoma
 d Genosom

3731 gentian
 f gentiane
 e genciana
 i genziana
 d Gentian; Enzianbitter

3732 geometric isomerism
 f isomérie géométrique
 e isomerismo geométrico
 i isomerismo geometrico
 d geometrische Isomerie

3733 geraniol
 f géraniol
 e geraniol
 i geraniolo
 d Geraniol

3734 geraniol butyrate
 f butyrate de géraniol
 e butirato de geraniol
 i butirrato geraniolico
 d Geraniolbutyrat

3735 geraniol formate
 f formiate de géraniol
 e formiato de geraniol
 i formiato di geraniolo
 d Geraniolformiat

3736 geranyl acetate
 f acétate de géranyle
 e acetato de geranilo
 i geranilacetato
 d Geranylacetat

3737 germ
 f germe
 e germen
 i germe
 d Spross

3738 germanium
 f germanium
 e germanio
 i germanio
 d Germanium

3739 germanium oxide
 f oxyde de germanium
 e óxido de germanio
 i ossido di germanio
 d Germaniumoxyd

* German silver → 5723

3740 germ carrier
 f porteur de germes
 e portador de gérmenes
 i portatore di germi
 d Bazillenträger; Keimträger

3741 germicide; microbicide
 f germicide
 e germicida
 i germicida
 d mikrobizides Mittel; Keimtötungsmittel

3742 germ plasm
 f germen; plasma germinal
 e plasma germinal
 i plasma germinale
 d Keimplasma

3743 giant chromosome
 f chromosome géant
 e cromosoma gigante
 i cromosoma gigantico
 d Riesenchromosom

3744 gibberellic acid
 f acide gibbérellique
 e ácido giberélico
 i acido gibberellico
 d Gibberellinsäure

3745 gigantocyte
 f gigantocyte
 e gigantocito
 i cellula gigante
 d Gigantozyt

3746 gilsonite
 f gilsonite
 e gilsonita

i gilsonite
d Gilsonit

3747 ginger
f gingembre
e ginebra
i zenzero
d Ingwer

3748 Girbotol process
f procédé Girbotol
e método de absorción Girbotol
i processo Girbotol
d Girbotol-Absorptionsverfahren

3749 gitoxin
f gitoxine
e gitoxina
i gitossina
d Gitoxin

3750 glacial acetic acid
f acide acétique cristallisable; vinaigre
glacial
e ácido acético glacial
i acido acetico cristallizzabile
d Eisessig; kristallisierte Essigsäure

3751 gland
f glande
e glándula
i ghiandola
d Drüse

3752 glaserite
f glasérite
e glaserita
i glaserite
d Glaserit

3753 glass
f verre
e vidrio
i vetro
d Glas

3754 glass cloth
f tissu de verre
e tela de fibra de vidrio
i tessuto di vetro
d Glasgewebe

* **glass electrode** → 3756

3755 glass fibre
f fibre de verre
e fibra de vidrio
i fibra di vetro
d Glasfaser

3756 glass half-cell; glass electrode
f demi-cellule en verre
e semicelda de vidrio
i elettrodo a vetro
d Glashalbzelle

3757 glass rod
f agitateur en verre
e varilla de vidrio
i bacchetta di vetro per agitare
d Rührstab; Glasstab

3758 glass tube
f tube en verre
e tubo de vidrio
i tubo di vetro
d Glasröhre

3759 glauberite
f glaubérite
e glauberita
i glauberite
d Glauberit

3760 Glauber's salt
f sel de Glauber
e sal de Glauber
i sale di Glauber
d Glaubersalz

3761 glaucodot; glaucodote
f glaucodot
e glaucodota
i glaucodote
d Glaukodot

3762 glauconite
f glauconite
e glauconita
i glauconite
d Glaukonit

3763 glaze
f glaçure
e vidriado
i smalto a vetrino
d Glasur

3764 glazing
f émaillage
e vidriado
i lucidatura
d Glasieren

3765 globule
f globule
e glóbulo
i globulo
d Globulus; Kügelchen

3766 globulins
f globulines
e globulinas
i globuline
d Globuline

3767 gloss
f lustre
e brillo; lustre
i lucentezza
d Glanz

3768 glossimeter
f brillancemètre
e brillancímetro
i misuratore della brillantezza
d Glanzmesser

3769 gloss oil
f huile vernic polic
e aceite para barnizar
i copalina
d Glanzöl

3770 gloss white
f blanc brillant
e blanco brillante
i vernice di fondo di solfato di barium e
  allumina
d Glanzweiß

3771 glost firing
f émaillage au feu
e cocción de vidriado
i cottura per smalto a vetrino
d Glasurbrand

3772 glost kiln
f four à émailler
e horno de vidriar
i forno per smalto a vetrino
d Glasurofen

3773 Glover tower
f glover; tour de Glover
e torre de Glover
i torre di Glover
d Gloverturm

3774 glucagon
f glucagon
e glucagón
i glucagone
d Glukagon

3775 gluconic acid
f acide gluconique
e ácido glucónico

i acido gluconico
d Gluconsäure

* glucosides → 3816

3776 glucuronolactone
f glucuronolactone
e glucuronolactona
i glucuronolattone
d Glucuronlacton

3777 glue
f colle
e cola; pasta adhesiva
i colla; sostanza glutinosa
d Leim; Klebstoff; Kleister

3778 glue penetration
f pénétration de colle
e penetración de la cola
i penetrazione della colla
d Ausschwitzen des Klebstoffes

3779 glutamic acid
f acide glutamique
e ácido glutámico
i acido glutammico
d Glutaminsäure

3780 glutamine
f glutamine
e glutamina
i glutammina
d Glutamin

3781 glutaraldehyde
f aldéhyde glutarique
e aldehído glutárico
i glutaraldeide
d Glutaraldehyd

3782 glutaric acid
f acide glutarique
e ácido glutárico
i acido glutarico
d Glutarsäure

3783 glutaric anhydride
f anhydride glutarique
e anhídrido glutárico
i anidride glutarica
d Glutarsäureanhydrid

3784 glutaronitrile
f glutaronitrile
e glutaronitrilo
i glutaronitrile
d Glutaronitril

3785 **gluten**
 f gluten; phytocolle
 e gluten
 i glutine
 d Gluten; Glutin

3786 **glyceraldehyde**
 f glycéraldéhyde
 e gliceraldehído
 i gliceraldeide
 d Glycerinaldehyd

3787 **glyceride**
 f glycéride
 e glicerido
 i gliceride
 d Glycerid

3788 **glycerin carbonate**
 f carbonate de glycérine
 e carbonato de glicerina
 i carbonato di glicerina
 d Glycerincarbonat

3789 **glycerol; glycerin(e)**
 f glycérol; glycérine
 e glicerina; glicerol
 i glicerina
 d Glycerin; Glyzerin

3790 **glycerol boriborate**
 f boriborate de glycérine
 e boriborato de glicerina
 i boriborato di glicerina
 d Glycerinboriborat

3791 **glycerol diacetate**
 f diacétate de glycérine
 e diacetato de glicerina
 i diacetato di glicerina
 d Glyceryldiacetat

3792 **glycerol ether acetate**
 f acétate de l'éther glycérique
 e acetato del éter de glicerina
 i etere glicerico acetato
 d Glycerinätheracetat

3793 **glycerol monoacetate**
 f monoacétate de glycérine
 e monoacetato de glicerina
 i monoacetato di glicerina
 d Glycerylmonoacetat

3794 **glycerol monolaurate**
 f monolaurate de glycérine
 e monolaurato de glicerina
 i monolaurato di glicerolo
 d Glycerinmonolaurat

3795 **glycerol monoricinoleate**
 f monoricinoléate de glycérine
 e monorricinoleato de glicerina
 i monoricinoleato di glicerolo
 d Glycerinmonoricinoleat

3796 **glycerol monostearate**
 f monostéarate de glycérine
 e monoestearato de glicerina
 i monostearato di glicerolo
 d Glycerinmonostearat

3797 **glycerol triacetate**
 f triacétate de glycérine
 e triacetato de glicerina
 i triacetato di glicerina; triacetina
 d Glyceryltriacetat

3798 **glycerophosphoric acid**
 f acide glycérophosphorique
 e ácido glicerofosfórico
 i acido glicerofosforico
 d Glycerophosphorsäure

3799 **glyceryl abietate**
 f abiétate de glycéryle
 e abietato de glicerilo
 i glicerolabietato
 d Glycerylabietat

3800 **glyceryl phthalate**
 f phtalate de glycéryle
 e ftalato de glicerilo
 i ftalato di glicerile; resina gliceroftalica
 d Glycerylphthalat

3801 **glyceryl triacetoxystearate**
 f triacétoxystéarate de glycéryle
 e triacetoxiestearato de glicerilo
 i triacetossistearato di glicerile
 d Glyceryltriacetoxystearat

3802 **glyceryl triacetylricinoleate**
 f triacétylricinoléate de glycéryle
 e triacetilricinoleato de glicerilo
 i triacetilricinoleato di glicerile
 d Glyceryltriacetylricinoleat

3803 **glyceryl tributyrate**
 f tributyrate de glycéryle
 e tributirato de glicerilo
 i tributirrato di glicerile
 d Glyceryltributyrat

3804 **glyceryl trihydroxystearate**
 f trihydroxystéarate de glycéryle
 e trihidroxiestearato de glicerilo
 i triidrossistearato di glicerile
 d Glyceryltrihydroxystearat

* **glyceryl trinitrate** → 5773

3805 **glyceryl tripropionate**
    f tripropionate de glycéryle
    e tripropionato de glicerilo
    i tripropionato di glicerile
    d Glyceryltripropionat

3806 **glyceryl triricinoleate**
    f triricinoléate de glycéryle
    e trirricinoleato de glicerilo
    i triricinoleato di glicerile
    d Glyceryltriricinoleat

3807 **glycidol**
    f glycidol; glycide
    e glicidol
    i glicidolo; glicide
    d Glycidol

3808 **glycine**
    f glycine
    e glicocola
    i glicocolla
    d Glykokol; Leimzucker

3809 **glycine ethyl ester hydrochloride**
    f chlorhydrate éthylestérique de glycine
    e hidrocloruro del etiléster de la glicina
    i cloridrato etilesterico di glicina
    d Glycinäthylesterhydrochlorid

3810 **glycogen**
    f glycogène
    e glucógeno
    i glicogeno
    d Glykogen

3811 **glycol dimercaptoacetate**
    f dimercaptoacétate de glycol
    e dimercaptoacetato de glicol
    i dimercaptoacetato di glicolo
    d Glykoldimercaptoacetat

3812 **glycolic acid**
    f acide glycolique
    e ácido glicólico
    i acido glicolico
    d Glykolsäure

3813 **glycolonitrile**
    f glycolonitrile
    e glicolonitrilo
    i glicolonitrile
    d Glykolonitril

3814 **glycol salicylate**
    f salicylate de glycol
    e salicilato de glicol
    i glicolsalicilato
    d Glykolsalicylat

3815 **glycosecretory**
    f glucosécrétoire
    e glicosecretorio
    i glicosecretorio
    d zuckersekretorisch

3816 **glycosides; glucosides**
    f glucosides
    e glicósidos; glucósidos
    i glucosidi
    d Glucoside; Glycoside

3817 **glycyrrhiza**
    f glycyrrhizine
    e glicirrina
    i glicirrizia
    d Glycyrrhizin; Süßholzzucker

3818 **glyodin**
    f glyodine
    e gliodina
    i gliodina
    d Glyodin

3819 **glyoxal**
    f glyoxal
    e glioxal
    i glyossale
    d Glyoxal

* **glyptal resins** → 339

3820 **goethite**
    f goethite
    e goetita
    i goethite
    d Goethit

3821 **gold**
    f or
    e oro
    i oro
    d Gold

3822 **gold bronze**
    f bronze doré
    e bronce dorado
    i bronzo dorato
    d Goldbronze

3823 **gold chloride**
    f chlorure d'or; aurichlorure
    e cloruro de oro
    i cloruro d'oro
    d Goldchlorid; Aurichlorid

3824 **gold-film glass**
 f verre à pellicule d'or
 e vidrio con una película de oro
 i vetro a pellicola d'oro
 d Goldfilmglas

3825 **gold hydroxide**
 f hydroxyde d'or
 e hidróxido de oro
 i idrossido d'oro
 d Goldhydroxyd

3826 **gold leaf**
 f or en feuilles
 e hoja de oro
 i foglia d'oro
 d Blattgold

3827 **gold oxide**
 f oxyde d'or
 e óxido de oro
 i ossido d'oro
 d Goldoxyd

3828 **gold potassium chloride**
 f chlorure d'or et de potassium
 e cloruro de oro y potasio
 i cloruro d'oro e potassio
 d Goldkaliumchlorid

3829 **gold size**
 f or couleur
 e cola para dorado
 i vernice a dorare
 d Goldleim

3830 **gold sodium chloride**
 f chlorure d'or et de sodium
 e cloruro de oro y sodio
 i cloruro di oro e sodio
 d Goldnatriumchlorid

3831 **gold sodium thiomalonate**
 f thiomalonate d'or et de sodium
 e tiomalonato de oro y sodio
 i tiomalonato d'oro e sodio
 d Goldnatriumthiomalonat

3832 **gold stoving varnish**
 f vernis doré à l'étuvage
 e barniz dorado a la estufa
 i doratura a fuoco
 d Goldeinbrennfirnis

3833 **gold tellurides**
 f tellurures d'or
 e telururos de oro
 i telluluri d'oro
 d Goldtellurüre

3834 **gold-tin purple**
 f pourpre de Cassius
 e púrpura de oro y estaño
 i porpora di Cassius
 d Goldpurpur

3835 **gold tribromide**
 f tribromure d'or
 e tribromuro de oro
 i tribromuro d'oro
 d Goldtribromid

3836 **Golgi apparatus**
 f appareil de Golgi
 e aparato de Golgi
 i apparato di Golgi
 d Golgi-Apparat

3837 **golgiolysis**
 f golgiolyse
 e golgiólisis
 i golgiolisi
 d Golgiolysis

3838 **gonad**
 f gonade
 e gónada
 i gonade
 d Gonade

3839 **gone**
 f gonie
 e gonio
 i gonio
 d Gone

3840 **gonocele**
 f spermatocèle
 e gonocele
 i spermatocele
 d Samenbruch

3841 **gonocyte**
 f gonocyte
 e gonócito
 i gonocita
 d Gonocyte

3842 **gonomere**
 f gonomère
 e gonómero
 i gonomero
 d Gonomere

3843 **gonosome**
 f gonosome
 e gonosoma
 i gonosoma
 d Gonosom

3844 **Gooch crucible**
  f creuset de Gooch
  e crisol de Gooch
  i crogiuolo di Gooch
  d Googscher Tiegel; Goochtiegel

3845 **Gossage's process**
  f procédé Gossage
  e proceso Gossage
  i processo Gossage
  d Gossagsches Verfahren

3846 **gossypol**
  f gossypol
  e gosipol
  i gossipolo
  d Gossypol

3847 **graduated pipette**
  f pipette graduée
  e pipeta de medida
  i pipetta di misurazione
  d Meßpipette

3848 **graft copolymer; graft polymer**
  f copolymère à greffe
  e copolímero de injerto
  i copolimero a innesto
  d Propfcopolymer

3849 **grahamite**
  f grahamite
  e grahamita
  i grahamite
  d Grahamit

3850 **graining**
  f veinage; marbrage
  e decoración imitando madera
  i verniciare simulando la grana del legno
  d Maserung

3851 **graining point**
  f point de cristallisation; point de
    granulation
  e punto de cristalización
  i punto di cristallizzazione
  d Granulierpunkt

3852 **gram-atom**
  f gramme-atome
  e átomo-gramo
  i grammoatomo
  d Grammatom

3853 **gram-equivalent**
  f gramme-équivalent
  e gramo-equivalente

  i grammo-equivalente
  d Grammäquivalent

3854 **gramicidin**
  f gramicidine
  e gramicidina
  i gramicidina
  d Gramicidin

3855 **gram-molecular volume**
  f volume d'un gramme-molécule
  e volumen gramomolecular
  i volume grammomolecolare
  d Grammolekülvolumen

3856 **gram-molecule**
  f molécule-gramme
  e molécula-gramo
  i grammo molecola
  d Grammolekül, Mol

3857 **Gram-negative**
  f Gram-négatif
  e Gram negativo
  i Gram-negativo
  d Gram-negativ

3858 **Gram-positive**
  f Gram-positif
  e Gram-positivo
  i Gram-positivo
  d Gram-positiv

3859 **granular ash**
  f cendre granulaire
  e ceniza granular
  i cenere granulare
  d körnige Asche

3860 **granulates**
  f granulés
  e granulados
  i granulati
  d Granulate, Körnchen

3861 **granule**
  f granule
  e gránulo
  i granulo
  d Korn

3862 **graphite; plumbago**
  f graphite; plombagine
  e grafito; plombagina
  i grafite; piombaggine
  d Graphit

3863 **gravimeter**
  f gravimètre

e gravímetro
i gravimetro
d Gravimeter

3864 gravimetric analysis
f analyse pondérale
e análisis gravimétrico
i analisi gravimetrica
d Gewichtsanalyse

3865 gravity filter
f filtre par gravité
e filtro de gravidad
i filtro a gravità
d Filter mit Doppelboden

* **greasiness** → 8608

3866 green acids
f acides verdiques
e ácidos verdes
i acidi verdi
d Grünsäuren

3867 green earth; terra verte
f terre verte; céladonite
e tierra verde
i terra verde; terra di Verona
d Grünerde

* **Greenland spar** → 2193

3868 green malt
f malt vert
e malta verde
i malto verde
d Grünmalz

3869 greenockite
f greenockite
e greenoquita
i greenochite
d Greenockit

3870 grid
f grille
e rejilla
i griglia
d Gitter

3871 Griess reagent
f réactif de Griess
e reactivo de Griess
i reagente di Griess
d Griessches Reagens

3872 Grignard reagents
f réactifs de Grignard
e reactivos de Grignard

i reagenti di Grignard
d Grignardsche Verbindungen

3873 Grillo-Schroeder process
f procédé Grillo-Schroeder
e método Grillo-Schroeder
i processo Grillo-Schroeder
d Grillo-Schroeder-Verfahren

3874 grindelia
f grindélia
e grindelia
i grindelia
d Grindelia

3875 grinning through
f se voyant au travers
e color del barniz anterior visto a través de
la nueva capa
i riaffioramento della vernice
d Durchscheinen; Durchschlagen

3876 grizzley; grizzly
f grille à barreaux
e barras paralelas de hierro para clasificar
roca partida
i griglia
d Gittersieb

3877 grog
f chamotte
e tierra cocida pulverizada; arcilla
calcinada
i mattoni rotti
d gemahlener Ton

3878 Gross-Almerode clay
f argile Gross-Almerode
e arcilla Gross-Almerode
i terra refrattaria
d Gross-Almerode-Ton

3879 grounds
f dépôt
e depósito; sedimento
i feccia
d Geläger

3880 ground state
f état normal
e estado fundamental
i stato fondamentale
d Grundzustand

3881 Grove cell
f élément de Grove
e elemento de Grove
i elemento di Grove
d Grove-Element

3882 guaiac
  f résine de guaïac
  e goma de guayaco
  i resina di guaiaco
  d Guajakharz

3883 guaiacol
  f guaïacol
  e guayacol
  i guaiacolo
  d Guayakol

3884 guanidine
  f guanidine
  e guanidina
  i guanidina
  d Guanidin

3885 guanidine carbonate
  f carbonate de guanidine
  e carbonato de guanidina
  i carbonato di guanidina
  d Guanidincarbonat

3886 guanidine nitrate
  f nitrate de guanidine
  e nitrato de guanidina
  i nitrato di guanidina
  d Guanidinnitrat

3887 guanine
  f guanine
  e guanina
  i guanina
  d Guanin

3888 guano
  f guano
  e guano
  i guano
  d Guano

3889 guanosine
  f guanosine
  e guanosina
  i guanosina
  d Guanosin

3890 guanylic acid
  f acide guanylique
  e ácido guanílico
  i acido guanilico
  d Guanylsäure

3891 guanyl nitrosoaminoguanylidene hydrazine
  f guanylnitrosoaminoguanylidènehydrazine
  e guanilnitrosaminoguanilidenhidracina
  i guanilidenidrazina di guanil-nitrosammina
  d Guanylnitrosoaminoguanylidinhydrazin

3892 guanyl nitrosoaminoguanyl tetracene
  f guanylnitrosoaminoguanyltétracène
  e guanilnitrosoaminoguaniltetraceno
  i guaniltetracene di guanilnitrosammina
  d Guanylnitrosoaminoguanyltetracin

3893 guanyl urea sulphate
  f sulfate de guanylurée
  e sulfate de guanilurea
  i solfato di guanile e urea
  d Guanylharnstoffsulfat

3894 Guggenheim process
  f procédé Guggenheim
  e proceso Guggenheim
  i processo Guggenheim
  d Guggenheim-Verfahren

3895 Guignet green; viridian green
  f vert Guignet; vert de chrome; oxyde de chrome hydraté
  e verde de Guignet; verde viridiano
  i verde di Guignet; verde di Verona
  d Guignetsgrün; Chromgrün; Chromhydroxyd

3896 Guild colorimeter
  f colorimètre de Guild
  e colorímetro de Guild
  i colorimetro Guild
  d Guildfarbmesser

3897 Guinea green
  f vert de Guinée
  e verde Guinea
  i verde Guinea
  d Guineagrün

3898 gum arabic
  f gomme arabique
  e goma arábica
  i gomma arabica
  d Gummiarabicum

3899 gummy
  f gommeux
  e gomoso
  i gommoso
  d gummös

3900 gum resins
  f résines naturelles
  e resinas naturales
  i gommaresina
  d Gummiharze

3901 gums
f gommes
e gomas
i gomme
d Gummi; Kautschuk

3902 gum spirit of turpentine
f essence de térébenthine
e aceite de trementina
i essenza di trementina
d Balsamterpentinöl

3903 gum tragacanth; tragacanth gum
f gomme adragante
e goma tragacanto
i gomma adragante
d Gummitragant; Tragantgummi

3904 guncotton; nitro-cotton
f coton-poudre
e algodón pólvora; nitroalgodón
i nitrocotone; fulmicotone
d Schießbaumwolle

3905 gun-powder
f poudre
e pólvora
i polvere nera
d Schießpulver

3906 Gurley densimeter
f densimètre de Gurley
e densímetro de Gurley
i densimetro Gurley
d Gurley-Dichtemesser

3907 gutta percha
f gutta-percha
e gutapercha
i guttaperga
d Guttapercha

3908 guttate
f gouttiforme
e maculado en gota
i maculato a forma di goccia
d gesprenkelt

3909 gynogenesis
f gynogénèse
e ginogénesis
i ginogenesi
d Gynogenese

3910 gypsite
f gypsite
e gipsita; yesita
i gipsite
d Gypsit

3911 gypsum
f gypse; plâtre
e yeso; sulfato de calcio
i gesso; solfato di calcio
d Gips; Calciumsulfat

3912 gyrate
f sinueux
e girado
i serpiginoso
d geschlängelt

3913 gyratory crusher
f broyeur giratoire; concasseur giratoire
e quebrantadora giratoria
i scagliuola; frantoio ruotante
d Kreiselbrecher; Glockenmühle

H

3914 Haber ammonia process
f procédé à l'ammoniac de Haber
e procedimiento al amoníaco de Haber
i processo Haber per la produzione dell'ammoniaca
d Haber-Ammoniakverfahren

3915 haemacytometer; haematimeter
f hématimètre
e hematímetro
i ematimetro
d Blutkörperchenzählapparat

3916 haemacytometry
f numération globulaire
e hematimetria
i emocitometria
d Blutkörperchenzählung

3917 haemagglutination
f hémo-agglutination
e hemoaglutinación
i emoagglutinazione
d Blutkörperchenagglutination

3918 haemagglutinin
f hémagglutinine
e hemaglutinina
i emoagglutinina
d Hämagglutinin

3919 haemalexin
f alexine du sang
e alexina de la sangre
i alessina del sangue
d Blutalexine

* **haematimeter** → 3915

3920 haematin
f hématine
e hematina
i ematina
d Hämatin

3921 haematite
f hématite
e hematita
i ematite
d Hämatit; Blutstein

3922 haematoblast
f hématoblaste
e hematoblasto
i ematoblasto
d Hämatoblast

3923 haematocyte
f globule sanguin
e hemacito
i ematocita
d Blutzelle

3924 haematopoiesis
f hématopoïèse
e hemopoyesis
i emopoiesi
d Hämatopoese

3925 haematoporphyrin
f hématoporphyrine
e hematoporfirina
i ematoporfirina
d Hämatoporphyrin

3926 haematoxylin
f hématoxyline
e hematoxilina
i ematossilina
d Hämatoxylin

3927 haemoculture
f hémoculture
e hemocultura
i emocoltura
d Hämokultur

3928 haemoglobin
f hémoglobine
e hemoglobina
i emoglobina
d Hämoglobin

3929 haemolysin
f hémolysine
e hemolisina
i emolisina
d Hämolysin

3930 haemolytic
f hémolitique
e hemolítico
i emolitico
d hämolitisch

* **haemotoxylon** → 4989

3931 hafnium
f hafnium
e hafnio
i afnio
d Hafnium

3932 halazone
f halazone
e halazona

   i alazone
   d Halazon

3933 half-cell
   f demi-cellule
   e semicelda
   i semicellula
   d Halbzelle

3934 half-chiasma
   f demi-chiasma
   e semiquiasma
   i mezzochiasma
   d Halbchiasma

3935 half-chromatid break
   f rupture des demi-chromatides
   e ruptura mediocromatidio
   i rottura dei mezzicromatidi
   d Halbchromatidenbruch

3936 half-chromatid fragmentation
   f fragmentation des demi-chromatides
   e fragmentación mediocromatidio
   i frammentazione dei mezzicromatidi
   d Halbchromatidenfragmentation

3937 half-life
   f période de demi-vie
   e período de vida media
   i mezza vita
   d Halbwertzeit

3938 half-time of exchange
   f demi-période d'échange
   e semiperíodo de intercambio
   i semiperiodo di scambio
   d Austauschhalbwertzeit

3939 halibut liver oil
   f huile de foie de flétan
   e aceite de hígado de halibut
   i olio di fegato di rombo
   d Heilbuttlebertran

3940 halide
   f halogénure
   e haluro; halogenuro; haloideo
   i aloide
   d Halogenid

3941 halite
   f halite
   e halita
   i alite
   d Halit; Steinsalz

3942 halloysite
   f halloysite

   e haloysita
   i alloisite
   d Halloysit

3943 Hall process
   f procédé Hall
   e método Hall
   i processo Hall
   d Hall-Verfahren

3944 halogenated derivative
   f dérivé halogéné
   e derivado halogenado
   i derivato alogenato
   d Halogenderivat

3945 halogenation
   f halogénation
   e halogenación
   i alogenazione
   d Halogenierung

3946 halogens
   f halogènes
   e halógenos
   i alogeni
   d Halogene

3947 haloid acids
   f hydracides halogéniques
   e ácidos haloideos
   i acidi aloidici
   d Halogenwasserstoffsäuren

3948 hammer mill
   f broyeur à marteaux
   e trituradora de martillos
   i molino a martelli
   d Hammermühle

3949 Hansa yellow
   f jaune Hansa
   e amarillo hansa
   i azogiallo
   d Hansagelb

3950 Hansgirg process
   f procédé Hansgirg
   e método Hansgirg
   i processo Hansgirg
   d Hansgirg-Verfahren

3951 haploid
   f haploïde
   e haploide
   i aploide
   d haploid

3952 haploid-sufficient
  f haploïde dominant
  e haploide-suficiente
  i aplosufficiente
  d haplosuffizient

3953 haploidy
  f haploïdie
  e haploidia
  i aploidismo
  d Haploidie

3954 haplomitosis
  f haplomitose
  e haplomitosis
  i aplomitosi
  d Haplomitose

3955 haplontic
  f haplontique
  e haplóntico
  i aplontico
  d haplontisch

3956 haplopolyploid
  f haplopolyploïde
  e haplopoliploide
  i aplopoliploide
  d haplopolyploid

3957 hard dry stage
  f tout à fait sec
  e estado duro y seco
  i stato duro e asciutto
  d vollständig trocken

3958 hardened fats
  f huiles durcies
  e grasas endurecidas
  i grassi induriti
  d gehärtete Fette

3959 hardened rosin
  f colophane durcie
  e colofonia endurecida
  i colofonia indurita
  d Hartharz

3960 hardeners
  f durcisseurs
  e agentes endurecedores
  i induritori
  d Härter

3961 hardening
  f durcissement
  e induración
  i indurimento
  d Verhärtung

3962 hard flow
  f grande dureté
  e mala fluidez
  i fluidità bassa
  d harter Fluß

3963 hard gloss paint
  f vernis dur brillant
  e barniz duro brillante
  i vernice dura lustra
  d Hartglanzfarbe

3964 hardness
  f dureté
  e dureza
  i durezza
  d Härte

3965 hard-paste porcelain
  f porcelaine dure
  e porcelana dura
  i porcellana dura
  d Hartporzellan

3966 hard water
  f eau dure
  e agua dura
  i acqua dura
  d harter Fluß

3967 Hargreaves process
  f procédé Hargreaves
  e método Hargreaves
  i processo Hargreaves
  d Hargreaves-Verfahren

3968 Harris process
  f procédé Harris
  e proceso Harris
  i processo Harris
  d Harris-Verfahren

3969 hartshorn
  f corne de cerf
  e cuerno de ciervo
  i soluzione acquosa di ammoniaca
  d Hirschhorn

3970 heat ageing inhibitor
  f antivieillisseur contre hautes
    températures
  e antienvejecidor contra temperaturas altas
  i antinvecchiante contro alte temperature
  d Alterungsschutzmittel gegen Wärme-
    einwirkung

3971 heat bodied oils
  f huiles cuites
  e aceites polimerizados por calor

i polimerizzati a caldo
d durch Hitze eingedickte Öle

3972 heat build-up
f échauffement accumulatif interne
e almacenamiento de calor interno
i sviluppo di calore interno
d Wärmeaufspeicherung

3973 heat exchanger
f échangeur de chaleur
e termointercambiador; termopermutador
i scambiatore di calore
d Wärmeaustauscher

3974 heat fastness
f stabilité sous l'effet de la chaleur
e firmeza al calor
i resistenza al calore
d Hitzebeständigkeit

3975 heating cable
f câble chauffant
e cable de calefacción
i cavo riscaldatore
d Heizkabel

3976 heating channel
f canal de chauffage
e canal de calefacción
i canale di riscaldamento
d Heizkanal

3977 heat of formation
f chaleur de formation
e calor de formación
i calore di formazione
d Bildungswärme

3978 heat of fusion
f chaleur de fusion
e calor de fusión
i calore di fusione
d Schmelzwärme

3979 heat of solution
f chaleur de dissolution
e calor de solución
i calore di diluizione
d Lösungswärme; Auflösungswärme

3980 heat-proof; heat resistant
f résistant à la chaleur
e resistente al calor
i resistente al calore
d wärmebeständig

3981 heat radiation
f radiation thermique

e radiación térmica
i radiazione termica; irradiazione di calore
d Wärmeausstrahlung

3982 heat recovery
f récupération thermique
e termorrecuperación
i ricupero di calore
d Wärmerückgewinn

* heat resistant → 3980

3983 heat sensitive
f thermosensible
e termosensible
i sensibile al calore
d wärmeempfindlich

3984 heat-sensitive paint
f peinture sensible à la chaleur
e barniz sensible al calor
i vernice sensibile al calore
d wärmeempfindliche Farbe

3985 heat treatment
f traitement thermique
e tratamiento térmico
i trattamento termico
d Wärmebehandlung

3986 heat-treatment bath
f four à bain
e horno de baño
i forno a bagno
d Wärmebehandlungsbad

3987 heavy atom
f atome lourd
e átomo pesado
i atomo pesante
d schweres Atom

3988 heavy element
f élément lourd
e elemento pesado
i elemento pesante
d schweres Element

3989 heavy element chemistry
f chimie des éléments lourds
e química de los elementos pesados
i chimica degli elementi pesanti
d Chemie der schweren Elemente

3990 heavy naphtha
f essence lourde
e nafta pesada
i benzina pesante
d Schwerbenzin

* heavy spar → 816

3991 heavy water
f eau lourde
e agua pesada
i acqua pesante
d schweres Wasser

3992 helio fast scarlet; post office red
f héliorouge
e rojo de estafeta; rojo de correos
i rosso scarlatto
d Scharlachrot

3993 heliotropin
f héliotropine
e heliotropina
i eliotropina
d Heliotropin

3994 helium
f hélium
e helio
i elio
d Helium

3995 helium pressure tank
f réservoir d'hélium sous pression
e tanque de presión de helio
i serbatoio d'elio a pressione
d Heliumdruckgefäß

3996 hellebore
f ellébore
e eléboro
i elleboro
d Nieswurz

3997 helleborein
f elléboréine
e eleboreína
i elleborina
d Helleborein

3998 Hellige comparator
f comparateur Hellige
e comparador Hellige
i comparatore Hellige
d Hellige-Kolorimeter

* helminthicide → 8730

3999 helvite
f helvite
e helvita
i elvite
d Helvit

4000 hemicellulose
f hémicellulose
e hemicelulosa
i emicellulosa
d Hemicellulose

4001 hemichromatidic
f hémichromatidique
e hemicromatídico
i emicromatidico
d halbchromatidisch

4002 hemihaploid
f hémihaploïde
e hemihaploide
i emiaploide
d hemihaploid

4003 hemimorphite
f hémimorphite
e hemimorfita
i calamina
d Hemimorphit

4004 hemixis
f hémixie
e hemixis
i emissi
d Hemixis

4005 hemizygous
f hémizygotique
e hemicigótico
i emizigotico
d hemizygot

4006 hemlock bark
f écorce de sapin-ciguë
e corteza de abeto
i corteccia dell'abete
d Hemlockrinde

4007 hemp
f chanvre
e cáñamo
i canapa
d Hanf

4008 hempseed oil
f huile de chanvre
e aceite de cáñamo
i olio di canapa
d Hanföl

4009 Henry's law
f loi d'Henry
e ley de Henry
i legge di Henry
d Henrysches Gesetz

4010 heparin
f héparine
e heparina
i eparina
d Heparin

4011 heptabarbital
f heptabarbital
e heptabarbital
i eptabarbital
d Heptabarbital

4012 heptachlor
f heptachlore
e heptacloro
i eptacloro
d Heptachlor

4013 heptadecanol
f heptadécanol
e heptadecanol
i eptadecanol
d Heptadecanol

4014 heptadecylglyoxalidine
f heptadécylglyoxalidine
e heptadecilglioxalidina
i eptadecilglioxalidina
d Heptadecylglyoxalidin

4015 heptafluorobutyric acid
f acide heptafluorobutyrique
e ácido heptafluobutírico
i acido eptafluorobutirrico
d Heptafluorbuttersäure

4016 heptanal
f heptanal
e heptanal
i eptanal
d Heptanal

4017 heptane
f heptane
e heptano
i eptano
d Heptan

4018 heptanol
f heptanol
e heptanol
i eptanolo
d Heptanol

4019 heptavalent
f heptavalent; septivalent
e heptavalente; septivalente
i eptavalente
d siebenwertig

4020 heptene
f heptène
e hepteno
i eptene
d Hepten

4021 heptoses
f heptoses
e heptosas
i eptosi
d Heptosen

4022 heptyl formate
f formiate d'heptyle
e formiato de heptilo
i formiato di eptile
d Heptylformiat

4023 herb
f herbe
e hierba
i erba
d Kraut

4024 herbicides
f herbicides
e herbicidas
i erbicidi
d Herbizide; Unkrautbekämpfungsmittel

4025 hereditary
f héréditaire
e hereditario
i ereditario
d erblich

4026 heroin
f héroïne
e heroína
i eroina
d Heroin

4027 Heroult furnace
f four Héroult
e horno Heroult
i forno Heroult
d Heroult-Ofen

4028 hesperidin
f hespéridine
e hesperidina
i esperidina
d Hesperidin

4029 Hess's law
f loi d'Hess
e ley de Hess
i legge di Hess
d Hessches Gesetz

4030 hetero-
Prefix denoting "different".
f hétéro-
e hetero-
i etero-
d Hetero-

4031 heterochromatic
f hétérochromatique
e heterocromático
i eterocromatico
d heterochromatisch

4032 heterochromatin
f hétérochromatine
e heterocromatina
i eterocromatina
d Heterochromatin

4033 heterochromatinosome
f hétérochromatinosome
e heterocromatinosoma
i eterocromatinosoma
d Heterochromatinosom

4034 heterochromatism
f hétérochromatisme
e heterocromatismo
i eterocromatismo
d Heterochromatie

4035 heterochromosome
f hétérochromosome
e heterocromosoma
i eterocromosoma
d Heterochromosom

4036 heterocyclic compounds
f composés hétérocycliques
e compuestos heterocíclicos
i composti eterociclici
d heterozyklische Verbindungen

4037 heterogamete
f hétérogamète
e heterogameto
i eterogamete
d Heterogamet

* **heterogametic** → 2606

4038 heterogamous
f hétérogame
e heterógamo
i eterogamico
d heterogam

4039 heteroicous
f hétéroécique
e heteroecio
i eteroico
d heteroezisch

4040 heterolecithal
f hétérolécithe
e heterolecito
i eterolecito
d heterolezithal

4041 heteromolybdates
f hétéromolybdates
e heteromolibdatos
i eteromolibdati
d Heteromolybdate

4042 heteropolar
f hétéropolaire
e heteropolar
i eteropolare
d heteropolar

4043 heteropolar bond
f liaison heteropolaire
e enlace heteropolar
i legame eteropolare
d heteropolare Bindung

4044 heteropolymer
f hétéropolymère
e heteropolímero
i eteropolimero
d Heteropolymer

4045 heterosomal
f hétérosomal
e heterosomal
i eterosomo
d heterosomal

4046 heterotopic
f hétérotopique
e hetcrotópico
i eterotopico
d heterotopisch

4047 heulandite
f heulandite; zéolithe feuilletée
e heulandita
i eulandite
d Heulandit; Blätterzeolith

4048 hexabromoethane
f hexabromoéthane
e hexabromoetano
i esabromoetano
d Hexabromäthan

4049 hexachlorobenzene
 f hexachlorobenzène
 e hexaclorobenceno
 i esaclorobenzene
 d Hexachlorbenzol

4050 hexachlorobutadiene
 f hexachlorobutadiène
 e hexaclorobutadieno
 i esaclorobutadiene
 d Hexachlorbutadien

4051 hexachlorocyclohexane
 f hexachlorocyclohexane
 e hexaclorociclohexano
 i esaclorocicloesano
 d Hexachlorcyclohexan

4052 hexachlorocyclopentadiene
 f hexachlorocyclopentadiène
 e hexaclorociclopentadieno
 i esaclorociclopentadiene
 d Hexachlorcyclopentadien

4053 hexachlorodiphenyl oxide
 f oxyde d'hexachlorodiphényle
 e óxido de hexaclorodifenilo
 i ossido di esaclorodifenile
 d Hexachlordiphenyloxyd

* hexachloroethane → 1449

4054 hexachloropropylene
 f hexachloropropylène
 e hexacloropropileno
 i esacloropropilene
 d Hexachlorpropylen

4055 hexadecane
 f hexadécane
 e hexadecano
 i esadecano
 d Hexadecan

4056 hexadecene
 f hexadécène
 e hexadeceno
 i esadecene
 d Hexadecen

4057 hexadecyl mercaptan
 f mercaptan d'hexadécyle
 e mercaptano de hexadecilo
 i mercaptano di esadecile
 d Hexadecylmercaptan

4058 hexadecyltrichlorosilane
 f hexadécyltrichlorosilane
 e hexadeciltriclorosilano

 i esadeciltriclorosilano
 d Hexadecyltrichlorsilan

4059 hexaethyl tetraphosphate
 f tétraphosphate d'hexaéthyle
 e tetrafosfato de hexaetilo
 i tetrafosfato di esaetile
 d Hexaäthyltetraphosphat

* hexahydric alcohol → 7668

4060 hexahydrobenzoic acid
 f acide hexahydrobenzoïque
 e ácido hexahidrobenzoico
 i acido esaidrobenzoico
 d Hexahydrobenzoesaure

4061 hexahydrophthalic acid
 f acide hexahydrophtalique
 e ácido hexahidroftálico
 i acido esaidroftalico
 d Hexahydrophthalsäure

4062 hexaldehyde
 f hexaldéhyde
 e hexaldehído
 i esaldeide
 d Hexaldehyd

4063 hexamethylenediamine
 f hexaméthylènediamine
 e hexametilendiamina
 i esametilenediammina
 d Hexamethylendiamin

4064 hexamethylenetetramine
 f hexaméthylènetétramine
 e hexametilentetramina
 i esametilentetrammina
 d Hexamethylentetramin; Utropin

4065 hexamethylphosphoric triamide
 f triamide hexaméthylphosphorique
 e triamida hexametilfosfórica
 i triammide esametilfosforica
 d Hexamethylphosphorsäuretriamid

4066 hexamine; urotropine
 f hexaméthylènetétramine; urotropine
 e hexametilenotetramina; urotropina
 i esametilentetrammina; urotropina
 d Hexamethylentetramin; Urotropin

4067 hexane
 f hexane
 e hexano
 i esano
 d Hexan

4068 hexanetriol
 f hexanetriol
 e hexanotriol
 i esanotriol
 d Hexantriol

4069 hexanitrodiphenyl amine
 f hexanitrodiphénylamine
 e hexanitrodifenilamina
 i esanitrodifenilammina
 d Hexanitrodiphenylamin

4070 hexanol
 f hexanol
 e hexanol
 i esanol
 d Hexanol

4071 hexavalent
 f hexavalent
 e hexavalente, sexivalente
 i esavalente
 d sechswertig

4072 hexene
 f hexène
 e hexeno
 i esene
 d Hexen

4073 hexenol
 f hexénol
 e hexenol
 i esenolo
 d Hexenol

4074 hexobarbital
 f hexobarbital
 e hexobarbital
 i acido dietilesobarbiturico
 d Hexobarbital

4075 hexokinase
 f hexokinase
 e hexoquinasa
 i esochinasi
 d Hexokinase

4076 hexoses
 f hexoses
 e hexosas
 i esosi
 d Hexosen

4077 hexyl acetate
 f acétate hexylique
 e acetato de hexilo
 i acetato di esile
 d Hexylacetat

4078 hexyl bromide
 f bromure hexylique
 e bromuro de hexilo
 i bromuro di esile
 d Hexylbromid

4079 hexylene glycol
 f hexylèneglycol
 e hexilenglicol
 i esilenglicolo
 d Hexylenglykol

4080 hexyl ether
 f éther hexylique
 e éter hexílico
 i etere esilico
 d Hexyläther

4081 hexyl mercaptan
 f mercaptan d'hexyle
 e mercaptano de hexilo
 i mercaptano esilico
 d Hexylmercaptan

4082 hexyl methacrylate
 f méthacrylate d'hexyle
 e metacrilato de hexilo
 i esilmetacrilato
 d Hexylmethacrylat

4083 hexylphenol
 f hexylphénol
 e hexilfenol
 i esilfenolo
 d Hexylphenol

4084 hexylresorcinol
 f hexylrésorcine
 e hexilresorcina
 i esilresorcinolo
 d Hexylresorcinol

4085 Hibbert cell
 f élement Hibbert
 e elemento de Hibbert
 i cella di Hibbert
 d Hibbertelement

4086 hibernation
 f hibernation
 e hibernación
 i ibernazione
 d Hibernation

4087 Hicks hydrometer
 f hydromètre Hicks
 e hidrómetro de Hicks
 i idrometro di Hicks
 d Hicksches Hydrometer

4088 hiding power
f pouvoir couvrant; capacité de couverture
e poder cubriente
i potere coprente
d Deckkraft

4089 high-boiling phenols
f phénols de point d'ébullition élevé
e fenoles de alto punto de ebullición
i fenoli ad alto punto di ebollizione
d hochsiedende Phenole

4090 high-flash solvent
f solvant à degré élevé d'inflammation
e disolvente de punto de inflamación alto
i solvente ad alto punto di infiammabilita
d Lösungsmittel mit hohem Flammpunkt

4091 high-frequency coagulation
f coagulation par courant à haute
  fréquence
e coagulación por alta frecuencia
i coagulazione ad alta frequenza
d Hochfrequenzkoagulation

4092 high-frequency vulcanization
f vulcanisation à haute fréquence
e vulcanización de alta frequencia
i vulcanizzazione ad alta frequenza
d Hochfrequenzvulkanisation

4093 high lights
f points brillants
e puntos brillantes
i punti brillanti
d Glanzpunkte

4094 high polymer
f haut polymère
e polímero elevado
i alto polimero
d Hochpolymerisat; Hochpolymer

4095 high viscous
f à viscosité élevée
e de alta viscosidad
i ad alta viscosità
d hoch viskös

4096 hindered settling
f décantation retardée
e sedimentación retardada
i sedimentazione non libera
d gestörte Sedimentation

4097 hippuric acid
f acide hippurique
e ácido hipúrico
i acido ippurico
d Hippursäure

4098 histamine
f histamine
e histamina
i istamina
d Histamin

4099 histamine phosphate
f phosphate d'histamine
e fosfato de histamina
i fosfato di istamina
d Histaminphosphat

* **histazylamine hydrochloride** → 8264

4100 histidine
f histidine
e histidina
i istidina
d Histidin

4101 histiocyte
f histiocyte
e histiocito
i istiocita
d Histiozyt

4102 histochemistry
f histochimie
e histoquímica
i istochimica
d Histochemie

4103 Hofmann's reaction
f réaction d'Hofmann
e reacción de Hofmann
i reazione di Hofmann
d Hofmannsche Reaktion

4104 holmium
f holmium
e holmio
i olmio
d Holmium

4105 holmium oxide
f oxyde de holmium
e óxido de holmio
i ossido di olmio
d Holmiumoxyd

4106 holoblastic cleavage
f clivage holoblastique
e segmentación holoblástica
i segmentazione oloblastica
d holoblastische Furchung

4107 holocrine
f holocrine
e holocrino
i olocrino
d holokrin

4108 homatropine
f homatropine
e homatropina
i omatropina
d Homatropin

4109 homenergic flow
f flux homénergétique
e flujo homenergético
i flusso omeoenergetico
d homenergische Strömung

4110 homentropic flow
f flux homentropique
e flujo homentrópico
i flusso omeoentropico
d homentropische Strömung

4111 homoallele
f homoallèle
e homoalelo
i omoallelo
d Homoallel

4112 homocaryon; homokarion
f homocaryon
e homocarión
i omocarion
d Homokaryon

4113 homocyclic
f homocyclique
e homocíclico
i omociclico
d homozyklisch

4114 homoeosis
f homéose
e homeosis
i omeosi
d Homöosis

4115 homoeostasis
f homéostasie
e homeostasis
i omeostasi
d Homöostasis

4116 homoeotic
f homéotique
e homeótico
i omeotico
d homöotisch

4117 homogametic
f homogamétique
e homogamético
i omogametico
d homogametisch

4118 homogenizing
f homogénisation
e homogenización
i omogeneizzazione
d Homogenisierung

* **homokarion** → 4112

4119 homolecithal
f homolécithe
e homolecito
i omolecitico
d homolezithal

4120 homologous series
f série homologue
e serie homóloga
i serie omologhe
d homologe Reihe

4121 homologues
f homologues
e homólogos
i omologhi
d homologe Verbindungen

4122 homomorphs
f homomorphes
e homomorfos
i omomorfi
d Homomorphe

4123 homonuclear molecule
f molécule homonucléaire
e molécula homonuclear
i molecola omonucleare
d homonukleares Molekül

4124 homoploidy
f homoploïdie
e homoploidia
i omoploidismo
d Homoploidie

4125 homopolar
f homopolaire
e homopolar
i omopolare
d gleichpolig

4126 homopolymer
f homopolymère
e homopolímero

i omopolimero
d Homopolymer

4127 Hooker cell
f élément d'Hooker
e célula de Hooker
i cella di Hooker
d Hooker-Zelle

4128 Hooke's law
f loi d'Hooke
e ley de Hooke
i legge di Hooke
d Hookesches Gesetz

4129 Hoopes process
f procédé Hoopes
e proceso de Hoopes
i processo di Hoopes
d Hoopesches Verfahren

4130 hopper vibrator
f vibrateur de trémie
e vibrador de tolva
i vibratore a tramoggia
d Trichterschüttelvorrichtung

4131 hops
f houblon
e lúpulo
i luppolo
d Hopfen

4132 hordenine
f hordénine
e hordenina
i ordenina
d Hordenin

4133 hormones
f hormones
e hormonas
i ormoni
d Hormone

* **horn quicksilver** → 1398

* **horn silver** → 1593

4134 horse-shoe shaped
f en fer à cheval
e en forma de herradura
i a ferro di cavallo
d abgeschrägt; hufeisenförmig

4135 host
f hôte
e huésped

i ospite
d Wirt

4136 hot atom chemistry
f chimie des atomes fortement excités
e química de los átomos muy excitados
i chimica degli atomi molto eccitati
d Chemie hochangeregter Atome

4137 hot blast
f vent chaud
e viento caliente
i corrente d'aria calda
d Heißluft; Heißwind

4138 hot-blast stove
f préchauffeur
e precalentador de aire
i preriscaldatore d'aria
d Winderhitzer; Vorwärmer

4139 hot-dip compound
f mélange anticorrosif de trempage
e mezcla por inmersión anticorrosiva
i miscela per immersione antiruggine
d Mischung für abstreubaren Tauch-
    überzug

4140 hot-dip galvanizing bath
f bain chaud de galvanisation
e baño caliente de galvanización
i bagno galvanico caldo
d Zinkbadofen

4141 hot-dip tinning bath
f bain chaud d'étamage
e baño caliente de estañado
i bagno galvanico caldo
d Zinnbadofen

4142 hot plate
f plaque de chauffage
e placa de calentamiento
i piastra di riscaldamento
d Heizplatte; Wärmeplatte

4143 hot-setting adhesive
f colle durcissable à chaud
e adhesivo en caliente
i adesivo termoindurente
d warmabbindender Klebstoff

4144 hot-wire anemometer
f anémomètre à fil chaud
e anemómetro térmico
i anemometro a filo caldo
d Hitzdrahtanemometer

4145 Houdry process
f procédé Houdry
e proceso Houdry
i processo Houdry
d Houdry-Verfahren

4146 Huber's reagent
f réactif d'Huber
e reactivo de Huber
i reagente di Huber
d Hubersches Reagenz

4147 humectant
f humectant
e humectante
i umettante
d Benetzungsmittel; Anfeuchter

4148 humidity
f humidité
e humedad
i umidità
d Feuchtigkeit

4149 humour
f humeur
e humor
i umore
d Flüssigkeit

4150 humus
f humus
e humus
i humus
d Humus

4151 hyalin
f hyaline
e hialina
i ialina
d Hyalin

4152 hyaluronic acid
f acide hyaluronique
e ácido hialurónico
i acido ialuronico
d Hyaluronsäure

4153 hyaluronidase
f hyaluronidase
e hialuronidasa
i ialuronidasi
d Hyaluronidase

4154 hybrid
f hybride
e híbrido
i ibrido
d Hybrid

4155 hybridization
f hybridation
e hibridación
i ibridazione
d Hybridisation

4156 hydantoin
f hydantoïne
e hidantoína
i idantoina
d Hydantoin

4157 hydnocarpic acid
f acide hydnocarpique
e ácido hidnocárpico
i acido idnocarpico
d Hydnocarpussäure

4158 hydrastine hydrochloride
f chlorhydrate d'hydrastine
e hidrocloruro de hidrastina
i cloridrato di idrastina
d Hydrastinhydrochlorid

* hydrated magnesium sulphate
→ 3082

4159 hydrates
f hydrates
e hidratos
i idrati
d Hydrate

4160 hydration
f hydratation
e hidratación
i idratazione
d Hydration

4161 hydraulic fluid
f fluide hydraulique
e fluido hidráulico
i olio per presse idrauliche
d Druckflüssigkeit; Drucköl

4162 hydraulic lime; lean lime
f chaux hydraulique
e cal hidráulica; cal magra
i calce idraulica
d Wasserkalk

4163 hydrazides
f hydrazides
e hidrazidas
i idrazidi
d Hydrazide

4164 hydrazine
f hydrazine

e hidracina
i idrazina
d Hydrazin

4165 hydrazine hydrate
f hydrate d'hydrazine
e hidrato de hidracina
i idrato di idrazina
d Hydrazinhydrat

4166 hydrazine sulphate
f sulfate d'hydrazine
e sulfato de hidracina
i solfato di idrazina
d Hydrazinsulfat

4167 hydrazobenzene
f hydrazobenzène
e hidrazobenceno
i benzene idrazoico
d Hydrazobenzol

4168 hydrazo compounds
f hydrazoïques
e compuestos hidrazoicos
i idrazoici; composti idrazoici
d Hydrazoverbindungen

4169 hydrazones
f hydrazones
e hidrazonas
i idrazoni
d Hydrazone

4170 hydrides
f hydrures
e hidruros
i idridi
d Hydride

4171 hydriodic acid
f acide iodhydrique
e ácido yodhídrico
i acido iodidrico
d Jodwasserstoffsäure

4172 hydrobromic acid
f acide hydrobromique
e ácido bromhídrico
i acido bromidrico
d Bromwasserstoffsäure; Hydrobromsäure

4173 hydrocarbon; carbon hydride
f hydrocarbure; carbure d'hydrogène
e hidrocarburo; hidruro de carbono
i idrocarburo; idruro di carbonio
d Kohlenwasserstoff

4174 hydrocarbon polymer
f hydrocarbure polymérique
e hidrocarburo polímero
i polimero idrocarburo
d polymerer Kohlenwasserstoff

4175 hydrocelluloses
f hydrocelluloses
e hidrocelulosas
i idrocellulose
d Hydrocellulosen

4176 hydrochloric acid; muriatic acid
f acide chlorhydrique; esprit de sel
e ácido clorhídrico; ácido muriático
i acido cloridrico; acido muriatico
d Salzsäure

4177 hydrocinnamic acid
f acide hydrocinnamique
e ácido hidrocinámico
i acido idrocinnamico
d Hydrozimtsäure

4178 hydrocortisone
f hydrocortisone
e hidrocortisona
i idrocortisone
d Hydrocortison

4179 hydrocracking
f hydrocraquage
e hidrofisuración
i piroscissione in presenza di idrogeno
d Hydrocracken

4180 hydrocyanic acid
f acide cyanhydrique
e ácido cianhídrico
i acido cianidrico; acido prussico
d Blausäure; Cyanwasserstoffsäure

4181 hydrofluoric acid
f acide hydrofluorique; acide fluorhydrique
e ácido fluorhídrico
i acido fluoridrico
d Fluorwasserstoffsäure

4182 hydroforming
f hydroforming
e hidroformación
i "reforming" con un catalizzatore di ossido di molibdeno
d Hydroforming

4183 hydroforming process
f procédé d'hydroformage
e proceso de hidroformación
i procedimento di "reforming" con un cata-

lizzatore di ossido di molibdeno
d Hydroforming

4184 hydrofuramide
f hydrofuramide
e hidrofuramida
i idrofurammide
d Hydrofuramid

4185 hydrogel
f hydrogel
e hidrogela
i idrogelo
d Hydrogel

4186 hydrogen
f hydrogène
e hidrógeno
i idrogeno
d Wasserstoff

4187 hydrogenated oils
f huiles hydrogénées
e aceites hidrogenados
i oli idrogenati
d gehärtete Öle

4188 hydrogenated rubber
f caoutchouc hydrogéné
e caucho hidrogenado
i gomma idrogenata
d Hydrokautschuk

4189 hydrogenation
f hydrogénation
e hidrogenación
i idrogenazione
d Hydrierung; Wasserstoffanlagerung

4190 hydrogenator
f appareil d'hydrogénation
e tanque hidrogenador
i idrogenatore
d Härtungskessel

4191 hydrogen bond
f liaison d'hydrogène
e enlace de hidrógeno
i legame d'idrogeno
d Wasserstoffbindung

4192 hydrogen bromide
f acide bromhydrique
e bromuro de hidrógeno
i acido bromidrico
d Bromwasserstoff

4193 hydrogen chloride
f acide chlorhydrique

e cloruro de hidrógeno
i acido cloridrico
d Chlorwasserstoff

4194 hydrogen cooling
f refroidissement par hydrogène
e enfriamiento por hidrógeno
i raffreddamento a circolazione d'idrogeno
d Wasserstoffkühlung

4195 hydrogen cyanide
f acide hydrocyanique
e cianuro de hidrógeno
i acido cianidrico
d Cyanwasserstoff

4196 hydrogen electrode
f électrode à hydrogène
e electrodo de hidrógeno
i eléttrodo a idrogeno
d Wasserstoffelektrode

4197 hydrogen embrittlement
f fragilité par l'hydrogène
e fragilidad por el hidrógeno
i fragilità dovuta all'idrogeno
d Wasserstoffsprödigkeit

4198 hydrogen fluoride
f acide hydrofluorique
e fluoruro de hidrógeno
i idrogeno fluorato
d Fluorwasserstoff

4199 hydrogen iodide
f acide iodhydrique
e yoduro de hidrógeno
i idrogeno iodato
d Jodwasserstoff

4200 hydrogen-like atom
f atome hydrogénoïde
e átomo hidrogenoide
i atomo idrogenoide
d wasserstoffähnliches Atom

4201 hydrogenous
f hydrogéné
e hidrogenado
i idrogenato
d wasserstoffhaltig

4202 hydrogen peroxide
f eau oxygénée
e peróxido de hidrógeno; agua oxigenada
i acqua ossigenata
d Wasserstoffsuperoxyd; Wasserstoff-
peroxyd

4203 hydrogen sulphide; sulphuretted hydrogen
f hydrogène sulfuré
e sulfuro de hidrógeno; hidrógeno sulfurado
i idrogeno solforato
d Schwefelwasserstoff

4204 hydrohalogenated product
f produit hydrohalogéné
e producto hidrohalogenado
i prodotto idroalogenato
d hydrohalogeniertes Produkt

4205 hydrolase
f hydrolase
e hidrolasa
i idrolasi
d Hydrolase

4206 hydrolysis
f hydrolyse
e hidrólisis
i idrolisi
d Hydrolyse

4207 hydrometallurgy
f hydrométallurgie
e hidrometalurgia
i idrometallurgia
d Hydrometallurgie; Naßmetallurgie

4208 hydrometer
f hydromètre; aréomètre
e hidrómetro; areómetro
i areometro
d Areometer; Senkwaage

4209 hydrophilic
f hydrophile
e hidrófilo
i idrofilo
d hydrophil

4210 hydrophobic
f hydrophobe
e hidrófobo
i idrofobo
d hydrophob

4211 hydroquinone
f hydroquinone
e hidroquinona
i idrochinone
d Hydrochinon

4212 hydroquinone benzyl ether
f éther benzylique d'hydroquinone
e benciléter de hidroquinona

i etere benzilico dell'idrochinone
d Hydrochinonbenzyläther

4213 hydroquinone dimethyl ether
f diméthyléther de l'hydroquinone
e dimetiléter de la hidroquinona
i dimetiletere di idrochinone
d Hydrochinondimethyläther

4214 hydroquinone hydrochloride
f chlorhydrate d'hydroquinone
e hidrocloruro de hidroquinona
i cloridrato di idrochinone
d Hydrochinonhydrochlorid

4215 hydroquinone monomethyl ether
f monométhyléther de l'hydroquinone
e monometiléter de la hidroquinona
i monometiletere di idrochinone
d Hydrochinonmonomethyläther

4216 hydrotropes
f hydrotropes
e hidrótropos
i idrotropi
d Hydrotrope

4217 hydrous
f hydraté
e hidroso; hidratado
i idratato
d wasserhaltig; wässerig

4218 hydroxyacetic acid
f acide hydroxyacétique
e ácido hidroxiacético
i acido idrossiacetico
d Hydroxyessigsäure

4219 hydroxyadipaldehyde
f hydroxyadipaldéhyde
e hidroxiadipaldehído
i idrossiadipaldeide
d Hydroxyadipaldehyd

4220 hydroxybenzaldehyde
f hydroxybenzaldéhyde
e hidroxibenzaldehído
i idrossibenzaldeide
d Hydroxybenzaldehyd

4221 hydroxybutyric acid
f acide hydroxybutyrique
e ácido hidroxibutírico
i acido idrossibutirrico
d Hydroxybuttersäure

4222 hydroxycitronellal
f hydroxycitronellal

e hidroxicitronelal
i idrossicitronellal
d Hydroxycitronellal

4223 hydroxydibenzofuran
f hydroxydibenzofuranne
e hidroxidibenzofurano
i idrossidibenzofurano
d Hydroxydibenzofuran

* **hydroxydimethol benzene** → 8883

4224 hydroxydiphenylamine
f hydroxydiphénylamine
e hidroxidifenilamina
i idrossidifenilammina
d Hydroxydiphenylamin

4225 hydroxyethylcellulose
f hydroxyéthylcellulose
e hidroxietilcelulosa
i idrossietilcellulosa
d Hydroxyäthylcellulose

4226 hydroxyethylenediamine
f hydroxyéthylènediamine
e hidroxietilendiamina
i idrossietilendiammina
d Hydroxyäthylendiamin

4227 hydroxyethylhydrazine
f hydroxyéthylhydrazine
e hidroxietilhidracina
i idrossietilidrazina
d Hydroxyäthylhydrazin

4228 hydroxyethyl piperazine
f hydroxyéthylpipérazine
e hidroxietilpiperazina
i idrossietilpiperazina
d Hydroxyäthylpiperazin

4229 hydroxyethyltrimethylammonium bicarbonate
f bicarbonate d'hydroxyéthyltriméthyl-ammonium
e bicarbonato de hidroxietiltrimetilamonio
i bicarbonato di idrossietiltrimetilammonio
d Hydroxyäthyltrimethylammonium-bicarbonat

4230 hydroxylamine
f hydroxylamine
e hidroxilamina
i idrossilammina
d Hydroxylamin

4231 hydroxylamine acid sulphate
f sulfate acide d'hydroxylamine

e sulfato ácido de hidroxilamina
i solfato acido di idrossilammina
d Hydroxylaminsäuresulfat

4232 hydroxylamine hydrochloride
f chlorhydrate d'hydroxylamine
e hidrocloruro de hidroxilamina
i cloridrato di idrossilammina
d Hydroxylaminhydrochlorid

4233 hydroxylamine sulphate
f sulfate d'hydroxylamine
e sulfato de hidroxilamina
i solfato di idrossilammina
d Hydroxylaminsulfat

4234 hydroxymercurichlorophenol
f hydroxymercurichlorophénol
e hidroximercuricloBrofenol
i idrossimercuriclorofenolo
d Hydroxymerkurichlorphenol

4235 hydroxymercuricresol
f hydroxymercuricrésol
e hidroximercuricresol
i idrossimercuricresol
d Hydroxymerkurikresol

4236 hydroxymercurinitrophenol
f hydroxymercurinitrophénol
e hidroximercurinitrofenol
i idrossimercurinitrofenolo
d Hydroxymerkurinitrophenol

4237 hydroxymethylbutanone
f hydroxyméthylbutanone
e hidroximetilbutanona
i idrossimetilbutanone
d Hydroxymethylbutanon

4238 hydroxynaphthoic acid
f acide hydroxynaphtoïque
e ácido hidroxinaftoico
i acido idrossinaftoico
d Hydroxynaphthoesäure

4239 hydroxynaphthoic anilide
f anilide hydroxynaphtoïque
e anilida hidroxinaftoica
i anilide idrossinaftoica
d Hydroxynaphthoesäureanilid

4240 hydroxynaphthoquinone
f hydroxynaphtoquinone
e hidroxinaftoquinona
i idrossinaftochinone
d Hydroxynaphthochinon

* **hydroxyphenylamine** → 8592

4241 hydroxyphenylglycine
f hydroxyphénylglycine
e hidroxifenilglicina
i idrossifenilglicina
d Hydroxyphenylglycin

4242 hydroxyphenylmercuric chloride
f chlorure hydroxyphénylmercurique
e cloruro hidroxifenilmercúrico
i cloruro idrossifenilmercurico
d Hydroxyphenylmerkurichlorid

4243 hydroxyproline
f hydroxyproline
e hidroxiprolina
i idrossiprolino
d Hydroxyprolin

4244 hydroxypropylglycerin
f hydroxypropylglycérine
e hidroxipropilglicerina
i idrossipropilglicerina
d Hydroxypropylglycerin

4245 hydroxypropyl toluidine
f hydroxypropyltoluidine
e hidroxipropiltoluidina
i idrossipropiltoluidina
d Hydroxypropyltoluidin

4246 hydroxypyridine oxide
f oxyde d'hydroxypyridine
e óxido de hidroxipiridina
i ossido di idrossipiridina
d Hydroxypyridinoxyd

4247 hydroxyquinoline
f hydroxyquinoléine
e hidroxiquinoleína
i idrossichinolina
d Hydroxychinolin

4248 hydroxyquinoline benzoate
f benzoate d'hydroxyquinoléine
e benzoato de hidroxiquinoleína
i benzoato di idrossichinolina
d Hydroxychinolinbenzoat

4249 hydroxyquinoline sulphate
f sulfate d'hydroxyquinoléine
e sulfato de hidroxiquinoleína
i solfato di idrossichinolina
d Hydroxychinolinsulfat

4250 hydroxystearyl alcohol
f alcool hydroxystéarylique
e alcohol hidroxiestearílico
i alcool idrossistearilico
d Hydroxystearylalkohol

4251 hydroxytitanium stearate
f stéarate d'hydroxytitane
e estearato de hidroxititanio
i stearato di idrossititanio
d Hydroxytitanstearat

4252 hydrozincite
f hydrozincite
e hidrocincita
i idrozincite
d Hydrozinkit; Zinkblüte

4253 hygrograph
f hygrographe
e higrógrafo; higrómetro registrador
i igrografo
d Hygrograph

4254 hygrometer; air-humidity indicator
f hygromètre
e higrómetro
i igrometro
d Hygrometer

4255 hygroscopic
f hygroscopique
e higroscópico
i igroscopico
d wasseraufnehmend; hygroskopisch

* hyoscine → 7256

4256 hyperchromasy
f hyperchromasie
e hipercromasia
i ipercromasia
d Hyperchromasie

4257 hyperpolyploidy
f hyperpolyploïdie
e hiperpoliploidia
i iperpoliploidismo
d Hyperpolyploidie

4258 hypertely
f hypertélie
e hipertelia
i ipertelia
d Hypertelie

4259 hypoadrenalemia
f hyposurrénalisme
e hipoadrenia
i iposurrenalismo
d Nebennierenunterfunktion

4260 hypochlorous acid
f acide hypochloreux
e ácido hipocloroso

   i acido ipocloroso
   d unterchlorige Säure; Unterchlorsäure

4261 hypochromaticity
   f hypochromaticité
   e hipocromaticidad
   i ipocromaticità
   d Hypochromatizität

4262 hypohaploid
   f hypohaploïde
   e hipohaploide
   i ipoaploide
   d hypohaploid

4263 hypohaploidy
   f hypohaploïdie
   e hipohaploidia
   i ipoaploidismo
   d Hypohaploidie

4264 hypophosphorous acid
   f acide hypophosphoreux
   e ácido hipofosforoso
   i acido ipofosforoso
   d Unterphosphorsäure; unterphosphorige
      Säure

4265 hypopituitarism
   f hypopituitarisme
   e hipopituitarismo
   i ipopituitarismo
   d Hypopituitarismus

4266 hypostasis
   f hypostase
   e hipostasia
   i ipostasi
   d Hypostasis

4267 hypostatic
   f hypostatique
   e hipostático
   i ipostatico
   d hypostatisch

I

4268 ice colours
f teintures azoïques
e colores al hielo; colorantes azoicos
i colori a ghiaccio
d Eisfarben

4269 Iceland spar
f spath d'Islande
e espato de Islandia
i spato d'Islanda
d isländischer Doppelspat

4270 ideal solution
f solution idéale
e solución ideal
i soluzione ideale
d ideale Lösung

4271 idiomutation
f idiomutation
e idiomutación
i idiomutazione
d Idiomutation

4272 idiopathic
f idiopathique
e idiopático
i idiopatico
d idiopathisch

4273 idioplasm
f idioplasma
e idioplasma
i plasma germinativo
d Idioplasma

4274 idiosome
f idiosome
e idiosoma
i idiosoma
d Idiosom

4275 Igewesky's solution
f liqueur d'Igewesky
e solución de Igewesky
i soluzione di Igewesky
d Igeweskysche Lösung

4276 ignitible
f inflammable
e combustible; inflamable
i accendibile; combustibile
d entzündbar

4277 ignition point
f point d'ignition
e punto de ignición
i punto di accensione
d Zündpunkt

4278 ilmenite
f ilménite
e ilmenita
i ilmenite
d Ilmenit; Eisentitan

4279 ilmenite black
f noir d'ilménite
e negro de ilmenita
i nero d'ilmenite
d Ilmenitschwarz

4280 imidazole
f imidazole; glyoxaline
e imidazol
i imidazolo
d Imidazol; Glyoxalin

4281 imides
f imides
e imidas
i immidi
d Imide

4282 iminobispropylamine
f iminobispropylamine
e iminobispropilamina
i iminobispropilammina
d Iminobispropylamin

4283 immersion lubricant
f graissage par trempage
e lubricación por baño de aceite
i lubrificazione a immersione
d Tauchschmierung

4284 immersion plating
f dépôt par immersion
e depósito por inmersión
i deposito per immersione
d Eintauchplattierung

4285 immiscible
f non miscible
e inmiscible
i non mescolabile
d unmischbar

4286 immiscible solutions
f solutions non miscibles
e soluciones inmiscibles
i soluzioni non mescolabili
d unmischbare Lösungen

4287 immunity
f immunité
e inmunidad
i immunità
d Immunität; Indifferenz

4288 immunization
f immunisation
e inmunización
i immunizzazione
d Immunisierung

4289 immunology
f immunologie
e inmunología
i immunologia
d Immunitätsforschung

4290 impact mixer
f broyeur centrifuge
e mezcladora centrífuga
i turboagitatore
d Turbomischer; Turbozerstäuber

4291 impact test
f essai de flexion par choc
e prueba de flexión por choque
i prova di piegamento per urto
d Schlagbiegeversuch

4292 impervious
f étanche; imperméable
e impermeable; impenetrable; hermético
i impermeabile
d undurchlässig

4293 impregnate v
f imprégner
e impregnar
i impregnare
d imprägnieren; tränken

4294 impregnation
f imprégnation
e impregnación
i impregnazione
d Imprägnieren; Tränken

4295 impsonite
f impsonite
e impsonita
i impsonite
d Impsonit

4296 impurity
f élément d'impurité
e elemento de impureza
i elemento d'impurezza
d Fremdatom

4297 impurity level
f niveau d'énergie d'impurité
e nivel de energía de impureza
i livello d'energia d'impurezza
d Fremdatomenergieniveau

4298 inassimilable
f inassimilable
e inasimilable
i inassimilabile
d nicht assimilierbar

4299 incidence
f incidence
e incidencia
i incidenza
d Incidenz

4300 inclusion
f inclusion
e inclusión
i inclusione
d Einschluß

4301 inclusion bodies
f inclusions cellulaires
e cuerpos de inclusión
i inclusioni cellulari
d Einschlußkörperchen

4302 inclusion complexes
f composés d'inclusion
e complejos de inclusión
i inclusioni
d Einschlußverbindungen

4303 incompatibility
f incompatibilité
e incompatibilidad
i incompatibilità
d Inkompatibilität

4304 incompatibility factor
f facteur d'incompatibilité
e factor de incompatibilidad
i fattore d'incompatibilità
d Inkompatibilitätsfaktor

4305 incompatibility genes
f gènes d'incompatibilité
e genes de incompatibilidad
i geni d'incompatibilità
d Inkompatibilitätsgene

4306 incompressible flow
f flux incompressible
e flujo incompresible
i flusso incompressibile
d unzusammendrückbares Fließen

4307 incompressible volume
f volume incompressible
e covolumen
i volume incompressibile
d unzusammendrückbares Volumen

4308 indamines
f indamines
e indaminas
i indamine
d Indamine

4309 indanthrene
f indanthrène
e indantreno
i indantrene
d Indanthren

4310 indene
f indène
e indeno
i indene
d Inden

4311 independent variable
f variable indépendante
e variable independiente
i variabile indipendente
d unabhängige Veränderliche

4312 Indian red
f rouge des Indes
e rojo de la India
i rosso d'India
d Indischrot; Persischrot

4313 Indian yellow
f jaune de cobalt; jaune des Indes
e amarillo de la India
i giallo d'India
d Indischgelb

4314 indicator
f indicateur
e indicador
i indicatore
d Indikator

4315 indices of crystal faces
f indices des facettes
e índices de las caras
i indici delle facce
d Kristallflächenindex

4316 indigo
f indigo bleu
e índigo
i indaco
d Indigo; Indigoblau

4317 indigo carmine
f carmin d'indigo
e carmín índigo
i carminio d'indaco
d Indigokarmin; Indigotin

4318 indirect arc furnace
f four à arc indirect
e horno de arco indirecto
i forno ad arco indiretto
d indirekter Lichtbogenofen

4319 indirect arc heating
f chauffage indirect par arc
e calefacción indirecta por arco
i riscaldamento indiretto ad arco
d indirekte Lichtbogenheizung

4320 indirect resistance furnace
f four à chauffage indirect par résistance
e horno de calentamiento indirecto por resistencia
i forno a riscaldamento indiretto a resistenza
d indirekter Widerstandsofen

4321 indirect resistance heating
f chauffage indirect par résistance
e calefacción indirecta por resistencia
i riscaldamento indiretto a resistenza
d indirekte Widerstandsheizung

4322 indium
f indium
e indio
i indio
d Indium

4323 indole
f indol
e indol
i indolo
d Indol

4324 indolebutyric acid
f acide indolbutyrique
e ácido indolbutírico
i acido indolobutirrico
d Indolbuttersäure

4325 indophenols
f indophénols
e indofenoles
i indofenoli
d Indophenole

4326 induced reaction
f réaction induite
e reacción inducida

i reazione indotta
d induzierte Reaktion

4327 induction channel furnace
  f four à induction à canal
  e horno de inducción con canal
  i forno a induzione a canale
  d Induktionsrinnenofen

4328 induction heating
  f chauffage par induction
  e calefacción por inducción
  i riscaldamento a induzione
  d induktive Heizung

4329 induction period
  f période d'induction
  e periodo de inducción
  i periodo d'induzione
  d Induktionsperiode

4330 induction type magnetic separator
  f séparateur magnétique à induction
  e separador magnético por inducción
  i separatore magnetico a induzione
  d induktiver Magnetscheider

4331 indulines
  f indulines
  e indulinas
  i induline
  d Induline

4332 industrial alcohol
  f alcool dénaturé
  e alcohol desnaturalizado
  i alcool denaturato
  d denaturierter Alkohol

4333 industrial chemistry
  f chimie industrielle
  e química industrial
  i chimica industriale
  d technische Chemie

4334 industrial methylated spirit
  f alcool dénaturé industriel
  e alcohol metilado industrial
  i alcool denaturato per uso industriale
  d denaturierter Alkohol für industrielle
     Zwecke

4335 inert
  f inerte
  e inerte
  i inerte
  d inert; träge

4336 inert gases
  f gaz inertes
  e gases inertes
  i gas inerti
  d Inertgase; Schutzgase

4337 inert gas purification
  f épuration de gaz inertes
  e depuración de gases inertes
  i depurazione di gas inerti
  d Edelgasreinigung

4338 infection
  f infection
  e infección
  i infezione
  d Infektion

4339 infestation
  f infestation
  e infestación
  i infestazione
  d Infestation

4340 inflammability
  f inflammabilité
  e inflamabilidad
  i infiammabilità
  d Brennbarkeit

4341 inflating agent
  f agent de gonflement
  e agente expansivo
  i agente dilatante
  d Blähmittel

4342 infraproteins
  f infraprotéines
  e infraproteínas
  i infraproteine
  d Infraproteine

4343 infra-red
  f infrarouge
  e infrarroja
  i infrarosso
  d infrarot

4344 infra-red radiation
  f radiation infrarouge
  e radiación infrarroja
  i radiazione nell'infrarosso
  d Infrarot-Strahlung

4345 infra-red radiation heating
  f chauffage par rayonnement infrarouge
  e calefacción por radiación infrarroja
  i riscaldamento a raggi infrarossi
  d Infrarot-Strahlungsheizung

4346 infuse v
f infuser
e poner en infusión
i fare un'infusione
d aufgießen

4347 infusion
f infusion
e infusión
i infusione
d Aufguß; Infusion

4348 inheritance
f hérédité
e herencia
i eredità
d Vererbung

4349 inhibiting gene
f gène inhibiteur
e gen inhibidor
i gene inibitore
d Inhibitorgen

4350 inhibition
f inhibition
e inhibición
i inibizione
d Inhibition; Inhibitorwirkung

4351 inhibitors
f inhibiteurs
e inhibidores
i inibitori
d Inhibitoren; Stabilisatoren

4352 initiator
f initiateur
e iniciador de reacción
i iniziatore di reazione
d Reaktionseinleiter

4353 inoculable
f inoculable
e inoculable
i inoculabile
d impfbar

4354 inoculate v
f inoculer
e inocular
i trapiantare
d abimpfen; okulieren

4355 inoculation
f inoculation
e inoculación
i inoculazione
d Impfung

4356 inorganic
f inorganique
e inorgánico
i inorganico
d anorganisch

4357 inorganic chemistry
f chimie minérale
e química inorgánica
i chimica inorganica
d anorganische Chemie

4358 inosine
f inosine
e inosina
i inosina
d Inosin

4359 inositol
f inositol
e inositol
i inositolo
d Inositol; Muskelzucker

4360 insecticide
f insecticide
e insecticida
i insetticida
d Insektizid; Insektenbekämpfungsmittel

4361 insectifuge
f insectifuge
e insectífugo
i insettifugo
d insektenvertreibendes Mittel

4362 insertional translocation
f translocation insertionnelle
e translocación de inserción
i traslocazione inserzionale
d insertionale Translokation

4363 insertion breakage
f cassure d'insertion
e ruptura de inserción
i rottura inserzionale
d Insertionsbruch

4364 insolubilize v
f rendre insoluble
e insolubilizar
i rendere insolubile
d unlöslich machen

4365 insoluble colourant
f colorant insoluble
e colorante insoluble
i colorante insolubile
d unlöslicher Farbstoff

4366 instrumentation
f instrumentation
e instrumentación
i strumentazione
d Instrumentierung

4367 insulating refractories
f réfractaires calorifuges
e ladrillos refractarios aislantes
i mattoni refrattari isolanti
d feuerfeste Steine

4368 insulating varnish
f vernis isolant
e barniz aislante
i vernice isolante
d Isolierlack

4369 insulin
f insuline
e insulina
i insulina
d Insulin

4370 intake
f apport
e aflujo
i afflusso
d Aufnahme

4371 interaction theory
f théorie de l'interaction
e teoría de la interacción
i teoria dell'interazione
d Wechselwirkungstheorie

4372 interband
f interbande
e entrebanda
i interbanda
d Zwischenscheibe

4373 interchange
f interchange
e intercambio
i interscambio
d Austausch

4374 interchange trisomic
f trisomique par interchange
e trisómico de intercambio
i trisomico d'interscambio
d "Interchange"-Trisome

4375 interchromomere
f interchromomère
e intercromómero
i intercromomero
d Interchromomer

4376 interchromosomal
f interchromosomique
e intercromosómico
i intercromosomico
d interchromosomisch

4377 intercooler
f refroidisseur intermédiaire
e termocambiador intermedio
i refrigeratore intermedio
d Zwischenkühler; Mitteldrucksammel-
flasche

4378 interface; interphase
f interface; surface de contact
e interfase
i faccia intermedia
d Phasengrenzfläche

4379 interfacial film
f film interfacial
e membrana interfacial
i strato interfacciale
d Grenzflächenfilm

4380 interfacial tension
f tension interfaciale
e tensión interfacial
i tensione interfacciale
d Grenzflächenspannung

4381 intergenic
f intergénique
e intergénico
i intergenico
d intergenisch

4382 intermediate
f produit intermédiaire
e producto intermedio
i prodotto intermedio
d Zwischenprodukt

4383 intermediate coupling
f couplage à valeur intermédiaire
e acoplamiento de valor intermedio
i accoppiamento a valore intermedio
d Mittelwertkupplung

4384 intermedin
f intermédine
e hormona cromatoforotrópica
i melanoforina
d Intermedin

4385 internal compensation
f compensation intérieure
e compensación interna

i compensazione interna
d intermolekulare Kompensation

4386 internal indicator
f indicateur intérieur
e indicador interior
i indicatore interno
d aufgelöster Indikator

4387 internal resistance
f résistance intérieure
e resistencia interna
i resistenza interna
d innerer Widerstand

* interphase → 4378

4388 interspecific
f interspécifique
e interespecífico
i interspecifico
d interspezifisch

4389 intragenic
f intragénique
e intragénico
i intragenico
d intragenisch

4390 intrahaploid
f intrahaploïde
e intrahaploide
i intraaploide
d Intrahaploid

4391 intramolecular force
f force intramoléculaire
e fuerza intramolecular
i forza intramolecolare
d intramolekülare Kraft

4392 intrinsic energy
f énergie intrinsèque
e energía intrínseca
i energia intrinseca
d Eigenenergie

4393 intrinsic viscosity
f viscosité intrinsèque
e viscosidad intrínseca
i viscosità intrinseca
d grundmolare Viskosität

4394 intumescence; swelling
f gonflement
e intumescencia
i tumescenza
d Intumeszenz

4395 intumescent coating
f revêtement intumescent
e revestimiento intumescente
i rivestimento intumescente
d anschwellender Überzug

4396 inulin
f inuline
e inulina
i inulina
d Inulin

4397 invariant
f invariant
e invariante
i invariante
d unveränderlich

4398 inversion
f inversion
e inversión
i inversione
d Inversion; Umkehrung

4399 invertase
f invertase
e invertasa
i invertasi
d Invertase

4400 invert sugar
f sucre inverti
e azúcar invertido
i zucchero invertito
d Invertzucker

4401 inviability
f inviabilité
e inviabilidad
i invitalità
d Lebensunfähigkeit

4402 in vitro
f in vitro
e in vitro
i in vitro
d in vitro

4403 in vivo
f in vivo
e in vivo
i in vivo
d in vivo

4404 iodargyrite
f iodargyrite
e yodargirita
i iodargirite
d Jodargyrit

4405 iodeosin
 f iodéosine
 e yodoeosina
 i iodeosina
 d Jodeosin

4406 iodic acid
 f acide iodique
 e ácido yódico
 i acido iodico
 d Jodsäure

4407 iodine
 f iode
 e yodo
 i iodio
 d Jod

4408 iodine monobromide
 f monobromure d'iode
 e monobromuro de yodo
 i monobromuro di iodio
 d Jodmonobromid

4409 iodine monochloride
 f monochlorure d'iode
 e monocloruro de yodo
 i monocloruro di iodio
 d Jodmonochlorid

4410 iodine number
 f indice d'iode
 e índice de yodo
 i indice di iodio
 d Jodzahl

4411 iodine tincture
 f teinture d'iode
 e tintura de yodo
 i tintura di iodio
 d Jodtinktur

4412 iodine trichloride
 f trichlorure d'iode
 e tricloruro de yodo
 i tricloruro di iodio
 d Jodtrichlorid

4413 iodized oil
 f huile iodée
 e aceite yodado
 i olio iodato
 d Jodöl

4414 iodoform
 f iodoforme
 e yodoformo
 i iodoformio
 d Jodoform

4415 iodosuccinimide
 f iodosuccinimide
 e yodosuccinimida
 i iodosuccinimmide
 d Jodsuccinimid

4416 ion
 f ion
 e ión
 i ione
 d Ion

4417 ion acceptor
 f accepteur d'ions
 e aceptador de iones
 i accettore d'ioni
 d Ionenakzeptor

4418 ion activity
 f activité ionique
 e actividad iónica
 i attività ionica
 d Ionenaktivität

4419 ion concentration
 f concentration ionique
 e concentración iónica
 i concentrazione ionica
 d Ionenkonzentration

4420 ion exchange
 f échange d'ions
 e intercambio de iones
 i scambio di ioni
 d Ionenaustausch

4421 ion exchange resins
 f résines à échanges d'ions
 e resinas para cambiar iones
 i resine scambiatriche di ioni
 d Ionenaustauschharze

4422 ion exclusion
 f exclusion d'ions
 e exclusión de iones
 i esclusione di ioni
 d Ionenausschluß

4423 ionic equilibrium
 f équilibre ionique
 e equilibrio iónico
 i equilibrio ionico
 d Ionengleichgewicht

4424 ionic mobility
 f mobilité ionique
 e movilidad iónica
 i mobilità ionica
 d Ionenbeweglichkeit

4425 ionic strength
 f force ionique
 e fuerza iónica
 i forza ionica
 d Ionenstärke

4426 ionium
 f ionium
 e ionio
 i ionio
 d Ionium

4427 ionization
 f ionisation
 e ionización
 i ionizzazione
 d Ionisation; Ionenspaltung; Ionisierung

4428 ionization dosemeter
 f ionimètre
 e ionómetro
 i dosimetro a ionizzazione
 d Ionisationsdosismesser

4429 ionization energy
 f énergie d'ionisation
 e energía de ionización
 i energia di ionizzazione
 d Ionisationsenergie

4430 ionization foaming
 f production de mousse par ionisation
 e alveolización por ionización
 i formazione di schiuma mediante ionizzazione
 d Schaumstoffherstellung durch Ionisation

4431 ionization potential
 f potentiel d'ionisation
 e potencial de ionización
 i potenziale d'ionizzazione
 d Ionisationspotential

4432 ionization pressure
 f pression d'ionisation
 e presión de ionización
 i pressione d'ionizzazione
 d Ionisierungsdruck

4433 ionizing medium
 f milieu ionisant
 e medio ionizante
 i mezzo ionizzante
 d ionisierendes Medium

4434 ionizing radiation
 f rayonnement ionisant
 e radiación ionizadora
 i radiazione ionizzatrice
 d ionisierende Strahlung

4435 ionogenic
 f ionogène
 e ionógeno
 i ionogeno
 d ionogen

4436 ionone
 f ionone
 e ionona
 i ionone
 d Jonon

4437 ipecacuanha
 f ipécacuanha
 e ipecacuana
 i ipecacuana
 d Ipecacuanha; Brechwurzel

4438 IR drop
 f chute de tension ohmique
 e caída de tensión óhmica
 i caduta di tensione ohmica
 d ohmscher Spannungsabfall

4439 iridic chloride
 f chlorure iridique
 e cloruro irídico
 i cloruro iridico
 d Iridiumchlorid

4440 iridium
 f iridium
 e iridio
 i iridio
 d Iridium

4441 iridium potassium chloride
 f chlorure d'iridium et de potassium
 e cloruro de iridio y potasio
 i cloruro di iridio e potassio
 d Iridiumkaliumchlorid

4442 iridium sesquioxide
 f sesquioxyde d'iridium
 e sesquióxido de iridio
 i sesquiossido di iridio
 d Iridiumsesquioxyd

* iridosmine → 6016

* Irish moss → 1475

4443 iron
 f fer
 e hierro

i ferro
d Eisen

4444 iron acetate liquor
f pyrolignite de fer
e licor de acetato de hierro
i soluzione di acetato di ferro
d Eisenacetatlauge

4445 iron buff
f jaune de fer
e amarillo de hierro
i idrossido ferrico
d Rostgelb

4446 iron-nickel storage battery
f accumulateur au fer-nickel
e acumulador de ferro-níquel
i accumulatore a ferro-nichel
d Nickel-Eisen-Akkumulator, Nife-Akkumulator

* **iron oxide** → **1953**

4447 iron oxide pigments
f pigments d'oxyde de fer
e pigmentos de óxido de hierro
i pigmenti all'ossido di ferro
d Eisenoxydpigmente

4448 iron oxide process
f procédé à l'oxyde de fer
e método del óxido de hierro
i processo all'ossido di ferro
d Eisenoxydverfahren

4449 iron oxide reds
f oxydes de fer rouge
e rojos de óxido de hierro
i rossi all'ossido di ferro
d Eisenoxydrot

4450 iron oxide yellows
f oxydes de fer jaunes
e amarillos de óxido de hierro
i gialli all'ossido di ferro
d Eisenoxydgelb

4451 iron pentacarbonyl
f fer pentacarbonyle
e hierro pentacarbonilo
i pentacarbonile di ferro
d Eisenpentacarbonyl

4452 iron potassium tartrate
f tartrate ferrico-potassique
e tartrato de hierro y potasio
i tartrato di ferro e potassio
d Eisenkaliumtartrat; Ferrokaliumtartrat

4453 iron pyrites
f pyrite jaune
e pirita de hierro
i pirite di ferro
d Eisenkies

4454 irreversible gel
f gel irréversible
e gel irreversible
i gel irreversibile
d irreversibles Gel

4455 irreversible process
f processus irréversible
e proceso irreversible
i processo irreversibile
d irreversibler Vorgang

4456 irreversible reaction
f reaction irreversible
e reacción irreversible
i reazione irreversibile
d irreversible Reaktion

4457 irritant
f irritant
e irritante
i irritante
d Reizmittel

4458 irrotational flow
f flux irrotationnel
e flujo irrotacional
i corrente irrotazionale
d wirbelfreie Strömung

4459 isatin
f isatine
e isatina
i isatina
d Isatin

4460 isethionic acid
f acide iséthionique
e ácido isetiónico
i acido isetionico
d Isäthionsäure

4461 isinglass
f colle de poisson
e colapez; cola de pescado
i colla di pesce
d Fischleim

4462 islands of isomerism
f îles d'isomérie
e islas de isomería
i isole d'isomeria
d Isomerieinsel

4463 iso-
Prefix signifying the isomer of a compound.
f iso-
e iso-
i iso-
d Iso-

4464 isoallele
f isoallèle
e isoalelo
i isoallelo
d Isoallel

4465 isoalloxazine
f isoalloxazine
e isoaloxacina
i isoallossazina
d Isoalloxazin

4466 isoamyl acetate
f acétate d'isoamyle
e acetato de isoamilo
i acetato di isoamile
d Isoamylacetat; Essigsäureamylester

4467 isoamyl alcohol
f alcool d'isoamyle
e alcohol isoamílico
i alcool di isoamile
d Isoamylalkohol

4468 isoamyl benzoate
f benzoate d'isoamyle
e benzoato de isoamilo
i benzoato di isoamile
d Isoamylbenzoat

4469 isoamyl benzyl ether
f isoamylbenzyléther
e isoamilbenciléter
i etere benzilico di isoamile
d Isoamylbenzyläther

4470 isoamyl butyrate
f butyrate d'isoamyle
e butirato de isoamilo
i bitirrato di isoamile
d Isoamylbutyrat; buttersaures Isoamyl

4471 isoamyl chloride
f chlorure d'isoamyle
e cloruro de isoamilo
i cloruro di isoamile
d Isoamylchlorid

4472 isoamylenes
f isoamylènes
e isoamilenos

i isoamilene
d Isoamylene

4473 isoamyl salicylate
f salicylate d'isoamyle
e salicilato de isoamilo
i salicilato di isoamile
d Isoamylsalicylat

4474 isoamyl valerate
f valérianate d'isoamyle
e valerianato de isoamilo
i valerato di isoamile
d Isoamylvalerianat

4475 isoautopolyploidy
f isoautopolyploïdie
e isoautopoliploidia
i isoautopoliploidismo
d Isoautopolyploidie

4476 isoborneol
f isobornéol
e isoborneol
i isoborneol
d Isoborneol

4477 isobornyl acetate
f acétate d'isobornyle
e acetato de isobornilo
i acetato di isobornile
d Isobornylacetat

4478 isobornyl salicylate
f salicylate d'isobornyle
e salicilato de isobornilo
i salicilato di isobornile
d Isobornylsalicylat

4479 isobornyl thiocyanoacetate
f thiocyanoacétate d'isobornyle
e tiocianoacetato de isobornilo
i tiocianoacetato di isobornile
d Isobornylthiocyanacetat

4480 isobutane
f isobutane
e isobutano
i isobutano
d Isobutan

4481 isobutene
f isobutène
e isobuteno
i isobuteno
d Isobuten

4482 isobutyl acetate
f acétate d'isobutyle

e acetato de isobutilo
i acetato di isobutile
d Isobutylacetat

4483 isobutyl alcohol
f alcool isobutylique
e alcohol isobutílico
i alcool isobutilico
d Isobutylalkohol

4484 isobutylamine
f isobutylamine
e isobutilamina
i isobutilammina
d Isobutylamin

4485 isobutyl aminobenzoate
f aminobenzoate d'isobutyle
e aminobenzoato de isobutilo
i amminobenzoato di isobutilo
d Isobutylaminobenzoat

4486 isobutyl benzoate
f benzoate d'isobutyle
e benzoato de isobutilo
i benzoato di isobutile
d Isobutylbenzoat

4487 isobutyl cinnamate
f cinnamate d'isobutyle
e cinamato de isobutilo
i cinnamato di isobutile
d Isobutylcinnamat

4488 isobutyl propionate
f propionate d'isobutyle
e propionato de isobutilo
i propionato di isobutile
d Isobutylpropionat

4489 isobutyl salicylate
f salicylate d'isobutyle
e salicilato de isobutilo
i salicilato di isobutile
d Isobutylsalicylat; Salicylsäureisobutyl-
   ester

4490 isobutylundecylenamide
f isobutylundécylénamide
e isobutilundecilenamida
i isobutilundecilenammide
d Isobutylundecylenamid

4491 isobutyraldehyde
f isobutyraldéhyde; aldéhyde isobutyrique
e aldehído isobutírico
i isobutiraldeide
d Isobutyraldehyd

4492 isobutyric acid
f acide isobutyrique
e ácido isobutírico
i acido isobutirrico
d Isobuttersäure

4493 isobutyric anhydride
f anhydride isobutyrique
e anhídrido isobutírico
i anidride isobutirrica
d Isobuttersäureanhydrid

4494 isobutyronitrile
f isobutyronitrile
e isobutironitrilo
i isobutirronitrile
d Isobutyronitril

4495 isocetyl laurate
f laurate d'isocétyle
e laurato de isocetilo
i laurato di isocetile
d Isocetyllaurat

4496 isocetyl myristate
f myristate d'isocétyle
e miristato de isocetilo
i miristato di isocetile
d Isocetylmyristat

4497 isocetyl stearate
f stéarate d'isocétyle
e estearato de isocetilo
i stearato di isocetile
d Isocetylstearat

4498 isochromatid
f isochromatide
e isocromatidio
i isocromatide
d Isochromatide

4499 isochromosome
f isochromosome
e isocromosoma
i isocromosoma
d Isochromosom

4500 isocyanates
f isocyanates
e isocianatos
i isocianati
d Isocyanate

4501 isocyanuric acid
f acide isocyanurique
e ácido isocianúrico
i acido isocianurico
d Isocyanursäure

4502 isocyclic compounds
f composés isocycliques
e compuestos isocíclicos
i composti isociclici
d isozyklische Verbindungen

4503 isodecaldehyde
f isodécaldéhyde
e isodecaldehído
i isodecaldeide
d Isodecaldehyd

4504 isodecanoic acid
f acide isodécanoïque
e ácido isodecanoico
i acido isodecanoico
d Isodecanoinsäure

4505 isodecanol
f isodécanol
e isodecanol
i isodecanol
d Isodecanol

4506 isodecyl chloride
f chlorure isodécylique
e cloruro de isodecilo
i cloruro di isodecile
d Isodecylchlorid

4507 isodecyl octyl adipate
f adipate isodécyloctylique
e adipato de isodecilo y octilo
i isodecilottiladipato
d Isodecyloctyladipat

4508 isodurene
f isodurène
e isodureno
i isodurene
d Isodurol

4509 isoelectric point
f point isoélectrique
e punto isoeléctrico
i punto isoelettrico
d isoelektrischer Punkt

4510 isoeugenol
f isoeugénol
e isoeugenol
i isoeugenol
d Isoeugenol

4511 isoeugenol ethyl ether
f éther éthylique de l'isoeugénol
e éter etílico del isoeugenol
i etere etilico di isoeugenol
d Isoeugenoläthyläther

4512 isogeny
f isogénie
e isogenia
i isogenia
d Isogenie

4513 isoheptane
f isoheptane
e isoheptano
i isoeptano
d Isoheptan

4514 isohexane
f isohexane
e isohexano
i isoesano
d Isohexan

4515 isolecithal
f isolécithe
e isolecito
i isolecito
d isolezithal

4516 isoleucine
f isoleucine
e isoleucina
i isoleucina
d Isoleucin

4517 isologues
f isologues
e isólogos
i isologhi
d Isologen

4518 isomer
f isomère
e isómero
i isomero
d Isomer

4519 isomerization
f isomérisation
e isomerización
i isomerizzazione
d Isomerisation; Isomerie

4520 isomorphic
f isomorphe
e isomórfico
i isomorfico
d isomorph

4521 isomorphism
f isomorphie; isomorphisme
e isomorfismo
i isomorfia
d Isomorphie; Gleichgestaltigkeit

4522 isonicotinic acid
 f acide isonicotinique
 e ácido isonicotínico
 i acido isonicotinico
 d Isonikotinsäure

4523 isooctane
 f isooctane
 e isooctano
 i isoottano
 d Isooctan

4524 isooctene
 f isooctène
 e isoocteno
 i isoottene
 d Isoocten

4525 isooctyl adipate
 f adipate isooctylique
 e adipato de isoocilo
 i isoottiladipato
 d Isooctyladipat

4526 isooctyl alcohol
 f alcool isooctylique
 e alcohol isooctílico
 i alcool isoottilico
 d Isooctylalkohol

4527 isooctyl isodecyl phthalate
 f phtalate d'isooctyl-isodécyle
 e ftalato de isooctiloisodecilo
 i isoottilisodecilftalato
 d Isooctylisodecylphthalat

4528 isooctyl palmitate
 f palmitate isooctylique
 e palmitato de isooctilo
 i palmitato isoottilico
 d Isooctylpalmitat

4529 isooctyl thioglycolate
 f thioglycolate isooctylique
 e tioglicolato de isooctilo
 i tioglicolato di isoottile
 d Isooctylthioglykolat

4530 isopentaldehyde
 f isopentaldéhyde
 e isopentaldehído
 i isopentaldeide
 d Isopentaldehyd

4531 isopentane
 f isopentane
 e isopentano
 i isopentano
 d Isopentan

4532 isopentanoic acid
 f acide isopentanoïque
 e ácido isopentanoico
 i acido isopentanoico
 d Isopentanoinsäure

4533 isophenous
 f isophène
 e isófeno
 i isofenoso
 d isophän

4534 isophorone
 f isophorone
 e isoforona
 i isoforone
 d Isophoron

4535 isophthalic acid
 f acide isophtalique
 e acido isoftálico
 i acido isoftalico
 d Isophthalsäure

4536 isophthaloyl chloride
 f bichlorure de métaphtaloyle
 e cloruro de isoftaloílo
 i cloruro di isoftaloile
 d Isophthaloylchlorid

4537 isoploidy
 f isoploïdie
 e isoploidia
 i isoploidismo
 d Isoploidie

4538 isopolyploid
 f isopolyploïde
 e isopoliploide
 i isopoliploide
 d isopolyploid

4539 isopral
 f isopral
 e isopral
 i isopral
 d Isopral

4540 isoprene
 f isoprène
 e isopreno
 i isoprene
 d Isopren

4541 isopropanolamine
 f isopropanolamine
 e isopropanolamina
 i isopropanolammina
 d Isopropanolamin

4542 isopropenylacetylene
f isopropénylacétylène
e isopropenilacetileno
i isopropenilacetilene
d Isopropenylacetylen

4543 isopropyl acetate
f acétate d'isopropyle
e acetato de isopropilo
i acetato di isopropile
d Isopropylacetat

4544 isopropyl alcohol
f alcool isopropylique
e alcohol isopropílico
i alcool isopropilico
d Isopropylalkohol

4545 isopropylamine
f isopropylamine
e isopropilamina
i isopropilammina
d Isopropylamin

4546 isopropylaminodiphenylamine
f isopropylaminodiphénylamine
e isopropilaminodifenilamina
i isopropilamminodifenilammina
d Isopropylaminodiphenylamin

4547 isopropylaminoethanol
f isopropylaminoéthanol
e isopropilaminoetanol
i isopropilamminoetanolo
d Isopropylaminoäthanol

4548 isopropyl antimonite
f antimonite d'isopropyle
e antimonita de isopropilo
i antimonite di isopropile
d Isopropylantimonit

4549 isopropyl bromide
f bromure d'isopropyle
e bromuro de isopropilo
i bromuro di isopropile
d Isopropylbromid

4550 isopropyl butyrate
f butyrate d'isopropyle
e butirato de isopropilo
i butirrato di isopropile
d Isopropylbutyrat

4551 isopropyl chloride
f chlorure d'isopropyle
e cloruro de isopropilo
i cloruro di isopropile
d Isopropylchlorid

4552 isopropyl ether
f éther isopropylique
e éter isopropílico
i etere di isopropile
d Isopropyläther

4553 isopropyl iodide
f iodure d'isopropyle
e yoduro de isopropilo
i ioduro di isopropile
d Isopropyljodid

4554 isopropyl mercaptan
f isopropylmercaptan
e isopropilmercaptano
i mercaptano di isopropile
d Isopropylmercaptan

4555 isopropyl methyl pyrazolyl dimethyl carbamate
f isopropylméthylpyrazolyldiméthyl-carbamate
e isopropilmetilpirazolildimetilcarbamato
i isopropilmetilpirazolildimetilcarbammato
d Isopropylmethylpyrazolyldimethyl-carbamat

4556 isopropyl myristate
f myristate d'isopropyle
e miristato de isopropilo
i isopropilmiristato
d Isopropylmyristat

4557 isopropyl oleate
f oléate d'isopropyle
e oleato de isopropilo
i isopropiloleato
d Isopropyloleat

4558 isopropyl palmitate
f palmitate d'isopropyle
e palmitato de isopropilo
i isopropilpalmitato
d Isopropylpalmitat

4559 isopropyl percarbonate
f percarbonate d'isopropyle
e percarbonato de isopropilo
i percarbonato di isopropile
d Isopropylpercarbonat

4560 isopropyl peroxydicarbonate
f peroxydicarbonate de isopropyle
e peroxidicarbonato de isopropilo
i isopropilperossibicarbonato
d Isopropylperoxydicarbonat

4561 isopropylphenol
f isopropylphénol

e isopropilfenol
i isopropilfenolo
d Isopropylphenol

4562 isopropyl phenylcarbamate
f phénylcarbamate d'isopropyle
e fenilcarbamato de isopropilo
i fenilcarbammato di isopropile
d Isopropylphenylcarbamat

4563 isopulegol
f isopulégol
e isopulegol
i isopulegolo
d Isopulegol

4564 Isoquinoline
f isoquinoléine
e isoquinoleina
i isochinolina
d Isochinolin

4565 isosafrole
f isosafrol
e isosafrol
i isosafrolo
d Isosafrol

4566 isosteric molecule
f molécule isostérique
e molécula isostérica
i molecola isosterica
d isosteres Molekül

4567 isosterism
f isostérie
e isosterismo
i isosterismo
d Isosterie

4568 isotactic
f isotactique
e isotáctico
i isotattico
d isotaktisch

4569 isothermal
f isotherme
e isotermo; isotérmico
i isotermico
d isotherm

4570 isothermal line
f isotherme
e línea isoterma
i linea isotermica
d Isotherme

4571 isotonic
f isotonique
e isotónico
i isotonico
d isotonisch

4572 isotope
f isotope
e isótopo
i isotopo
d Isotop

4573 isotopic dilution analysis
f analyse par dilution isotopique
e análisis por dilución isotópica
i analisi per diluizione isotopica
d Isotopenverdünnungsanalyse

4574 isotropic body
f substance isotropique
e substancia isotrópica
i costanza isotropica
d isotrope Substanz

4575 isotypy
f isotypie
e isotipia
i isotipia
d Isotypie

4576 isovaleraldehyde
f isovaléraldéhyde
e aldehído isovaleriánico
i isovaleraldeide
d Baldrianaldehyd; Isovalerianaldehyd;
   Isovaleraldehyd

4577 isovaleric acid
f acide isovalérianique
e ácido isovaleriánico
i acido isovalerico
d Isovaleriansäure

4578 itaconic acid
f acide itaconique
e ácido itacónico
i acido itaconico
d Itaconsäure

4579 ivory black
f noir d'ivoire
e negro de marfil
i nero d'avorio
d Elfenbeinschwarz; Spodium

J

4580 jacket
f chemise
e camisa
i camicia
d Gehäuse

4581 jacket cooling
f refroidissement par chemise réfrigérante
e enfriamiento por camisa
i raffreddamento in camicia
d Mantelkühlung

4582 jacketed
f à double chemise
e a doble pared
i a doppia parete
d doppelwandig

4583 jade
f jade
e jade
i giada
d Jade

4584 jalap resin
f résine de jalap
e resina de jalapa
i gialappa
d Jalapenharz

4585 jamesonite
f jamésonite
e jamesonita
i giamesonite
d Jamesonit; Querspießglanz

4586 Japan
f vernis du Japon
e laca del Japón
i lacca del Giappone
d Japanlack

4587 Japan wax
f cire du Japon
e cera del Japón
i cera di sommacco
d Japanwachs

4588 jar
f bac
e recipiente
i contenitore; vaso
d Zellenkasten

4589 jasmine oil
f essence de jasmin
e aceite de jazmín
i essenza di gelsomino
d Jasminöl

4590 jaune brilliant
f jaune brillant
e amarillo brillante
i giallo brillante
d Brillantgelb

4591 jaune d'or
f jaune d'or
e amarillo de oro
i giallo d'oro
d Goldgelb

4592 Javelle water
f eau de Javel
e agua de Javel
i acqua di Javel
d Eau de Javelle; Javellesche Lauge

4593 jaw crusher
f concasseur à mâchoires
e machacadora de mordazas;
  quebrantadora de mandíbulas
i frantoio a mascella
d Backenbrecher

4594 jet
f jet
e azabache
i carbone a lunga fiamma
d Pechkohle

4595 jet mill
f broyeur à jet
e trituradora de chorro
i mescolatore a getto
d Strahlmühle

4596 Jolly's apparatus
f appareil de Jolly
e aparato de Jolly
i apparecchio di Jolly
d Jollyscher Apparat

4597 Joule's law
f loi de Joule
e ley de Joule
i legge di Joule
d Joulesches Gesetz

4598 Joule-Thomson effect
f effet Joule
e efecto Joule
i effetto Joule-Thomson
d Joule-Effekt

4599 juniper oil
   f essence de genièvre
   e aceite de enebro
   i essenza di ginepro
   d Wachholderöl

K

4600 kainite
 f kaïnite
 e kainita
 i kainite
 d Kainit

4601 kaolin; china clay
 f kaolin; terre à porcelaine
 e caolín; arcilla
 i caolino
 d Kaolin; Porzellanerde

4602 kaolinite
 f kaolinite
 e caolinita
 i caolinite
 d Kaolinit

4603 kapok
 f kapok
 e capoc
 i kapok
 d Bombaxwolle; Kapok

4604 kappa-factor
 f facteur kappa
 e factor kappa
 i fattore cappa
 d Kappa-Faktor

4605 karaya gum
 f gomme de karaya
 e goma de karaya
 i gomma di karaya
 d Karaya-Gummi

4606 Karl Fischer reagent
 f réactif de Karl Fischer
 e reactivo de Karl Fischer
 i reagente di Karl Fischer
 d Karl Fischer Reagenz

4607 karyogamy
 f caryogamie
 e cariogamia
 i cariogamia
 d Karyogamie

4608 karyolysis
 f caryolyse
 e cariólisis
 i cariolisi
 d Karyolysis

4609 karyomere
 f caryomère

 e cariómero
 i cariomero
 d Karyomer

4610 karyosome
 f caryosome
 e cariosoma
 i cariosoma
 d Karyosom

4611 karyotin
 f caryotine
 e cariotina
 i cariotina
 d Karyotin

4612 kauri
 f kauri
 e goma de kauri
 i resina di kauri
 d Kauriharz

4613 Keller furnace
 f four Keller
 e horno Keller
 i forno Keller
 d Keller-Ofen

4614 kelp
 f varec
 e quelpo
 i laminaria
 d Kelp

4615 keratin
 f kératine
 e queratina
 i cheratina
 d Keratin; Hornstoff

4616 kermes
 f kermès
 e quermes
 i chermes
 d Kermes

4617 kermesite; red antimony
 f kermésite; antimoine rouge
 e quermesita
 i kermesite; antimonio rosso
 d Kermesit; Antimonblende;
   Antimonzinnober

4618 kernite
 f kernite
 e kernita
 i kernite
 d Kernit; Rasorit

4619 kerogen
 f kérogène
 e querógeno
 i kerogene
 d Kerogen

4620 kerosene; kerosine *(US)*
 f kérosène; pétrole lampant
 e keroseno; petróleo de lámpara
 i cherosene; petrolio raffinato
 d Kerosin; Leuchtpetroleum

4621 ketene
 f cétène
 e cetena
 i chetene
 d Keten

4622 ketobenzotriazine
 f kétobenzotriazine
 e cetobenzotriazina
 i chetobenzotriazina
 d Ketobenzotriazin

4623 ketones
 f cétones
 e cetonas
 i chetoni
 d Ketone

4624 ketonimine dyestuffs
 f teintures de kétonimine
 e colorantes de cetonimina
 i sostanze coloranti alle chetonimina
 d Ketoniminfarbstoffe

4625 ketoses
 f cétoses
 e cetosas
 i chetosi
 d Ketosen

4626 ketoximes
 f cétoximes
 e cetoximas
 i chetossime
 d Ketoxime

4627 kettle
 f marmite; bouilleur
 e marmita; caldereta
 i caldaia
 d Kochkessel; Kessel

4628 Keyes process
 f procédé Keyes
 e proceso Keyes
 i processo Keyes
 d Keyesches Verfahren

4629 kibbler
 f égrugeoir
 e trituradora
 i frantoio
 d Zerkleinerungsmaschine

4630 Kick's law
 f loi de Kick
 e ley de Kick
 i legge di Kick
 d Kicksches Gesetz

4631 kidney ore
 f minerai en rognons; roussier
 e variedad de hematites roja arriñonada
 i ematite da masse reniformi
 d Nierenerz

 * kieselguhr ⟶ 2489

4632 kieserite
 f kiesérite
 e kieserita
 i kieserite
 d Kieserit

4633 kilderkin
 f barril (80 litres)
 e barrilito
 i bariletto; fusticino
 d Fäßchen (80 Liter)

4634 kiln
 f touraille; séchoir
 e tostadero; secadero
 i essiccatoio
 d Darre; Trockner

4635 kiln *v*
 f dessécher
 e desecar
 i torrefare
 d ausdarren

4636 kiln-dry *v*
 f donner le coup de feu
 e tostar
 i essiccare; torrefare
 d abdarren; darren

4637 kiln floor
 f plateau de touraillage
 e plato de tostadero
 i graticcio per essiccamento
 d Abdarrhorde; Darrboden

4638 kiln malt
 f malt touraillé
 e malta tostada

i malto essiccato
d Darrmalz

4639 kiln temperature
f température de four
e temperatura de tostación
i temperatura della fornace
d Abdarrtemperatur

4640 kinematic viscosity
f viscosité cinématique
e viscosidad cinemática
i viscosità cinematica
d kinematische Viskosität

4641 kinetochore
f kinétochore
e cinetócoro
i cinetocoro
d Kinetochor

* **king's blue** → 1897

4642 king's yellow
f jaune royal
e amarillo real
i giallo re
d Königsgelb

4643 kinoplasm
f cinoplasme
e cinoplasma
i cinoplasma
d Kinoplasma

4644 kinosome
f cinosome
e cinosoma
i cinosoma
d Kinosom

4645 Kipps' apparatus
f appareil de Kipp
e aparato de Kipp
i apparecchio di Kipp
d Kippscher Apparat

4646 Kjeldahl flask
f flacon de Kjeldahl
e frasco de Kjeldahl
i matraccio di Kjeldahl
d Kjeldahl-Kolben

4647 Kjeldahl's method
f méthode de Kjeldahl
e método de Kjeldahl
i metodo Kjeldahl
d Kjeldahlsche Methode

4648 knife-edge *(of a balance)*
f couteau de balance
e cuchillo de balanza
i coltello
d Schneide

4649 knotting
f vernis à masquer les nœuds
e barniz de goma laca
i vernice alla gomma lacca
d schwerer Schellack

4650 Knowles cell
f élément Knowles
e elemento Knowles
i pila di Knowles
d Knowles-Element

4651 kojic acid
f acide kojique
e ácido cójico
i acido cogico
d Kojinsäure

4652 konimeter; konometer
f conimètre
e conímetro
i conimetro
d Konimeter; Staubmesser

4653 Kroll process
f procédé Kroll
e método Kroll
i processo Kroll
d Kroll-Verfahren

4654 krypton
f krypton
e criptón
i cripto
d Krypton

4655 kyanizing
f kyanisation
e kianización
i impregnazione del legname con sublimato corrosivo
d Kyanisieren

L

4656 Labarraque's solution
f eau de Labarraque
e solución de Labarraque
i acqua di Labarraque
d Labarraquesche Flüssigkeit

4657 labdanum oil
f essence de labdanum
e aceite de ládano
i olio essenziale di labdano
d Labdanumöl

4658 label v
f marquer
e marcar; rotular; titular
i contrassegnare (con etichetta)
d bezeichnen, etikettieren

4659 labelled compound
f composé marqué
e compuesto marcado
i composto marcato
d markierte Verbindung

4660 labile
f labile
e lábil; inestable
i labile
d labil

4661 lability
f labilité
e labilidad
i labilità
d Labilität

4662 laboratory
f laboratoire
e laboratorio
i laboratorio
d Laboratorium; Labor

4663 laboratory bench
f table de laboratoire
e mesa de laboratorio
i tavolo da lavoro
d Laboratoriumtisch

4664 laboratory test
f essai de laboratoire
e ensayo de laboratorio; prueba de
   laboratorio
i prova di laboratorio
d Laboratoriumsversuch

4665 labyrinth seal
f scellement au labyrinthe
e cierre de laberinto
i guarnizione a labirinto
d Labyrinthdichtung

4666 lac
f laque
e laca
i lacca
d Lack

4667 lachrymatory
f lacrymogène
e lacrimógeno
i lacrimogeno
d tränenerregend

4668 lacmoid
f lacmoïde
e lacmoïde
i lacmoide
d Lakmoid

4669 lacquer
f vernis; laque
e barniz; laca
i lacca; vernice
d Lack

4670 lactalbumins
f lactalbumines
c lactalbúminas
i lattalbumine
d Lactalbumine

4671 lactam
f lactame
e lactama
i lattame
d Lactam

4672 lactase
f lactase
e lactasa
i lattasi
d Lactase

4673 lactate
f lactate
e lactato
i lattato
d Lactat; milchsaures Salz

4674 lacteous
f laiteux
e lácteo
i latteo
d milchig

4675 lactic acid
f acide lactique
e ácido láctico
i acido lattico
d Milchsäure

* **lactobiose** → **4681**

4676 lactobutyrometer
f lactobutyromètre
e lactobutirómetro
i lattobutirrometro
d Lactobutyrometer

4677 lactoflavin; vitamin B$_2$
f lactoflavine
e lactoflavina
i lattoflavina
d Laktoflavin

4678 lactonitrile
f lactonitrile
e lactonitrilo
i lattonitrile
d Lactonitril

4679 lactonization
f lactonisation
e lactonización
i lattonizzazione
d Lactonbildung

4680 lactophenine
f lactophénine
e lactofenina
i lattofenina
d Lactophenin

4681 lactose; lactobiose; milk sugar
f lactose; sucre de lait
e lactosa
i lattosio
d Lactose; Milchzucker

4682 ladle
f poche de coulée
e caldero de colada
i siviera
d Gießpfanne

4683 ladle crane
f grue de coulée
e grúa de cucharón
i martinello a siviera
d Gießpfannenkran

4684 laevorotatory
f lévogyre
e levorrotatorio; levógiro

i levogiro; levorotativo
d linksdrehend

4685 laevulic acid
f acide lévulique
e ácido levúlico
i acido levulico
d Lävulinsäure

4686 laevulose
f lévulose
e levulosa
i levuloso
d Lävulose; Schleimzucker

4687 lag *v*
f revêtir
e revestir
i rivestire
d ummanteln

4688 lager beer
f bière de fermentation basse
e cerveza de fermentación baja
i birra normale
d Lagerbier

4689 lagging
f enrobage d'isolation thermique
e revestimiento aislador térmico
i rivestimento
d Ummantelung; Verschalung

4690 laitance
f croûte
e lechada de cemento
i crosta
d Betonschaum

4691 lake
f laque
e laca colorante
i lacca
d Pigmentfarbe

4692 lake orange
f orangé pour laque
e laca anaranjada
i pigmento arancione
d Lackorange

4693 Lalande cell
f élément de Lalande
e pila de Lalanda
i pila di Lalande
d Lalande-Element

4694 lamella
f lamelle

e laminilla
i lamella
d Lamelle; Blättchen

4695 lamellar
f lamellaire
e laminar
i lamellare
d lamellenförmig

4696 lamina
f lamelle
e lámina
i lamina
d Lamelle; Blättchen

4697 laminar
f lamellaire
e laminar; laminada
i lamellare; laminare
d lamellenförmig

4698 laminar flow
f flux laminaire
e flujo laminar
i flusso laminare
d laminare Strömung

4699 laminate *v*
f laminer
e laminar; exfoliar
i laminare
d lamellieren; walzen

4700 laminated cloth; laminated fabric
f stratifié-tissu
e estratificado a base de tejido
i laminato di tessuto
d Hartgewebe

4701 laminating
f stratification
e laminación
i laminazione
d Schichtstoffherstellung

4702 laminating resin
f résine pour stratifiés
e resina para laminados
i resina per laminati
d Laminierharz

4703 lamination
f strate
e capa
i strato
d Lage

4704 lamination coating
f doublage par extrusion-laminage
e recubrimiento con una máquina de extrusión
i rivestimento con estrusore
d Beschichten über die Schneckenpresse

4705 lamp
f lampe
e lámpara
i lampada
d Lampe

4706 lamp black; Paris black
f noir de lampe; noir de fumée
e negro de humo; negro de lámpara; negro
   de Paris
i parofumo
d Lampenschwarz; Lampenruß

4707 lampbrush chromatid
f chromatide plumeuse
e cromatidio plumoso
i cromatide piumosa
d Lampenbürstenchromatid

4708 lampbrush chromosome
f chromosome plumeux
e cromosoma plumoso
i cromosoma piumoso
d Lampenbürstenchromosom

4709 lanatoside C
f lanatoside C
e lanatósido C
i lanatoside C
d Lanatosid C

4710 lanceolate
f lancéolé
e lanceolado
i lanceolato
d lanzenförmig

4711 Landsberger apparatus
f appareil de Landsberger
e aparato de Landsberger
i apparecchio di Landsberger
d Landsberger-Apparat

4712 langbeinite
f langbéinite
e langbeinita
i langbeinite
d Langbeinit

4713 lanolin; wool fat
f lanoléine; lanoline; graisse de laine
e lanolina; manteca de lana

i lanolina; grasso di lana
d Lanolin; Wollfett

4714 lanosterol
f lanostérol
e lanosterol
i lanosterolo
d Lanosterin

4715 lanthanide series; lanthanides
f lanthanides
e lantánidos
i lantanidi
d Lanthanide; seltene Erden

4716 lanthanum
f lanthane
e lantano
i lantanio
d Lanthan

4717 lanthanum nitrate
f nitrate de lanthane
e nitrato de lantano
i nitrato di lantanio
d Lanthannitrat

4718 lanthanum oxide
f oxyde de lanthane
e óxido de lantano
i ossido di lantanio
d Lanthanoxyd

4719 lanthionine
f lanthionine
e lantionina
i lantionina
d Lanthionin

4720 lapis lazuli
f lapis lazuli
e lapislázuli
i lapislazzuli
d Lapis lazuli

4721 lard oil
f huile de saindoux
e aceite de manteca de cerdo
i olio di lardo
d Lardöl; Specköl

4722 latent heat
f chaleur latente
e calor latente
i calore latente
d latente Wärme; bleibende Wärme

4723 latent image
f image latente

e imagen latente
i immagine latente
d latentes Bild

4724 lateral chiasma
f chiasma latéral
e quiasma lateral
i chiasma laterale
d Lateralchiasma

4725 latex
f latex
e látex
i lattice
d Latex; Kautschukmilch; Milchsaft

4726 latex-adhesive
f adhésif au latex
e cola de látex
i adesivo da lattice
d Latexleim

4727 lather
f mousse
e espuma
i schiuma
d Schaum

4728 lather *v*
f savonner
e producir espuma
i ricoprire di spuma
d schäumen

4729 lather booster
f exalteur de mousse
e agente de mejoración de espuma
i esaltatore di schiuma
d Schaumverbesserer

4730 lather collapse
f chute de la mousse
e rompimiento de la espuma
i caduta della schiuma
d Zusammenbruch des Schaumes

4731 lather value
f indice de mousse
e índice espumante
i indice di schiuma
d Schaumzahl

4732 lattice girder
f poutre à treillis
e viga de celosía
i travatura a traliccio
d Gitterträger; Fachwerkträger

4733 laudanine
  f laudanine
  e laudanina
  i laudanina
  d Laudanin

4734 laudanosine
  f laudanosine
  e laudanosina
  i laudanosina
  d Laudanosin

4735 laudanum
  f laudanum
  e láudano
  i laudano
  d Laudanon

  [*] **laughing gas** → 5802

4736 laundry blue
  f bleu fixe
  e anzulete; azul de lavandería
  i azzurro di lavanderia
  d Waschblau

4737 laurel
  f laurier
  e laurel
  i lauro
  d Lorbeer

4738 laurel oil
  f huile de graines de laurier
  e aceite de laurel
  i olio di lauro
  d Lorbeeröl

4739 lauric acid
  f acide laurique
  e ácido láurico
  i acido laurico
  d Laurinsäure

4740 lauric alcohol
  f alcool laurique
  e alcohol láurico
  i alcool laurinico
  d Laurylalkohol

4741 lauroyl chloride
  f chlorure de lauroyle
  e cloruro de lauroílo
  i cloruro di lauroile
  d Lauroylchlorid

4742 lauroyl peroxide
  f peroxyde de lauroyle
  e peróxido de lauroílo

  i perossido di lauroile
  d Lauroylperoxyd

4743 lauryl alcohol
  f alcool laurylique
  e alcohol laurílico
  i alcool laurilico
  d Laurylalkohol

4744 lauryl aldehyde
  f aldéhyde laurylique
  e aldehído laurílico
  i aldeide laurilica
  d Laurylaldehyd

4745 lauryl chloride
  f chlorure de lauryle
  e cloruro de laurilo
  i cloruro di laurile
  d Laurylchlorid

4746 lauryl mercaptan
  f laurylmercaptan
  e laurilmercaptano
  i mercaptano di laurile
  d Laurylmercaptan

4747 lauryl methacrylate
  f méthacrylate de lauryle
  e metacrilato de laurilo
  i metacrilato di laurile
  d Laurylmethacrylat

4748 lauryl pyridinium chloride
  f chlorure de lauryle et pyridium
  e cloruro de laurilo y piridinio
  i cloruro di laurile e piridinio
  d Laurylpyridiniumchlorid

4749 lavender
  f lavande
  e espliego
  i lavanda
  d Lavendelblüten

4750 lavender oil
  f essence de lavande
  e aceite de lavanda; aceite de espliego
  i olio essenziale de lavanda
  d Lavendelöl

  [*] **lavender spike oil** → 7710

4751 lavender water
  f eau de lavande
  e agua de colonia
  i acqua di lavanda
  d Lavendelwasser

4752 **lawrentium**
  f lawrencium
  e lawrencio
  i lawrencio
  d Lawrencium

4753 **laxative**
  f laxatif
  e laxante
  i lassativo
  d Abführmittel

4754 **lazulite**
  f lazulite; pierre d'azur; lapis lazuli
  e lazulita
  i lazulite
  d Lazulit; Lasurspat

4755 **leached rubber**
  f "leached rubber"
  e caucho mojado
  i "leached rubber"
  d Bleicher

  * **leaching** → 4982

  * **leaching plant** → 4983

4756 **leach liquor**
  f solution de lixivation
  e solución de lixiviación
  i liscivia
  d Laugenlösung

4757 **lead**
  f plomb
  e plomo
  i piombo
  d Blei

4758 **lead acetate; sugar of lead**
  f acétate de plomb; sucre de saturne
  e acetato de plomo; azúcar de plomo
  i acetato di piombo; zucchero di saturno
  d Bleiacetat; Bleizucker

4759 **lead-acid battery**
  f accumulateur au plomb
  e acumulador de plomo
  i accumulatore a piombo
  d Bleiakkumulator

4760 **lead antimonate**
  f antimoniate de plomb
  e antimoniato de plomo
  i antimoniato di piombo
  d Bleiantimonat

4761 **lead arsenate**
  f arséniate de plomb
  e arseniato de plomo
  i arseniato di piombo
  d Bleiarsenat

4762 **lead arsenite**
  f arsénite de plomb
  e arsenito de plomo
  i arsenito di piombo
  d Bleiarsenit

4763 **lead atomizer**
  f pulvérisateur de plomb
  e pulverizador de plomo
  i polverizzatore di piombo
  d Bleizerstäuber

4764 **lead azide**
  f azothydrure de plomb
  e acida de plomo
  i aziomide di piombo
  d Bleiazid

4765 **lead bath**
  f bain de plomb
  e baño de plomo
  i bagno di piombo
  d Bleibad

4766 **lead bath furnace**
  f four à bain de plomb
  e horno de baño de plomo
  i fornace a bagno di piombo
  d Bleibadofen

4767 **lead borate**
  f borate de plomb
  e borato de plomo
  i borato di piombo
  d Bleiborat

4768 **lead borosilicate**
  f borosilicate de plomb
  e borosilicato de plomo
  i borosilicato di piombo
  d Bleiborsilikat

4769 **lead buckle**
  f disque perforé de plomb
  e disco perforado de plomo
  i disco perforato di piombo
  d perforierte Bleiplatte

4770 **lead-burning**
  f soudage autogène du plomb
  e soldadura autógena del plomo
  i unione per fusione; saldatura autogena

del piombo
d autogene Schweißung

* **lead carbonate, basic ~ → 827**

4771 lead chloride
f chlorure de plomb
e cloruro de plomo
i cloruro di piombo
d Bleichlorid; Chlorblei

4772 lead chromate
f chromate de plomb
e cromato de plomo
i cromato di piombo
d Bleichromat

4773 lead chrome
f chrome au plomb
e cromato de plomo
i cromato di piombo
d Bleichromat

4774 lead dioxide
f peroxyde de plomb
e dióxido de plomo
i biossido di piombo
d Bleidioxyd

4775 lead disilicate; lead frit
f disilicate de plomb
e disilicato de plomo
i bisilicato di piombo
d Bleidisilikat

4776 lead driers
f siccatifs au plomb
e secadores al plomo
i essiccanti al piombo
d Bleisikkative

4777 leaded bronze
f bronze au plomb
e bronce plomoso
i bronzo al piombo
d Bleibronze

4778 leaded zinc oxide
f oxyde de zinc plombifère
e pigmento compuesto de óxido de cinc y sulfato básico de plomo
i ossido di zinco piombifero
d Mischoxyd

4779 lead formate
f formiate de plomb
e formiato de plomo
i formiato di piombo
d Bleiformiat

* **lead frit → 4775**

4780 leadhillite
f leadhillite
e leadhillita
i leadhillite
d Bleihillit

4781 lead hydroxide
f hydroxyde de plomb
e hidróxido de plomo
i idrossido di piombo
d Bleihydroxyd

4782 lead iodide
f iodure de plomb
e yoduro de plomo
i ioduro di piombo
d Bleijodid; Jodblei

4783 lead linoleate
f linoléate de plomb
e linoleato de plomo
i linoleato di piombo
d Bleilinoleat

4784 lead molybdate
f molybdate de plomb
e molibdato de plomo
i molibdato di piombo
d Bleimolybdat

4785 lead mononitroresorcinate
f mononitrorésorcinate de plomb
e mononitrorresorcinato de plomo
i mononitroresorcinato di piombo
d Bleimononitroresorcinat

* **lead monoxide → 4940**

4786 lead naphthalenesulphonate
f naphtalène-sulfonate de plomb
e naftalenosulfonato de plomo
i naftalenesolfonato di piombo
d Bleinaphthalinsulfonat

4787 lead naphthenate
f naphténate de plomb
e naftenato de plomo
i naftenato di piombo
d Bleinaphthenat

4788 lead nitrate
f nitrate de plomb
e nitrato de plomo
i nitrato di piombo
d Bleinitrat; Bleisalpeter

4789 lead oleate
f oléate de plomb
e oleato de plomo
i oleato di piombo
d Bleioleat

* **lead oxychloride** → 1495

4790 lead paint
f peinture à base de plomb
e pintura a base de plomo
i minio
d Bleifarbe

4791 lead peroxide
f peroxyde de plomb
e peróxido de plomo
i perossido di piombo
d Bleioxyd; Bleiperoxyd

* **lead phosphate, dibasic** ~ → 2498

4792 lead phosphite
f phosphite de plomb
e fosfito de plomo
i fosfito di piombo
d Bleiphosphit

* **lead poisoning** → 6479

4793 lead resinate
f résinate de plomb
e resinato de plomo
i resinato di piombo
d Bleiresinat; harzsaures Blei

4794 lead rubber
f caoutchouc plombeux
e goma plombífera
i gomma piombifera
d Bleigummi

4795 lead salt-ether method
f méthode de sels de plomb-éther
e método de sales de plomo-éter
i metodo di sali di piombo-etere
d Bleisalz-Äthermethode

4796 lead sesquioxide
f sesquioxyde de plomb
e sesquióxido de plomo
i sesquiossido di piombo
d Bleisesquioxyd

4797 lead silicate
f silicate de plomb
e silicato de plomo
i silicato di piombo
d Bleisilikat; Bleiglas

4798 lead stannate
f stannate de plomb
e estannato de plomo
i stannato di piombo
d Bleistannat

4799 lead stearate
f stéarate de plomb
e estearato de plomo
i stearato di piombo
d Bleistearat

4800 lead sulphate
f sulfate de plomb
e sulfato de plomo
i solfato di piombo
d Bleisulfat

* **lead sulphate, basic** ~ → 828

4801 lead sulphide
f sulfure de plomb
e sulfuro de plomo
i solfuro di piombo
d Bleisulfid; Bleiglanz

4802 lead tetracetate
f tétracétate de plomb
e tetraacetato de plomo
i tetraacetato di piombo
d Bleitetraacetat

4803 lead tetraethyl
f tétraéthyle de plomb
e tetraetilo de plomo
i tetraetile di piombo
d Bleitetraäthyl

4804 lead thiocyanate
f thiocyanate de plomb
e tiocianato de plomo
i tiocianato di piombo
d Bleithiocyanat

4805 lead titanate
f titanate de plomb
e titanato de plomo
i titanato di piombo
d Bleititanat

4806 lead trinitroresorcinate
f trinitrorésorcinate de plomb
e trinitrorresorcinato de plomo
i trinitroresorcinato di piombo
d Bleitrinitroresorcinat

4807 lead tungstate
f tungstate de plomb
e tungstato de plomo

i tungstato di piombo
d Bleiwolframat

4808 lead vanadate
f vanadate de plomb
e vanadato de plomo
i vanadiato di piombo
d Bleivanadat

4809 leakage water
f eau provenant des fuites
e agua perdida por fugas
i acqua perduta per fughe
d Sickerwasser

* **lean lime** → 4162

4810 lean mixture (UK)
f mélange pauvre
e mezcla pobre
i miscela povera
d mageres Gemisch

4811 lean solvent
f solvant pauvre
e solvente pobre
i solvente magro
d dünnes Lösungsmittel

* **lear** → 4819

4812 leather black
f noir pour cuir
e negro por cuero
i nero per cuoio
d Lederschwarz

4813 leather-cloth
f simili-cuir
e tela cuero
i similcuioio
d Kunstleder

4814 leather packing
f garniture en cuir
e guarnición de cuero
i guarnizione di cuoio
d Lederdichtung

4815 Leblanc process
f procédé Leblanc
e proceso de Leblanc
i procedimento Leblanc
d Leblanc-Verfahren

4816 Le Chatelier-Braun principle
f principe de Le Chatelier
e principio de Le Chatelier

i principio di Le Chatelier
d Le Chatelier-Prinzip

4817 lecithin; lecithol
f lécithine
e lecitina
i lecitina
d Lecithin

4818 Leclanché cell
f pile de Leclanché
e pila de Leclanché
i pila di Leclanché
d Leclanché-Element

4819 lehr; lear
f fourneau à recuire
e horno de recocer continuo
i forno a canale
d Kühlofen

4820 Leipzig yellow
f jaune de chrome; jaune de Leipzig
e amarillo de cromo
i giallo di cromo
d Leipzigergelb

4821 lemon chrome
f jaune citron
e cromo limón
i giallo limone
d Zitronengelb

4822 lemon-grass oil
f essence de lemongrass
e esencia de lemon-gras
i olio essenziale di Cymbopogon
d Lemongrasöl; Grasöl; Zitronengrasöl

4823 lemon oil
f essence de citron
e esencia de limón
i essenza di limone
d Zitronenöl

4824 lens
f lentille
e lente
i lente
d Linse

4825 lenticular
f lenticulaire
e lenticular
i lenticolare
d linsenförmig

4826 leonite
f léonite

e leonita
i leonite
d Leonit

4827 lepidine
  f lépidine
  e lepidina
  i lepidina
  d Lepidin

4828 lepidolite
  f lépidolite
  e lepidolita
  i lepidolite
  d Lepidolith; Lithiumglimmer

4829 leptocyte
  f leptocyte
  e leptocito
  i leptocita
  d Leptozyt

4830 leptometer
  f leptomètre
  e leptómetro
  i lettometro
  d Leptometer

4831 leptonema
  f leptonème
  e leptonema
  i leptonema
  d Leptonema

4832 leptotene
  f leptotène
  e leptoteno
  i leptotene
  d Leptotän

4833 lessen *v*
  f diminuer
  e aminorar; disminuir; rebajar; reducir
  i diminuire; ridurre
  d abschwächen; vermindern

4834 leucine
  f leucine
  e leucina
  i leucina
  d Leucin

4835 leucite
  f leucite
  e leucita
  i leucite
  d Leucit

4836 leuco-base
  f leucobase
  e leucobase
  i leucobase
  d Leukobase

4837 leuco-compounds
  f leucodérivés
  e leucocompuestos
  i leucocomposti
  d Leuko-Derivate; Leuko-Verbindungen

4838 leucocyte
  f leucocyte
  e leucocito
  i leucocita
  d Leukozyt

* **leucoline** → **6875**

4839 level *v*
  f égaliser
  e igualar; producir un color uniforme
  i livellare
  d abgleichen

4840 lever safety-valve
  f soupape de sûreté à levier
  e válvula de seguridad de contrapeso
  i valvola di sicurezza a contrappeso
  d Hebelsicherheitsventil

4841 levigation
  f lévigation
  e levigación
  i levigazione
  d Abschlemmen; Dekantieren

4842 lewisite; chlorovinyl dichlorarsine
  f léwisite
  e lewisita
  i lewisite
  d Lewisit; Chlorvinyldichlorarsin

4843 Lewis process
  f procédé Lewis
  e método de Lewis
  i procedimento di Lewis
  d Lewis-Verfahren

4844 liberate *v*
  f libérer
  e liberar
  i liberare
  d befreien

4845 liberator tank
  f cuve libératrice
  e cuba de liberación

i cella di esaurimento
d Entmetallisierungsbad

4846 lichen
  f lichen
  e liquen
  i lichene
  d Flechte

4847 lichenin
  f lichénine
  e liquenina
  i lichenina
  d Lichenin

4848 licorice
  f réglisse
  e regaliz
  i liquirizia
  d Lakritze

4849 Liebig condenser
  f condensateur Liebig
  e condensador de Liebig
  i refrigerante di Liebig
  d Liebig-Kühler

4850 light alloy
  f alliage léger
  e aleación ligera
  i lega leggera
  d Leichtlegierung

4851 light element
  f élément léger
  e elemento liviano
  i elemento leggero
  d leichtes Element

4852 light element chemistry
  f chimie des éléments légers
  e química de los elementos livianos
  i chimica degli elementi leggeri
  d Chemie der leichten Elemente

4853 light fastness; light resistance
  f résistance à la lumière
  e resistencia al deslucimiento
  i stabilità alla luce
  d Lichtbeständigkeit

4854 light-fog
  f brouillard lumineux
  e niebla luminosa
  i nebbia luminosa
  d Lichtnebel

4855 light fraction
  f fraction légère

e fracción ligera
i frazione leggera
d leichte Fraktion

4856 lighting gas
  f gaz d'éclairage
  e gas de alumbrado
  i gas per illuminazione
  d Leichtgas

4857 light metal
  f métal léger
  e metal ligero
  i metallo leggero
  d Leichtmetall

4858 light-negative; photographic negative
  f photorésistant
  e fotonegativo; fotorresistente
  i negativo
  d lichtnegativ; lichtwiderstandsfähig

4859 light oils
  f huiles légères
  e aceites ligeros
  i oli leggeri
  d Leichtöle

4860 light-piping
  f amenée de la lumière
  e alumbrado por los bordes
  i conduttura della luce
  d Lichtleitung

4861 light-relay
  f relais photoélectrique
  e relé fotoeléctrico
  i valvola a luce variabile
  d Lichtrelais

* **light resistance → 4853**

4862 light-sensitive
  f sensible à la lumière
  e fotosensible
  i sensibile alla luce
  d lichtempfindlich

4863 light solvent
  f solvant à bas point d'ébullition
  e solvente de bajo punto de ebullición
  i solvente a basso punto di ebollizione
  d leicht siedendes Lösungsmittel

4864 light source
  f source lumineuse
  e fuente luminosa
  i sorgente di luce
  d Lichtquelle

4865 light stability agent
f agent de stabilité à la lumière
e agente de estabilidad a la acción de la luz
i stabilizzante alla luce
d Lichtschutzmittel

4866 light transmittance
f transmittance
e transmisión de la luz
i trasmissione della luce
d Lichtdurchlässigkeit

4867 light-weight concrete
f béton leger
e hormigón ligero
i cemento di basso peso
d Leichtbeton

4868 ligneous
f ligneux
e leñoso
i legnoso
d holzig

4869 lignification
f lignification
e lignificación
i lignificazione
d Verholzung

4870 lignin
f lignine
e lignina
i lignina
d Lignin; Holzfaserstoff

4871 lignite; brown coal
f lignite; houille brune
e lignito
i lignite
d Braunkohle

4872 lignite coke
f coke de lignite
e coque de lignito
i coke di lignite
d Braunkohlenkoks

4873 lignite oil
f huile de lignite
e aceite de lignito
i olio di lignite
d Braunkohlenöl

4874 lignocelluloses
f lignocelluloses
e lignocelulosas
i lignocellulose
d Holzfaserstoffe

4875 lignoceric acid
f acide lignocérique
e ácido lignocerínico
i acido lignocerico
d Lignocerinsäure

4876 ligroin
f ligroïne
e ligroína
i ligroina
d Ligroin

4877 lime
f chaux
e cal
i calce
d Kalk

4878 lime blue
f bleu de Brême
e azul a la cal
i azzurro di calce
d Kalkblau

4879 lime defecation
f défécation au lait de chaux
e clarificación con lechada de cal
i defecazione della calce
d nasse Scheidung

4880 limed rosin
f résine durcie
e colofonia endurecida
i resina calcinata
d gehärtetes Colophonium

4881 lime hydrated; slack lime; slaked lime
f chaux délitée; chaux éteinte
e cal hidratada
i calce spenta
d gelöschter Kalk

4882 lime kiln
f four à chaux
e horno de cal
i fornace da calce
d Kalkbrennofen

4883 lime oil
f essence de limon
e aceite de lima
i olio essenziale di tiglio
d Limonenöl

4884 lime red
f rouge de chaux
e rojo de cal
i rosso di calce
d Kalkrot

4885 lime soap
 f savon de chaux
 e jabón de cal; jabón cálcico
 i sapone di calcio
 d Kalkseife

4886 lime-soda process
 f procédé chaux-soude
 e método cal-sosa
 i processo calce-soda
 d Kalksodaverfahren

4887 limestone
 f pierre à chaux; calcaire
 e caliza; piedra caliza
 i pietra calcarea; calcare
 d Kalkstein

4888 limewash
 f lait de chaux; blanc de chaux; badigeon
 e lechada de cal
 i latte di calce
 d Kalktünche

4889 lime-water
 f eau de chaux
 e agua de cal
 i latte di calce; acqua di calce
 d Kalkwasser

4890 lime yellow
 f jaune de chaux
 e amarillo de cal
 i giallo di calce
 d Kalkgelb

4891 liming
 f chaulage
 e tratamiento en un baño de cal
 i trattamento alla calce
 d Kalken

4892 limonite
 f limonite
 e limonita
 i limonite
 d Limonit; Braunerz

4893 limpidity
 f limpidité
 e limpidez
 i limpidità
 d Helle; Klarheit

4894 limy
 f calcaire
 e calizo; cálcico
 i calcareo
 d kalkhaltig

4895 linaloe oil
 f essence de linaloe
 e esencia de lináloe
 i olio essenziale di linaloe
 d Linaloeöl

4896 linalool
 f linalol
 e linalol
 i linalolo
 d Linalool

4897 linalyl acetate
 f acétate de linalyle
 e acetato de linalilo
 i acetato di linalile
 d Linalylacetat

4898 Linde process
 f procédé Linde
 e proceso Linde
 i processo Linde
 d Linde-Verfahren

4899 linear chromosome
 f chromosome linéaire
 e cromosoma lineal
 i cromosoma lineare
 d lineares Chromosom

4900 linear expansion
 f dilatation linéaire
 e dilatación lineal
 i espansione lineare
 d Längenausdehnung

4901 linear macromolecule
 f macromolécule linéaire
 e macromolécula lineal
 i macromolecola lineare
 d lineares Makromolekül

4902 linear molecular chain
 f chaîne moléculaire linéaire
 e cadena molecular lineal
 i catena molecolare lineare
 d unverzweigte Molekülkette

4903 linear polymer
 f polymère linéaire
 e polímero lineal
 i polimero lineare
 d lineares Polymer

4904 linear superpolymer
 f superpolymère linéaire
 e superpolímero lineal
 i superpolimero lineare
 d lineares Superpolymer

4905 liniment
 f liniment
 e linimento
 i linimento
 d Liniment; Salbe; Einreibungsmittel

4906 linkage
 f liaison
 e enlace
 i legame
 d Bindung

4907 linnaeite
 f linnéite
 e lineíta
 i linneite
 d Linneit

4908 linoleate driers
 f siccatifs au linoléate
 e secadores al linoleato
 i essiccatori al linoleato
 d Linoleattrockenstoffe

4909 linoleic acid
 f acide linoléique
 e ácido linoleico
 i acido linoleico
 d Linolsäure

4910 linolein
 f linoléine
 e linoleína
 i linoleina
 d Linolen

4911 linolenic acid
 f acide linolénique
 e ácido linolénico
 i acido linolenico
 d Linolensäure

4912 linolenyl alcohol
 f alcool linolénylique
 e alcohol linolenílico
 i alcool linolenilico
 d Linolenylalkohol

4913 linoleyl alcohol
 f alcool linoléylique
 e alcohol linoleílico
 i alcool linoleilico
 d Linoleylalkohol

4914 linoleyltrimethylammonium bromide
 f bromure de linoléyltriméthylammonium
 e bromuro de linoleiltrimetilamonio
 i bromuro linoleiltrimetilammonico
 d Linoleyltrimethylammoniumbromid

4915 linoxyn
 f linoxyne
 e linoxina
 i linossina
 d Linoxyn

4916 linseed
 f graine de lin
 e linaza
 i linosa
 d Leinsamen

4917 linseed oil
 f huile de lin
 e aceite de linaza
 i olio di semi di lino
 d Leinöl; Baumöl

4918 linters
 f linters
 e linters; borra de algodón
 i cascami di cotone
 d Linters

* liothyronine sodium → **8485**

4919 lip *(of a ladle)*
 f bec de la poche
 e boca del caldero
 i becco della siviera
 d Pfannenausguß

4920 liparoid
 f gras
 e liparoideo
 i grasso
 d fettähnlich

4921 lipases
 f lipases
 e lipasas
 i lipasi
 d Lipasen

4922 lipid
 f lipoïde
 e lípido
 i lipoide
 d Lipoid

4923 lipidosis
 f lipidose
 e lipídosis
 i lipoidosi
 d Lipoidose

4924 lipochromes
 f lipochromes
 e lipocromos

i lipocromi
d Lipochrome

* **lipoclasis** → 4926

4925 lipocyte
f lipocyte
e lipocito
i lipocita
d Lypozyt

4926 lipolysis; lipoclasis
f lipolyse
e lipólisis
i lipolisi
d Lipolyse; Fettspaltung

4927 lipophore
t xanthophore
e lipoforo
i ⸬⸬⸬⸬⸬⸬⸬⸬
d Chromatophor

4928 lipotropic agent
f agent lipotropique
e agente lipotrópico
i agente lipotropico
d lipotropisches Mittel

4929 lipoxidase
f lipoxydase
e lipoxidasa
i lipossidasi
d Lipoxydase

4930 liquate *v*
f rendre liquide
e licuar
i rendere liquido
d verflüssigen

4931 liquation
f liquation
e licuación
i liquazione
d Seigerung

4932 liquefaction
f liquéfaction
e licuefacción
i liquefazione
d Verflüssigung

4933 liquid detergent
f détergent liquide; lessive liquide
e detergente líquido
i detergente liquido
d flüssiges Waschmittel

4934 liquid driers
f siccatifs liquides
e secantes líquidos de pinturas
i essiccatori liquidi
d flüssige Trockenstoffe

4935 liquid level indicator
f limnimètre
e indicador de nivel
i misuratore di livello
d Flüssigkeitstandanzeiger

4936 liquid/liquid extraction
f extraction par partage
e extracción por repartición
i estrazione per ripartizione
d Flüssig/Flüssig-Extraktion

4937 liquid/liquid interface
f interface liquide/liquide
e interfase de dos líquidos que son inm
i interfaccia liquido/liquido
d Grenzfläche flüssig/flüssig

4938 liquid rosin
f colophane liquide
e colofonia líquida
i resina liquida
d Tallöl

4939 liquor *(brewing)*
f eau de brassage
e agua de cocimiento
i acqua per birra
d Brauwasser

4940 litharge; lead monoxide
f litharge; oxyde de plomb
e litargirio; monóxido de plomo
i litargirio; monossido di piombo
d Bleiglätte; Bleioxyd

4941 lithia; lithium monoxide
f lithine; oxyde de lithium
e monóxido de litio
i monossido di litio
d Lithion

4942 lithium
f lithium
e litio
i litio
d Lithium

4943 lithium alcoholates
f alcoolates de lithium
e alcoholatos de litio
i alcoolati di litio
d Lithiumalkoholate

4944 lithium aluminate
 f aluminate de lithium
 e aluminato de litio
 i alluminato di litio
 d Lithiumaluminat

4945 lithium aluminium hydride
 f hydrure de lithium et d'aluminium
 e hidruro de aluminio y litio
 i idruro di alluminio e litio
 d Lithiumaluminiumhydrid

4946 lithium amide
 f lithiumamide
 e litioamida
 i ammide di litio
 d Lithiumamid

4947 lithium borohydride
 f borohydrure de lithium
 e borohidruro de litio
 i boroidruro di litio
 d Lithiumborhydrid

4948 lithium bromide
 f bromure de lithium
 e bromuro de litio
 i bromuro di litio
 d Lithiumbromid

4949 lithium carbonate
 f carbonate de lithium
 e carbonato de litio
 i carbonato di litio
 d Lithiumcarbonat

4950 lithium chlorate
 f chlorate de lithium
 e clorato de litio
 i clorato di litio
 d Lithiumchlorat

4951 lithium chloride
 f chlorure de lithium
 e cloruro de litio
 i cloruro di litio
 d Chlorlithium; Lithiumchlorid

4952 lithium chromate
 f chromate de lithium
 e cromato de litio
 i cromato di litio
 d Lithiumchromat

4953 lithium citrate
 f citrate de lithium
 e citrato de litio
 i citrato di litio
 d Lithiumcitrat

4954 lithium cobaltite
 f cobaltite de lithium
 e cobaltita de litio
 i cobaltite di litio
 d Lithiumkobaltit

4955 lithium fluophosphate
 f fluophosphate de lithium
 e fluofosfato de litio
 i fluofosfato di litio
 d Lithiumfluophosphat

4956 lithium fluoride
 f fluorure de lithium
 e fluoruro de litio
 i fluoruro di litio
 d Lithiumfluorid

4957 lithium hydride
 f hydrure de lithium
 e hidruro de litio
 i idruro di litio
 d Lithiumhydrid

4958 lithium hydroxide
 f hydroxyde de lithium; lithine caustique
 e hidróxido de litio
 i idrossido di litio
 d Lithiumhydroxyd; Lithionhydrat;
   Lithiumhydrat

4959 lithium hypochlorite
 f hypochlorite de lithium
 e hipoclorito de litio
 i ipoclorito di litio
 d Lithiumhypochlorit

4960 lithium iodide
 f iodure de lithium
 e yoduro de litio
 i ioduro di litio
 d Lithiumjodid; Jodlithium

4961 lithium manganite
 f manganite de lithium
 e manganita de litio
 i manganite di litio
 d Lithiummanganit

4962 lithium metasilicate
 f métasilicate de lithium
 e metasilicato de litio
 i metasilicato di litio
 d Lithiummetasilikat

4963 lithium molybdate
 f molybdate de lithium
 e molibdato de litio

i molibdato di litio
d Lithiummolybdat

* **lithium monoxide** → 4941

4964 lithium nitrate
f nitrate de lithium
e nitrato de litio
i nitrato di litio
d Lithiumnitrat

4965 lithium peroxide
f peroxyde de lithium
e peróxido de litio
i perossido di litio
d Lithiumperoxyd

4966 lithium ricinoleate
f ricinoléate de lithium
e ricinoleato de litio
i ricinoleato di litio
d Lithiumricinoleat

4967 lithium salicylate
f salicylate de lithium
e salicilato de litio
i salicilato di litio
d Lithiumsalicylat

4968 lithium stearate
f stéarate de lithium
e estearato de litio
i stearato di litio
d Lithiumstearat

4969 lithium tetraborate
f tétraborate de lithium
e tetraborato de litio
i tetraborato di litio
d Lithiumtetraborat

4970 lithium titanate
f titanate de lithium
e titanato de litio
i titanato di litio
d Lithiumtitanat

4971 lithium zirconate
f zirconate de lithium
e zirconato de litio
i zirconato di litio
d Lithiumzirconat

4972 lithocholic acid
f acide lithocholique
e ácido litocólico
i acido litocolico
d Lithocholsäure

4973 lithographic oils; litho oils
f huiles lithographiques
e aceites litográficos
i oli litografici
d lithographische Öle

4974 lithographic varnish
f vernis lithographique
e barniz litográfico
i vernice litografica
d lithographischer Lack

* **litho oils** → 4973

4975 lithopone
f lithopone
e litopón
i litopone
d Lithopone, Schwefelzinkweiß

4976 litmus
f tournesol
e tornasol
i tornasole
d Lackmus

4977 litmus paper; turnsol paper
f papier de tournesol
e papel tornasol
i carta al tornasole
d Lackmuspapier

4978 litre
f litre
e litro
i litro
d Liter

4979 livering
f épaississage
e espesamiento
i impolmonimento; indurimento
d Eindicken

4980 liver of sulphur
f foie de soufre
e azufre hepático
i fegato di zolfo
d Schwefelleber

4981 lixiviate *v*
f lessiver
e lixiviar
i liscivare
d auslaugen

4982 lixiviation; leaching
f lixiviation; lessivage
e lixiviación; lejioción

i lisciviazione
d Laugen; Auslaugen; Laugung

4983 lixiviation plant; leaching plant
f installation à lixiviation
e instalación de lixiviación
i impianto di lisciviazione
d Laugenanlage

4984 loading funnel
f trémie de remplissage
e embudo
i imbuto di caricamento
d Füllstutzen; Fülltrichter

4985 loam
f terre glaise
e barro de fundición; arcilla plástica
i terra grassa; argilla da fonderia
d Lehm

4986 loam-core
f noyau en terre
e macho de arcilla
i nucleo di terra grassa
d Lehmkern

4987 lobelia
f lobélie
e lobelia
i lobelia
d Lobelie

4988 lobeline
f lobéline
e lobelina
i lobelina
d Lobelin

4989 logwood; haemotoxylon
f bois de campêche; bois de Brésil
e palo campeche
i campeggio
d Blauholz; Kampecheholz

4990 long malt
f malt vert
e malta verde
i malto verde
d Grünmalz

4991 Lovibond tintometer
f colorimètre de Lovibond
e tintómetro de Lovibond
i tintometro Lovibond
d Lovibond-Kolorimeter

4992 low foamer; low sudser
f détergent peu moussant

e detergente poco espumante
i detergente poco schiumogeno
d wenig schäumendes Waschmittel

4993 low grade
f à basse teneur
e de poca ley; de calidad inferior; pobre
i a basso tenore
d geringhaltig

* **low sudser** → **4992**

4994 low-temperature accelerator
f accélérateur agissant à basse température
e acelerador de temperatura baja
i accelerante a bassa temperatura
d Beschleuniger für niedrige Temperaturen

4995 low-temperature carbonization
f carbonisation à basse température
e carbonización a baja temperatura
i carbonizzazione a bassa temperatura
d Schwelung; Tiefverkokung; Urdestillation

4996 low-temperature coke
f coke à basse température
e coque obtenido a baja temperatura
i coke prodotto a bassa temperatura
d Schwelkoks

4997 low-temperature oxidation
f oxydation à basse température
e oxidación a baja temperatura
i ossidazione a bassa temperatura
d Tieftemperaturoxydation

4998 lubricant
f lubrifiant; agent de lubrification
e lubricante
i lubrificante
d Schmiermittel; Schmierstoff

4999 lubricating greases
f graisses lubrifiantes
e grasas lubricantes
i grassi lubrificanti
d Schmierfette

5000 lubricating oils
f huiles de graissage
e aceites
i oli lubrificanti
d Schmieröle; Maschinenöle

5001 lucidification
f éclaircissement
e lucidificación

i chiarificazione
d Hellwerden

5002 lukewarm
f tiède
e tibio
i tepido
d handwarm

5003 lumen
f lumen
e lumen
i lumen
d Lumen

5004 luminescence
f luminescence
e luminiscencia
i luminescenza
d Lumineszenz; Leuchten

5005 luminescent pigments
f pigments luminescents
e colorantes luminescentes
i pigmenti luminescenti
d fluoreszierende Pigmente

5006 luminous flux
f flux lumineux
e flujo luminoso
i flusso luminoso
d Lichtstrom

5007 luminous intensity
f intensité lumineuse
e intensidad luminosa
i intensità luminosa
d Beleuchtungsstärke

5008 luminous paint
f peinture luminescente
e pintura luminosa
i vernice fosforescente
d Leichtfarbe; phosphoreszierende Farbe

5009 Lunge nitrometer
f nitromètre de Lunge
e nitrómetro de Lunge
i nitrometro Lunge
d Lunge-Nitrometer

5010 lupanine
f lupanine
e lupanina
i lupanina
d Lupanin

* **lupetazine** → **2705**

5011 lupulin
f lupuline
e lupulina
i lupulina
d Lupulin

5012 lustre
f éclat; lustrage
e lustre; brillo
i splendore
d Glanz

5013 lustreless
f terne
e sin brillo; mate
i opaco
d glanzlos

5014 lustrous
f brillant
e lustroso; satinado; brillante
i brillante
d glänzend

5015 lutecium
f lutécium
e lutecio
i lutezio
d Lutetium

5016 lutein
f lutéine
e luteína
i luteina
d Lutein

5017 lutidine
f lutidine
e lutidina
i lutidina
d Lutidin

5018 luting
Attaching preformed decoration to
pottery with liquid clay slip before firing.
f applicage
e fijación de adornos a la pieza antes de la
cochura
i applicazione lutare
d Kitten

5019 luting
Sealing of vessels or the like with clay,
cement or other substances.
f masticage
e mastiqueación
i tamponatura
d Kitten; Verkitten

5020 lycopodium
f poudre de lycopode; soufre végétal
e licopodio
i licopodio
d Lycopodium

5021 lye
f lessive
e lejía
i lescivia
d Lauge

5022 lymph
f lymphe
e linfa
i linfa
d Lymphe

5023 lymphoblast
f lymphoblaste
e linfoblasto
i linfoblasto
d Lymphoblast

5024 lyolysis
f lyolyse
e liólisis
i liolisi
d Lyolyse

5025 lyophilic
f lyophile; aisément soluble
e liofílico
i liofilo
d lyophil

5026 lyophobic
f insoluble; lyophobe
e liófobo
i liofobo
d lyophob

5027 lysine
f lysine
e lisina
i lisina
d Lysin

5028 lysis
f lyse
e lisis
i lisi
d Lyse

5029 lysogenic
f lysogénique
e lisogénico
i lisogenico
d lysogen

5030 lysogenization
f lysogénisation
e lisogenisación
i lisogenizzazione
d Lysogenisierung

5031 lysozyme
f lysozime
e lisozima
i lisozima
d Lysozym

5032 lytic
f lytique
e lítico
i litico
d lytisch

M

5033 mace oil
  f huile de muscade
  e aceite de macis
  i essenza di miristica
  d Macisöl; Muskatblütenöl

5034 macerate *v*
  f faire macérer
  e macerar
  i macerare
  d einweichen

5035 Mackey test
  f épreuve de Mackey
  e prueba de Mackey
  i prova di Mackey
  d Mackey-Versuch

5036 macro-axis
  f macro-axe
  e eje mayor
  i macroasse
  d Makroachse

5037 macromolecular dispersion
  f dispersion macromoléculaire
  e dispersión macromolecular
  i dispersione macromolecolare
  d makromolekulare Lösung

5038 macromolecule
  f macromolécule
  e macromolécula
  i macromolecola
  d Makromolekül

5039 macronuclear regeneration
  f régénération macronucléaire
  e regeneración macronuclear
  i rigenerazione macronucleare
  d Makronukleusregeneration

5040 macronucleus
  f macronucléus
  e macronúcleo
  i macronucleo
  d Makronukleus

5041 macroscopic
  f macroscopique
  e macroscópico
  i macroscopico
  d makroskopisch

5042 macrospore
  f macrospore
  e macrospora
  i macrospora
  d Makrospore

5043 macrostructure
  f macrostructure
  e macroestructura
  i macrostruttura
  d Makrostruktur; Grobgefüge

5044 madder
  f garance
  e rubia
  i robbia
  d Krapp

5045 madder lake
  f laque de garance
  e lacas de la rubia
  i lacca di robbia
  d Färberröte; Krapplack

5046 madescent
  f humescent
  e madescente
  i umettato
  d angefeuchtet

5047 magnesite
  f magnésite
  e magnesita
  i magnesite
  d Magnesit; Bitterspat

5048 magnesium
  f magnésium
  e magnesio
  i magnesio
  d Magnesium

5049 magnesium acetate
  f acétate de magnésium
  e acetato magnésico
  i acetato di magnesio
  d Magnesiumacetat; essigsaures Magnesium

5050 magnesium ammonium phosphate
  f phosphate d'ammonium et de magnésium
  e fosfato magnésico-amónico
  i fosfato di ammonio e magnesio
  d Magnesiumammoniumphosphat

5051 magnesium arsenate
  f arséniate de magnésium
  e arseniato magnésico
  i arseniato di magnesio
  d Magnesiumarsenat

5052 magnesium borate
f borate de magnésium
e borato magnésico
i borato di magnesio
d Magnesiumborat

5053 magnesium-boron fluoride
f fluorure de magnésium et de bore
e fluoruro de magnesio y boro
i fluoruro di boro e magnesio
d Magnesiumborofluorid

5054 magnesium bromate
f bromate de magnésium
e bromato magnésico
i bromato di magnesio
d Magnesiumbromat

5055 magnesium carbonate
f carbonate de magnésium; carbonate de
  magnésie
e carbonato magnésico
i carbonato di magnesio
d Magnesiumcarbonat; kohlensaures
  Magnesium

5056 magnesium cell
f pile au magnésium
e pila de magnesio
i pila al magnesio
d Magnesiumelement

5057 magnesium chlorate
f chlorate de magnésium
e clorato magnésico
i clorato di magnesio
d Magnesiumchlorat

5058 magnesium chloride
f chlorure de magnésium
e cloruro magnésico
i cloruro di magnesio
d Magnesiumchlorid; Chlormagnesium

5059 magnesium citrate
f citrate de magnésium
e citrato magnésico
i citrato di magnesio
d Magnesiumcitrat

5060 magnesium fluoride
f fluorure de magnésium
e fluoruro magnésico
i fluoruro di magnesio
d Magnesiumfluorid

5061 magnesium fluosilicate
f fluosilicate de magnésium
e fluosilicato magnésico

i fluosilicato di magnesio
d Magnesiumfluosilikat

5062 magnesium formate
f formiate de magnésium
e formiato magnésico
i formiato di magnesio
d Magnesiumformiat

5063 magnesium glycerophosphate
f glycérophosphate de magnésium
e glicerofosfato magnésico
i glicerofosfato di magnesio
d Magnesiumglycerophosphat

5064 magnesium hydroxide
f hydroxyde de magnésium; hydrate de
  magnésie
e hidróxido magnésico
i idrossido di magnesio
d Magnesiumhydroxyd

5065 magnesium methylate
f méthylate de magnésium
e metilato magnésico
i metilato di magnesio
d Magnesiummethylat

5066 magnesium nitrate
f nitrate de magnésium
e nitrato magnésico
i nitrato di magnesio
d Magnesiumnitrat

5067 magnesium oleate
f oléate de magnésium
e oleato magnésico
i oleato di magnesio
d Magnesiumoleat

5068 magnesium oxide
f magnésie; oxyde de magnésium
e magnesia; óxido magnésico
i magnesia; ossido di magnesio
d Magnesia; Magnesiumoxyd

5069 magnesium palmitate
f palmitate de magnésium
e palmitato magnésico
i palmitato di magnesio
d Magnesiumpalmitat

5070 magnesium perborate
f perborate de magnésium
e perborato magnésico
i perborato di magnesio
d Magnesiumperborat

5071 magnesium perchlorate
f perchlorate de magnésium
e perclorato magnésico
i perclorato di magnesio
d Magnesiumperchlorat

5072 magnesium permanganate
f permanganate de magnésium
e permanganato magnésico
i permanganato di magnesio
d Magnesiumpermanganat

5073 magnesium peroxide
f peroxyde de magnésium; superoxyde de magnésium
e peróxido magnésico
i perossido di magnesio
d Magnesiumperoxyd;
   Magnesiumsuperoxyd

* **magnesium phosphate, dibasic ~**
   → 2499

* **magnesium phosphate, monobasic ~ → 5567**

* **magnesium phosphate, tribasic ~ → 8395**

5074 magnesium ricinoleate
f ricinoléate de magnésie
e ricinoleato magnésico
i ricinoleato di magnesio
d Magnesiumricinoleat

5075 magnesium salicylate
f salicylate de magnésie
e salicilato magnésico
i salicilato di magnesio
d Magnesiumsalicylat

5076 magnesium silicate
f silicate de magnésium
e silicato magnésico
i silicato di magnesio
d Magnesiumsilikat

5077 magnesium stearate
f stéarate de magnésie
e estearato magnésico
i stearato di magnesio
d Magnesiumstearat

5078 magnesium sulphate
f sulfate de magnésium; sel anglais
e sulfato magnésico
i solfato di magnesio
d Magnesiumsulfat; Bittersalz

* **magnesium sulphate, hydrated ~**
   → 3082

5079 magnesium sulphite
f sulfite de magnésium
e sulfito magnésico
i solfito di magnesio
d Magnesiumsulfit

5080 magnesium trisilicate
f trisilicate de magnésium
e trisilicato magnésico
i trisilicato di magnesio
d Magnesiumtrisilikat

5081 magnesium tungstate
f tungstate de magnésium
e tungstato magnésico
i tungstato di magnesio
d Magnesiumwolframat

5082 magnetic pulley
f poulie magnétique
e polea magnética
i puleggia magnetica
d magnetische Rolle

5083 magnetic separator
f séparateur magnétique
e separador magnético
i separatore magnetico
d Magnetscheider

5084 magnetic vibrator
f vibreur magnétique
e vibrador magnético
i convogliatore a vibrazione magnetica
d elektromagnetischer Vibrator

5085 magnetism
f magnétisme
e magnetismo
i magnetismo
d Magnetismus

5086 magnetite
f magnétite
c magnetita
i magnetite
d Magnetit; Magneteisenstein

5087 magnetite black
f noir de magnétite
e negro de magnetita
i nero di magnetite
d Magnetitschwarz

5088 magnetization
f aimantation

e imantación
i magnetizzazione
d Magnetisierung

5089 magnetometer
f magnétomètre
e magnetómetro
i magnetometro
d Magnetometer

5090 magnetostriction
f magnétostriction
e magnetoestricción
i magnetostrizione
d Magnetostriktion

5091 magnification
f grossissement
e aumento
i ingrandimento
d Vergrößerung

5092 magnifying glass
f loupe
e lupa
i lente d'ingrandimento
d Lupe

5093 magnifying power
f grossissement
e aumento
i potenza di ingrandimento
d Vergrößerung

5094 magnitude
f grandeur
e magnitud
i grandezza
d Größe

5095 major gene
f gène majeur
e gen mayor
i gene maggiore
d Hauptgen

5096 make-up water
f eau d'appoint
e agua de relleno; agua adicional
i acqua di integrazione
d Zusatzwasser

5097 malabsorption
f défaut d'absorption
e absorción defectuosa
i difetto di assorbimento
d mangelhafte Absorption

5098 malachite
f malachite
e malaquita
i malachite
d Kupferspat; Malachit

5099 malachite green
f vert malachite
e verde malaquita
i verde malachite
d Malachitgrün

5100 malassimilation
f assimilation incomplète
e asimilación incompleta
i assimilazione incompleta
d schlechte Assimilation

5101 malathion
f malathion
e malatión
i malathion
d Malathion

5102 male die
f moule mâle
e punzón
i stampo maschio
d Patrize

5103 maleic acid
f acide maléique
e ácido maleico
i acido maleico
d Maleinsäure

5104 maleic anhydride
f anhydride maléique
e anhídrido maleico
i anidride maleica
d Maleinsäureanhydrid

5105 maleic hydrazide
f hydrazide maléique
e hidracida del ácido maleico
i idrazide maleica
d Maleinsäurehydrazid

5106 maleic resins
f résines maléiques
e resinas maleicas
i resine maleiche
d Maleinharze

5107 malic acid
f acide malique
e ácido málico
i acido malico
d Apfelsäure

5108 malleable
  f malléable
  e maleable
  i malleabile
  d verformbar

5109 malonic acid
  f acide malonique
  e ácido malónico
  i acido malonico
  d Malonsäure

5110 malt
  f malt
  e malta
  i malto
  d Malz

5111 malt *v*
  f malter
  e maltear
  i maltare
  d mälzen

5112 maltase
  f maltase
  e maltasa
  i maltasi
  d Maltase

  * **malt couch** → 5116

5113 malthenes
  f malthènes
  e maltenos
  i malteni
  d Malthene

5114 malthouse
  f malterie
  e maltería
  i malteria
  d Mälzerei

5115 maltose
  f maltose; sucre de malt
  e maltosa
  i maltosio
  d Maltose; Malzzucker

5116 malt piece; malt couch
  f tas de malt
  e pila de malta
  i mucchio
  d Malzhaufen

5117 mandarin oil
  f essence de mandarine
  e esencia de mandarina

  i essenza di mandarino
  d Mandarinenöl

5118 mandelic acid
  f acide mandélique
  e ácido mandélico
  i acido mandelico
  d Mandelsäure

5119 mandrel test
  f épreuve de flexibilité
  e prueba del mandril
  i prova di flessibilità
  d Biegeprobe

5120 maneb; manganese ethylene-bisdithiocarbamate
  f maneb
  e maneb
  i maneb
  d Maneb

5121 manganese
  f manganèse
  e manganeso
  i manganese
  d Mangan

5122 manganese acetate
  f acétate de manganèse
  e acetato de manganeso
  i acetato di manganese
  d Manganacetat; essigsaures Mangan

5123 manganese black
  f bioxyde de manganèse
  e dióxido de manganeso
  i biossido di manganese
  d Mangandioxyd

5124 manganese borate
  f borate de manganèse
  e borato de manganeso
  i borato di manganese
  d Manganborat

5125 manganese brown
  f marron de manganèse
  e marrón de manganeso
  i marrone di manganese
  d Manganbister

5126 manganese butyrate
  f butyrate de manganèse
  e butirato de manganeso
  i butirrato di manganese
  d Manganbutyrat

5127 **manganese carbonate**
  f carbonate de manganèse; blanc de
    manganèse
  e carbonato de manganeso
  i carbonato di manganese
  d Mangancarbonat; kohlensaures Mangan

5128 **manganese citrate**
  f citrate de manganèse
  e citrato de manganeso
  i citrato di manganese
  d Mangancitrat

5129 **manganese dioxide**
  f bioxyde de manganèse
  e dióxido de manganeso
  i biossido di manganese
  d Mangandioxyd; Mangansuperoxyd

5130 **manganese driers**
  f siccatifs au manganèse
  e desecadores de manganeso
  i essiccatori al manganese
  d Mangantrockenstoffe

  * **manganese ethylene-
    bisdithiocarbamate** → 5120

5131 **manganese gluconate**
  f gluconate de manganèse
  e gluconato de manganeso
  i gluconato di manganese
  d Mangangluconat

5132 **manganese glycerophosphate**
  f glycérophosphate de manganèse
  e glicerofosfato de manganeso
  i glicerofosfato di manganese
  d Manganglycerophosphat

5133 **manganese green**
  f vert de manganèse
  e verde de manganeso
  i verde di manganese
  d Mangangrün

5134 **manganese hypophosphite**
  f hypophosphite de manganèse
  e hipofosfito de manganeso
  i ipofosfito di manganese
  d Manganhypophosphit

5135 **manganese linoleate**
  f linoléate de manganèse
  e linoleato de manganeso
  i linoleato di manganese
  d Manganlinoleat

5136 **manganese naphthenate**
  f naphténate de manganèse
  e naftenato de manganeso
  i naftenato di manganese
  d Mangannaphthenat

5137 **manganese oleate**
  f oléate de manganèse
  e oleato de manganeso
  i oleato di manganese
  d Manganoleat

5138 **manganese resinate**
  f résinate de manganèse
  e resinato de manganeso
  i resinato di manganese
  d Manganresinat

5139 **manganic fluoride**
  f fluorure de manganèse
  e fluoruro mangánico
  i fluoruro di manganese
  d Manganfluorid

5140 **manganic hydroxide**
  f hydroxyde de manganèse
  e hidróxido mangánico
  i idrossido di manganese
  d Manganhydroxyd; Manganbraun

5141 **manganite**
  f manganite
  e manganita
  i mantanite
  d Manganit

5142 **manganous chloride**
  f chlorure manganeux
  e cloruro manganoso
  i cloruro manganoso
  d Manganchlorür

5143 **manganous nitrate**
  f nitrate manganeux
  e nitrato manganoso
  i nitrato manganoso
  d Manganonitrat

5144 **manganous oxide**
  f oxyde de manganèse
  e óxido manganoso
  i ossido manganoso
  d Manganoxyd

5145 **manganous sulphate**
  f sulfate manganeux
  e sulfato manganoso
  i solfato manganoso
  d Manganosulfat; Mangansulfat

5146 mangle
f calandre
e máquina secadora
i mangano
d Mangel

5147 man hole
f trou d'homme; trou de visite
e agujero de visita; agujero de hombre
i botola; passaggio d'ispezione
d Mannloch

5148 man-made fibre
f fibre artificielle
e fibra artificial
i fibra artificiale
d Chemiefaser; Kunstfaser

5149 manna
f manne
e maná
i manna
d Manna

5150 mannitol
f mannitol; sucre de manne
e manitol
i mannitolo
d Mannit; Mannazucker

5151 mannitol hexanitrate
f hexanitrate de mannitol
e hexanitrato de manitol
i esanitrato di mannitolo
d Mannithexanitrat

5152 mannose
f mannose
e manosa
i mannosio
d Mannose

5153 manometer
f manomètre
e manómetro
i manometro
d Druckmesser

5154 mantle fibres
f fibres du fuseau
e fibras del manto
i mantello del fuso
d Mantelfasern

5155 manufacturing process
f procédé de fabrication
e proceso de fabricación
i processo di fabbricazione
d Herstellungsverfahren

5156 manure
f engrais
e abono; fertilizante
i concime
d Dünger

* **M.A.R.** → 5467

5157 Marathon-Howard process
f procédé Marathon-Howard
e método Marathon-Howard
i processo Marathon-Howard
d Marathon-Howard-Verfahren

5158 marcasite
f marcasite
e marcasita
i marcassite
d Markasit

5159 margarine
f margarine
e margarina
i margarina
d Margarin

5160 margarite
f margarite
e margarita
i margarite
d Margarit

5161 marginal stability
f stabilité marginale
e estabilidad marginal
i stabilità marginale
d Grenzwertestabilität

5162 marihuana
f haschisch
e marijuana
i marihuana
d Marihuana

5163 marjoram oil
f essence de marjolaine
e aceite de mejorana
i olio di maggiorana
d Majoranöl

* **marrow** → 1110

* **marsh gas** → 5284

5164 martensite
f martensite
e martensita
i martensite
d Martensit

5165 mash
Kneaded or crushed material.
f produit malaxé
e amasijo; pasta
i macerato
d Brei

5166 mash
Mixture of coarse malt and hot water.
f maische; trempe
e mosto
i miscela; tempera
d Maische

5167 mash *v*
f pétrir
e amasar
i macerare
d zerquetschen

5168 masher
f hydrateur
e cuba mezcladora; premezclador
i promescolatore
d Maischepfanne

5169 mashing
f brassage
e braceaje
i macerazione
d Maischen

5170 mash tun
f cuve-matière
e tina de mezcla; cuba de mosto
i tino di miscela
d Maischbottich

5171 mass
f masse
e masa
i massa
d Masse

5172 mass abundance
f concentration en pourcentage de poids
e concentración en porcentajes de pesos
i concentrazione in percentuali di pesi
d Konzentration in Gewichtsprozenten

5173 mass action
f action de masse
e acción de masa
i azione di massa
d Massenwirkung

5174 mass balance
f bilan matière
e balance de materia

i bilancio di materia
d Materialbilanz

5175 massicot
f massicot
e masicote
i massicot
d Massicot

5176 mass spectrometer
f spectromètre de masse
e espectrómetro de masa
i spettrometro di massa
d Massenspektrometer

5177 mass velocity
f densité de courant massique
e densidad de corriente de masa
i densità di corrente di massa
d Mengenflußdichte

5178 masterbatch
f mélange maître
e mezcla básica
i partita maestra
d Vorgemisch

5179 master gauge
f comparateur
e calibre de comparación
i comparatore
d Vergleichslehre

5180 mastic
f mastic
e mástico
i mastice
d Kitt; Mastix; Gummimastiche

5181 masticator
f masticateur
e masticador
i macchina masticatrice
d Mastikator

5182 mastic gum
f gomme mastic
e almáciga
i gomma mastice
d Mastix; Gummimastiche

5183 matched metal dies
f moule et contre-moule rigides
e molde y contramolde rígidos
i stampo a controstampo rigido
d starre Formhälften

5184 material well
f espace de contraction

e cavidad para el material
i cavità per il materiale
d Füllraum

5185 matrix
f matrice
e matriz
i matrice
d Matrize

5186 matrix stickiness
f viscosité de la matrice
e viscosidad de la matriz
i viscosità della matrice
d Matrixverklebung

5187 matt dip
f bain de matage
e baño de matcado
i bagno di mordenzatura
d Mattbrenne

5188 matte
f matte
e mata
i metallina
d Stein

5189 maturation
f maturation
e maduración
i maturazione
d Alterung; Reifung

5190 maturation division
f division de maturation
e división de maturación
i divisione maturativa
d Reifungsteilung

5191 mauvein
f mauvéine
e mauveína
i mauvina
d Mauvein

5192 maximum allowable concentration
f concentration maximum admissible
e concentración máxima admisible
i concentrazione massima consentita
d maximale Arbeitsplatzkonzentration

5193 measuring cylinder
f vase gradué
e tubo graduado
i bicchiere graduato
d Meßglas

5194 measuring flask
f ballon gradué
e matraz graduado
i matraccio tarato
d Meßkolben

5195 measuring glass
f verre gradué
e vaso graduado
i cilindro graduato
d Meßglas

5196 measuring rule
f règle graduée
e regla graduada
i riga di misurazione
d Strichmaß

5197 measuring tank
f jaugeur
e cuba de medida
i serbatoio di misura
d Meßbehälter

5198 mechanical electroplating
f galvanoplastie mécanique
e galvanoplastia mecánica
i galvanoplastica meccanica
d Plattierung mit bewegten Kathoden

5199 mechanical foaming
f production mécanique de mousse
e alveolación mecánica
i formazione meccanica di schiuma
d mechanische Schaumstoffherstellung

5200 mechanical press
f presse mécanique
e prensa mecánica
i pressa meccanica
d mechanische Presse

5201 mechanical stage
f platine à chariot
e platina de ajuste
i piatto regolabile
d verstellbarer Objektivträger

5202 mechanical stripping
f dépouillage mécanique
e eliminación mecánica
i degalvanizzazione meccanica
d Abziehen; Abschleifen

5203 mechanical woodpulp
f cellulose de bois
e pasta de madera desintegrada
mecánicamente

i pasta di legno ottenuta meccanicamente
d Holzzellstoff

5204 meconic acid
f acide méconique
e ácido mecónico
i acido meconico
d Meconinsäure

5205 mediator
f intermédiaire
e intercesor
i portatore
d Vermittler

* **medullocell** → **5625**

5206 megasporocyte
f mégasporocyte
e megasporocito
i megasporocito
d Megasporocyte

5207 meiocyte
f méiocyte
e meiocito
i meiocito
d Meiocyte

5208 meiosome
f méiosome
e meiosoma
i meiosoma
d Meiosom

5209 Meker burner
f bec Méker
e mechero Méker
i becco Méker
d Méker-Brenner

5210 melamine
f mélamine
e melamina
i melamina
d Melamin

5211 melamine-formaldehyde resins
f résines à base de mélamine-
   formaldéhyde
e resinas de melamina y formaldehído
i resine alla melamina e formaldeide
d Melamin-Formaldehydharze

5212 melamine resins
f résines à la mélamine
e resinas de melamina
i resine melaminiche
d Melaminharze

5213 melanin
f mélanine
e melanina
i melanina
d Melanin

5214 meldometer
f meldomètre
e meldómetro
i meldometro
d Meldometer; Schmelzpunktmesser

5215 melissic acid
f acide mélissique
e ácido melísico
i acido melissico
d Melissinsäure

5216 mellitose
f mélitose
e melitosa
i raffinosio
d Melitose

5217 melt flow index
f indice de fluage
e índice de fusión
i indice di fusione
d Schmelzindex

* **melting** → **3646**

5218 melting point
f point de fusion
e punto de fusión
i punto di fusione
d Schmelzpunkt

5219 melting section
f zone de plastification
e zona de plastificación
i zona di fusione
d Plastifizierzone

5220 mendelevium
f mendelévium
e mendelevio
i mendelevio
d Mendelevium

5221 menthanediamine
f menthanediamine
e mentanodiamina
i mentanediammina
d Menthandiamin

5222 menthol
f menthol
e mentol

i mentolo
d Menthol

5223 menthone
f menthone
e mentona
i mentone
d Menthon

5224 menthyl acetate
f acétate de menthyle
e acetato de mentilo
i acetato di mentile
d Menthylacetat

5225 mercaptans; thiols
f mercaptans; thioalcools
e mercaptanos
i mercaptani
d Mercaptane; Thioalkohole

5226 mercaptobenzothiazole
f mercaptobenzothiazol
e mercaptobenzotiazol
i mercaptobenzotiazolo
d Mercaptobenzothiazol

5227 mercaptoethanol
f mercaptoéthanol
e mercaptoetanol
i mercaptoetanolo
d Mercaptoäthanol

5228 mercaptothiazoline
f mercaptothiazoline
e mercaptotiazolina
i mercaptotiazolina
d Mercaptothiazolin

5229 mercuric ammonium chloride
f chlorure double de mercure et
d'ammonium
e cloruro mercúrico-amónico
i cloruro di mercurio-ammonico
d Merkuriammoniumchlorid

5230 mercuric arsenate
f arséniate mercurique
e arseniato mercúrico
i arseniato mercurico
d Merkuriarsenat

5231 mercuric barium iodide
f iodure de mercure et de barium
e yoduro mercúrico-bárico
i ioduro mercurico di bario
d Merkuribariumjodid

5232 mercuric chloride
f chlorure mercurique
e cloruro mercúrico
i cloruro mercurico
d Merkurichlorid; Quecksilberchlorid

5233 mercuric cyanide
f cyanure mercurique
e cianuro mercúrico
i cianuro mercurico
d Quecksilbercyanid; Cyanquecksilber

5234 mercuric fluoride
f fluorure mercurique
e fluoruro mercúrico
i fluoruro mercurico
d Quecksilberfluorid

5235 mercuric iodide
f iodure mercurique
e yoduro mercúrico
i ioduro mercurico
d Jodquecksilber; Quecksilberjodid

5236 mercuric nitrate
f nitrate mercurique
e nitrato mercúrico
i nitrato mercurico
d Merkurinitrat; Quecksilberoxydnitrat

5237 mercuric oleate
f oléate mercurique
e oleato mercúrico
i oleato mercurico
d Quecksilberoleat

* **mercuric oxide, red ~ → 6983**

* **mercuric oxide, yellow ~ → 8899**

5238 mercuric oxycyanide
f oxycyanure de mercure
e oxicianuro mercúrico
i ossicianuro mercurico
d Quecksilberoxycyanid

5239 mercuric potassium iodide
f iodure de mercure et de potassium
e yoduro mercúrico-potásico
i ioduro mercurico di potassio
d Quecksilberkaliumjodid

5240 mercuric stearate
f stéarate de mercure
e estearato mercúrico
i stearato mercurico
d Quecksilberstearat

5241 mercuric sulphate
  f sulfate mercurique
  e sulfato mercúrico
  i solfato mercurico
  d Merkurisulfat; Quecksilbersulfat

5242 mercuric sulphide
  f sulfure mercurique
  e sulfuro mercúrico
  i solfuro mercurico
  d Quecksilbersulfid

* mercuric sulphide, red ~ → 6984

5243 mercuric thiocyanate
  f thiocyanate de mercure
  e tiocianato mercúrico
  i tiocianato mercurico
  d Merkurirhodanid

5244 mercurous chloride
  f chlorure mercureux
  e cloruro mercurioso
  i cloruro mercuroso; calomelano
  d Merkurochlorid; Quecksilberchlorür

5245 mercurous chromate
  f chromate mercureux
  e cromato mercurioso
  i cromato mercuroso
  d Merkurochromat

5246 mercurous nitrate
  f nitrate mercureux
  e nitrato mercurioso
  i nitrato mercuroso
  d Quecksilberoxydulnitrat

5247 mercurous sulphate
  f sulfate mercureux
  e sulfato mercurioso
  i solfato mercuroso
  d Merkurosulfat; Quecksilberoxydulsulfat

5248 mercury; quicksilver
  f mercure
  e mercurio
  i mercurio
  d Quecksilber

5249 mercury electrolytic cell
  f cellule électrolytique au mercure
  e celda electrolítica de mercurio
  i cella elettrolitica a mercurio
  d Quecksilberzelle

5250 mercury fulminate
  f fulminate de mercure
  e fulminato de mercurio

  i fulminato di mercurio
  d Knallquecksilber

5251 mercury naphthenate
  f naphténate de mercure
  e naftenato de mercurio
  i naftenato di mercurio
  d Quecksilbernaphthenat

5252 mercury oxide cell
  f pile à oxyde de mercure
  e pila de óxido de mercurio
  i pila a ossido di mercurio
  d Quecksilberoxydelement

5253 mercury vapour lamp
  f lampe à vapeur de mercure
  e lámpara de vapor de mercurio
  i lampada a vapori di mercurio
  d Quecksilberdampflampe

5254 mercury vapour rectifier
  f redresseur à vapeur de mercure
  e rectificador de vapor de mercurio
  i raddrizzatore a vapore di mercurio
  d Quecksilberdampfgleichrichter

5255 meroblastic cleavage
  f clivage méroblastique
  e escisión meroblástica
  i scissione meroblastica
  d meroblastische Furchung

5256 mesh
  f maille
  e malla
  i maglia
  d Masche

5257 mesitylene
  f mésitylène
  e mesitileno
  i mesitilene
  d Mesitylen

5258 mesityl oxide
  f oxyde de mésityle
  e óxido de mesitilo
  i ossido di mesitile
  d Mesityloxyd

5259 mesomerism
  f mésomérie
  e mesomería
  i mesomeria
  d Mesomerie

5260 meson
  f méson

e mesón
i mesone
d Meson

5261 mesothorium-I
f mésothorium-I
e mesotorio-I
i mesotorio-I
d Mesothorium-I

5262 metabolic nucleus
f noyau métabolique
e núcleo metabólico
i nucleo metabolico
d Arbeitskern

5263 metabolism
f métabolisme
e metabolismo
i metabolismo
d Stoffwechsel

5264 metabolite
f métabolite
e metabolita
i metabolito
d Stoffwechselprodukt

* **metaformaldehyde** → 8524

5265 metaldehyde
f métaldéhyde
e metaldehído
i metaldeide
d Metaldehyd

5266 metal distribution ratio
f rapport de distribution du métal
e relación de distribución del metal
i rapporto di distribuzione del metallo
d Niederschlagsverteilungsverhältnis

5267 metal fog; metal mist
f brouillard métallique
e niebla metálica
i nebbia metallica
d Metallnebel

5268 metallic fibre
f fibre métallique
e fibra metalizada
i fibra metallica
d metallisierter Faden

5269 metallization
f métallisation
e metalización
i metallizzazione
d Metallspritzverfahren; Metallisieren

5270 metallized dyes
f teintures métallisées
e colorantes metalizados
i coloranti metallizzati
d metallisierte Farbstoffe

5271 metalloid
f métalloïde
e metaloide
i metalloide
d Halbmetall; Metalloid

5272 metallo-organic
f organométallique
e metalorgánico
i organicometallico
d organometallisch

5273 metallo-organic pigment
f pigment semi minéral
e pigmento metalorgánico
i pigmento metallo-organico
d metall-organisches Pigment

5274 metallurgical coke
f coke métallurgique
e coque metalúrgico
i coke metallurgico
d Hochofenkoks; Zechenkoks

5275 metallurgy
f métallurgie
e metalurgia
i metallurgia
d Metallurgie; Hüttenkunde

* **metal mist** → 5267

5276 metamorphism
f métamorphisme
e metamorfismo
i metamorfismo
d Metamorphic

5277 metanilic acid
f acide métanilique
e ácido metanílico
i acido metanilico
d Metanilsäure

5278 metaphase
f métaphase
e metafase
i metafase
d Metaphase

5279 metastable
f métastable
e metaestable

i metastabile
d metastabil

5280 methacrolein; methacrylaldehyde
f méthacroléine
e metacroleína
i metacroleina
d Methacrolein

5281 methacrylate esters
f esters de méthacrylate
e ésteres de metacrilato
i esteri al metacrilato
d Methacrylatester

5282 methacrylatochromic chloride
f chlorure méthacrylatochromique
e cloruro metacrilatocrómico
i cloruro metacrilatocromico
d Methacrylatochromichlorid

5283 methacrylic acid
f acide méthacrylique
e ácido metacrílico
i acido metacrilico
d Methacrylsäure

5284 methane; marsh gas
f méthane; gas des marais; grisou
e metano; gas de pantano
i metano; gas delle paludi
d Methan; Sumpfgas; Grubengas

5285 methanediamine
f méthanediamine
e metanodiamina
i metanodiammina
d Methandiamin

5286 methane hydroperoxide
f hydroperoxyde de méthane
e hidroperóxido de metano
i idroperossido di metano
d Methanhydroperoxyd

5287 methanol; methyl alcohol; colonial spirit
f méthanol; alcool méthylique
e metanol; alcohol de metilo; alcohol colonial
i metanolo; alcool metilico
d Methanol; Methylalkohol

5288 methionine
f méthionine
e metionina
i metionina
d Methionin

* **methoxsalen** → **8877**

5289 methoxyacetic acid
f acide méthoxyacétique
e ácido metoxiacético
i acido metossiacetico
d Methoxyessigsäure

5290 methoxyacetophenone
f méthoxyacétophénone
e metoxiacetofenona
i metossiacetofenone
d Methoxyacetophenon

5291 methoxybutanol
f méthoxybutanol
e metoxibutanol
i metossibutanolo
d Methoxybutanol

5292 methoxychlor
f méthoxychlore
e metoxicloro
i metossicloro
d Methoxychlor

5293 methoxyethyl acetyl ricinoleate
f ricinoléate acétométhoxyéthylique
e acetilricinoleato de metoxietilo
i metossietilacetilricinoleato
d Methoxyäthylacetylricinoleat

5294 methoxyethylmercury acetate
f acétate méthoxyéthylmercurique
e acetato metoxietilmercúrico
i acetato metossietilmercurico
d Methoxyäthylquecksilberacetat

5295 methoxyethyl oleate
f oléate de méthoxyéthyle
e oleato de metoxietilo
i oleato metossietilico
d Methoxyäthyloleat

5296 methoxyethyl ricinoleate
f ricinoléate de méthoxyéthyle
e ricinoleato de metoxietilo
i metossietilricinoleato
d Methoxyäthylricinoleat

5297 methoxyethyl stearate
f stéarate de méthoxyéthyle
e estearato de metoxietilo
i stearato metossietilico
d Methoxyäthylstearat

5298 methoxymethylpentanol
f méthoxyméthylpentanol
e metoximetilpentanol
i metossimetilpentanolo
d Methoxymethylpentanol

5299 methoxymethylpentanone
f méthoxyméthylpentanone
e metoximetilpentanona
i metossimetilpentanone
d Methoxymethylpentanon

5300 methoxypropylamine
f méthoxypropylamine
e metoxipropilamina
i metossipropilammina
d Methoxypropylamin

5301 methoxytriglycol acetate
f acétate de méthoxytriglycol
e acetato de metoxitriglicol
i acetato di metossitriglicolo
d Methoxytriglykolacetat

5302 methyl abietate
f abiétate de méthyle
e abietato de metilo
i abietato di metile
d Methylabietat

5303 methylacetanilide
f méthylacétanilide
e metilacetanilida
i metilacetanilide
d Methylacetanilid

5304 methyl acetate
f acétate de méthyle
e acetato de metilo
i acetato di metile
d Methylacetat; essigsaures Methyl

5305 methyl acetoacetate
f acéto-acétate de méthyle
e acetoacetato de metilo
i acetoacetato di metile
d Methylacetoacetat

5306 methylacetone
f méthylacétone
e metilacetona
i acetone di metile
d Methylaceton

5307 methylacetophenone
f méthylacétophénone
e metilacetofenona
i metilacetofenone
d Methylacetophenon

5308 methylacetylene
f méthylacétylène
e metilacetileno
i metilacetilene
d Methylacetylen

5309 methyl acetylricinoleate
f acétylricinoléate de méthyle
e acetilricinoleato de metilo
i acetilricinoleato di metile
d Methylacetylricinoleat

5310 methyl acetylsalicylate
f méthylacétylsalicylate
e acetilsalicilato de metilo
i acetilsalicilato di metile
d Methylacetylsalicylat

5311 methyl acrylate
f acrylate de méthyle
e acrilato de metilo
i acrilato di metile
d Methylacrylat

5312 methylal; formal
f méthylal
e metilal
i metilal
d Methylal

* **methyl alcohol** → 5287

5313 methylallyl chloride
f chlorure de méthylallyle
e cloruro de metilalilo
i cloruro di metilallile
d Methylallylchlorid

5314 methylaluminium sesquibromide
f sesquibromure de méthylaluminium
e sesquibromuro de metilaluminio
i sesquibromuro di metilalluminio
d Methylaluminiumsesquibromid

5315 methylamine
f méthylamine
e metilamina
i metilammina
d Methylamin

5316 methylaminophenol
f méthylaminophénol
e metilaminofenol
i amminofenolo metilico
d Methylaminophenol

5317 methylamyl acetate
f acétate de méthylamyle
e acetato de metilamilo
i acetato di metilamile
d Methylamylacetat

5318 methylamyl alcohol
f alcool méthylamylique
e alcohol metilamílico

i alcool metilamilico
d Methylamylalkohol

5319 methylamyl carbinol
f méthylamylcarbinol
e metilamilcarbinol
i metilamilcarbinolo
d Methylamylcarbinol

5320 methylamyl ketone
f méthylamylcétone
e metilamilcetona
i metilamilchetone
d Methylamylketon

5321 methylaniline
f méthylaniline
e metilanilina
i metilanilina
d Methylanilin

5322 methylanthracene
f méthylanthracène
e metilantraceno
i metilantracene
d Methylanthracen

5323 methyl anthranilate
f anthranilate de méthyle
e antranilato de metilo
i antranilato di metile
d Methylanthranilat

5324 methylanthraquinone
f méthylanthraquinone
e metilantraquinona
i metilantrachinone
d Methylanthrachinon

5325 methyl arachidate
f arachidate de méthyle
e araquidato de metilo
i arachidato di metile
d Methylarachidat

5326 methylated spirits
f alcool dénaturé
e alcohol metilado
i alcool denaturato
d denaturierter Alkohol

5327 methyl behenate
f béhénate de méthyle
e behenato de metilo
i beenato di metile
d Methylbehenat

5328 methyl benzoate
f benzoate de méthyle

e benzoato de metilo
i benzoato di metile
d Methylbenzoat

5329 methyl benzoylbenzoate
f benzoylbenzoate de méthyle
e benzoilbenzoato de metilo
i benzoilbenzoato di metile
d Methylbenzoylbenzoat

5330 methylbenzylamine
f méthylbenzylamine
e metilbencilamina
i metilbenzilammina
d Methylbenzylamin

5331 methylbenzyldiethanolamine
f méthylbenzyldiéthanolamine
e metilbencildietanolamina
i metilbenzildietanolammina
d Methylbenzyldiäthanolamin

5332 methylbenzyldimethylamine
f méthylbenzyldiméthylamine
e metilbencildimetilamina
i metilbenzildimetilammina
d Methylbenzyldimethylamin

5333 methylbenzyl ether
f éther méthylbenzylique
e éter metilbencílico
i etere metilbenzilico
d Methylbenzyläther

5334 methyl bromide
f bromure de méthyle
e bromuro de metilo
i bromuro di metile
d Methylbromid; Brommethyl

5335 methyl bromoacetate
f bromo-acétate de méthyle
e bromoacetato de metilo
i bromoacetato di metile
d Methylbromacetat

5336 methylbutanol
f méthylbutanol
e metilbutanol
i metilbutanolo
d Methylbutanol

5337 methylbutene
f méthylbutène
e metilbuteno
i metilbutene
d Methylbuten

5338 methylbutenol
 f méthylbuténol
 e metilbutenol
 i metilbutenolo
 d Methylbutenol

5339 methylbutyl benzene
 f méthylbutylbenzène
 e metilbutilbenceno
 i metilbutilbenzene
 d Methylbutylbenzol

5340 methyl butyl ketone; propylacetone
 f méthylbutylcétone
 e metilbutilcetona
 i metilbutilchetone
 d Methylbutylketon

5341 methylbutynol
 f méthylbutynol
 e metilbutinol
 i metilbutinolo
 d Methylbutynol

5342 methyl butyrate
 f butyrate de méthyle
 e butirato de metilo
 i metilbutirrato
 d Methylbutyrat

5343 methyl caprate
 f caprate de méthyle
 e caprato de metilo
 i caprato di metile
 d Methylcaprat

5344 methyl caprylate; methyl octanoate
 f caprylate de méthyle
 e caprilato de metilo
 i caprilato di metile
 d Methylcaprylat

5345 methyl carbonate
 f carbonate de méthyle
 e carbonato de metilo
 i carbonato di metile
 d Methylcarbonat

5346 methylcellulose
 f méthylcellulose
 e metil celulosa
 i metilcellulosa
 d Methylcellulose

5347 methyl cerotate
 f cérotate de méthyle
 e cerotato de metilo
 i cerotato di metile
 d Methylcerotat

5348 methyl chloride
 f chlorure de méthyle
 e cloruro de metilo
 i cloruro di metile
 d Methylchlorid

5349 methyl chloroacetate
 f chloracétate de méthyle
 e cloroacetato de metilo
 i cloroacetato di metile
 d Methylchloracetat

5350 methyl chloroformate
 f chloroformiate de méthyle
 e cloroformiato de metilo
 i cloroformiato di metile
 d Methylchlorformiat

5351 methyl chlorophenoxypropionic acid
 f acide méthyl chlorophénoxypropionique
 e ácido metilclorofenoxipropiónico
 i acido metilclorofenossipropionico
 d Methylchlorphenoxypropionsäure

5352 methyl chlorosulphonate
 f sulfonate d'éthyle chloré
 e clorosulfonato de metilo
 i clorosolfonato di metile
 d Chlorsulfonsäuremethylester

5353 methyl cinnamate
 f cinnamate de méthyle
 e cinamato de metilo
 i cinnamato di metile
 d Methylcinnamat

5354 methylclothiazide
 f méthylclothiazide
 e metilclotiazida
 i metilclotiazide
 d Methylclothiazid

5355 methylcoumarin
 f méthylcoumarine
 e metilcumarina
 i metilcumarina
 d Methylkumarin

5356 methyl cyanoacetate
 f cyanoacétate de méthyle
 e cianoacetato de metilo
 i cianoacetato di metile
 d Methylcyanacetat

5357 methyl cyanoformate
 f cyanoformiate de méthyle
 e cianoformiato de metilo
 i cianoformiato di metile
 d Methylcyanformiat

5358 methylcyclohexane
 f méthylcyclohexane
 e metilciclohexano
 i metilcicloesano
 d Methylcyclohexan

5359 methylcyclohexanol
 f méthylcyclohexanol
 e metilciclohexanol
 i metilcicloesanolo
 d Methylcyclohexanol

5360 methylcyclohexanone
 f méthylcyclohexanone
 e metilciclohexanona
 i metilcicloesanone
 d Methylcyclohexanon; Methylanon

5361 methylcyclohexanone glyceryl acetal
 f glycérylacétal de méthylcyclohexanone
 e glicerilacetal de la metilciclohexanona
 i metilcicloesanoneglicerilacetale
 d Methylcyclohexanonglycerylacetal

5362 methyl cyclohexene carboxaldehyde
 f méthylcyclohexènecarboxaldéhyde
 e carboxaldehído de metilciclohexeno
 i carbossialdeide di metilcicloesene
 d Methylcyclohexencarboxaldehyd

5363 methylcyclohexylamine
 f méthylcyclohexylamine
 e metilciclohexilamina
 i metilcicloesilammina
 d Methylcyclohexylamin

5364 methylcyclohexyl isobutyl phthalate
 f phtalate de méthylcyclohexylisobutyle
 e ftalato de metilciclohexilisobutilo
 i metilcicloesilisobutilftalato
 d Methylcyclohexylisobutylphthalat

5365 methylcyclopentadiene dimer
 f dimère de méthylcyclopentadiène
 e dímero de metilciclopentadieno
 i dimero del metilciclopentadiene
 d Methylcyclopentadiendimer

5366 methylcyclopentane
 f méthylcyclopentane
 e metilciclopentano
 i metilciclopentano
 d Methylcyclopentan

5367 methyl dichloroacetate
 f dichloracétate de méthyle
 e dicloroacetato de metilo
 i dicloroacetato di metile
 d Methyldichloracetat

5368 methyl dichlorostearate
 f dichlorostéarate de méthyle
 e dicloroestearato de metilo
 i diclorostearato di metile
 d Methyldichlorstearat

5369 methyldiethanolamine
 f méthyldiéthanolamine
 e metildietanolamina
 i metildietanolammina
 d Methyldiäthanolamin

5370 methyldioxolane
 f méthyldioxolane
 e metildioxolano
 i biossolano di metile
 d Methyldioxolan

5371 methylenebisacrylamide
 f méthylènebisacrylamide
 e metilenbisacrilamida
 i metilenebisacrilammide
 d Methylenbisacrylamid

5372 methylene blue
 f bleu de méthylène
 e azul de metileno
 i blu di metilene
 d Methylenblau

5373 methylene bromide
 f bromure de méthylène
 e bromuro de metileno
 i bromuro di metilene
 d Methylenbromid

5374 methylene chloride
 f chlorure de méthylène
 e cloruro de metileno
 i cloruro di metilene
 d Methylenchlorid; Dichlormethan

5375 methylenedianiline
 f méthylènedianiline
 e metilendianilina
 i metilenedianilina
 d Methylendianilin

5376 methylene ditannin
 f méthylèneditannine
 e metilenditanino
 i ditannino di metilene
 d Methylenditannin

5377 methylene iodide
 f iodure de méthylène
 e yoduro de metileno
 i ioduro di metilene
 d Methylenjodid

* **methylene-protocatechuic aldehyde**
→ **6413**

5378 methylergonovine maleate
f maléate de méthylergonovine
e maleato de metilergonovina
i maleato di metilergonovina
d Methylergonovinmaleat

5379 methyl ethyl ketone
f méthyléthylcétone
e metiletilcetona
i metiletilchetone
d Methyläthylketon

5380 methylethylpyridine
f méthyléthylpyridine
o metiletilpiridina
i metiletilpiridina
u Methylathylpyridin

5381 methylformanilide
f méthylformanilide
e metilformanilida
i metilformanilide
d Methylformanilid

5382 methyl formate
f formiate de méthyle
e formiato de metilo
i formiato di metile
d Methylformiat

5383 methylfuran
f méthylfuranne
e metilfurano
i metilfurano
d Methylfuran

5384 methyl furoate
f furoate de méthyle
e furoato de metilo
i furoato di metile
d Methylfuroat

5385 methylglucamine diatrizoate
f diatrizoate de méthylglucamine
e diatrizoato de metilglucamina
i diatrizoato di metilglucamina
d Methylglucamindiatrizoat

5386 methyl glucoside
f méthylglucoside
e metilglucósido
i metilglucoside
d Methylglucosid

5387 methyl heneicosanoate
f hénéicosanoate

e heneicosanoato de metilo
i eneicosanoato di metile
d Methylheneikosanoat

5388 methyl heptadecanoate
f heptadécanoate de méthyle
e heptadecanoato de metilo
i eptadecanoato di metile
d Methylheptadecanoat

5389 methylheptane
f méthylheptane
e metilheptano
i metileptano
d Methylheptan

5390 methylheptenone
f méthylheptenone
e metilheptenona
i metileptenone
d Methylheptenon

5391 methylhexane
f méthylhexane
e metilhexano
i metilesano
d Methylhexan

5392 methylhexaneamine
f méthylhexaneamine
e metilhexanoamina
i metilesanammina
d Methylhexanamin

5393 methyl hexyl ketone
f méthylhexylcétone
e metilhexilcetona
i metilesilchetone
d Methylhexylketon

5394 methylhydrazine
f méthylhydrazine
e metilhidracina
i metilidrazina
d Methylhydrazin

5395 methylhydroxybutanone
f méthylhydroxybutanone
e metilhidroxibutanona
i metilidrossibutanone
d Methylhydroxybutanon

5396 methyl iodide
f iodure de méthyle
e yoduro de metilo
i ioduro di metile
d Methyljodid; Jodmethyl

5397 methylionone
f méthylionone
e metilionona
i metilionone
d Methyljonon

5398 methyl isoamyl ketone
f méthylisoamylcétone
e metilisoamilcetona
i metilisoamilchetone
d Methylisoamylketon

5399 methyl isobutyl ketone
f méthylisobutylcétone
e metilisobutilcetona
i metilisobutilchetone
d Methylisobutylketon

5400 methylisoeugenol
f méthylisoeugénol
e metilisoeugenol
i metilisoeugenolo
d Methylisoeugenol

5401 methyl isopropenyl ketone
f méthylisopropénylcétone
e metilisopropenilcetona
i metilisopropenilchetone
d Methylisopropenylketon

5402 methyl lactate
f lactate de méthyle
e lactato de metilo
i lattato di metile
d Methyllactat

5403 methyl laurate
f laurate de méthyle
e laurato de metilo
i laurato di metile
d Methyllaurat; Laurinsäuremethylester

5404 methyl lauroleate
f lauroléate de méthyle
e lauroleato de metilo
i lauroleato di metile
d Methyllauroleat

5405 methyl lignocerate
f lignocérate de méthyle
e lignocerato de metilo
i lignocerato di metile
d Methyllignocerat

5406 methyl linoleate
f linoléate de méthyle
e linoleato de metilo
i linoleato di metile
d Methyllinoleat

5407 methylmagnesium bromide
f bromure de méthyle et de magnésium
e bromuro de metilmagnesio
i bromuro di metilmagnesio
d Methylmagnesiumbromid

5408 methylmagnesium iodide
f iodure de méthyle et de magnésium
e yoduro de metilmagnesio
i ioduro di metilmagnesio
d Methylmagnesiumjodid

5409 methyl mercaptan
f méthylmercaptan
e metilmercaptano
i metilmercaptano
d Methylmercaptan

5410 methyl methacrylate
f méthacrylate de méthyle
e metacrilato de metilo
i metacrilato di metile
d Methylmethacrylat

5411 methyl morpholine
f méthylmorpholine
e metilmorfolina
i metilmorfolina
d Methylmorpholin

5412 methyl myristate
f myristate de méthyle
e miristato de metilo
i miristato di metile
d Methylmyristat

5413 methyl myristoleate
f myristoléate de méthyle
e miristoleato de metilo
i miristoleato di metile
d Methylmyristoleat

5414 methylnaphthalene
f méthylnaphtalène
e metilnaftaleno
i metilnaftalene
d Methylnaphthalen

5415 methyl naphthyl ketone
f méthylnaphtylcétone
e metilnaftilcetona
i metilnaftilchetone
d Methylnaphthylketon

5416 methyl nitrate
f nitrate de méthyle
e nitrato de metilo
i nitrato di metile
d Methylnitrat

5417 methyl nonadecanoate
 f nonadécanoate de méthyle
 e nonadecanoato de metilo
 i nonadecanoato di metile
 d Methylnonadecanoat

5418 methyl nonanoate
 f nonanoate de méthyle
 e nonanoato de metilo
 i nonanoato di metile
 d Methylnonanoat

5419 methylnonylacetaldehyde
 f méthylnonylacétaldéhyde
 e metilnonilacetaldehído
 i metilnonilacetaldeide
 d Methylnonylacetaldehyd

5420 methyl nonyl ketone
 f méthylnonylcétone
 e metilnonilcetona
 i metilnonilchetone
 d Methylnonylketon

 * **methyl octanoate** → 5344

5421 methylol dimethylhydantoin
 f méthyloldiméthylhydantoïne
 e metiloldimetilhidantoína
 i dimetilidantoina di metilolo
 d Methyloldimethylhydantoin

5422 methyl oleate
 f oléate de méthyle
 e oleato de metilo
 i oleato di metile
 d Methyloleat

5423 methylol urea
 f méthylolurée
 e metilolurea
 i metilolurea
 d Methylolharnstoff

5424 methylorthoanisidine
 f méthylorthoanisidine
 e metilortoanisidina
 i metilortoanisidina
 d Methylorthoanisidin

5425 methyl palmitate
 f palmitate de méthyle
 e palmitato de metilo
 i palmitato di metile
 d Methylpalmitat

5426 methyl palmitoleate
 f palmitoléate de méthyle
 e palmitoleato de metilo

 i palmitoleato di metile
 d Methylpalmitoleat

5427 methyl pentachlorostearate
 f pentachlorostéarate de méthyle
 e pentacloroestearato de metilo
 i metilpentaclorostearato
 d Methylpentachlorstearat

5428 methyl pentadecanoate
 f pentadécanoate de méthyle
 e pentadecanoato de metilo
 i pentadecanoato di metile
 d Methylpentadecanoat

5429 methylpentadiene
 f méthylpentadiène
 e metilpentadieno
 i metilpentadiene
 d Methylpentadien

5430 methylpentaldehyde
 f méthylpentaldéhyde
 e metilpentaldehído
 i metilpentaldeide
 d Methylpentaldehyd

5431 methylpentane
 f méthylpentane
 e metilpentano
 i metilpentano
 d Methylpentan

5432 methylpentanediol
 f méthylpentanediol
 e metilpentanodiol
 i metilpentanediolo
 d Methylpentandiol

5433 methylpentene
 f méthylpentène
 e metilpenteno
 i metilpentene
 d Methylpenten

5434 methylpentynol
 f méthylpentynol
 e metilpentinol
 i metilpentinolo
 d Methylpentynol

5435 methyl phenylacetate
 f acétate de méthylphényle
 e acetato de fenilmetilo
 i acetato di fenilmetile
 d Methylphenylacetat

5436 methyl phenyldichlorosilane
 f méthylphényldichlorosilane

e metilfenildiclorosilano
i fenildiclorosilano di metile
d Methylphenyldichlorsilan

5437 methylphloroglucinol
f méthylphloroglucine
e metilfloroglucina
i metilfloroglucinolo
d Methylphloroglycin

5438 methylphosphoric acid
f acide méthylphosphorique
e ácido metilfosfórico
i acido metilfosforico
d Methylphosphorsäure

5439 methyl phthalyl ethyl glycolate
f méthyl-phtalyl-éthyl-glycolate
e metilftalilglicolato de etilo
i metilftaliletilglicolato
d Methylphthalyläthylglykolat

5440 methylpiperazine
f méthylpipérazine
e metilpiperacina
i metilpiperazina
d Methylpiperazine

5441 methyl propionate
f propionate de méthyle
e propionato de metilo
i propionato di metile
d Methylpropionat

5442 methyl propyl ether; metopryl
f éther méthylpropylique
e éter metilpropílico
i etere metilpropilico
d Methylpropyläther

5443 methyl propyl ketone
f méthylpropylcétone
e metilpropilcetona
i metilpropilchetone
d Methylpropylketon

5444 methylpyrrole
f méthylpyrrole
e metilpirrol
i metilpirrolo
d Methylpyrrol

5445 methylpyrrolidone
f méthylpyrrolidone
e metilpirrolidona
i metilpirrolidone
d Methylpyrrolidon

5446 methyl red
f rouge de méthyle
e rojo de metilo
i rosso di metile; metilrosso
d Methylrot

5447 methyl ricinoleate
f ricinoléate de méthyle
e ricinoleato de metilo
i ricinoleato di metile
d Methylricinoleat

5448 methyl salicylate
f salicylate de méthyle
e salicilato de metilo
i salicilato di metile
d Methylsalicylat

5449 methyl stearate
f stéarate de méthyle
e estearato de metilo
i stearato di metile
d Methylstearat

5450 methylstyrene
f méthylstyrolène
e metilestireno
i metilstirene
d Methylstyrol

* **methylsuccinic acid** → 6840

* **methylsulphonal** → 7933

5451 methyltaurine
f méthyltaurine
e metiltaurina
i metiltaurina
d Methyltaurin

5452 methyltetrahydrofuran
f méthyltétrahydrofuranne
e metiltetrahidrofurano
i metiltetraidrofurano
d Methyltetrahydrofuran

5453 methyl thiouracil
f méthylthiouracile
e metiltiouracilo
i metiltiouracile
d Methylthiouracil

5454 methyltrichlorosilane
f méthyltrichlorosilane
e metiltriclorosilano
i metiltriclorosilano
d Methyltrichlorsilan

5455 methyl tridecanoate
f tridécanoate de méthyle
e tridecanoato de metilo
i tridecanoato di metile
d Methyltridecanoat

5456 methyl undecanoate
f undécanoate de méthyle
e undecanoato de metilo
i undecanoato di metile
d Methylundecanoat

5457 methylvinyldichlorosilane
f méthylvinyldichlorosilane
e metilvinildiclorosilano
i metilvinildiclorosilano
d Methylvinyldichlorsilan

5458 methyl vinyl ether
f éther méthylvinilique
e metilviniléter
i ètere metilvinilico
d Methylvinyläther

5459 methyl vinylpyridine
f méthylvinylpyridine
e metilvinilpiridina
i metilvinilpiridina
d Methylvinylpyridin

5460 methyl violet; Paris violet
f violet de méthyle
e violeta de metilo; violeta de París
i violetto di metile
d Methylviolett

* metopryl → 5442

5461 miasma
f miasme
e miasma
i miasma
d Miasma

5462 mica
f mica
e mica
i mica
d Glimmer

5463 mica board
f mica aggloméré
e cartón de mica
i cartone di mica
d glimmerhaltige Schichtplatte

5464 micaceous iron oxide
f oxyde de fer micacé
e óxido de hierro micáceo
i ossido di ferro micaceo
d Eisenglimmer

5465 micelle
f micelle
e micela
i micella
d Mizelle

5466 microanalysis
f microanalyse
e microanálisis
i microanalisi
d Mikroanalyse

5467 microanalytical reagent; M.A.R.
f réactif microanalytique
e reactivo microanalítico
i reagente microanalitico
d mikroanalytisches Reagens

5468 microbalance
f microbalance
e microbalanza
i microbilancia
d Mikrowaage

5469 microballoons
f microballons
e microbalones
i microsfere
d kleine, gasgefüllte Hohlräume

5470 microbe
f microbe
e microbio
i microbo
d Mikrobe

* microbicide → 3741

5471 microbiology
f microbiologie
e microbiología
i microbiologia
d Mikrobiologie

5472 microcellular rubber
f caoutchouc microcellulaire
e goma microcelular
i gomma microcellulare
d Mikrozellgummi

5473 microchemistry
f microchimie
e microquímica
i microchimica
d Mikrochemie

5474 microchromosome
f microchromosome
e microcromosoma
i microcromosoma
d Mikrochromosom

5475 microcosmic salt
f sel microcosmique; phosphate sodio-
ammonique
e sal microcósmica
i sale microcosmico; fosfato di sodio
ammonico
d Phosphorsalz

5476 microfoam rubber
f mousse de latex microporeux
e espuma de látex microporosa
i gomma piuma microporosa
d Mikroschaumgummi

5477 microglia
f microglie
e microglia
i microglia
d Mikroglie

5478 microgliacyte
f microgliacyte
e microgliacito
i microgliacito
d Mikrogliazyt

5479 microheterochromatic
f microhétérochromatique
e microheterocromático
i microeterocromatico
d mikroheterochromatisch

5480 microlite
f microlite
e microlita
i microlite
d Mikrolith

5481 micromanometer
f micromanomètre
e micromanómetro
i micromanometro
d Mikromanometer

5482 micrometer
f micromètre
e micrómetro
i micrometro
d Mikrometer

5483 micron
f micron
e micrón

i micron
d Mikron

5484 micronucleus
f micronucléus
e micronúcleo
i micronucleo
d Mikronukleus

5485 microporous
f microporeux
e microporoso
i microporoso
d mikroporös

5486 microscope slide
f porte-objet
e portaobjeto
i portaoggetti
d Objektglas

5487 microscopic concentration
f concentration microscopique
e concentración microscópica
i concentrazione microscopica
d mikroskopische Konzentration

5488 migration
f migration
e migración
i migrazione
d Wanderung

5489 migration fastness
f résistance à la migration de couleur
e resistencia a la migración de color
i resistenza alla migrazione di colori
d Beständigkeit gegen Farbwanderung

5490 mild
f doux
e suave
i leggero
d mild

5491 milking booster
f survolteur
e convertidor para carga de acumuladores
i sulvoltrice per carica batterie
d Spannungserhöher

5492 milk of magnesia
f magnésie blanche; magnésie anglaise
e lechada de magnesia
i magnesia
d Magnesia

* milk sugar → 4681

5493 mill
f moulin; broyeur
e molino
i frantumatore; molino
d Mühle

5494 milled glass fibre
f fibre de verre broyée
e fibra de vidrio triturado
i fibra di vetro macinato
d gemahlene Glasfaser

5495 Mills-Packard chamber
f chambre de plomb Mills-Packard
e cámara de plomo Mills-Packard
i camera di Mills-Packard
d Mills-Packardsche Bleikammer

5496 mimetite
f mimétite; mimétèse; mimótòno
e mimetita
i mimesite; mimotito
d Mimetesit; Grünbleierz

5497 mineral black
f noir minéral
e negro mineral
i nero minerale
d Mineralschwarz

5498 mineral charcoal
f fusain
e carbón vegetal mineralizado
i carbone minerale
d Steinkohle

5499 mineral orange
f orangé minéral
e anaranjado mineral
i ossido di minio
d Mineralorange

5500 mineral pigment
f pigment minéral
e pigmento mineral
i pigmento minerale
d Mineralfarbstoff

5501 mineral rubber
f caoutchouc minéral
e caucho mineral
i gomma minerale
d Mineralkautschuk

5502 mineral violet
f violet minéral
e violeta mineral
i violetto minerale
d Mineralviolett

* **mineral wax** → 6063

5503 mineral white
f gypse; pierre à plâtre
e blanco mineral
i bianco minerale
d Mineralweiß; Permanentweiß

5504 mineral wool
f laine de scorie
e lana mineral
i lana di scorie
d Schlackenwolle

5505 minimum ionization
f ionisation minimale
e ionización mínima
i ionizzazione minima
d minimale Ionisation

5506 minium; red lead
f minium
e minio
i minio; rosso di piombo
d Minium; Bleirot; Bleimennige

5507 mirabilite
f mirabilite
e mirabilita
i mirabilite
d Mirabilit

* **mirbane oil** → 5755

* **mispickel** → 671

5508 mitosome
f mitosome
e mitosoma
i mitosoma
d Mitosom

5509 mitotic
f mitotique
e mitótico
i mitotico
d mitotisch

5510 mitotic poison
f poison mitotique
e tóxico mitótico
i veleno mitotico
d Mitosegift

5511 mixed glue
f colle préparée
e adhesivo mezclado
i colla preparata
d Klebemischung

5512 mixed polyelectrode potential
f tension mixte d'une polyélectrode
e tensión mixta de un polielectrodo
i potenziale misto di elettrodo multiplo
d Mischpotential

5513 mixer
f mélangeur
e mezclador
i mescolatrice
d Mischer

5514 mixochromosome
f mixochromosome
e mixocromosoma
i mixocromosoma
d Mixochromosom

5515 mixture
f mélange
e mezcla
i mescolanza; miscela
d Gemisch; Mischung

5516 mobilometer
f viscosimètre
e viscosímetro
i mobilometro
d Viskositätsmesser

5517 mock-up
f maquette; modèle
e maqueta
i prototipo sperimentale
d Attrappe; Baumodell

5518 modacrylic fibre
f fibre modacrylique
e fibra modacrílica
i fibra modacrilica
d Modacrylfaser

5519 modifier complex
f complexe modificateur
e complejo modificador
i complesso modificatore
d "modifier complex"

5520 modulus of elasticity
f module d'élasticité
e módulo de elasticidad
i modulo di elasticità
d Elastizitätsmodul

5521 Mohr's clip
f pince
e pinza de Mohr
i stringitubi a morsetto
d Quetschhahn

5522 Mohs' scale
f échelle de Mohs
e escala de Mohs
i scala di Mohs
d Mohssche Härteskala

5523 moisture content
f teneur en humidité
e contenido en humedad
i contenuto d'umidità
d Feuchtigkeitsgehalt

5524 moisture loss
f perte d'humidité
e pérdida de humedad
i perdità d'umidità
d Feuchtigkeitsverlust

5525 moisture vapour transmission
f perméabilité à l'humidité
e permeabilidad para al vapor de agua
i emissione di vapore d'acqua
d Wasserdampfabgabe

5526 molar conductance
f conductance molaire
e conductancia molar
i conduttanza molare
d molare Leitfähigkeit

5527 molar quantities
f quantités molaires
e cantidades molares
i quantità molari
d Molargrößen

5528 molar solution
f solution molaire
e solución molar
i soluzione molare
d molare Lösung

5529 molasses; treacle
f mélasse
e melazas
i molasse
d Melasse

5530 mole
f mole
e mol
i grammo-molecola
d Mol

5531 molecular abundance
f abondance moléculaire
e abundancia molecular
i abbondanza molecolare
d molekulare Häufigkeit

5532 molecular association
f association moléculaire
e asociación molecular
i associazione molecolare
d molekulare Assoziation

5533 molecular bond
f liaison moléculaire
e enlace molecular
i legame molecolare
d Molekülbindung

5534 molecular collision
f choc moléculaire
e choque molecular
i urto molecolare
d molekularer Stoß

5535 molecular depression of freezing point
f dépression moléculaire du point de
  congélation
e depresión molecular del punto de
  congelación
i depressione molecolare del punto di
  congelamento
d molekulare Gefrierpunktverlagerung

5536 molecular diffusion
f diffusion moléculaire
e difusión molecular
i diffusione molecolare
d Molekulardiffusion

5537 molecular effusion
f effusion moléculaire
e efusión molecular
i effusione molecolare
d Molekularausströmung

5538 molecular excitation
f excitation moléculaire
e excitación molecular
i eccitazione molecolare
d Molekülanregung

5539 molecular formula
f formule moléculaire
e fórmula molecular
i formula molecolare
d Molekularformel

5540 molecular heat
f chaleur moléculaire
e calor molecular
i calore molecolare
d Molarwärme

5541 molecular mass
f masse moléculaire

e masa molecular
i massa molecolare
d Molekülmasse

5542 molecular rotation
f rotation moléculaire
e rotación molecular
i rotazione molecolare
d Molekulardrehung

* **molecular sieves → 8913**

5543 molecular solution
f solution moléculaire
e solución molecular
i soluzione molecolare
d molekulare Lösung

5544 molecular spiral
f spirale moléculaire
e espiral molecular
i spirale molecolare
d Molekularspirale

5545 molecular structure
f structure de molécule
e estructura molecular
i struttura di molecola
d Molekularstruktur

5546 molecular volume
f volume moléculaire
e volumen molecular
i volume molecolare
d Normalvolumen

5547 molecular weight
f poids moléculaire
e peso molecular
i massa molecolare
d Molekulargewicht

5548 molecule
f molécule
e molécula
i molecola
d Molekül

5549 molybdate orange
f orangé de molybdate
e anaranjado de molibdato
i arancio di molibdato
d Molybdatorange

5550 molybdenite
f molybdénite
e molibdenita
i molibdenite
d Molybdänit; Molybdänglanz

5551 molybdenum
f molybdène
e molibdeno
i molibdeno
d Molybdän

5552 molybdenum disulphide
f bisulfure de molybdène
e bisulfuro de molibdeno
i bisolfuro di molibdeno
d Molybdändisulfid

5553 molybdenum pentachloride
f pentachlorure de molybdène
e pentacloruro de molibdeno
i pentacloruro di molibdeno
d Molybdänpentachlorid

5554 molybdenum sesquioxide
f sesquioxyde de molybdène
e sesquióxido de molibdeno
i sesquiossido di molibdeno
d Molybdänsesquioxyd

5555 molybdenum trioxide
f trioxyde de molybdène
e trióxido de molibdeno
i triossido di molibdeno
d Molybdäntrioxyd

5556 molybdite
f molybdine; molybdite
e molibdita
i molibdite
d Molybdit

5557 monatomic
f monatomique
e monoatómico
i monatomico
d einatomig

5558 monazite
f monazite
e monacita
i monacite
d Monazit; Turnerit

5559 Mond gas
f gaz de Mond
e gas Mond
i gas Mond
d Mondgas

5560 Mond process
f procédé Mond
e proceso Mond
i processo Mond
d Mond-Verfahren

5561 mongolism
f mongolisme
e mongolismo
i mongolismo
d Mongolismus

5562 monitoring
f contrôle
e control; supervisión
i controllo
d Kontrolle; Überwachung

5563 monobasic
f monobasique
e monobásico
i monobasico
d monobasisch

5564 monobasic acid
f acide monobasique
e ácido monobásico
i acido monobasico
d einbasische Säure

5565 monobasic ammonium phosphate
f phosphate d'ammonium monobasique
e fosfato amónico monobásico
i fosfato di ammonio monobasico
d Ammoniumphosphat; Ammoniummonophosphat

5566 monobasic calcium phosphate
f phosphate calcique
e fosfato cálcico monobásico
i fosfato di calcio
d Monocalciumphosphat

5567 monobasic magnesium phosphate
f phosphate monobasique de magnésium
e fosfato monobásico de magnesio
i fosfato monobasico di magnesio
d einbasisches Magnesiumphosphat

5568 monobasic potassium phosphate
f phosphate monopotassique
e fosfato potásico monobásico
i fosfato potassico monobasico
d Kaliumphosphat; Monokaliumphosphat

5569 monobasic sodium phosphate
f phosphate monobasique de sodium
e fosfato sódico monobásico
i fosfato monobasico di sodio
d Mononatriumphosphat

5570 monoblast
f monoblaste
e monoblasto

i monoblasto
d Monoblast

5571 monocellular
f unicellulaire
e monocelular
i unicellulare
d einzellig

5572 monoclinic
f monoclinique
e monoclínico
i monoclinico
d monoklin

5573 monodisperse
f monodispersé
e monodisperso
i monodisperso
d monodispers

5574 monoecy
f monoécie
e monoecia
i monoicismo
d Monoezie

5575 monofilament; monofil
f monofilament
e monofilamento
i monofilamento
d Einzelfaden

5576 monomer
f monomère
e monómero
i monomero
d Monomer

5577 monomery
f monomérie
e monomería
i monomeria
d Monomerie

5578 monomolecular layer
f couche monomoléculaire
e capa monomolecular
i strato monomolecolare
d monomolekulare Schicht

5579 monomorphous
f monomorphe
e monomorfo
i monomorfo
d monomorph

5580 monosaccharoses; monoses
f monosaccharoses; monoses

e monosacarosas; monosas
i monosaccarosi; monosi
d Monosaccharosen; Monosen

5581 monosome
f monosome
e monosoma
i monosoma
d Monosom

5582 monotropic
f monotropique
e monotrópico
i monotropico
d monotrop

5583 monovalent; univalent
f monovalent
e monovalente
i monovalente
d einwertig

5584 montan wax
f cire de lignite
e cera montana
i cera montana; cera di lignite
d Montanwachs

5585 montmorillonite
f montmorillonite
e montmorillonita
i montmorillonite
d Montmorillonit

5586 mordant
f mordant
e mordiente
i mordente
d Beize; Beizmittel

5587 morin
f morin
e morina
i morina
d Morin

5588 morphine
f morphine
e morfina
i morfina
d Morphin

5589 morphism
f morphisme
e morfismo
i morfismo
d Morphismus

5590 morpholine
f morpholine
e morfolina
i morfolina
d Morpholin

5591 morphology
f morphologie
e morfología
i morfologia
d Morphologie

5592 morphoplasm
f morphoplasme
e morfoplasma
i morfoplasma
d Substanz des Zellretikulums

5593 mortar
f mortier
e mortero
i mortaio
d Mörser

5594 mother blank
f plaque-support de feuille de départ
e placa soporte de hoja de arranque
i piastra di supporto dei lamierini catodici
d Mutterblechkathodenblech

5595 mother liquor
f liqueur-mère
e aguas madres
i soluzione madre
d Mutterlauge

5596 motility
f mobilité; motilité
e motilidad
i motilità
d Beweglichkeit

5597 mottle
f marbrure
e jaspeado
i screziatura
d Marmormuster

5598 mottling
f marbrure; diaprure; tiqueture
e jaspeado; reticulado
i screziatura
d Marmorieren

5599 mould
f moule
e molde
i stampo
d Form; Preßform

5600 mould cavity
f cavité d'un moule
e cavidad del molde
i matrice
d Matrize

5601 moulding cycle
f cycle de moulage
e ciclo de moldeo
i ciclo di formatura
d Preßdauer

5602 mouth blowpipe
f chalumeau
e soplete de boca
i cannello ferruminatorio
d Lötrohr

5603 mucic acid
f acide mucique
e ácido múcico
i acido mucico
d Schleimsäure

5604 mucilage
f mucilage
e mucilago
i mucillagine
d Mucilago; Pflanzenschleim; Schleimstoff

5605 mucus
f mucus
e moco
i muco
d Schleim

5606 mud
f boue
e barro; lodo; fango
i fango
d Schlamm

5607 muffle
f moufle
e mufla
i muffola
d Muffel

5608 muffle furnace
f fourneau à moufles
e horno de mufla
i forno a muffola
d Muffelofen

5609 muffle kiln
f four à moufle
e horno de copela; horno de esmaltar
i fornace a muffola
d Muffelofen

5610 mullite
 f mullite
 e mullita
 i mullite
 d Mullit

5611 multi-deck screen
 f crible à étages multiples
 e criba de pisos múltiples
 i vaglio a setacci multipli
 d Etagensieb

5612 multiple chiasma
 f chiasma multiple
 e quiasma múltiplo
 i chiasma multiplo
 d multiples Chiasma

5613 multiple-effect evaporator
 f ~~évaporateur à multiple effet~~
 e evaporador de efecto múltiple
 i evaporatore ad effetto multiplo
 d Mehrfachverdampfapparat

5614 multiple-screw extruder
 f boudineuse à vis multiple
 e extruidor dotado de varios tornillos
 i estrusore a vite multipla
 d Mehrschneckenpresse

5615 multiple system
 f système multiple
 e sistema múltiple
 i sistema multiplo
 d monopolare Schaltung

5616 multivalent
 f multivalent
 e multivalente
 i multivalente
 d mehrwertig; vielwertig

 * **mu oil** → **8568**

 * **muriatic acid** → **4176**

5617 muscovite
 f muscovite
 e muscovita
 i muscovite
 d Muskovit; weißer Glimmer

 * **musk root** → **7949**

5618 must
 f moût; vin doux
 e mosto
 i mosto
 d Most

5619 mustard oil
 f sénévol; essence de moutarde
 e aceite de mostaza
 i olio di senape
 d Senföl

5620 musty
 f à odeur de moisi
 e enmohecido
 i muffo
 d muffig

5621 mutation
 f mutation
 e mutación
 i mutazione
 d Mutation

5622 mutator gene
 f gène-mutateur
 e ~~gen mutador~~
 i gene mutatore
 d Mutatorgen

5623 mycelium
 f mycélium; mycélion; blanc de champignon
 e micelio
 i micelio
 d Myzelium

5624 myeloblast
 f myéloblaste
 e mieloblasto
 i mieloblasto
 d Myeloblast

5625 myelocyte; medullocell
 f myélocyte
 e mielocito
 i mielocita
 d Markzelle

5626 myoblast
 f myoblaste
 e mioblasto
 i mioblasto
 d Myoblast

5627 myoglobin
 f myoglobine
 e mioglobina
 i mioglobina
 d Myoglobin

5628 myristic acid
 f acide myristique
 e ácido mirístico

i acido miristico
d Myristinsäure

5629 myristoyl peroxide
f peroxyde de myristoyle
e peróxido de miristoílo
i perossido di miristoilo
d Myristoylperoxyd

5630 myristyl alcohol
f alcool myristylique
e alcohol miristílico
i alcool di miristile
d Myristylalkohol

5631 myrrh oil
f essence de myrrhe
e aceite de mirra
i olio di mirra
d Myrrhenöl

N

5632 nacre
f nacre
e nácar; madreperla
i madreperla
d Perlmutter

5633 nacreous
f nacré
e nacarado; nacarino
i madreperlaceo
d perlmutterartig

* **nagyagite → 1019**

5634 naphtha
f naphta
e nafta
i nafta
d Naphtha

5635 naphthalene
f naphtalène; naphtaline
e naftaleno
i naftalene; naftalina
d Naphthalin

5636 naphthalene black
f noir acide
e negro ácido
i nero di naftalene
d Säureschwarz

5637 naphthalenedisulphonic acid
f acide bisulfonique de naphtalène
e ácido naftalendisulfónico
i acido naftalendisolfonico
d Naphthalindisulfonsäure

5638 naphthalenesulphonic acid
f acide sulfonique de naphtalène
e ácido naftalensulfónico
i acido naftalensolfonico
d Naphthalinsulfonsäure

5639 naphthionic acid
f acide naphtionique
e ácido naftiónico
i acido naftionico
d Naphthionsäure

5640 naphthol
f naphtol
e naftol
i naftolo
d Naphthol

5641 naphtholdisulphonic acid
f acide bisulfonique de naphtol
e ácido naftoldisulfónico
i acido naftoldisolfonico
d Naphtholdisulfonsäure

5642 naphthoquinone
f naphtoquinone
e naftoquinona
i naftochinone
d Naphthochinon

5643 naphthylamine
f naphtylamine
e naftilamina
i naftilammina
d Naphthylamin

5644 naphthylamine disulphonic acid
f acide naphtylamindisulfonique
e ácido naftilaminodisulfónico
i acido naftilammindisolfonico
d Naphthylamindisulfonsäure

5645 naphthylamine sulphonic acid
f acide naphtylaminsulfonique
e ácido naftilaminosulfónico
i acido naftilamminsolfonico
d Naphthylaminsulfonsäure

5646 naphthylamine trisulphonic acid
f acide naphtylamintrisulfonique
e ácido naftilaminotrisulfónico
i acido naftilammintrisolfonico
d Naphthylamintrisulfonsäure

5647 naphthylenediamine
f naphtylène-diamine
e naftilendiamina
i naftilenediammina
d Naphthylendiamin

5648 naphthyl ethyl ether
f éther naphtyléthylique
e éter naftiletílico
i etere etilnaftilico
d Naphthyläthyläther

5649 naphthyl methylcarbamate
f naphtylméthylcarbamate
e naftilmetilcarbamato
i metilcarbammato naftilico
d Naphthylmethylcarbamat

5650 naphthyl methyl ether
f éther naphtylméthylique
e éter naftilmetílico
i etere naftilmetilico
d Naphthylmethyläther

5651 naphthylthiourea
 f naphtylthiourée
 e naftiltiourea
 i naftiltiourea
 d Naphthylthioharnstoff

5652 Naples yellow; galliolino
 f jaune de Naples
 e amarillo de Nápoles
 i giallo di Napoli
 d Neapelgelb

5653 narceine
 f narcéine
 e narceína
 i narceina
 d Narcein

5654 narcotic
 f narcotique
 e narcótico
 i narcotico
 d Betäubungsmittel; Rauschgift

5655 narcotine
 f narcotine
 e narcotina
 i narcotina
 d Narcotin; Opian

5656 native
 f natif
 e nativo
 i vergine
 d bergfein; gediegen

5657 Natta catalyst
 f catalyseur Natta
 e catalizador Natta
 i catalizzatore di Natta
 d Nattascher Katalysator

5658 natural ageing
 f vieillissement naturel
 e envejecimiento natural
 i invecchiamento naturale
 d natürliches Altern

5659 natural cement
 f ciment naturel
 e cemento natural
 i cemento naturale
 d natürlicher Zement

5660 natural element
 f élément naturel
 e elemento natural
 i elemento naturale
 d natürliches Element

5661 natural fibres
 f fibres naturelles
 e fibras naturales
 i fibre naturali
 d Naturfasern

5662 natural gas
 f gaz naturel
 e gas natural
 i gas naturale
 d Erdgas; Leuchtgas

5663 natural resins
 f résines naturelles
 e resinas naturales
 i resine naturali
 d Naturharze

5664 nauseous
 f nauséabond
 e nauseabundo
 i nauseante
 d ekelhaft; widerlich

5665 neatsfoot oil
 f huile de pieds de bœuf
 e aceite de pie de buey
 i olio di piede di bue
 d Rinderklauenöl

5666 necrobiosis
 f nécrobiose
 e necrobiosis
 i necrobiosi
 d Nekrobiose

5667 necrobiotic
 f nécrobiotique
 e necrobiótico
 i necrobiotico
 d nekrobiotisch

5668 needle-shaped
 f aciculaire
 e acicular; aguzado
 i aghiforme
 d nadelförmig

5669 needle valve
 f robinet à pointeau
 e válvula de aguja
 i valvola a spillo
 d Nadelventil

5670 negative catalysis
 f catalyse négative
 e catálisis negativa
 i catalisi negativa
 d negative Katalyse

5671 negative catalyst
 f catalyseur négatif
 e catalizador negativo
 i catalizzatore negativo
 d negativer Katalysator

5672 negative electrode
 f électrode négative
 e electrodo negativo
 i elettrodo negativo
 d negative Elektrode

5673 negative plate
 f plaque négative
 e placa negativa
 i piastra negativa
 d negative Platte

5674 negative proton
 f antiproton
 e protón negativo
 i protone negativo
 d Antiproton

5675 Nelson cell
 f élément de Nelson
 e célula de Nelson
 i pila di Nelson
 d Nelson-Element

5676 nematicide
 f destructeur de nématodes
 e nematicida
 i nematicida
 d Nematodenmittel

5677 neoarsphenamine
 f néoarsphénamine
 e neoarsfenamina
 i neoarsfenammina
 d Neoarsphenamin

5678 neocentromere
 f néocentromère
 e neocentrómero
 i neocentromero
 d Neocentromer

5679 neocinchophen
 f néocinchophène
 e neocincófeno
 i neocincofene
 d Neocinchophen

5680 neodymium
 f néodymium
 e neodimio
 i neodimio
 d Neodym

5681 neohexane
 f néohexane
 e neohexano
 i neoesano
 d Neohexan

5682 neomycin
 f néomycine
 e neomicina
 i neomicina
 d Neomycin

5683 neon
 f néon
 e neón
 i neon
 d Neon

5684 neoprene
 f néoprène
 e neopreno
 i neoprene
 d Neopren

5685 neostigmine
 f néostigmine
 e neostigmina
 i neostigmina
 d Neostigmin

5686 nephelite
 f néphéline
 e nefelita
 i nefelite
 d Nephelin; Fettstein

5687 nephelometer
 f opacimètre
 e nefelómetro
 i nefelometro
 d Trübungsmesser

5688 neptunium
 f neptunium
 e neptunio
 i nettunio
 d Neptunium

5689 nerol
 f nérol
 e nerol
 i nerol
 d Nerol

5690 nerolidol
 f nérolidol
 e nerolidol
 i nerolidolo
 d Nerolidol

5691 neroli oil; orange flower oil
f essence de néroli; essence de fleurs
d'orangers
e aceite de neroli
i essenza di neroli
d Neroliöl

5692 nesidioblasts
f cellules insulaires
e células insulares
i cellule insulari
d Inselzellen

5693 Nessler's reagent
f réactif de Nessler
e reactivo de Nessler
i reagente di Nessler
d Nessler-Reagens

5694 nest of tubes
f faisceau tubulaire
e haz tubular; haz de tubos
i fascio di tubi
d Rohrbündel

5695 net cooling
f refroidissement en circuit
e enfriamiento continuo
i raffreddamento a rete in circuito
d Netzkühlung

5696 net heating
f chauffage en circuit
e calefacción continua
i riscaldamento a rete in circuito
d Netzheizung

5697 net weight
f poids net
e peso neto
i peso netto
d Nettogewicht

5698 neurine
f neurine
e neurina
i neurina
d Neurin

5699 neurogenic
f neurogène
e neurógeno
i neurogeno
d neurogen

5700 neurogliocyte
f cellule de la neuroglie
e célula glial

i cellula gliale
d Neurogliazelle

5701 neurokeratin
f neurokératine
e neuroqueratina
i neurocheratina
d Neurokeratin

5702 neurolysis
f neurolyse
e neurólisis
i neurolisi
d Neurolyse

5703 neutral
f neutre
e neutro
i neutro
d neutral

5704 neutralization
f neutralisation
e neutralización
i neutralizzazione
d Neutralisierung

5705 neutralize v
f neutraliser
e neutralizar
i neutralizzare
d neutralisieren

5706 neutral molecule
f molécule neutre
e molécula neutra
i molecola neutra
d neutrales Molekül

* **neutral red** → 8332

5707 neutral solution
f solution neutre
e solución neutra
i soluzione neutra
d neutrale Lösung

5708 Neville and Winther's acid
f acide de Neville et Winther
e ácido de Neville y Winther
i acido di Neville e Winther
d Neville-und-Winthersche Säure

5709 Newtonian flow
f flux newtonien
e flujo Newtoniano
i flusso newtoniano
d Newtonsche Strömung

* niacin → 5729

* niacinamide → 5727

5710 niccolite
f nickéline
e nicolita
i nichelina
d Nickelin; Arsennickel

5711 nick
f encoche
e muesca
i sbocconcellatura
d Kerbe; Nute

5712 nickel
f nickel
e níquel
i nichel
d Nickel

5713 nickel acetate
f acétate de nickel
e acetato de níquel
i acetato di nichel
d Nickelacetat

5714 nickel ammonium chloride
f chlorure de nickel ammoniacal
e cloruro de níquel y amonio
i cloruro di nichel e ammonio
d Nickelammoniumchlorid

5715 nickel ammonium sulphate
f sulfate de nickel ammoniacal
e sulfato de níquel y amonio
i solfato di nichel e ammonio
d Nickelammoniumsulfat

5716 nickel arsenate
f arséniate de nickel
e arseniato de níquel
i arseniato di nichel
d Nickelarsenat

* nickel bloom → 566

5717 nickel carbonate
f carbonate de nickel
e carbonato de níquel
i carbonato di nichel
d Nickelcarbonat

5718 nickel carbonyl
f nickel carbonyle
e níquel carbonilo
i nichel carbonile
d Nickelcarbonyl; Nickeltetracarbonyl

5719 nickel chloride
f chlorure de nickel
e cloruro de níquel
i cloruro di nichel
d Nickelchlorid

5720 nickel dibutyldithiocarbamate
f dibutyldithiocarbamate de nickel
e dibutilditiocarbamato de níquel
i dibutilditiocarbammato di nichel
d Nickeldibutyldithiocarbamat

5721 nickel nitrate
f nitrate de nickel
e nitrato de níquel
i nitrato di nichel
d Nickelnitrat

5722 nickel plating
f nickelage
e niquelado electrolítico
i nichelatura
d Nickelüberzug; Vernickeln

5723 nickel silver; German silver
f maillechort; argent allemand
e alpaca; metal blanco; plata de níquel;
  maillechort
i lega di rame, zinco e nichel; argentone
d Neusilber; Argenton

5724 nickel steel
f acier au nickel
e acero al níquel
i acciaio al nichel
d Nickelstahl

5725 nickel sulphate
f sulfate de nickel
e sulfato de níquel
i solfato di nichel
d Nickelvitriol; Nickelsulfat

5726 Nicol prism
f nicol; prisme de Nicol
e prisma de Nicol; nicol
i prisma di Nicol
d Nicolsches Prisma; Nikol

5727 nicotinamide; niacinamide
f nicotinamide
e nicotinamida
i nicotinammide
d Nikotinamid

5728 nicotine
f nicotine
e nicotina

i nicotina
d Nicotin; Nikotin

5729 nicotinic acid; niacin
f acide nicotinique
e ácido nicotínico
i acido nicotinico
d Nicotinsäure

5730 nidation
f nidation
e nidación
i annidamento
d Einnistung

5731 nigrosin
f nigrosine
e nigrosina
i nigrosina
d Nigrosin

5732 nikethamide; aminocordine
f nikéthamide
e niquetamida
i nichetamide
d Nikethamid

5733 niobium; columbium
f niobium
e niobio
i niobio
d Niob; Niobium

5734 niobium carbide
f carbure de niobium
e carburo de niobio
i carburo di niobio
d Niobiumcarbid

5735 nip
f cylindres presseurs
e par de cilindros de calandria
i rulli di compressione
d Walzenpaar

5736 nipple
f mamelon
e niple
i nipplo; raccordo filettato
d Nippel

5737 nitrate *v*
f nitrer
e nitrar
i nitrare
d nitrieren

5738 nitrate
f nitrate

e nitrato
i nitrato
d Nitrat

* **nitrating apparatus** → 5740

5739 nitration
f nitration
e nitración
i nitrazione
d Nitrierung

5740 nitrator; nitrating apparatus
f nitreur
e nitradora
i nitratore
d Nitrierapparat; Nitriertopf

* **nitre** → 6630

5741 nitric acid
f acide nitrique
e ácido nítrico
i acido nitrico
d Salpetersäure

5742 nitric ester
f ester nitrique
e éster nítrico
i estere nitrico
d Salpetersäureester

5743 nitric oxide
f oxyde azotique
e óxido nítrico
i ossido nitrico
d Stickstoffoxyd

5744 nitrides
f nitrures
e nitruros
i nitruri
d Nitride

5745 nitriding
f nitruration
e nitruración
i nitrurazione
d Nitrierhärtung

5746 nitrification
f nitrification
e nitrificación
i nitrurazione
d Nitrierung

5747 nitrify *v*
f nitrifier
e nitrificar

i nitrificare
d nitrieren

5748 nitrile
  f nitrile
  e nitrilo
  i nitrile
  d Nitril

5749 nitrite
  f nitrite
  e nitrito
  i nitrito
  d Nitrit

5750 nitroacetanilide
  f nitroacétanilide
  e nitroacetanilida
  i nitroacetanilide
  d Nitroacetanilid

5751 nitroaminophenol
  f nitroaminophénol
  e nitroaminofenol
  i nitroamminofenolo
  d Nitroaminophenol

5752 nitroaniline
  f nitroaniline
  e nitroanilina
  i nitroanilina
  d Nitroanilin

5753 nitroanisole
  f nitroanisole
  e nitroanisol
  i nitroanisolo
  d Nitroanisol

5754 nitrobenzaldehyde
  f nitrobenzaldéhyde
  e nitrobenzaldehído
  i nitrobenzaldeide
  d Nitrobenzaldehyd

5755 nitrobenzene; mirbane oil
  f nitrobenzène
  e nitrobenceno
  i nitrobenzène
  d Nitrobenzol

5756 nitrobenzeneazoresorcinol
  f nitrobenzèneazorésorcinol
  e nitrobencenoazorresorcina
  i nitrobenzeneazoresorcinolo
  d Nitrobenzolazoresorcinol

5757 nitrobenzoic acid
  f acide nitrobenzoïque

  e ácido nitrobenzoico
  i acido nitrobenzoico
  d Nitrobenzoesäure

5758 nitrobenzoyl chloride
  f chlorure de nitrobenzoyle
  e cloruro de nitrobenzoílo
  i cloruro di nitrobenzoile
  d Nitrobenzoylchlorid

5759 nitrobenzoyl cyanide
  f cyanure de nitrobenzoyle
  e cianuro de nitrobenzoílo
  i cianuro di nitrobenzoile
  d Nitrobenzoylcyanid

5760 nitrobiphenyl
  f nitrobiphényle
  e nitrobifenilo
  i nitrobifenile
  d Nitrodiphenyl

5761 nitrocellulose
  f nitrocellulose
  e nitrocelulosa
  i nitrocellulosa
  d Nitrocellulose

  * nitro-cotton → 3904

5762 nitrodiphenylamine
  f nitrodiphénylamine
  e nitrodifenilamina
  i nitrodifenilammina
  d Nitrodiphenylamin

5763 nitroethane
  f nitroéthane
  e nitroetano
  i nitroetano
  d Nitroäthan

5764 nitrofurantoin
  f nitrofurantoïne
  e nitrofurantoína
  i nitrofurantoina
  d Nitrofurantoin

5765 nitrofurazone
  f nitrofurazone
  e nitrofurazona
  i nitrofurazone
  d Nitrofurazon

5766 nitrogen; azote
  f azote
  e nitrógeno
  i azoto
  d Stickstoff

5767 nitrogenase
f nitrogénase
e nitrogenasa
i nitrogenasi
d Nitrogenase

* nitrogen dioxide → 5770

5768 nitrogen fixation
f azotation
e fijación del nitrógeno
i assimilazione dell'azoto
d Bindung des atmosphärischen Stickstoffes

5769 nitrogenous
f azoté
e nitrogenado
i azotato
d stickstoffhaltig

5770 nitrogen peroxide; nitrogen dioxide
f peroxyde d'azote
e peróxido de nitrógeno
i perossido di azoto
d Stickstoffdioxyd

5771 nitrogen trichloride
f trichlorure d'azote
e tricloruro de nitrógeno
i tricloruro di azoto
d Stickstofftrichlorid

5772 nitrogen trifluoride
f trifluorure d'azote
e trifluoruro de nitrógeno
i trifluoruro di azoto
d Stickstofftrifluorid

5773 nitroglycerine; glyceryl trinitrate
f nitroglycérine
e nitroglicerina
i nitroglicerina
d Nitroglycerin

5774 nitroguanidine
f nitroguanidine
e nitroguanidina
i nitroguanidina
d Nitroguanidin

5775 nitromersol
f nitromersol
e nitromersol
i nitromersolo
d Nitromersol

5776 nitromethane
f nitrométhane
e nitrometano
i nitrometano
d Nitromethan

5777 nitron
f nitron
e nitrón
i nitron
d Nitron

5778 nitronaphthalene
f nitronaphtaline
e nitronaftaleno
i nitronaftalene
d Nitronaphthalin

5779 nitroparacresol
f nitroparacrésol
e nitroparacresol
i nitroparacresolo
d Nitroparacresol

5780 nitroparaffins
f nitroparaffines
e nitroparafinas
i nitroparaffine
d Nitroparaffine

5781 nitrophenetole
f nitrophénétol
e nitrofenetol
i nitrofenetolo
d Nitrophenetol

5782 nitrophenide
f nitrophénide
e nitrofenida
i nitrofenide
d Nitrophenid

5783 nitrophenol
f nitrophénol
e nitrofenol
i nitrofenolo
d Nitrophenol

5784 nitrophenylacetic acid
f acide nitrophénylacétique
e ácido nitrofenilacético
i acido nitrofenilacetico
d Nitrophenylessigsäure

5785 nitropropane
f nitropropane
e nitropropano
i nitropropano
d Nitropropan

5786 nitrosalicylic acid
  f acide nitrosalicylique
  e ácido nitrosalicílico
  i acido nitrosalicilico
  d Nitrosalicylsäure

5787 nitroso compounds
  f composés nitrosés
  e compuestos nitrosos
  i composti nitrosi
  d Nitrosoverbindungen

5788 nitrosodimethylaniline
  f nitrosodiméthylaniline
  e nitrosodimetilanilina
  i nitrosodimetilanilina
  d Nitrosodimethylanilin

5789 nitrosodiphenylamine
  f nitrosodiphénylamin
  e nitrosodifenilamina
  i nitrosodifenilammina
  d Nitrosodiphenylamin

5790 nitrosoguanidine
  f nitrosoguanidine
  e nitrosoguanidina
  i nitrosoguanidina
  d Nitrosoguanidin

5791 nitrosonaphthol
  f nitrosonaphtol
  e nitrosonaftol
  i nitrosonaftolo
  d Nitrosonaphthol

5792 nitrosophenol
  f nitrosophénol
  e nitrosofenol
  i nitrosofenolo
  d Nitrosophenol

5793 nitroso rubber
  f caoutchouc nitrosé
  e caucho nitroso
  i gomma nitrosa
  d Nitrosokautschuk

5794 nitrostarch
  f nitroamidon; nitrate d'amidon
  e nitroalmidón
  i nitroamido
  d Nitrostärkemehl

5795 nitrostyrene
  f nitrostyrène
  e nitroestireno
  i nitrostirene
  d Nitrostyrol

5796 nitrosulphathiazole
  f nitrosulfathiazol
  e nitrosulfatiazol
  i nitrosolfatiazolo
  d Nitrosulphathiazol

5797 nitrosyl chloride
  f chlorure de nitrosyle
  e cloruro de nitrosilo
  i cloruro di nitrosile
  d Nitrosylchlorid

5798 nitrotoluene
  f nitrotoluène
  e nitrotolueno
  i nitrotoluolo
  d Nitrotoluol

5799 nitrotoluidine
  f nitrotoluidine
  e nitrotoluidina
  i nitrotoluidina
  d Nitrotoluidin

5800 nitrotrifluoromethylbenzonitrile
  f nitrotrifluorométhylbenzonitrile
  e nitrotrifluometilbenzonitrilo
  i nitrotrifluorometilbenzonitrile
  d Nitrotrifluormethylbenzonitril

5801 nitrourea
  f nitrourée
  e nitrourea
  i nitrourea
  d Nitroharnstoff

5802 nitrous oxide; laughing gas
  f oxyde nitreux; gaz hilarant
  e óxido nitroso; gas hilarante
  i ossido nitroso
  d GMI-Stoff; Stickoxydul; Lachgas

5803 nitroxylene
  f nitroxylène
  e nitroxileno
  i nitroxileno
  d Nitroxylen

5804 nivenite
  f nivénite
  e nivenita
  i nivenite
  d Nivenit

5805 nobelium
  f nobélium
  e nobelio
  i nobelio
  d Nobelium

5806 noble gases
f gaz nobles; gaz inertes
e gases nobles
i gas nobili; gas inerti
d Edelgase; Inertgase

5807 noble metals
f métaux nobles
e metales nobles
i metalli nobili
d Edelmetalle

5808 nocuity
f nocivité
e nocividad
i nocività
d Schädlichkeit

5809 nodal
f nodal
e nodal
i nodale
d Knoten-

5810 nodule
Small round swelling.
f nodule
e nódulo
i nodulo
d Knollen

5811 nodules
Round excrescences formed on the
cathode during deposition.
f nodules
e nódulos
i noduli
d Klümpchen

5812 non-actinic
f inactinique
e inactínico
i inattinico
d unaktinisch

5813 nonadecane
f nonadécane
e nonadecano
i nonadecano
d Nonadecan

5814 nonaldecanoic acid
f acide nonaldécanoïque
e ácido nonaldecanoico
i acido nonaldecanoico
d Nonaldecansäure

5815 nonanal
f nonanal

e nonanal
i nonanal
d Nonanal

5816 nonane
f nonane
e nonano
i nonano
d Nonan

5817 non-centrifugal sugar
f sucre non cristallisé
e azúcar no centrífugo
i zucchero non cristallizzato
d nicht kristallisierter Zucker

5818 non-conductor
f non conducteur
e aislador; no conductor
i isolante
d Nichtleiter-

5819 non-congealable
f non congelable
e incongelable
i incongelabile
d nicht gefrierbar

5820 non-crawling
f résistant au fluage
e no sujeto al arrugamiento
i resistente a repellenza
d nicht verlaufend

5821 non-destructive test
f essai non destructif
e ensayo no destructivo
i prova non distruttiva
d zerstörungsfreie Prüfung

5822 non-drying oil
f huile non siccative
e aceite no secante
i olio non essiccabile
d nicht trocknendes Öl

5823 non-explosive
f inexplosible
e inexplosible; indetonante
i non esplosivo
d explosionssicher

5824 non-ferrous
f non ferreux
e no ferroso
i non ferroso
d nicht eisenhaltig

5825 non-ferrous metal
f métal non ferreux
e metal no ferroso
i metallo non ferroso
d Nichteisenmetall

5826 non-flammable
f ininflammable
e ininflamable
i non infiammabile
d unentzündlich; unentflammbar

5827 non-lined construction
f montage sans habillage
e montaje sin recubrimiento
i montaggio senza rivestimento
d wickellose Bauweise

5828 non-magnetic
f amagnétique
e amagnético; antimagnético
i antimagnetico; paramagnetico
d nicht magnetisch

5829 non-miscible
f non miscible
e inmiscible
i immiscibile
d unmischbar

* **nonoic acid → 6156**

5830 nonoses
f nonoses
e nonosas
i nonosi
d Nonosen

5831 non-polar
f non polaire
e no polar
i non polare
d nicht polar

5832 non-porous
f non poreux
e aporoso; imporoso
i non poroso
d nicht porös

5833 non-return valve; one-way valve
f soupape de retenue
e válvula de retención
i valvola di non ritorno
d Rückstromventil

5834 non-volatile
f non volatil
e involátil

i non volatile
d nicht flüchtig

5835 nonyl acetate
f acétate de nonyle
e acetato de nonilo
i acetato di nonile
d Nonylacetat

5836 nonyl alcohol
f alcool nonylique
e alcohol nonílico
i alcool nonílico
d Nonylalkohol

5837 nonylamine
f nonylamine
c nonilamina
i nonilammina
d Nonylamin

5838 nonylbenzene
f nonylbenzène
e nonilbenceno
i nonilbenzene
d Nonylbenzol

5839 nonylene
f nonylène
e nonileno
i nonilene
d Nonylen

5840 nonyl lactone
f nonyllactone
e nonil-lactona
i nonillattone
d Nonyllacton

5841 nonyl phenol
f nonylphénol
e nonilfenol
i nonilfenolo
d Nonylphenol

5842 no-pressure resin
f résine durcissant sans pression
e resina endurecible sin presión
i resina a contatto
d drucklos härtendes Harz

5843 noradrenalin
f noradrénaline
e noradrenalina
i noradrenalina
d Noradrenalin

* **norephedrine → 6311**

5844 **norleucine**
 f norleucine
 e norleucina
 i norleucina
 d Norleucin

5845 **norm**
 f norme
 e norma
 i norma
 d Norm

5846 **normal**
 f normal
 e normal
 i normale
 d normal

5847 **normalizing**
 f normalisation
 e normalización
 i normalizzazione
 d Ausglühen

5848 **normal salts**
 f sels normaux
 e sales normales
 i sali normali
 d Normalsalze

5849 **normal solution**
 f solution normale
 e solución normal
 i soluzione normale
 d Normallösung

5850 **normocyte**
 f normocyte
 e normocito
 i normocita
 d Normozyt

5851 **notch-sensitivity**
 f sensibilité à l'entaille
 e sensibilidad al efecto de entalladura
 i sensibilità all'intaglio
 d Kerbempfindlichkeit

5852 **novobiocin; streptonivicin**
 f novobiocine
 e novobiocina
 i novobiocina
 d Novobiocin

5853 **noxious**
 f nocif
 e nocivo; dañino
 i nocivo; dannoso
 d schädlich

5854 **nozzle**
 f buse
 e tobera; boquilla
 i boccaglio; ugello
 d Düse

5855 **nuclear**
 f nucléaire
 e nuclear
 i nucleare
 d Kern-

5856 **nuclear association**
 f association nucléaire
 e asociación nuclear
 i associazione nucleare
 d Kernassoziation

5857 **nuclear cap**
 f capsule nucléaire
 e casquete nuclear
 i capsula nucleare
 d Kernkappe

5858 **nuclear chemistry**
 f chimie nucléaire
 e química nuclear
 i chimica nucleare
 d Kernchemie

5859 **nuclear dimorphism**
 f dimorphisme nucléaire
 e dimorfismo nuclear
 i dimorfismo nucleare
 d Kerndimorphismus

5860 **nuclear disruption**
 f dislocation nucléaire
 e desgarradura nuclear
 i dislocamento nucleare
 d Kernbruch

5861 **nuclear division**
 f division nucléaire
 e división nuclear
 i divisione nucleare
 d Kernteilung

5862 **nuclear fragmentation**
 f fragmentation nucléaire
 e fragmentación nuclear
 i frammentazione nucleare
 d Kernfragmentation

5863 **nuclear interaction**
 f action mutuelle nucléaire
 e interacción nuclear
 i interazione nucleare
 d Kernwechselwirkung

5864 nuclear isomerism
f isomérie nucléaire
e isomería nuclear
i isomeria nucleare
d Kernisomerie

5865 nuclear sap
f suc nucléaire
e jugo nuclear
i succo nucleare
d Kernsaft

5866 nuclear technology; nucleonics
f technologie nucléaire
e tecnología nuclear
i tecnologia nucleare
d Kerntechnologie

5867 nucleic acids
f acides nucléiques
e ácidos nucleiços
i acidi nucleici
d Nukleinsäuren

5868 nuclein
f nucléine
e nucleína
i nucleina
d Nuklein

5869 nucleination
f nucléisation
e nucleinación
i nucleinazione
d Nukleinisierung

5870 nucleoalbumin
f nucléoalbumine
e nucleoalbúmina
i nucleoalbumina
d Nukleoalbumin

5871 nucleolar associated chromatin
f chromatine nucléolaire associée
e cromatina nucleolar asociada
i cromatina nucleolare associata
d "nucleolar associated chromatin"

5872 nucleolar constriction
f constriction nucléolaire
e constricción nucleolar
i costrizione nucleolare
d Nukleolareinschnürung

5873 nucleolar fragmentation
f fragmentation nucléolaire
e fragmentación nucleolar
i frammentazione nucleolare
d Nukleolusfragmentation

5874 nucleolus
f nucléole
e nucléolo
i nucleolo
d Nukleolus

5875 nucleolus sickle-stage
f paranucléole
e paranucléolo
i paranucleolo
d Paranukleolus

* **nucleonics** → 5866

5876 nucleoplasmatic ratio
f rapport nucléoplasmatique
e relación nucleoplasmática
i rapporto nucleoplasmatico
d Kern-Plasma-Relation

5877 nucleoprotein
f nucléoprotéine
e nucleoproteína
i nucleoproteina
d Nukleoprotein

5878 nucleotid
f nucléotide
e nucleótida
i nucleotide
d Nukleotid

5879 nucleus
f noyau
e núcleo
i nucleo
d Kern

5880 nucleus crystal
f amorce de cristallisation
e núcleo de cristalización
i germe cristallino
d Keimkristall

5881 nugget
f pépite
e pepita
i pepita
d Klumpen

5882 nullisomic
f nullisomique
e nulisómico
i nullisomico
d nullisom

5883 null valence
f valence zéro
e valencia nula

i valenza nulla
d Nullvalenz

5884 nux vomica
f noix vomique
e nuez vómica
i noce vomica
d Brechnuß

5885 nylon
f nylon
e nylon; nilón
i nailon; nylon
d Nylon

5886 nylon bristles
f poils de nylon
e cerdas de nylon
i setole di nailon
d Nylonbürsten

5887 nylon fibre
f fibre de nylon
e fibra de nylon
i fibra di nailon
d Nylonfaser

5888 nylon monofilaments
f monofilaments de nylon
e monofilamentos de nylon
i monofilamenti di nailon
d Nyloneinzelfäden

5889 nylon moulding powders
f poudres à mouler de nylon
e polvos de moldear de nylon
i polveri per la stampatura del nailon
d Nylonpreßpulver

5890 nylon yarn
f fil de nylon
e hilo de nylon
i filo di nailon
d Nylongarn

5891 nytril
f nytril
e nitrilo
i nitrile
d Nytril

O

5892 oak bark
 f écorce de chêne
 e corteza de roble
 i corteccia di quercia
 d Eichenrinde

5893 object-glass
 f objectif
 e objetivo
 i obiettivo
 d Objektiv

5894 objective
 f objectif
 e objetivo
 i obiettivo
 d Objektiv

5895 obsidian
 f obsidiane; obsidienne
 e obsidiana
 i agata dell'Islanda
 d Obsidian; Glasachat

5896 obturation
 f obturation
 e obturación
 i otturazione
 d Verstopfung

5897 occlude *v*
 f occlure
 e ocluir; absorber
 i assorbire; occludere
 d absorbieren; okkludieren

5898 occluded gas
 f gaz occlus
 e gas ocluso; gas encerrado
 i gas occluso
 d eingeschlossenes Gas; okkludiertes Gas

5899 occlusion
 f absorption; occlusion
 e oclusión; absorción
 i assorbimento; occlusione
 d Absorption; Okklusion

5900 ochre
 f ocre
 e ocre
 i ocra
 d Ocker; Mennig

5901 octadecane
 f octadécane

 e octadecano
 i ottadecano
 d Octadecan

5902 octadecene
 f octadécène
 e octadeceno
 i ottadecene
 d Octadecen

5903 octadecenyl aldehyde
 f aldéhyde octadécénylique
 e aldehído octadecenílico
 i aldeide di ottadecenile
 d Octadecenylaldehyd

5904 octahedral
 f octaédrique
 e octaédrico
 i ottaedrico
 d achtflächig; oktaedrisch

5905 octahedron
 f octaèdre
 e octaedro
 i ottaedro
 d Oktaeder

5906 octamethyl pyrophosphoramide
 f octaméthyl-pyrophosphoramide
 e octametilpirofosforamida
 i pirofosforammide di ottametile
 d Octamethylpyrophosphoramid

5907 octanal
 f octanal
 e octanal
 i aldeide caprilica
 d Octanal

5908 octane
 f octane
 e octano
 i ottano
 d Octan; Oktan

5909 octane number; octane rating
 f indice d'octane
 e número de octano
 i numero di ottano
 d Oktanwert; Oktanzahl

5910 octanol
 f octanol
 e octanol
 i alcool ottoico
 d Octanol

5911 octanoyl chloride
f chlorure d'octanoyle
e cloruro de octanoílo
i cloruro di caprilile
d Octanoylchlorid

5912 octavalent
f octavalent
e octovalente
i ottovalente
d achtwertig

5913 octene
f octène
e octeno
i caprilene; ottene
d Octen

5914 octoses
f octoses
e octosas
i ottosi
d Octosen

5915 octyl acetate
f acétate d'octyle
e acetato de octilo
i acetato di ottile
d Octylacetat

5916 octylamine
f octylamine
e octilamina
i ammina ottilica
d Octylamin

5917 octylbicycloheptene dicarboximide
f dicarboximide d'octylbicycloheptène
e dicarboximida de octilbiciclohepteno
i dicarbossimide di ottilbicicloeptene
d Octylbicycloheptendicarboximid

5918 octyl bromide
f bromure d'octyle
e bromuro de octilo
i bromuro di ottile
d Octylbromid

5919 octyl decyl adipate
f adipate d'octyldécyle
e adipato de octildecilo
i adipato di ottildecile
d Octyldecyladipat

5920 octyl decyl phthalate
f phtalate d'octyldécyle
e ftalato de octildecilo
i ftalato di ottildecile
d Octyldecylphthalat

5921 octylene glycol titanate
f titanate d'octylèneglycol
e titanato de octilenglicol
i glicol titanato di ottilene
d Octylenglykoltitanat

5922 octylene oxide
f oxyde d'octylène
e óxido de octileno
i ossido di ottilene
d Octylenoxyd

5923 octyl iodide
f iodure octyle
e yoduro de octilo
i ioduro di ottile
d Octyljodid

5924 octyl mercaptan
f octylmercaptan
e octilmercaptano
i ottilmercaptano
d Octylmercaptan

5925 octyl methacrylate
f méthacrylate d'octyle
e metacrilato de octilo
i metacrilato di ottile
d Octylmethacrylat

5926 octyl phenol
f octylphénol
e octilfenol
i ottilfenolo
d Octylphenol

5927 octylphenoxy polyethoxyethanol
f octylphénoxy-polyéthoxyéthanol
e octilfenoxipolietoxietanol
i ottilfenossipolietossietanolo
d Octylphenoxypolyäthoxyäthanol

5928 ocular
f oculaire
e ocular
i oculare
d Okular

5929 odoriferous
f odorant; odoriférant
e odorante; odorífero
i odorifero
d riechend

5930 odorimetry
f odorimètre
e odorimetría
i odorimetria
d Geruchsmessung

5931 odourless
f inodore
e inodoro
i inodoro
d geruchlos

5932 oenometer; vinometer
f oenomètre; vinomètre
e enómetro
i enometro
d Weinmesser

* **oestradiol** → 3117

* **oestriol** → 3120

* **oestrogen** → 3121

* **oestrogenic** → 3122

* **oestrone** → 3123

5933 ohmic overvoltage
f surtension ohmique
e sobretensión óhmica
i sovratensione ohmica
d ohmscher Spannungsabfall

5934 oil absorption
f absorption d'huile
e absorción de aceite
i presa d'olio
d Ölaufnahme

5935 oil bath
f bain d'huile
e baño de aceite
i bagno d'olio
d Ölbad

5936 oil cleaner
f épurateur d'huile
e purificador de aceite
i purificatore d'olio
d Ölreiniger

5937 oil film
f film d'huile
e película de aceite; film de aceite
i pellicola d'olio
d Ölhäutchen

5938 oil filter
f filtre d'huile
e filtro de aceite
i filtro dell'olio
d Ölfilter

5939 oil firing
f chauffage au mazout
e caldeo con petróleo
i riscaldamento a nafta
d Ölfeuerung

5940 oil immersion test
f essai d'immersion dans l'huile
e ensayo de inmersión en aceite
i prova d'immersione in olio
d Öleintauchprobe

5941 oil length
f proportion d'huile
e galones de aceite por 100 libras de pigmento
i galloni di olio per 1000 litri di resina
d Ölgehalt

5942 oil-mist lubrication
f dispersion d'huile par brouillard
e lubrificación con neblina de aceite
i lubrificazione nebulizzata
d Ölschmierung

* **oil of vitriol** → 7945

5943 oil paint
f peinture à l'huile
e pintura al óleo; pintura al aceite
i pittura all'olio
d Ölfarbe

5944 oil paste
f concentré
e pasta de aceite
i vernici a olio
d Konzentrat

5945 oil pressure gauge
f jauge de pression d'huile
e manómetro del aceite
i manometro dell'olio
d Schmierstoffdruckmesser

5946 oil-reactive resin
f résine oléo-active
e resina que reacciona con el aceite secante
i resina reattiva in fase oleosa
d ölreaktives Harz

5947 oil resistant
f résistant à l'huile
e oleorresistente
i resistente all'olio
d ölbeständig

5948 oil shale
  f schiste bitumineux
  e esquisto aceitoso
  i scisto bituminoso
  d Ölschiefer

5949 oil-soluble resin
  f résine oléosoluble
  e resina soluble en aceite
  i resina solubile in olio
  d öllösliches Harz

5950 ointment
  f onguent
  e ungüento
  i unguento
  d Salbe

5951 ointment base
  f excipient de pommades
  e excipiente por pomadas
  i eccipiente per unguenti
  d Salbengrundlage

5952 old fustic
  f maclure
  e fustete viejo
  i giallo vegetale usato con merdente
     metallico
  d Gelbholz

5953 oleaginous
  f oléagineux
  e oleaginoso; aceitoso
  i oleaginoso
  d ölhaltig

5954 oleamide
  f oléamide
  e oleamida
  i oleoammide
  d Oleamid

5955 oleandomycin
  f oléandomycine
  e oleoandomicina
  i oleandomicina
  d Oleandomycin

  * **olefiant gas** → 3180

5956 olefin fibres
  f fibres oléfines
  e fibras de olefina
  i fibre olefine
  d Olefinfasern

5957 olefins
  f oléfines; alcènes

  e olefinas
  i olefine
  d Olefine

5958 oleic acid
  f acide oléique
  e ácido oleico
  i acido oleico
  d Ölsäure

5959 olein
  f oléine
  e oleína
  i oleina
  d Olein

5960 oleometer
  f oléomètre
  e oleómetro
  i oleometro
  d Oleometer; Ölmesser

5961 oleoresin
  f oléorésine
  e oleorresina
  i oleoresina
  d Ölharz

5962 oleoyl chloride
  f chlorure d'oléoyle
  e cloruro de oleoílo
  i cloruro di oleoile
  d Oleoylchlorid

  * **oleum** → 3624

5963 oleyl alcohol
  f alcool oléylique
  e alcohol oleílico
  i alcool oleilico
  d Oleylalkohol

5964 olivenite
  f olivénite
  e olivenita
  i olivenite
  d Olivenit

5965 olive oil
  f huile d'olive
  e aceite de oliva
  i olio d'oliva
  d Olivenöl

5966 olivine
  f chrysolithe; olivine
  e olivino
  i olivina
  d Olivin

5967 oncogenous
f tumorigène
e oncógeno
i oncogeno
d tumorbildend

5968 one-fluid cell
f pile à un liquide
e pila de un líquido
i pila a un solo elettrolito
d Flüssigkeitselement

5969 one-stage resin
f résol
e resol
i resolo
d Resol

* **one-way valve** → 5833

5970 "on site" oxygen plant
f générateur d'oxygène "de chantier"
e instalación de oxígeno "en el sitio"
i impianto di produzione dell'ossigeno "in situ"
d eigene Sauerstoffanlage

5971 on stream
f en marche
e en marcha
i in marcia
d in Betrieb

5972 ontogenesis
f ontogénie
e ontogenía; ontogénesis
i ontogenia; ontogenesi
d Ontogenie

5973 oocyte
f ovocyte
e oocito
i ovocita
d Eimutterzelle

5974 oosome
f oosome
e oosoma
i oosoma
d Oosom

5975 ootid
f ovotide
e oótida
i ootidio
d Ootide

5976 oozing
f suintement

e rezumamiento; rezume
i trasudamento
d Durchsickern

5977 opacity
f opacité; intransparence
e opacidad
i opacità
d Deckkraft; Undurchsichtigkeit

5978 opalescence
f opalescence
e opalescencia
i opalescenza
d Schiller

5979 opalescent
f opalescent
e opalescente
i opalescente
d opaleszent

5980 opaline
f opaline
e vidrio opalino
i vetro opalino
d Opalin

5981 opaque
f opaque
e opaco
i opaco
d undurchsichtig

5982 opiates
f opiates
e opiatos
i oppiati
d Opiate

5983 opium
f opium
e opio
i oppio
d Opium

5984 opposition factor
f facteur d'opposition
e factor de oposición
i fattore d'opposizione
d Oppositionsfaktor

5985 optical activity
f activité optique
e actividad óptica
i attività ottica
d optische Aktivität

5986 optical axis
 f axe optique
 e eje óptico
 i asse ottico
 d Linsenachse

5987 optical bleaches
 f décolorants optiques
 e blanqueadores ópticos
 i imbiancatori ottici
 d optische Bleichmittel

5988 optical isomers
 f isomères optiques
 e isómeros ópticos
 i isomeri ottici
 d optische Isomere

5989 optical plastics
 f plastiques optiques
 e plásticos ópticos
 i plastiche per l'ottica
 d optische Kunststoffe

5990 optical pyrometer
 f pyromètre optique
 e pirómetro óptico
 i pirometro ottico
 d optisches Pyrometer

5991 orange chromes
 f orangé de chrome
 e pigmentos anaranjados de cromato de
   plomo
 i arancione di cromo
 d Chromorange

 * **orange flower oil** → 5691

5992 orange lead
 f rouge au plomb
 e plomo anaranjado
 i arancio al piombo
 d Orangemennige

5993 orange peeling
 f formation de cratères
 e superficie rugosa
 i bucciatura
 d Kraterbildung

5994 orange shellac
 f laque en écailles orangé
 e laca anaranjada
 i gommalacca arancione
 d Orangeschellack

5995 ordinate
 f ordonnée

 e ordenada
 i ordinata
 d Ordinate

5996 ore
 f minerai
 e mineral
 i minerale
 d Erz

5997 ore dressing
 f préparation mécanique des minerais
 e beneficio de minerales
 i preparazione meccanica del minerale
 d Erzaufbereitung

5998 Orford process
 f procédé Orford
 e proceso Orford
 i processo Orford
 d Orford-Verfahren

5999 organic chemistry
 f chimie organique
 e química orgánica
 i chimica organica
 d organische Chemie

6000 organic pigments
 f pigments organiques
 e pigmentos orgánicos
 i pigmenti organici
 d organische Pigmente

6001 organometallic compounds
 f composés organométalliques
 e compuestos organometálicos
 i composti organometallici
 d Organometalle

6002 organosol
 f organosol
 e organosol
 i organosolo
 d Organosol

6003 orientate *v*
 f orienter
 e orientar
 i orientare
 d recken; richten

6004 orientation
 f orientation
 e orientación
 i orientamento
 d Orientierung

6005 orifice
  f orifice
  e orificio; abertura
  i orificio
  d Mündung; Öffnung

6006 ornithine
  f ornithine
  e ornitina
  i ornitina
  d Ornithin

6007 orphenadrine hydrochloride
  f chlorhydrate d'orphénadrine
  e hidrocloruro de orfenadrina
  i cloridrato di orfenadrina
  d Orphenadrinhydrochlorid

6008 orpiment
  f orpiment, orpin
  e oropimente
  i orpimento
  d Operment; gelbes Arsensulfid

6009 orthochromatic
  f orthochromatique
  e ortocromático
  i ortocromatico
  d orthochromatisch

6010 orthohelium
  f orthohélium
  e ortohelio
  i ortoelio
  d Orthohelium

6011 orthohydrogen
  f orthohydrogène
  e ortohidrógeno
  i ortoidrogeno
  d Orthowasserstoff

6012 orthophosphorous acid
  f acide orthophosphoreux
  e ácido ortofosforoso
  i acido ortofosforoso
  d Orthophosphorigsäure

6013 orthoploidy
  f orthoploïdie
  e ortoploidia
  i ortoploidismo
  d Orthoploidie

6014 orthorhombic
  f orthorhombique
  e ortorrómbico
  i ortorombico
  d orthorhombisch

* oscine → 7257

6015 osmic acid
  f acide osmique
  e ácido ósmico
  i acido osmico
  d Osmiumsäure

6016 osmiridium; iridosmine
  f osmiride; iridosmine
  e osmiridio; iridosmina
  i osmiridio; iridosmina
  d Osmiridium; Iridosmium

6017 osmium
  f osmium
  e osmio
  i osmio
  d Osmium

6018 osmometer
  f osmomètre
  e osmómetro
  i osmómetro
  d Osmometer

6019 osmosis
  f osmose
  e ósmosis
  i osmosi
  d Osmose

6020 osmotic pressure
  f pression osmotique
  e presión osmótica
  i pressione osmotica
  d osmotischer Druck

6021 ossification
  f ossification
  e osificación
  i ossificazione
  d Ossifikation

6022 osteoblast
  f ostéoblaste
  e osteoblasto
  i osteoblasto
  d Osteoblast

6023 Ostwald U-tube
  f tube en U d'Ostwald
  e tubo en U de Ostwald
  i tubo Ostwald a U
  d Ostwaldsche U-Röhre

6024 Othmer process
  f procédé Othmer
  e proceso Othmer

i processo Othmer
d Othmersches Verfahren

6025 ouabain
f ouabaïne
e uabaína
i strofantina
d Ouabain

6026 outcrop
f affleurement
e afloramiento; crestón
i affioramento
d Ausbiß; Aufschluß

6027 output
f rendement
e producción
i produzione
d Leistung

6028 oven
f four
e horno
i forno
d Ofen

6029 oven-dry weight
f poids sec absolu
e peso absoluto
i peso secco assoluto
d Darregewicht

6030 overall enrichment per stage
f facteur d'enrichissement par étage
e factor de enriquecimiento por etapa
i fattore d'arricchimento per stadio
d Gesamtanreicherungsfaktor je Stufe

6031 overcuring
f surcuisson
e curación excesiva
i indurimento eccessivo
d übermäßiges Härten

6032 overflow
f trop-plein
e rebose; desbordamiento; derrame
i traboccamento
d Überlaufen

6033 overhaul *v*
f réviser
e revisar; reparar
i revisionare
d überholen

6034 overhead product
f distillat de tête

e destilado de cabeza
i distillato di testa
d Kopfdestillat

6035 overmasticated compound; overmilled compound
f mélange trop mastiqué
e mezcla sobremasticada
i miscela sopraplastificata
d übermastizierte Mischung

6036 oversize
f refus
e material que no pasa por la criba
i residuo di vagliatura
d Überkorn

6037 overvoltage
f surtension
e sobretensión
i sovratensione
d Überspannung

6038 ovum
f œuf
e huevo
i uovo
d Ovum

6039 oxalates
f oxalates
e oxalatos
i ossalati
d Oxalate

6040 oxalic acid
f acide oxalique
e ácido oxálico
i acido ossalico
d Oxalsäure

6041 oxamide
f oxamide
e oxamida
i ossammide
d Oxamid

6042 oxidant
f oxydant
e oxidante
i ossidante
d Oxydationsmittel

6043 oxidase
f oxydase
e oxidasa
i ossidasi
d Oxydase

6044 oxidation
f oxydation
e oxidación
i ossidazione
d Oxydation

6045 oxidation-reduction indicators
f indicateurs d'oxydoréduction
e indicadores de oxidación-reducción
i indicatori di ossidazione e riduzione
d Oxydations-Reduktionsindikatoren

6046 oxides
f oxydes
e óxidos
i ossidi
d Oxyde

6047 oxidized bitumen
f bitume oxydé
e betún soplado
i bitume ossidato
d geblasenes Bitumen

* **oxidized turpentine** → 3359

6048 oxidizing agent
f agent oxydant
e agente de oxidación; oxidante
i agente ossidante; ossidante
d Oxydationsmittel

6049 oxidizing flame
f flamme oxydante; feu d'oxydation
e llama oxidante
i fiamma ossidante
d Oxydationsflamme; oxydierende Flamme

6050 oximes
f oximes
e oximas
i ossimi
d Oxime

6051 oxtriphylline
f oxtriphylline
e oxtrifilina
i ossitrifillina
d Oxtriphyllin

6052 oxycelluloses
f oxycelluloses
e oxicelulosas
i ossicellulose
d Oxycellulosen

6053 oxydiethylenebenzothiazole sulphenamide
f sulfénamide d'oxydiéthylène-benzothiazol

e oxidietilenbenzotiazolsulfenamida
i solfanamide di ossidietilenebenzotiazolo
d Oxydiäthylenbenzothiazolsulphenamid

6054 oxygen
f oxygène
e oxígeno
i ossigeno
d Sauerstoff

6055 oxyhaemoglobin
f oxyhémoglobine
e oxihemoglobina
i ossiemoglobina
d Oxyhämoglobin

6056 oxykrinin
f sécrétine
e oxicrinina
i secretina
d Sekretin

6057 oxymethandrolone
f oxyméthandrolone
e oximetandrolona
i ossimetandrolone
d Oxymethandrolon

6058 oxymorphone hydrochloride
f chlorhydrate d'oxymorphone
e hidrocloruro de oximorfona
i cloridrato di ossimorfone
d Oxymorphonhydrochlorid

6059 oxyphenonium bromide
f bromure d'oxyphénonium
e bromuro de oxifenonio
i bromuro di ossifenonio
d Oxyphenoniumbromid

6060 oxyproline
f oxyproline
e oxiprolina
i ossiprolina
d Oxyprolin

6061 oxytetracycline
f oxytétracycline
e oxitetraciclina
i ossitetraciclina
d Oxytetracyclin

6062 oxytocin
f oxytocine
e oxitocina
i ossitocina
d Oxytocin

6063 ozokerite; mineral wax
  f ozokérite; cire minérale
  e ozoquerita; cera mineral
  i ozocherite; cera minerale
  d Ozokerit; Mineralwachs

6064 ozone
  f ozone
  e ozono
  i ozono
  d Ozon

6065 ozonides
  f ozonides
  e ozonidas
  i ozonidi
  d Ozonide

P

6066 packed column
f colonne garnie
e columna de relleno
i colonna a riempimento
d Füllkörpersäule

6067 packed tower
f tour de percolation
e torre de rectificación de relleno
i torre a riempimento
d Füllkörpersäule

6068 packing
f garniture
e empaquetadura; relleno
i guarnitura; carica
d Dichtungsmittel; Füllung

6069 packing factor
f facteur d'empaquetage
e factor de empaque
i fattore d'impacchettamento
d Packungsfaktor

6070 paint
f peinture
e pintura
i pittura
d Farbe

6071 pair alleles
f allèles couplés
e pares alelomórficos
i alleli appaiati
d Pseudoallele

6072 pairing
f division en deux parties égales
e división por pares
i divisione in due parti uguali
d Halbieren

6073 palladinized asbestos
f amiante palladiée
e amianto paladinizado
i amianto al palladio
d Palladiumasbest

6074 palladious iodide
f iodure palladeux
e yoduro paladioso
i ioduro palladioso
d Palladiumjodür

6075 palladium
f palladium

e paladio
i palladio
d Palladium

6076 palladium chloride
f chlorure de palladium
e cloruro de paladio
i cloruro di palladio
d Palladiumchlorid

6077 palladium nitrate
f nitrate de palladium
e nitrato de paladio
i nitrato di palladio
d Palladiumnitrat

6078 palladium oxide
f oxyde de palladium
e óxido de paladio
i ossido di palladio
d Palladiumoxyd

6079 palmarosa oil
f essence de palma rosa
e aceite de palmarosa
i essenza di palmarosa
d Palmarosaöl

6080 palmitic acid
f acide palmitique
e ácido palmítico
i acido palmitico
d Palmitinsäure

6081 palmitoleic acid
f acide palmitoléique
e ácido palmitoleico
i acido palmitoleico
d Palmitoleinsäure

6082 palm nut oil
f huile de palmiste
e aceite de nuez de palma
i olio di noce di palma
d Palmkernöl

0083 pan *(of a balance)*
f plateau de balance
e platillo de balanza
i piattello della bilancia
d Waagschale

6084 panallele
f panallèle
e panalelo
i panallelo
d Panallel

6085 pan breeze
f scories
e cisco de cok
i scorie di coke
d Grus

6086 panchromatic
f panchromatique
e pancromático
i pancromatico
d panchromatisch

6087 pancreatin
f pancréatine
e pancreatina
i pancreatina
d Pankreatin

6088 pancytolysis
f pancytolyse
e pancitólisis
i pancitolisi
d Panzytolyse

6089 pangen
f pangène
e pángene
i pangene
d Pangen

6090 pangenosome
f pangénosome
e pangenosoma
i pangenosoma
d Pangenosom

6091 panmictic
f panmictique
e panmíctico
i panmittico
d panmiktisch

6092 pan mill
f broyeur à meules verticales
e molino de rodillos
i molazza a ruote verticali
d Kollergang; Kollermühle

6093 pantothenic acid
f acide pantothénique
e ácido pantoténico
i acido pantotenico
d Pantothensäure

6094 papain
f papaïne
e papaína
i papaina
d Papain

6095 papaver
f pavot
e papaverácea
i papavero
d Mohn

6096 papaverine
f papavérine
e papaverina
i papaverina
d Papaverin

6097 paper-lined construction
f montage au papier
e montaje recubierto de papel
i montaggio con carta impregnata
d Papierscheiderbauweise

6098 paper pulp
f pâte à papier
e pasta papelera
i pasta di legno
d Papiermasse; Papierbrei

6099 Papin's digester
f marmite de Papin; digesteur de Papin
e digestor de Papín
i pentola di Papin
d Papinscher Topf

6100 papyraceous
f papyracé
e papiráceo
i papiraceo
d papierartig

6101 para-
Prefix denoting an isomeric form of di-
substitution product derived from
benzene.
f para-
e para-
i para-
d Para-

6102 parachromatin
f parachromatine
e paracromatina
i paracromatina
d Parachromatin

6103 parachromatosis
f changement de couleur
e paracromia
i paracromatosi
d Farbverlust

6104 parachute
f écumeur automatique

e espumadera
i separatore del lievito
d Abheber

6105 paraffins
f paraffines
e parafinas
i paraffine
d Paraffine

6106 paraffin wax
f cire de paraffine
e cera de parafina
i cera di paraffina
d Paraffinwachs

6107 paraformaldehyde; polyoxymethylene
f paraformaldehyde
e paraformaldehído
i paraformaldeide
d Paraformaldehyd

6108 parahelium
f parahélium
e parahelio
i paraelio
d Parahelium

6109 parahydrogen
f parahydrogène
e parahidrógeno
i paraidrogeno
d Parawasserstoff

6110 paraldehyde
f paraldéhyde
e paraldehído
i paraldeide
d Paraldehyd

6111 parallax
f parallaxe
e paralaje
i parallasse
d Parallaxe

6112 parallel connexion
f montage en parallèle
e montaje en paralelo
i collegamento in parallelo
d Parallelschaltung

6113 paramagnetic
f paramagnétique
e paramagnético
i paramagnetico
d paramagnetisch

6114 parameiosis
f paraméiose
e parameiosis
i parameiosi
d Parameiose

6115 parameter
f paramètre
e parámetro
i parametro
d Parameter

6116 paramethadione
f paraméthadione
e parametadiona
i parametadione
d Paramethadion

6117 paramictic
f paramictique
e paramíctico
i paramittico
d paramiktisch

6118 paramitosis
f paramitose
e paramitosis
i paramitosi
d Paramitose

6119 paranuclein
f paranucléine
e paranucleína
i paranucleina
d Paranuklein

6120 parasiticide
f parasiticide
e parasiticida
i parassiticida
d parasitentötendes Mittel

6121 paratrophic
f paratrophe
e parátrofo
i paratrofo
d paratroph

6122 parboil *v*
f échauder; faire bouiller légèrement
e hervir parcialmente
i sobbollire
d brühen; nicht garkochen

6123 parchment
f parchemin
e pergamino
i pergamena
d Pergament

6124 parent element
 f élément père
 e elemento original
 i elemento padre
 d Ausgangselement

 * **Paris black** → 4706

 * **Paris green** → 3018

 * **Parisian blue** → 6782

 * **Paris violet** → 5460

6125 Paris white
 f blanc de Meudon; blanc d'Espagne
 e blanco de París
 i bianco di Parigi
 d Pariserweiß

6126 Paris yellow
 f jaune de Paris
 e amarillo de París
 i giallo di Parigi
 d Parisergelb

6127 Park's process
 f procédé Parke
 e proceso de Parke
 i processo di Parke
 d Parkerisieren

6128 partial chiasma
 f chiasma partiel
 e quiasma parcial
 i chiasma parziale
 d Partialchiasma

6129 partial plating
 f dépôt limité
 e depósito limitado
 i deposito limitato
 d Teilgalvanisierung

6130 particle
 f particule
 e partícula
 i particella
 d Teilchen

6131 particle shape
 f forme de particules
 e forma de las partículas
 i forma della particella
 d Teilchenform

6132 parting
 f séparation
 e desplatación; separación de la plata y oro

 i separazione
 d Scheiden

6133 parting agent
 f agent de démoulage
 e agente para facilitar la extracción del molde
 i agente di separazione
 d Gleitmittel

6134 passive
 f passif
 e pasivo; inerte; inactivo
 i passivo
 d passiv

6135 passivity
 f passivité
 e pasividad
 i passività
 d Passivität

6136 passivization
 f passivation
 e pasivación
 i passivazione
 d Passivierung

6137 paste
 Material forming the body of porcelain.
 f pâte
 e pasta; engrudo
 i pasta
 d Tonmasse

6138 paste
 Electrolyte containing gelatinized layer lying against the negative electrode.
 f pâte
 e pasta
 i pasta gelificante
 d Paste

6139 pasted plate
 f plaque à oxyde rapporté
 e placa de óxido empastado
 i piastra a ossidi riportati
 d pastierte Platte

6140 paste paint
 f pigment en pâte
 e pintura en pasta
 i colorante in pasta
 d Farbpaste

6141 paste resin
 f résine pour pâte
 e resina para pastas

i resina per paste
d Pastenharz

6142 pasteurization
f pasteurisation
e pasteurización
i pastorizzazione
d Pasteurisierung

6143 patent specification
f description de brevet
e especificación de patente
i descrizione del brevetto
d Patentschrift

6144 path coefficient
f "path-coefficient"
e coeficiente de trayectoria
i coefficiente di traiettoria
d Pfadkoeffizient

6145 pattern of damage
f type de lésion
e modelo de daño
i modello di danno
d Schädigungsmuster

6146 Pattinson process
f pattinsonage
e proceso Pattinson
i processo di Pattinson
d Pattinsonieren

6147 Pauling concentrator
f concentrateur Pauling
e torre de relleno concentradora
i concentratore di Pauling
d Pauling-Eindicker

6148 peak
f maximum
e valor máximo
i valore massimo
d Höchstwert

6149 peak load
f charge de pointe
e carga máxima
i carico massimo
d Spitzenbelastung

6150 pearlescent pigment
f pigment nacré
e pigmento nacarado
i pigmento perlaceo
d perlmutterfarbiges Pigment

6151 pearlite
f perlite

e perlita
i perlite
d Perlit

* pear oil → 496

6152 pebble mill
f broyeur à boulets; moulin à galets
e molino de bolas
i molino a biglie
d Kugelmühle

6153 pectins
f pectines
e pectinas
i pectine
d Pektine

6154 peeling
f écaillage
e escamado
i scagliatura; sfogliatura
d Abblättern

6155 peep hole
f regard
e mirilla; ventanilla
i foro di spia
d Schauloch

6156 pelargonic acid; nonoic acid
f acide pélargonique; acide nonylique
e ácido pelargónico; ácido nonoico
i acido pelargonico; acido nonoico
d Pelargonsäure

6157 pelargonyl chloride
f chlorure de pélargonyle
e cloruro de pelargonilo
i cloruro di pelargonile
d Pelargonylchlorid

6158 pellet
f pastille
e perdigón; pídora; bolita
i pasticca
d Tablette

6159 pelletierine tannate
f tannate de pelletiérine
e tanato de peletierina
i tannato di pelletierina
d Pelletierintannat

6160 pelleting
f pastillage
e formación de partículas pequeñas
i appallottolamento
d Tablettieren

6161 pellicle
f pellicule à autosupport
e emulsión autosoportada
i emulsione autoportante
d selbsttragende Emulsion

6162 pellucid
f transparent
e diáfano; transparente; pelúcido
i trasparente
d durchsichtig

6163 penetrometer
f pénétromètre
e penetrómetro
i penetrometro
d Penetrationsmesser; Härtemesser

6164 penicillin
f pénicilline
e penicilina
i penicillina
d Penicillin

6165 penicillinase
f pénicillinase
e penicilinasa
i penicillinasi
d Penicillinase

6166 penicillin-fast
f pénicilline-résistant
e resistente a la penicilina
i resistente alla penicillina
d penicillinresistent

6167 pennyroyal oil
f essence de menthe Pouliot
e aceite de poleo
i essenza di mentha pulegium
d Pennyroyalöl

6168 pentaborane
f pentaborane
e pentaborano
i pentaborano
d Pentaboran

6169 pentachloroethane
f pentachloréthane
e pentacloroetano
i pentacloroetano
d Pentachloräthan

6170 pentachloronitrobenzene
f pentachloronitrobenzène
e pentacloronitrobenceno
i pentacloronitrobenzene
d Pentachlornitrobenzol

6171 pentachlorophenol
f pentachlorophénol
e pentaclorofenol
i pentaclorofenolo
d Pentachlorphenol

6172 pentadecane
f pentadécane
e pentadecano
i pentadecano
d Pentadecan

6173 pentadecanoic acid
f acide pentadécanoïque
e ácido pentadecanoico
i acido pentadecanoico
d Pentadecansäure

6174 pentaerythritol
f pentaérythritol
e pentaeritrita; pentaeritritol
i pentaeritritolo
d Pentaerythrit

6175 pentaerythritol tetranitrate
f tétranitrate de pentaérythritol
e tetranitrato de pentaeritrita
i tetranitrato di pentaeritritolo
d Pentaerythrittetranitrat

6176 pentane
f pentane
e pentano
i pentano
d Pentan

6177 pentanediol
f pentandiol
e pentanodiol
i pentanediolo
d Pentandiol

* pentanepentol → 8886

6178 pentanol
f pentanol
e pentanol
i pentanolo
d Pentanol

6179 pentasomic
f pentasomique
e pentasómico
i pentasomico
d pentasom

6180 pentatriacontane
f pentatriacontane
e pentatriacontano

i pentatriacontano
d Pentatriakontan

6181 pentavalent; quinquivalent
f pentavalent
e pentavalente
i pentavalente
d fünfwertig

* pentene → 501

6182 pentlandite
f pentlandite
e pentlandita
i pentlandite
d Pentlandit

6183 pentobarbital
f pentobarbital
e pentobarbital
i pentobarbital
d Pentobarbital

6184 pentolite
f pentolite
e pentolita
i pentolite
d Pentolit

6185 pentose
f pentose
e pentosa
i pentosio
d Pentose

6186 pentylenetetrazol
f pentylènetétrazol
e pentilentetrazol
i pentilenetetrazolo
d Pentylentetrazol

6187 peppermint oil
f essence de menthe
e aceite de menta piperita
i essenza di menta
d Pfefferminzöl

6188 pepsin
f pepsine
e pepsina
i pepsina
d Pepsin

6189 peptization
f peptisation
e peptización
i peptizzazione
d Peptisieren

6190 peptize *v*
f peptiser
e peptizar
i peptizzare
d verteilen

6191 peptizer
f agent peptisant
e agente peptizante
i agente peptizzante
d Peptisiermittel

6192 peptone
f peptone
e peptona
i peptone
d Pepton

6193 peracetic acid
f acide peracétique
e ácido peracético
i acido peracetico
d Peressigsäure

6194 peracid
f peracide
e perácido
i peracido
d Persäure

6195 perchlorates
f perchlorates
e percloratos
i perclorati
d Perchlorate

6196 perchlorether
f perchloréther
e percloréter
i etere perclorico
d Perchloräther

6197 perchloric acid
f acide perchlorique
e ácido perclórico
i acido perclorico
d Perchlorsäure

6198 perchloroethylene
f perchloréthylène
e percloroetileno
i percloroetilene
d Perchloräthylen

6199 perchloromethyl mercaptan
f perchlorométhylmercaptan
e perclorometilmercaptano
i perclorometilemercaptan
d Perchlormethylmercaptan

6200 perchloryl fluoride
f fluorure de perchloryle
e fluoruro de perclorilo
i fluoruro di perclorile
d Perchlorylfluorid

6201 percolate v
f percoler; filtrer; passer
e percolar; filtrar; rezumar
i filtrare
d durchsickern lassen

6202 percolation
f percolation; suintement
e percolación; infiltración
i filtrazione; percolazione
d Durchsickern

6203 perfect gas
f gaz parfait
e gas ideal
i gas perfetto
d ideales Gas

6204 performance test
f essai de valeur d'emploi
e ensayo de valor de uso
i prova di comportamente delle
  caratteristiche nell'uso
d Gebrauchsfähigkeitsprüfung

6205 performic acid
f acide performique
e ácido perfórmico
i acido performico
d Perameisensäure

6206 pericarp
f péricarpe
e pericarpio
i pericarpio
d Fruchtschale

6207 periclase
f périclase
e periclasa
i periclasi
d Periklas

6208 perikinetic
f périkinétique
e pericinético
i pericinetico
d perikinetisch

6209 periodic acid
f acide periodique
e ácido peryódico

i acido periodico
d Perjodsäure

6210 periodic system
f classification périodique
e sistema periódico
i sistema periodico
d periodisches System

6211 periodic table
f table périodique
e tabla periódica
i tavola periodica
d periodische Tabelle

6212 periplasma
f périplasme
e periplasma
i periplasma
d Periplasma

6213 perishing of rubber
f dégradation du caoutchouc
e descomposición de goma
i degradazione della gomma
d Unbrauchbarwerden des Gummis

6214 perlite
f perlite
e perlita
i perlite
d Perlit

6215 permanent deformation; permanent set
f déformation permanente
e deformación permanente
i deformazione permanente
d Dauerveränderung; bleibende
  Veränderung

6216 permanent hardness
f crudité permanente
e dureza permanente
i durezza permanente
d bleibende Härte

* **permanent set** → 6215

6217 permanganates
f permanganates
e permanganatos
i permanganati
d Permanganate

6218 permeability; perviousness
f perméabilité
e permeabilidad; penetrabilidad
i permeabilità
d Durchlässigkeit; Permeabilität

6219 permeation
f imprégnation; pénétration
e permeación; impregnación; penetración
i permeazione
d Imprägnierung; Durchdringung;
  Tränkung

6220 peroxidase
f peroxydase
e peroxidasa
i perossidasi
d Peroxydase

6221 peroxides
f peroxydes
e peróxidos
i perossidi
d Peroxyde

6222 perphenazine
f perphénazine
e perfenacina
i perfenazina
d Perphenazin

6223 Persian Gulf red
f rouge persan
e rojo de Persia
i rosso di Persia
d Persischrot

6224 Peruvian balsam
f baume du Pérou
e bálsamo peruano
i balsamo del Perú
d Perubalsam

6225 pervious
f perméable
e pervio
i pervio
d durchlässig

* **perviousness → 6218**

6226 pestle
f pilon de mortier
e mano de mortero
i pestello
d Reiber; Stößer

6227 Peterson concentrator
f concentrateur de Peterson
e concentrador de Peterson
i concentratore di Peterson
d Peterson-Eindicker

6228 Petri dish
f boîte de Pétri

e placa de Petri
i scatola di Petri
d Petri-Schale

6229 petrolatum; petroleum jelly
f pétrolatum; vaseline
e petrolato; vaselina
i petrolato; vasellina
d Petrolatum; Vaseline

6230 petroleum
f pétrole
e petróleo
i petrolio
d Petroleum

6231 petroleum greases
f graisses de pétrole
e grasas del petróleo
i grassi di fondi di raffineria
d Petroleumfette

* **petroleum jelly → 6229**

6232 petroleum waxes
f cires de pétrole
e ceras del petróleo
i paraffine
d Petroleumwachse

6233 pH
f pH
e pH
i pH
d pH

6234 pH value
f valeur pH
e valor pH
i valore pH
d pH-Wert

6235 phage splitting
f rupture phagique
e ruptura de fago
i rottura del fago
d Phagenspaltung

6236 phagocyte
f phagocyte
e fagocito
i fagocita
d Phagozyt

6237 phagocytosis
f phagocytose
e fagocitosis
i fagocitosi
d Phagozytose

6238 **pharmacodynamics**
  f pharmacodynamie
  e farmacodinámica
  i farmacodinamica
  d Pharmakodynamik

6239 **phase**
  f phase
  e fase
  i fase
  d Phase

6240 **phase boundary**
  f couche limite de phases
  e capa límite de fases
  i strato limite di fasi
  d Phasengrenzschicht

6241 **phase relationship**
  f règle des phases
  e regla de las fases
  i regola della fasi
  d Phasenregel

6242 **phase reversal of emulsion**
  f inversion du type d'émulsion
  e inversión de fase del tipo de emulsión
  i emulsione reversibile
  d Emulsionsentmischung

6243 **phase transition**
  f transition de phase
  e transición de fase
  i transizione di fase
  d Phasenübergang

6244 **phenacemide**
  f phénacémide
  e fenacemida
  i fenacemmide
  d Phenacemid

6245 **phenakite; phenacite**
  f phénacite
  e fenacita
  i fenachite
  d Phenakit

6246 **phenanthrene**
  f phénanthrène
  e fenantreno
  i fenantrene
  d Phenanthren

6247 **phenanthrenequinone**
  f phénanthrène-quinone
  e fenantrenoquinona
  i fenantrenechinone
  d Phenanthrenchinon

6248 **phenarsazine chloride**
  f chlorure de phénarsazine
  e cloruro de fenarsacina
  i cloruro di fenarsazina
  d Diphenylaminchlorarsin

6249 **phenazine**
  f phénazine
  e fenacina
  i fenazina
  d Phenazin

6250 **phenetidine**
  f phénétidine
  e fenetidina
  i fenetidina
  d Phenetidin

6251 **phenindamine tartrate**
  f tartrate de phénindamine
  e tartrato de fenindamina
  i tartrato di fenindammina
  d Phenindamintartrat

6252 **phenobarbital**
  f phénobarbital
  e fenobarbital
  i fenobarbital
  d Phenobarbital

6253 **phenocline**
  f phénocline
  e fenoclino
  i fenoclino
  d Phänocline

6254 **phenol; carbolic acid**
  f phénol; acide phénylique
  e fenol; ácido carbólico
  i fenolo; acido carbolico
  d Phenol; Carbolsäure

6255 **phenoldisulphonic acid**
  f acide phénoldisulfonique
  e ácido fenoldisulfónico
  i acido fenoldisolfonico
  d Phenoldisulfonsäure

6256 **phenol-formaldehyde resins**
  f résines de phénolformaldéhyde
  e resinas de fenolformaldehído
  i resine fenolformaldeiche
  d Phenol-Formaldehyd-Harze

6257 **phenolic cement**
  f ciment phénolique
  e cemento fenólico
  i cemento fenolico
  d Phenolharzklebstoff

6258 phenolic plastics
f phénolplastes
e plásticos fenólicos
i materie plastiche fenoliche; fenoplasti
d Phenolharzkunststoffe; Phenolplaste

6259 phenolic resins
f résines phénoliques
e resinas fenólicas
i resine fenoliche
d Phenolharze

6260 phenology
f phénologie
e fenología
i fenologia
d Phänologie

6261 phenolphthalein
f phénolphtaléine
e fenolftaleína
i fenolftaleina
d Phenolphthalein

6262 phenol red; phenolsulphonphthalein
f rouge de phénol; fénolsulfophtaléine
e rojo de fenol; fenolsulfoftaleína
i rosso di fenolo; fenolsolfoftaleina
d Phenolrot; Phenolsulfophthalein

6263 phenol resins
f résines phénoliques
e resinas fenólicas
i resine fenoliche
d Phenolharze

6264 phenolsulphonic acid
f acide phénolsulfonique
e ácido fenolsulfónico
i acido fenolsolfonico
d Phenolsulfonsäure

* **phenolsulphonphthalein** → 6262

6265 phenomenon
f phénomène
e fenómeno
i fenomeno
d Erscheinung; Phänomen

6266 phenothiazine
f phénothiazine
e fenotiacina
i fenotiazina
d Phenothiazin

6267 phenotype
f phénotype
e fenótipo

i fenotipo
d Phänotyp

6268 phenotypic
f phénotypique
e fenotípico
i fenotipico
d phänotypisch

6269 phenoxyacetic acid
f acide phénoxyacétique
e ácido fenoxiacético
i acido fenossiacetico
d Phenoxyessigsäure

6270 phenoxybenzamine
f phénoxybenzamine
e fenoxibenzamina
i fenossibenzammina
d Phenoxybenzamin

6271 phenoxypropanediol
f phénoxypropandiol
e fenoxipropanodiol
i fenossipropanediolo
d Phenoxypropandiol

* **phenpromethamine** → 6314

6272 phentolamine hydrochloride
f chlorhydrate de phentolamine
e hidrocloruro de fentolamina
i cloridrato di fentolammina
d Phentolaminhydrochlorid

6273 phenylacetaldehyde
f phénylacétaldéhyde
e fenilacetaldehído
i fenilacetaldeide
d Phenylacetaldehyd

6274 phenylacetamide
f phénylacétamide
e fenilacetamida
i fenilacetammide
d Phenylacetamid

6275 phenyl acetate
f acétate de phényle
e acetato de fenilo
i acetato di fenile
d Phenylacetat

6276 phenylacetic acid
f acide phénylacétique
e ácido fenilacético
i acido fenilacetico
d Phenylessigsäure

6277 **phenylaminonaphthol sulphonic acid**
  f acide phénylaminonaphtolsulfonique
  e ácido fenilaminonaftolsulfónico
  i acido fenilamminonaftolsolfonico
  d Phenylaminonaphtholsulfonsäure

6278 **phenylarsonic acid**
  f acide phénylarsonique
  e ácido fenilarsónico
  i acido fenilarsonico
  d Phenylarsinsäure

6279 **phenylbutazone**
  f phénylbutazone
  e fenilbutazona
  i fenilbutazone
  d Phenylbutazon

6280 **phenylbutynol**
  f phénylbutynol
  e fenilbutinol
  i fenilbutinolo
  d Phenylbutynol

6281 **phenylcarbethoxypyrazolone**
  f phénylcarbéthoxypyrazolone
  e fenilcarbetoxipirazolona
  i fenilcarbetossipirazolona
  d Phenylcarbäthoxypyrazolon

6282 **phenylcarbylamine chloride**
  f chlorure de phénylcarbylamine
  e cloruro de fenilcarbilamina
  i cloruro di fenilcarbilammina
  d Phenylcarbylaminchlorid

6283 **phenyl chloride**
  f chlorure de phényle
  e cloruro de fenilo
  i cloruro di fenile
  d Chlorbenzol

6284 **phenylcyclohexane**
  f phénylcyclohexane
  e fenilciclohexano
  i fenilcicloesano
  d Phenylcyclohexan

6285 **phenylcyclohexanol**
  f phénylcyclohexanol
  e fenilciclohexanol
  i fenilcicloesanolo
  d Phenylcyclohexanol

6286 **phenyl diethanolamine**
  f phényldiéthanolamine
  e fenildietanolamina
  i fenildietanolammina
  d Phenyldiäthanolamin

6287 **phenyl dimethylurea**
  f phényldiméthylurée
  e fenildimetilurea
  i fenildimetilurea
  d Phenyldimethylharnstoff

6288 **phenylenediamine**
  f phénylènediamine
  e fenilendiamina
  i fenilendiammina
  d Phenylendiamin

6289 **phenylethanolamine**
  f phényléthanolamine
  e feniletanolamina
  i feniletanolammina
  d Phenyläthanolamin

6290 **phenylethyl acetate**
  f acétate de phényléthyle
  e acetato de feniletilo
  i acetato di feniletile
  d Phenyläthylacetat

6291 **phenylethylacetic acid**
  f acide phényléthylacétique
  e ácido feniletilacético
  i acido feniletilacetico
  d Phenyläthylessigsäure

6292 **phenylethyl alcohol**
  f alcool phényléthylique
  e alcohol feniletílico
  i alcool feniletilico
  d Phenyläthylalkohol

6293 **phenylethyl anthranilate**
  f anthranilate de phényléthyle
  e antranilato de feniletilo
  i antranilato di feniletile
  d Phenyläthylanthranilat

  * **phenylethylene** → 7871

6294 **phenylethylethanolamine**
  f phényléthyléthanolamine
  e feniletiletanolamina
  i feniletiletanolammina
  d Phenyläthyläthanolamin

6295 **phenylethyl isobutyrate**
  f isobutyrate de phényléthyle
  e isobutirato de feniletilo
  i isobutirrato di feniletile
  d Phenyläthylisobutyrat

6296 **phenylethyl propionate**
  f propionate de phényléthyle
  e propionato de feniletilo

i propionato di feniletile
d Phenyläthylpropionat

6297 phenylhydrazine
   f phénylhydrazine
   e fenilhidracina
   i fenilidrazina
   d Phenylhydrazin

6298 phenyl isocyanate
   f isocyanate de phényle
   e isocianato de fenilo
   i isocianato di fenile
   d Phenylisocyanat

  * phenyl mercaptan → 8247

6299 phenylmercuric acetate
   f acétate phénylmercurique
   e acetato fenilmercúrico
   i acetato fenilmercurico
   d Phenylmerkuriacetat

6300 phenylmercuric borate
   f borate phénylmercurique
   e borato fenilmercúrico
   i borato fenilmercurico
   d Phenolmerkuriborat

6301 phenylmercuric chloride
   f chlorure phénylmercurique
   e cloruro fenilmercúrico
   i cloruro fenilmercurico
   d Phenylmerkurichlorid

6302 phenylmercuric hydroxide
   f hydroxyde phénylmercurique
   e hidróxido fenilmercúrico
   i idrossido fenilmercurico
   d Phenylmerkurihydroxyd

6303 phenylmercuric naphthenate
   f naphténate phénylmercurique
   e naftenato fenilmercúrico
   i naftenato fenilmercurico
   d Phenylmerkurinaphthenat

6304 phenylmercuric propionate
   f propionate phénylmercurique
   e propionato fenilmercúrico
   i propionato fenilmercurico
   d Phenylmerkuripropionat

6305 phenylmercuriethanolammonium acetate
   f acétate de phénylmercuriéthanol-
      ammonium
   e acetato fenilmercurietanolamónico
   i acetato fenilmercurietanolammonico
   d Phenylmerkuriäthanolammoniumacetat

6306 phenylmethylethanolamine
   f phénylméthyléthanolamine
   e fenilmetiletanolamina
   i fenilmetiletanolammina
   d Phenylmethyläthanolamin

6307 phenylmorpholine
   f phénylmorpholine
   e fenilmorfolina
   i fenilmorfolina
   d Phenylmorpholin

6308 phenylnaphthylamine
   f phénylnaphtylamine
   e fenilnaftilamina
   i fenilnaftilammina
   d Phenylnaphthylamin

6309 phenylphenol
   f phénylphénol
   e fenilfenol
   i fenilfenolo
   d Phenylphenol

6310 phenylpiperazine
   f phénylpipérazine
   e fenilpiperacina
   i fenilpiperazina
   d Phenylpiperazin

**6311 phenylpropanolamine hydrochloride;
     norephedrine**
   f chlorhydrate de phénylpropanolamine
   e hidrocloruro de fenilpropanolamina
   i cloridrato di fenilpropanolammina
   d Phenylpropanolaminhydrochlorid;
      Norephedrin

6312 phenylpropyl acetate
   f acétate de phénylpropyle
   e acetato de fenilpropilo
   i acetato di fenilpropile
   d Phenylpropylacetat

6313 phenylpropyl aldehyde
   f aldéhyde phénylpropylique
   e aldehído fenilpropílico
   i aldeide fenilpropilica
   d Phenylpropylaldehyd

**6314 phenylpropylmethylamine;
     phenpromethamine**
   f phénylpropylméthylamine
   e fenilpropilmetilamina
   i fenilpropilmetilammina
   d Phenylpropylmethylamin

6315 phenyltrichlorosilane
   f phényltrichlorosilane

  e feniltriclorosilano
  i feniltriclorosilano
  d Phenyltrichlorsilan

6316 phlogistic
  f inflammatoire
  e flogístico
  i infiammatorio
  d entzündlich

6317 phloroglucinol
  f phloroglucine
  e floroglucinol
  i floroglucinolo
  d Phloroglucin

6318 phonochemistry
  f phonochimie
  e fonoquímica
  i fonochimica
  d Phonochemie

6319 phorone
  f phorone
  e forona
  i forone
  d Phoron

6320 phosgene; carbonyl chloride
  f phosgène
  e fosgeno
  i fosgene
  d Phosgen

6321 phosphate rock
  f roche phosphatée
  e roca de fosfato
  i fosforite
  d Calciumphosphat

6322 phosphates
  f phosphates
  e fosfatos
  i fosfati
  d Phosphate

  * **phosphatide** → 6326

6323 phosphide
  f phosphure
  e compuesto de fósforo y metal
  i fosfuro
  d Phosphid

6324 phosphine
  f phosphine
  e fosfina
  i fosfina
  d Phosphin

6325 phosphites
  f phosphites
  e fosfitos
  i fosfiti
  d Phosphite

6326 phospholipid; phosphatide
  f phospholipide; phosphatide
  e fosfolípido; fosfátido
  i fosfolipide; fosfatide
  d Phospholipid; Phosphatid

6327 phosphomolybdic acid
  f acide phosphomolybdique
  e ácido fosfomolíbdico
  i acido fosfomolibdico
  d Phosphormolybdänsäure

6328 phosphonium bases
  f bases de phosphonium
  e bases de fosfonio
  i basi di fosfonio
  d Phosphoniumbasen

6329 phosphonium iodide
  f iodure de phosphonium
  e yoduro de fosfonio
  i ioduro di fosfonio
  d Phosphoniumjodid

6330 phosphopenia
  f phosphoropénie
  e fosfopenia
  i mancanza di acido fosforico
  d Phosphormangel

6331 phosphor-bronzes
  f bronzes phosphoreux
  e bronces fosforosos
  i bronzi fosforosi
  d Phosphorbronzen

6332 phosphorescence
  f phosphorescence
  e fosforescencia
  i fosforescenza
  d Phosphoreszenz

6333 phosphorescent pigments
  f pigments phosphorescents
  e pigmentos fosforescentes
  i pigmenti fosforescenti
  d phosphoreszierende Pigmente;
    Leuchtstoffe

6334 phosphoric acid
  f acide phosphorique
  e ácido fosfórico

i acido fosforico
d Phosphorsäure

6335 phosphorous acid
f acide phosphoreux
e ácido fosforoso
i acido fosforoso
d Phosphorigsäure

6336 phosphors
f corps phosphorescents
e fósforos
i fosfori
d Phosphore; Leuchtschirmsubstanzen

6337 phosphorus
f phosphore
e fósforo
i fosforo
d Phosphor

6338 phosphorus oxychloride
f oxychlorure de phosphore
e oxicloruro de fósforo
i ossicloruro di fosforo
d Phosphoroxychlorid

6339 phosphorus pentabromide
f pentabromure de phosphore
e pentabromuro de fósforo
i pentabromuro di fosforo
d Phosphorpentabromid

6340 phosphorus pentachloride
f pentachlorure de phosphore
e pentacloruro de fósforo
i pentacloruro di fosforo
d Phosphorpentachlorid

6341 phosphorus pentasulphide
f pentasulfure de phosphore
e pentasulfuro de fósforo
i pentasolfuro di fosforo
d Phosphorpentasulfid

6342 phosphorus pentoxide
f pentoxyde de phosphore
e pentóxido de fósforo
i pentossido di fosforo
d Phosphorpentoxyd

6343 phosphorus sesquisulphide
f sesquisulfure de phosphore
e sesquisulfuro de fósforo
i sesquisolfuro di fosforo
d Phosphorsubsulfid

6344 phosphorus tribromide
f tribromure de phosphore

e tribromuro de fósforo
i tribromuro di fosforo
d Phosphorbromür

6345 phosphorus trichloride
f trichlorure de phosphore
e tricloruro de fósforo
i tricloruro di fosforo
d Phosphortrichlorid

6346 phosphotungstic acid
f acide phosphotungstique
e ácido fosfotúngstico
i acido fosfotungstico
d Phosphorwolframsäure

6347 phosphuranylite
f phosphuranylite
e fosfuranilite
i fosfuranilite
d Phosphuranylit

6348 photocatalysis
f photocatalyse
e fotocatálisis
i fotocatalisi
d Photokatalyse

* **photocell → 6351**

6349 photochemistry
f photochimie
e fotoquímica
i fotochimica
d Photochemie

6350 photodissociation
f photodissociation
e fotodisociación; fotodesdoblamiento
i fotodissociazione
d Photozersetzung

6351 photoelectric cell; photocell
f cellule photoélectrique
e célula fotoeléctrica
i cellula fotoelettrica
d Photozelle

6352 photographic developers
f révélateurs photographiques
e reveladores fotográficos
i materiale per sviluppo fotografico
d Entwickler

* **photographic negative → 4858**

6353 photolysis
f photolyse
e fotólisis

i fotolisi
d Photolyse

6354 photometer
f photomètre
e fotómetro
i fotometro
d Photometer; Lichtmesser

6355 photophoresis
f photophorèse
e fotofóresis
i fotoforesi
d Photophorese

6356 photopolymer
f photopolymère
e fotopolímero
i fotopolimero
d Photopolymer

6357 photoreactivation
f photoréactivation
e fotoreactivación
i fotoriattivazione
d Photoreaktivierung

6358 photosensitive
f photosensible
e sensible a la luz
i sensibile alla luce
d lichtempfindlich

6359 photosynthesis
f photosynthèse
e fotosíntesis
i fotosintesi
d Photosynthese

6360 phragmoplast
f phragmoplaste
e fragmoplasto
i fragmoplasto
d Phragmoplast

6361 phragmosome
f phragmosome
e fragmosoma
i fragmosoma
d Phragmosom

6362 phthalamide
f phtalamide
e ftalamida
i ftalammide
d Phthalamid

6363 phthaleins
f phtaléines

e ftaleínas
i ftaleine
d Phthaleine

6364 phthalic acid
f acide phtalique
e ácido ftálico
i acido ftalico
d Phthalsäure

6365 phthalic anhydride
f anhydride phtalique
e anhídrido ftálico
i anidride ftalica
d Phthalsäureanhydrid

6366 phthalic resins
f résines phtaliques
e resinas ftálicas
i resine ftaliche
d Phthalatharze

6367 phthalimide
f phtalimide
e ftalimida
i ftalimmide
d Phthalimid

6368 phthalonitrile
f phtalonitrile
e ftalonitrilo
i ftalonitrile
d Phthalonitril

6369 phthalylsulphacetamide
f phtalylsulfacétamide
e ftalilsulfacetamida
i ftalilsolfacetammide
d Phthalylsulfacetamid

6370 phthalylsulphathiazole
f phtalylsulfathiazol
e ftalilsulfatiazol
i ftalilsolfatiazolo
d Phthalylsulfathiazol

6371 physical scale of atomic weights
f table physique de poids atomiques
e tabla física de pesos atómicos
i tavola fisica di pesi atomici
d Atomgewichte

6372 physiochemistry
f physiochimie
e fisioquímica
i fisiochimica
d Physiochemie

6373 physiochromatin
 f physiochromatine
 e fisiocromatina
 i fisiocromatina
 d Physiochromatin

6374 physostigmine
 f physostigmine; ésérine
 e fisostigmina
 i fisostigmina
 d Physostigmin

6375 phytic acid
 f acide phytique
 e ácido fítico
 i acido fitico
 d Phytinsäure

6376 phytochemical
 f phytochimique
 e fitoquímico
 i fitochimico
 d pflanzenchemisch

6377 phytol
 f phytol
 e fitol
 i fitolo
 d Phytol

6378 phytonadione
 f phytonadione
 e fitonadiona
 i fitonadione
 d Phyllochinon

6379 phytosterols
 f phytostérols
 e fitosteroles
 i fitosteroli
 d Phytosterine

6380 pickle
 f décapant
 e decapante
 i bagno di decapaggio
 d Beize

6381 pickle *v*
 f décaper
 e decapar al ácido
 i decapare
 d beizen

6382 pickling
 f décapage
 e decapado con ácido
 i decapaggio
 d Beizen

6383 pickling inhibitor
 f inhibiteur de décapage
 e inhibidor de decapado
 i inibitore di decapaggio
 d Beizzusatz

6384 pickling pond
 f bac de décapage
 e recipiente de cristalización
 i vasca di decapaggio
 d Kristallisationsbecken

6385 picoline
 f picoline
 e picolina
 i picolina
 d Picolin

6386 picramic acid
 f acide picramique
 e ácido picrámico
 i acido picramico
 d Pikraminsäure

6387 picric acid; trinitrophenol
 f acide picrique
 e ácido pícrico
 i acido picrico
 d Pikrinsäure

6388 picrolonic acid
 f acide picrolonique
 e ácido picrolónico
 i acido picrolonico
 d Pikrolonsäure

6389 picrotoxin
 f picrotoxine
 e picrotoxina
 i picrotossina
 d Pikrotoxin

6390 piezochemistry
 f piézochimie
 e piezoquímica
 i piezochimica
 d Piezochemie; Kristallchemie

6391 piezoelectric effect
 f effet piézoélectrique
 e efecto piezoeléctrico
 i effetto piezoelettrico
 d piezoelektrischer Effekt

6392 piezometer
 f piézomètre
 e piezómetro
 i piezometro
 d Piezometer

6393 pigment
 f pigment
 e pigmento
 i pigmento
 d Pigment

6394 pigment paste
 f pigment en pâte
 e pasta de pigmento
 i pasta di pigmento; pigmento in pasta
 d Pigmentpaste

6395 pilbarite
 f pilbarite
 e pilbarita
 i pilbarite
 d Pilbarit

6396 pilocarpine
 f pilocarpine
 e pilocarpina
 i pilocarpina
 d Pilocarpin

6397 pilot cell
 f élément pilote
 e elemento piloto
 i elemento pilota
 d Prüfzelle

6398 pilot plant
 f installation d'essai; usine pilote
 e instalación piloto; instalación de ensayo
 i impianto pilota
 d Versuchsanlage

6399 pinch-cock
 f pince de Mohr
 e abrazadera de compresión
 i pinza per tubi di gomma
 d Quetschhahn

6400 pinch effect
 f effet de pincement
 e restricción
 i effetto di contrazione
 d Einschnürungseffekt

6401 pinene
 f pinène
 e pineno
 i pinene
 d Pinen

6402 pine oil
 f essence de pin
 e aceite de pino
 i olio essenziale di pino
 d Pineoil; Tannenöl

6403 pine tar
 f goudron de pin; goudron de bois de sapin
 e alquitrán de pino
 i catrame di pino
 d Kienteer

6404 pinguid
 f gras
 e aceitoso; grasiento
 i unto; untuoso
 d fettig

6405 pinholing
 f piquage
 e superficie picada
 i butteramento
 d Porenbildung; Kraterbildung

6406 pipeclay triangle
 f triangle en fil de fer
 e triángulo de alambre
 i triangolo di filo metallico
 d Drahtdreieck

6407 piperazine
 f pipérazine
 e piperacina
 i piperazina
 d Piperazin

6408 piperazine hydrate
 f hydrate de pipérazine
 e hidrato de piperacina
 i idrato di piperazina
 d Piperazinhydrat

6409 piperazine oestrone sulphate
 f sulfate de pipérazine-oestrone
 e sulfato de piperacina estrona
 i solfato di estrone e piperazina
 d Piperazinöstronsulfat

6410 piperidine
 f pipéridine
 e piperidina
 i piperidina
 d Piperidin

6411 piperidinoethanol
 f pipéridinoéthanol
 e piperidinoetanol
 i piperidinoetanolo
 d Piperidinäthanol

6412 piperocaine hydrochloride
 f chlorhydrate de pipérocaïne
 e hidrocloruro de piperocaína
 i cloridrato di piperocaina
 d Piperocainhydrochlorid

**6413 piperonal; methylene-protocatechuic
aldehyde**
 f pipéronal
 e piperonal
 i piperonal
 d Piperonal

6414 piperoxane hydrochloride
 f chlorhydrate de pipéroxane
 e hidrocloruro de piperoxano
 i cloridrato di piperossano
 d Piperoxanhydrochlorid

6415 piperylene
 f pipérylène
 e piperileno
 i piperilene
 d Piperylen

6416 pipe still
 f alambic tubulaire
 e alambique de tubos recalentados
 i distillatore tubolare
 d Röhrendestillationsofen

6417 pipette
 f pipette
 e pipeta
 i pipetta
 d Pipette

6418 pipette stand
 f étagère à pipettes
 e estante para pipetas
 i portapipette
 d Pipettenständer

6419 pitch
 f brai; poix
 e pez; brea
 i pece
 d Pech; Teer

6420 pitch *v*
 f levurer
 e inocular con levadura
 i inoculare lievito
 d anstellen (Hefe)

6421 pitchblende; uraninite
 f pechblende; uraninite
 e pechblenda; uraninita
 i pechblenda; blenda picea; uraninite
 d Pechblende; Uraninit

6422 pitching rate
 f dose de levure
 e dosis de levadura

 i aggiunta di lievito
 d Hefegabe

6423 pitching yeast
 f levain
 e levadura de siembra
 i lievito di semenza
 d Anstellhefe

6424 Pitot tube
 f tube de Pitot
 e tubo de Pitot
 i tubo di Pitot
 d Pitotrohr; Staurohr

6425 pits
 f piqûres
 e picaduras
 i figure di attacco
 d Narben; Poren

6426 pitting
 f formation de piqûres
 e corrosión alveolar superficial
 i pittatura
 d Grübchenbildung

6427 pituicyte
 f cellule hypophysaire
 e pituicito
 i cellula ipofisaria posteriore
 d Hypophysenzelle

6428 pituitary
 f pituitaire
 e pituitario
 i pituitario
 d pituitär

6429 planetary mixer
 f agitateur planétaire
 e mezcladora de paletas planetarias
 i mescolatore planetario
 d Planetenrührwerk

6430 planetary stirring machine
 f agitateur planétaire
 e agitador planetario
 i agitatore planetario
 d Planetenrührwerk

6431 planogamete
 f planogamète
 e planogameto
 i planogamete
 d Planogamet

6432 plantation white
 f sucre colonial

e azúcar colonial; azúcar blanco
i zucchero bianco
d Kolonialzucker; Weißzucker

6433 Planté plate
f plaque Planté
e placa Planté
i piastra Planté
d Großoberflächenplatte

6434 plaque
f plaque
e placa
i placca
d Plaque

6435 plasma
The clear, liquid part of blood.
f plasma
e plasma
i plasma
d Plasma

6436 plasma
Bright-green translucent variety of
chalcedony, used as a gem.
f plasma; calcédoine vert foncé
e calcedonia verde brillante
i plasma
d Plasma

6437 plasma balance
f équilibre de plasma
e equilibrio de plasma
i equilibrio di plasma
d Plasmagleichgewicht

6438 plasmachromatin
f plasmachromatine
e plasmacromatina
i plasmacromatina
d Plasmachromatin

6439 plasmagenes
f plasmagènes
e plasmagenes
i plasmageni
d Plasmagene

6440 plasma gun
f canon à plasma
e cañon de plasma
i cannone a plasma
d Plasmakanone

6441 plasma instability
f instabilité de plasma
e inestabilidad de plasma

i instabilità di plasma
d Plasmainstabilität

6442 plasma pressure
f pression du plasma
e presión del plasma
i pressione del plasma
d Plasmadruck

6443 plasma propulsion
f propulsion par plasma
e propulsión por plasma
i propulsione per plasma
d Plasmaantrieb

6444 plasma purity
f pureté du plasma
e pureza del plasma
i purezza del plasma
d Plasmareinheit

6445 plasmasome
f plasmasome
e plasmasoma
i plasmasoma
d Leukozytenkörnelung

6446 plasmid
f plasmide
e plasmidio
i plasmidio
d Plasmid

6447 plasmodium
f plasmodium
e plasmodium
i plasmodio
d Plasmodium

6448 plasmolysis
f plasmolyse
e plasmólisis
i plasmolisi
d Plasmolyse

6449 plasmon
f plasmone
e plasmón
i plasmon
d Plasmon

6450 plasmosome
f plasmosome
e plasmosoma
i plasmosoma
d Plasmosom

6451 plaster of Paris
f plâtre de moulage

e yeso de París
i gesso di Parigi
d Gips; gebrannter Gips

6452 plastic flow
f écoulement plastique
e flujo plástico
i flusso plastico
d plastisches Fließen

6453 plasticity
f plasticité
e plasticidad
i plasticità
d Plastizität

6454 plasticizer
f plastifiant
e plastificante
i plastificante
d Plastifizierer

6455 plasticizing capacity
f capacité de plastification
e capacidad de plastificación
i capacità plastificante
d Verflüssigungsleistung

6456 plastics
f plastiques; matière plastique
e plásticos
i materia plastica
d Kunststoffe

6457 plastid
f plastide
e plastidio
i plastidio
d Plastid

6458 plastidome
f plastidome
e plastidomio
i plastidoma
d Plastidom

6459 plastimeter
f plastimètre
e plastómetro
i plastometro
d Plastimeter

6460 plastisol
f plastisol
e plastisol
i plastisol
d Plastisol

6461 plastome
f plastome
e plastoma
i plastoma
d Plastom

6462 plastomer
f plastomère
e plastómero
i plastomero
d Plastomer

6463 plastometer
f plastomètre; consistomètre
e plastómetro; consistómetro
i plastometro; consistometro
d Plastometer; Konsistenzmesser

6464 plastosome
f plastosome
e plastosoma
i plastosoma
d Plastosom

6465 platiculture
f culture sur plaques
e platicultivo
i coltura su piastre
d Plattenkultur

6466 plating
f clichage
e clisado
i placcatura
d Plattieren

6467 plating generator; plating dynamo
f générateur pour galvanoplastie
e dínamo para galvanostegia
i generatore per galvanotecnica
d Plattiergenerator

6468 plating rack
f support d'accrochage
e soporte de los cátodos
i sospensioni
d Einhängegestell

6469 platinized asbestos
f amiante platinée
e amianto platinado
i amianto platinado
d Platinasbest

6470 platinum
f platine
e platino
i platino
d Platin

6471 platinum barium cyanide
 f cyanure de platine et de baryum
 e cianuro platinoso-bárico
 i cianuro di bario e platino
 d Platinbariumcyanid

6472 platinum black
 f noir de platine
 e negro de platino
 i nero di platino
 d Platinschwarz

6473 platinum chloride
 f chlorure de platine
 e cloruro de platino
 i cloruro di platino
 d Platinchlorid

6474 platinum sponge
 f mousse de platine
 e esponja de platino
 i platino spugnoso
 d Platinschwamm

6475 pleochroism
 f pléochroïsme
 e pleocroísmo
 i pleocroismo
 d Pleochroismus

6476 pliability
 f flexibilité
 e flexibilidad; plegabilidad
 i flessibilità
 d Biegsamkeit

6477 pliable
 f flexible
 e flexible
 i flessibile
 d biegsam

6478 plodder
 f emboutisseuse
 e extrusionador
 i imbutitrice
 d Peloteuse

 * **plumbago** → 3862

6479 plumbism; lead poisoning
 f empoisonnement au plomb
 e saturnismo
 i saturnismo
 d Bleivergiftung

6480 plumbosolvency
 f plumbo-solubilisation
 e plumbosolvencia
 i capacità di soluzione del piombo
   dell'acqua potabile
 d Bleilöslichkeit

6481 plumose
 f duveteux
 e en plumas
 i piumoso
 d fedrig

6482 plutonium
 f plutonium
 e plutonio
 i plutonio
 d Plutonium

6483 pneumatic dryer
 f séchoir à gaz
 e secadora neumática
 i essiccatore pneumatico
 d Gastrockner

6484 pocket-type plate
 f plaque à pochettes
 e placa de alveolos
 i piastra a taschette
 d Taschenplatte

6485 Podbielniak contactor
 f spirale contactrice de Podbielniak
 e espiral contactora de Podbielniak
 i spirale contattrice di Podbielniak
 d Kontaktspirale nach Podbielniak

6486 Poggendorf cell
 f élément de Poggendorf
 e elemento de Poggendorf
 i pila di Poggendorf
 d Poggendorf-Element

6487 poison
 f poison
 e veneno
 i veleno
 d Gift

 * **polar body** → 6499

6488 polar bond
 f liaison polaire
 e electrovalencia
 i legame polare
 d polare Verbindung

6489 polarimeter
 f polarimètre
 e polarímetro
 i polarimetro
 d Polarisationsapparat

6490 polarization
  f polarisation
  e polarización
  i polarizzazione
  d Polarisation

6491 polarization potential
  f potentiel de polarisation
  e potencial de polarización
  i potenziale di polarizzazione
  d Polarisationspotential

6492 polarized ionic bond
  f liaison ionique polarisée
  e enlace iónico polarizado
  i legame ionico polarizzato
  d polarisierte Ionenbindung

6493 polarizer
  f polarisant
  e polarizador
  i polarizzante
  d Polarisator

6494 polar liquid
  f liquide polaire
  e líquido polar
  i liquido polare
  d polare Flüssigkeit

6495 poliovirus
  f virus de la poliomyélite
  e poliovirus
  i virus della poliomielite
  d Poliomyelitisvirus

6496 polishing bob
  f disque à polir
  e disco de fieltro pulidor
  i disco per pulitrice
  d Schwabbelscheibe

6497 polishing mop
  f disques à polir
  e discos de fieltro pulidores
  i disco per lucidatrice
  d Poliermop; Flanellpolierscheiben

6498 polishing varnish; rubbing varnish
  f vernis de polissage
  e barniz que puede pulirse
  i vernice per lucidare
  d Polierlack

6499 polocyte; polar body
  f globule polaire
  e polocito; glóbulo polar
  i polocita
  d Polocyte; Polkörperchen

6500 polonium
  f polonium
  e polonio
  i polonio
  d Polonium

6501 poly-
  Prefix indicating "many".
  f poly-
  e poli-
  i poli-
  d Poly-

6502 polyacrylamide
  f polyacrylamide
  e poliacrilamida
  i poliacrilammide
  d Polyacrylamid

6503 polyacrylate
  f polyacrylate
  e poliacrilato
  i poliacrilato
  d Polyacrylat

6504 polyacrylic acid
  f acide polyacrylique
  e ácido poliacrílico
  i acido poliacrilico
  d Polyacrylsäure

6505 polyalkane
  f polyalcane
  e polialcano
  i polialcane
  d Polyalkan

6506 polyallomer
  f polyallomère
  e polialómero
  i polialomero
  d Polyallomer

6507 polyamide
  f polyamide
  e poliamida
  i poliammide
  d Polyamid

6508 polyamide resins
  f résines de polyamide
  e resinas poliamídicas
  i resine poliammidiche
  d Polyamidharz

6509 polyamine-methylene resin
  f résine de polyamine-méthylène
  e resina de poliamina-metileno

i resina poliammino-metilenica
d Polyaminmethylenharz

6510 polyaminotriazoles
f polyaminotriazoles
e poliaminotriazoles
i poliamminotriazoli
d Polyaminotriazole

6511 polybasic
f polybasique
e polibásico
i polibasico
d polybasisch; mehrbasisch

6512 polybasite
f polybasite
e polibasita
i polibasite
d Polybasit; Eugenglanz

6513 polybutadiene
f polybutadiène
e polibutadieno
i polibutadiene
d Polybutadien

6514 polybutadiene-acrylic acid copolymer
f copolymère de l'acide polybutadiène-acrylique
e copolímero del ácido polibutadien-acrílico
i copolimero dell'acido acrilico-polibutadienico
d Polybutadienacrylsäurecopolymer

6515 polybutylenes; polybutenes; polyisobutylenes
f polybutylènes; polybutènes
e polibutilenos
i polibutileni
d Polybutylene; Polybutene

6516 polychloroprene
f polychloroprène
e policloropreno
i policloroprene
d Polychloropren

6517 polychlorotrifluoroethylene
f polychlorotrifluoroéthylène
e policlorotrifluoroetileno
i policlorotrifluoroetilene
d Polychlortrifluoräthylen

6518 polychondric
f polychondrique
e policóndrico
i policondrico
d polychondrisch

6519 polycondensate
f polycondensat
e policondensado
i policondensato
d Polykondensat

6520 polycondensation
f polycondensation
e policondensación
i policondensazione
d Polykondensation

6521 polycyclic
f polycyclique
e policíclico
i policiclico
d polyzyklisch

6522 polydihydroperfluorobutyl acrylate
f acrylate de polydihydroperfluorobutyle
e acrilato de polidihidroperfluobutilo
i acrilato di polidiidroperfluorobutile
d Polydihydroperfluorbutylacrylat

6523 polydisperse
f polydispersé
e polidisperso
i polidisperso
d polydispers

6524 polyester
f polyester
e poliéster
i poliestere
d Polyester

6525 polyester amide
f polyester-amide
e amida poliéster
i ammide poliesterica
d Polyesteramid

6526 polyester fibre
f fibre de polyester
e fibra poliéster
i fibra poliesterica
d Polyesterfaser

6527 polyester resins
f résines polyesters
e resinas poliésteres
i resine poliesteriche
d Polyesterharze

* **polyester rubber** → **6558**

6528 polyether foams
f mousses de polyéther
e espumas de poliéter

i schiume polieteriche
d Polyätherschäume

6529 polyethylene; polythene
f polyéthylène
e polietileno
i polietilene
d Polyäthylen

6530 polyethylene glycol
f polyéthylèneglycol
e polietilenglicol
i glicolo polietilenico
d Polyäthylenglykol

6531 polyethylene terephthalate
f téréphtalate de polyéthylène
e tereftalato de polietileno
i tereftalato di polietilene
d Polyäthylenterephthalat

6532 polyformaldehydes
f polyformaldéhydes
e poliformaldehídos
i poliformaldeidi
d Polyformaldehyde

6533 polyglycol distearate
f distéarate de polyglycol
e diestearato de poliglicol
i distearato di poliglicolo
d Polyglykoldistearat

6534 polyglycols
f polyglycols
e poliglicoles
i poliglicoli
d Polyglykole

6535 polyhexamethyleneadipamide
f polyhexaméthylène-adipamide
e polihexametilenadipamida
i poliesametileneadipammina
d Polyhexamethylenadipamid

  * **polyisobutylenes** → **6515**

6536 polyisoprene
f polyisoprène
e polisopreno
i polisoprene
d Polyisopren

6537 polymer
f polymère
e polímero
i polimero
d Polymer

6538 polymeric
f polymérique
e polimérico
i polimerico
d polymer

6539 polymeric chromosome
f chromosome polymère
e cromosoma polímero
i cromosoma polimerico
d Polymerchromosom

6540 polymerization
f polymérisation
e polimerización
i polimerizzazione
d Polymerisation

6541 polymerized oils
f huiles polymérisées
e aceites polimerizados
i oli polimerizzati
d polymerisierte Öle

6542 polymer structure
f structure polymère
e estructura de polímeros
i struttura polimera
d Polymerstruktur

6543 polymethacrylates
f polyméthacrylates
e polimetacrilatos
i polimetacrilati
d Polymethacrylate; Polymethacrylsäure-
ester

6544 polymorphism
f polymorphisme
e polimorfismo
i polimorfismo
d Polymorphismus; Vielgestaltigkeit

6545 polymyxins
f polymyxines
e polimixinas
i polimissine
d Polymyxine

6546 polyolefins
f polyoléfines
e poliolefinas
i poliolefine
d Polyolefine

6547 polyoses
f polyoses
e poliosas

i poliosi
d Polyosen

* **polyoxymethylene** → **6107**

6548 polyoxypropylene glycol ethylene oxide
f oxyde de polyoxypropylène et éthylèneglycol
e óxido de etileno de polioxipropilenglicol
i ossido etilenico del glicolpoliossi-
propilenico
d Polyoxypropylenglykoläthylenoxyd

6549 polyoxypropylene glycols
f glycols de polyoxypropylène
e glicoles de polioxipropileno
i glicoli di poliossipropilene
d Polyoxypropylenglykole

6550 polyploidy
f polyploïdie
e poliploidia
i poliploidismo
d Polyploidie

6551 polypropylene
f polypropylène
e polipropileno
i polipropilene
d Polypropylen

6552 polystyrene
f polystyrène
e poliestireno
i polistirene
d Polystyrol

6553 polysulphide rubber
f caoutchouc de polysulfure
e caucho de polisulfuro
i gomma al polisolfuro
d Polysulfidkautschuk

6554 polyterpene resins
f résines de polyterpène
e resinas de politerpeno
i resine politerpeniche
d Polyterpinharze

6555 polytetrafluoroethylene
f polytétrafluoroéthylène
e politetrafluoetileno
i politetrafluoroetilene
d Polytetrafluoräthylen

* **polythene** → **6529**

6556 polyurethane foam
f mousse de polyuréthane

e espuma de poliuretano
i espanso di poliuretano
d Polyurethanschaumstoff

6557 polyurethane resins
f résines de polyuréthane
e resinas de poliuretano
i resine di poliuretano
d Polyurethanharze

6558 polyurethane rubber; polyester rubber
f caoutchouc d'uréthane
e caucho de poliuretano
i gomma al poliuretano
d Polyurethankautschuk

6559 polyvinyl acetal resins
f résines acétalpolyvinyliques
e resinas de acetalpolivinilo
i resine acetalpoliviniliche
d Polyvinylacetalharze

6560 polyvinyl acetals
f acétals polyvinyliques
e acetales de polivinilo
i acetali polivinilici
d Polyvinylacetale

6561 polyvinyl acetate
f acétate de polyvinyle
e acetato de polivinilo
i acetato di polivinile
d Polyvinylacetat

6562 polyvinyl alcohol
f alcool polyvinylique
e alcohol polivinílico
i alcool polivinilico
d Polyvinylalkohol

6563 polyvinyl butyral
f butyral de polyvinyle
e butiral de polivinilo
i butirrale di polivinile
d Polyvinylbutyral

6564 polyvinyl carbazole
f polyvinylcarbazol
e polivinilcarbazol
i carbazolo di polivinile
d Polyvinylcarbazol

6565 polyvinyl chloride
f chlorure de polyvinyle
e cloruro de polivinilo
i cloruro di polivinile
d Polyvinylchlorid

6566 polyvinyl dichloride
  f dichlorure de polyvinyle
  e dicloruro de polivinilo
  i bicloruro di polivinile
  d Polyvinyldichlorid

6567 polyvinyl ethyl ether
  f polyéther éthylvinylique
  e poliviniletiléter
  i poliviniletiletere
  d Polyvinyläthyläther

6568 polyvinyl fluoride
  f fluorure de polyvinyle
  e fluoruro de polivinilo
  i fluoruro di polivinile
  d Polyvinylfluorid

6569 polyvinyl formals
  f formals de polyvinyle
  e polivinilformals
  i formali di polivinile
  d Polyvinylformale

6570 polyvinylidene chloride
  f chlorure de polivinylidène
  e cloruro de polivinilideno
  i cloruro di polivinilidene
  d Polyvinylidenchlorid

6571 polyvinylidene fluoride
  f fluorure de polyvinylidène
  e fluoruro de polivinilideno
  i fluoruro di polivinilidene
  d Polyvinylidenfluorid

6572 polyvinyl isobutyl ether
  f éther polyvinylisobutylique
  e polivinilisobutiléter
  i etere isobutilpolivinilico
  d Polyvinylisobutyläther

6573 polyvinyl methyl ether
  f polyvinylméthyléther
  e polivinilmetiléter
  i etere metilpolivinilico
  d Polyvinylmethyläther

6574 polyvinylpyrrolidone
  f polyvinylpyrrolidone
  e polivinilpirrolidona
  i polivinilpirrolidone
  d Polyvinylpyrrolidon

6575 Pompeian blue
  f bleu pompéin
  e azul pompeyano
  i blu pompeiano
  d Pompejanischblau

6576 Pompey red
  f rouge pompéin
  e rojo pompeyano
  i rosso pompeiano
  d Pompejanischrot

6577 porcelain crucible
  f creuset en porcelaine
  e crisol de porcelana
  i crogiuolo di porcellana
  d Porzellantiegel

6578 pore water
  f eau interstitielle
  e agua de los poros
  i acqua contenuta nei pori
  d Porenwasser

6579 porometer
  f poromètre
  e porímetro
  i porometro
  d Porometer

6580 porosimeter
  f porosimètre
  e porosímetro
  i porosimetro
  d Porosimeter

6581 porosity
  f porosité
  e porosidad
  i porosità
  d Porosität

6582 porous
  f poreux
  e poroso
  i poroso
  d porös

6583 porous pot
  f récipient poreux
  e marmita porosa
  i vaso poroso
  d poröses Gefäß

6584 porphyrins
  f porphyrines
  e porfirinas
  i porfirine
  d Porphyrine

6585 positional alleles
  f allèles de position
  e alelos de posición
  i alleli di posizione
  d Positionsallele

6586 positive electrode
f électrode positive
e electrodo positivo
i elettrodo positivo
d positive Elektrode

6587 positive plate
f plaque positive
e placa positiva
i piastra positiva
d positive Platte

* **possuolana** → **6668**

* **post office red** → **3992**

6588 pot
f creuset
e crisol
i crogiuolo
d Tiegel

6589 potable
f potable
e potable
i potabile
d trinkbar

* **potash** → **6607**

6590 potash bulbs
f appareil à potasse
e recipiente para solución potásica
i apparato per potassa
d Kaliapparat

6591 potassium
f potassium
e potasio
i potassio
d Kalium

6592 potassium abietate
f abiétate de potassium
e abietato potásico
i abietato di potassio
d Kaliumabietat

6593 potassium acetate
f acétate de potassium
e acetato de potasio
i acetato di potassio
d Kaliumacetat

6594 potassium acid saccharate
f saccharate acide de potassium
e sacarato potásico ácido
i saccarato acido di potassio
d Kaliumsäuresaccharat

6595 potassium alum
f alum de potasse
e alumbre potásico
i solfato idrato di alluminio e potassio
d Kaliumalaun

6596 potassium aluminate
f aluminate de potassium
e aluminato potásico
i alluminato di potassio
d Kaliumaluminat

6597 potassium aluminium fluoride
f fluorure de potassium et aluminium
e fluoruro potásico-alumínico
i fluoruro di alluminio e potassio
d Kaliumaluminiumfluorid

6598 potassium arsenate
f arséniate de potassium
e arseniato potásico
i arseniato potassico
d Kaliumarsenat

6599 potassium arsenite
f arsénite de potasse
e arsenito potásico
i arsenito potassico
d Kaliumarsenit

6600 potassium bicarbonate
f bicarbonate de potassium
e bicarbonato de potasio
i bicarbonato di potassio
d Kaliumbicarbonat; doppelkohlensaures
Kalium

6601 potassium bioxalate
f bioxalate de potassium
e bioxalato potásico
i biossalato potassico
d Kaliumbioxalat; Kleesalz

6602 potassium bisulphate
f bisulfate de potassium
e bisulfato potásico
i bisolfato potassico
d Kaliumbisulfat; schwefelsaures Kalium

6603 potassium bisulphite
f bisulfite de potassium
e bisulfito potásico
i bisolfito potassico
d Kaliumbisulfit

6604 potassium bitartrate
f bitartrate de potassium; tartrate acide de
potassium
e bitartrato potásico

i bitartrato potassico
d Kaliumbitartrat; Weinsteinrahm;
   weinsaures Kalium

6605 potassium bromate
f bromate de potassium
e bromuro potásico
i bromato potassico
d Kaliumbromat; bromsaures Kalium

6606 potassium bromide
f bromure de potassium
e bromuro potásico
i bromuro potassico
d Kaliumbromid; Bromkalium

6607 potassium carbonate; potash
f carbonate de potassium; potasse
e carbonato de potasio
i carbonato di potassio; potassa
d Kaliumcarbonat; Pottasche

6608 potassium chlorate
f chlorate de potassium; sel de Berthollet
e clorato potásico
i clorato di potassio
d Kaliumchlorat; chlorsaures Kalium

6609 potassium chloride
f chlorure de potassium
e cloruro de potasio
i cloruro di potassio
d Kaliumchlorid

6610 potassium chlorochromate
f chlorochromate de potassium
e clorocromato potásico
i clorocromato di potassio
d Kaliumchlorchromat

6611 potassium chloroplatinate
f chloroplatinate de potassium
e cloroplatinato potásico
i cloroplatinato di potassio
d Kaliumchlorplatinat

6612 potassium chromate
f chromate neutre de potassium
e cromato potásico
i cromato potassio
d Kaliumchromat

6613 potassium citrate
f citrate de potassium
e citrato potásico
i citrato potassico
d Kaliumcitrat

6614 potassium cyanate
f cyanate de potassium
e cianato potásico
i cianato potassico
d Kaliumcyanat

6615 potassium dichromate
f bichromate de potassium
e dicromato potásico
i bicromato potassico
d Kaliumdichromat

6616 potassium ferricyanide
f ferricyanure de potassium
e ferricianuro potásico
i ferricianuro potassico
d Kaliumferricyanid; Kaliumeisencyanid

6617 potassium ferrocyanide
f ferrocyanure de potassium
e ferrocianuro potásico
i ferrocianuro potassico
d Kaliumferrocyanid; Kaliumeisencyanür

6618 potassium fluoride
f fluorure de potassium
e fluoruro potásico
i fluoruro potassico
d Kaliumfluorid

6619 potassium fluosilicate
f fluosilicate de potassium
e fluosilicato potásico
i fluosilicato potassico
d Kaliumfluosilikat

6620 potassium gluconate
f gluconate de potassium
e gluconato potásico
i gluconato potassico
d Kaliumgluconat

6621 potassium hydroxide
f hydroxyde de potassium
e hidróxido potásico
i idrossido di potassio
d Kaliumhydroxyd

6622 potassium iodate
f iodate de potassium
e yodato potásico
i iodato di potassio
d Kaliumjodat

6623 potassium iodide
f iodure de potassium
e yoduro potásico
i ioduro di potassio
d Kaliumjodid

6624 potassium magnesium sulphate
 f sulfate de potassium et magnésium
 e sulfato potásico-magnésico
 i solfato di magnesio e potassio
 d Kaliummagnesiumsulfat

6625 potassium manganate
 f manganate de potassium
 e manganato potásico
 i manganato di potassio
 d Kaliummanganat

6626 potassium metabisulphite
 f métabisulfite de potassium
 e metabisulfito potásico
 i metabisolfito di potassio
 d Kaliummetabisulfit

6627 potassium metaphosphate
 f métaphosphate de potasse
 e metafosfato potásico
 i metafosfato di potassio
 d Kaliummetaphosphat

6628 potassium molybdate
 f molybdate de potassium
 e molibdato potásico
 i molibdato di potassio
 d Kaliummolybdat

6629 potassium naphthenate
 f naphténate de potassium
 e naftenato potásico
 i naftenato di potassio
 d Kaliumnaphthenat

6630 potassium nitrate; saltpetre; nitre
 f nitrate de potassium; salpêtre; nitre
 e nitrato potásico; nitro; salitre
 i nitrato di potassio; nitro; salnitro
 d Kaliumnitrat; Kalisalpeter; Salpeter

6631 potassium nitrite
 f nitrite de potassium
 e nitrito potásico
 i nitrito di potassio
 d Kaliumnitrit

6632 potassium oleate
 f oléate de potassium
 e oleato potásico
 i oleato di potassio
 d Kaliumoleat

6633 potassium osmate
 f osmiate de potassium
 e osmiato potásico
 i osmiato potassico
 d osmiumsaures Kalium

6634 potassium oxalate
 f oxalate de potassium
 e oxalato potásico
 i ossalato potassico
 d Kaliumoxalat

6635 potassium percarbonate
 f percarbonate de potassium
 e percarbonato potásico
 i percarbonato di potassio
 d Kaliumpercarbonat

6636 potassium perchlorate
 f perchlorate de potassium
 e perclorato potásico
 i perclorato di potassio
 d Kaliumperchlorat

6637 potassium periodate
 f periodate de potassium
 e peryodato potásico
 i periodato di potassio
 d Kaliumperjodat

6638 potassium permanganate
 f permanganate de potassium
 e permanganato potásico
 i permanganato potassico
 d Kaliumpermanganat

6639 potassium peroxide
 f peroxyde de potassium
 e peróxido potásico
 i perossido di potassio
 d Kaliumperoxyd

6640 potassium persulphate
 f persulfate de potassium
 e persulfato potásico
 i persolfato di potassio
 d Kaliumpersulfat

* **potassium phosphate, dibasic ~**
 → **2500**

* **potassium phosphate, monobasic ~**
 → **5568**

* **potassium phosphate, tribasic ~**
 → **8396**

6641 potassium polysulphide
 f polysulfure de potassium
 e polisulfuro potásico
 i polisolfuro di potassio
 d Kaliumpolysulfid

6642 potassium pyrophosphate
 f pyrophosphate de potasse

e pirofosfato potásico
i pirofosfato di potassio
d Kaliumpyrophosphat

6643 potassium quadroxalate; salts of Sorrel
f quadroxalate de potassium; sels d'oseille
e quadroxalato potásico; sal de acederas
i quadrossalato di potassio; sale di acetosella
d Kaliumquadroxalat; Sorrelsalze

6644 potassium selenate
f séléniate de potassium
e seleniato potásico
i seleniato di potassio
d Kaliumselenat

6645 potassium silicate; water glass
f silicate de potassium; verre soluble
e silicato potásico; vidrio soluble
i silicato di potassio
d Kaliumsilikat; Wasserglas

6646 potassium sodium tartrate
f tartrate de potassium et sodium
e tartrato potásico-sódico
i tartrato di sodio e potassio
d Kaliumnatriumtartrat; Natronweinstein

6647 potassium stannate
f stannate de potassium
e estannato potásico
i stannato potassico
d Kaliumstannat

6648 potassium stearate
f stéarate de potassium
e estearato potásico
i stearato di potassio
d Kaliumstearat

6649 potassium sulphate
f sulfate de potassium
e sulfato potásico
i solfato di potassio
d Kaliumsulfat

6650 potassium sulphide
f sulfure de potassium
e sulfuro potásico
i solfuro di potassio
d Kaliumsulfid

6651 potassium sulphite
f sulfite de potassium
e sulfito potásico
i solfito di potassio
d Kaliumsulfit

6652 potassium sulphocarbonate
f sulfocarbonate de potassium
e sulfocarbonato potásico
i solfocarbonato di potassio
d Kaliumsulfocarbonat

6653 potassium tartrate
f tartrate de potassium
e tartrato potásico
i tartrato di potassio
d Kaliumtartrat

6654 potassium titanate
f titanate de potassium
e titanato potásico
i titanato di potassio
d Kaliumtitanat

6655 potassium tripolyphosphate
f tripolyphosphate de potassium
e tripolifosfato potásico
i tripolifosfato potassico
d Kaliumtripolyphosphat

6656 potassium undecylenate
f undécylénate de potassium
e undecilenato potásico
i undecilenato di potassio
d Kaliumdecylenat

6657 potato spirit
f eau-de-vie de pommes de terre
c aguardiente de patatas
i acquavite di patate
d Kartoffelspiritus

6658 potcher *(papermaking)*
f cuvier mélangeur
e mezcladora
i olandese imbiancatrice
d Mischer

6659 potential difference
f différence de potentiel
e diferencia de potencial
i differenza di potenziale
d Potentialdifferenz; Potentialunterschied

6660 potential gradient
f gradient de potentiel
e gradiente potencial
i gradiente potenziale
d Potentialgradient

6661 potentiometer
f potentiomètre
e potenciómetro
i potenziometro
d Potentiometer

6662 potion
f potion
e poción
i pozione
d Arzneitrank

6663 pot still
f alambic
e alambique caldeado directamente por la llama
i distillatrice a crogiolo
d Destillationsapparat

6664 powder adhesive
f colle en poudre
e adhesivo in polvo
i adesivo in polvere
d Klebepulver

6665 powder base
f excipient de poudres
e excipiente por polvos
i eccipiente per polvere
d Pudergrundlage

6666 powder metallurgy
f métallurgie des poudres
e pulvimetalurgia; pulvimetalogía
i metallurgia delle polveri
d Sintermetallurgie

6667 powdery
f pulvérulent
e pulverulento
i polverulento; friabile
d pulvrig

6668 pozzolana; possuolana; puzzolana
f pozzolane
e pozolana
i pozzolana
d Pozzolanerde

6669 pozzolana cement
f ciment de pozzolane
e cemento de pozolana
i cemento alla pozzolana
d Pozzolanzement

6670 pramoxine hydrochloride
f chlorhydrate de pramoxine
e hidrocloruro de pramoxina
i cloridrato di pramossina
d Pramoxinhydrochlorid

6671 praseodymium
f praséodyme
e praseodimio

i praseodimio
d Praseodymium

6672 precipitable
f précipitable
e precipitable
i precipitabile
d ausfällbar

6673 precipitate
f précipité
e precipitado
i precipitato
d Niederschlag

6674 precipitate v
f précipiter
e precipitar
i precipitare
d niederschlagen

6675 precipitated driers
f siccatifs précipités
e desecantes precipitados
i essiccatori precipitati
d gefällte Trockenmittel

6676 precipitation
f précipitation
e precipitación
i precipitazione
d Niederschlag

6677 precision balance
f balance de précision
e balanza de precisión
i bilancia di precisione
d Präzisionswaage

6678 precision instrument
f instrument de précision
e instrumento de precisión
i strumento di precisione
d Präzisionsinstrument

6679 precooling
f prérefroidissement
e preenfriado; prerrefrigerado
i prerefrigerazione
d Vorkühlung

6680 prednisolone
f prednisolone
e prednisolona
i prednisolone
d Prednisolon

6681 prednisone
f prednisone

e prednisona
i prednisone
d Prednison

6682 predrier
f préséchoir
e presecador; presecadora
i preessicatore
d Vortrockner

6683 preformed precipitate
f précipité préformé
e precipitado preformado
i precipitato preformato
d vorgefällter Niederschlag

6684 pregnanediol
f prégnandiol
e pregnanodiol
i pregnanodiolo
d Pregnandiol

6685 pregnenolone
f prégnénolone
e pregnenolona
i pregnenolone
d Pregnenolon

6686 preheater
f préchauffeur
e precalentador
i preriscaldatore
d Vorwärmer

6687 preheating
f préchauffage
e precalentamiento
i preriscaldamento
d Vorwärmen

6688 prehnitene; prenitol
f préhnitène
e prehniteno; preniteno
i prenitene
d Prehnitol

6689 preimmunization
f préimmunisation
e preinmunización
i preimmunizzazione
d frühzeitige Immunisierung

6690 premixing
f prémélange
e mezcla preliminar
i premiscelazione
d Vormischung

* **prenitol → 6688**

6691 preprophage
f préprophage
e preprofago
i preprofago
d Präprophage

6692 preprophase inhibitor
f inhibiteur de la préprophase
e inhibidor de la preprofase
i inibitore della preprofase
d Präprophase-Inhibitor

6693 preservative
f préservateur
e preservativo
i preservativo
d Vorbeugungsmittel

6694 pressing
f moulage à la presse
e moldeo a la prensa; prensado
i stampaggio
d Pressen

6695 pressure cooker
f marmite à pression
e autoclave
i autoclave
d Druckpfanne

6696 pressure reducing valve
f détendeur
e válvula reductora de presión
i valvola regolatrice di pressione
d Druckreduzierventil

6697 pretreatment
f traitement préalable
e tratamiento preliminar
i trattamento preliminare
d Vorbehandlung

6698 priceite
f pricéite
e priceita
i priceite
d Priceit

* **primaclone → 6706**

6699 primaquine phosphate
f phosphate de primaquine
e fosfato de primaquina
i fosfato di primachina
d Primachinphosphat

6700 primary
f produit de base
e producto primario

i prodotto primario
d Primärprodukt

6701 primary amide
f amide primaire
e amida primaria
i ammide primaria
d primäres Amid

6702 primary cell
f pile; élément primaire
e pila primaria
i pila primaria
d Primärelement; Batterie

6703 primary current ratio
f rapport de courant primaire
e relación de corriente primaria
i rapporto di corrente primaria
d Primärstromverhältnis

6704 prime *v (pumps)*
f amorcer
e cebar
i innescare
d auffüllen

6705 primer
f amorce
e cápsula fulminante; detonador
i innesco
d Zünder

6706 primidone; primaclone
f primidone
e primidona
i primidone
d Primidon

6707 priming paint
f peinture d'impression
e pintura de imprimación
i vernice di fondo
d Vorstreichfarbe

6708 primrose chromes
f chromes de primevère
e cromos primavera
i cromi di primula
d blaßgelbe Pigmente

6709 primuline
f primuline
e primulina
i primulina
d Primulin

6710 primuline red
f rouge de primuline

e rojo de primulina
i rosso di primulina
d Primulinrot

6711 prisometer
f prisomètre
e prisómetro
i prisometro
d Prisometer

6712 probarbital sodium
f probarbital sodique
e probarbital sódico
i probarbital sodico
d Probarbitalnatrium

6713 probe
f sonde
e sonda
i sonda
d Sonde

6714 procainamide hydrochloride
f chlorhydrate de procaïnamide
e hidrocloruro de procainamida
i cloridrato di procainammide
d Procainamidhydrochlorid

**6715 procaine hydrochloride; ethocaine
hydrochloride**
f chlorhydrate de procaïne
e hidrocloruro de procaína
i cloridrato di procaina
d Procainhydrochlorid

6716 process
f procédé; traitement
e proceso; método; procedimiento
i processo; procedimento
d Herstellungsverfahren

6717 process *v*
f traiter; usiner
e procesar
i trattare; lavorare
d behandeln; verarbeiten

6718 processability
f aptitude à subir un processus
e elaborabilidad
i lavorabilità
d Verarbeitbarkeit

6719 processing
f usinage
e trabajo
i procedimento
d Bearbeitung

6720 process steam
  f vapeur utilisée pour des usages
    industriels
  e vapor para uso industrial
  i vapore per uso industriale
  d Betriebsdampf

6721 process water
  f eau de processus
  e agua de elaboración; aguas residuales
  i acqua di lavorazione
  d Betriebswasser

6722 prochlorperazine dimaleate
  f dimaléate de prochlorpérazine
  e dimaleato de proclorperacina
  i dimaleato di proclorperazina
  d Prochlorperazindimaleat

6723 prochromosome
  f prochromosome
  e procromosoma
  i procromosoma
  d Prochromosom

6724 producer gas
  f gaz de gasogène; gaz pauvre
  e gas pobre; gas de gasógeno
  i gas di gasogeno
  d Generatorgas

6725 product
  f produit
  e producto
  i prodotto
  d Erzeugnis; Produkt

6726 product of combustion
  f produit de combustion
  e producto de combustión
  i prodotto della combustione
  d Verbrennungsprodukt

6727 profibrinolysin
  f plasminogène
  e profibrinolisina
  i profibrinolisina
  d Plasminogen

6728 proflavine
  f proflavine
  e proflavina
  i proflavina
  d Proflavin

6729 progesterone
  f progestérone
  e progesterona

  i progesterone
  d Progesteron

6730 prolactin
  f prolactine
  e prolactina
  i prolactina
  d Prolaktin

6731 prolamin
  f prolamine
  e prolamina
  i prolammina
  d Prolamin

6732 proline
  f proline
  e prolina
  i prolina
  d Prolin

6733 promazine hydrochloride
  f chlorhydrate de promazine
  e hidrocloruro de promacina
  i cloridrato di promazina
  d Promazinhydrochlorid

6734 prometaphase
  f prométaphase
  e prometafase
  i prometafase
  d Prometaphase

6735 promethazine hydrochloride
  f chlorhydrate de prométhazine
  e hidrocloruro de prometacina
  i cloridrato di prometazina
  d Promethazinhydrochlorid

6736 promethium
  f prométhéum
  e promecio; prometio
  i prometio
  d Promethium

6737 promoter
  f activeur
  e activador
  i attivatore
  d Aktivierungsmittel; Beschleuniger;
    Promotor

6738 pronucleolus
  f pronucléole
  e pronucléolo
  i pronucleolo
  d Pronukleolus

6739 pronucleus
f pronucléus
e pronúcleo
i pronucleo
d Pronukleus

* **propadiene** → 351

6740 propane
f propane
e propano
i propano
d Propan

6741 propanol
f propanol
e propanol
i propanolo
d Propanol

6742 propantheline bromide
f bromure de propanthéline
e bromuro de propantelina
i bromuro di propantelina
d Propanthelinbromid

6743 propargyl alcohol
f alcool propargylique
e alcohol propargílico
i alcool propargilico
d Propargylalkohol

* **propenal** → 143

6744 prophage
f prophage
e profago
i profago
d Prophage

6745 prophase
f prophase
e profase
i profase
d Prophase

6746 prophylaxis
f prophylaxie
e profilaxis
i profilassi
d Prophylaxe

6747 propiolactone
f propiolactone
e propiolactona
i propiolattone
d Propiolacton

6748 propionaldehyde
f propionaldéhyde
e propionaldehído
i aldeide propionica
d Propionaldehyd

6749 propionic acid
f acide propionique
e ácido propiónico
i acido propionico
d Propionsäure

6750 propionic anhydride
f anhydride propionique
e anhídrido propiónico
i anidride propionica
d Propionsäureanhydrid

6751 propiophenone
f propiophénone
e propiofenona
i propiofenone
d Propiophenon

6752 proportion pump
f pompe de dosage
e bomba de dosificación
i pompa a erogazione proporzionale
d Dosierpumpe

6753 propyl acetate
f acétate de propyle
e acetato de propilo
i acetato propilico
d Propylacetat

* **propylacetone** → 5340

6754 propyl alcohol
f alcool propylique
e alcohol propílico
i alcool propilico
d Propylalkohol

6755 propylamine
f propylamine
e propilamina
i propilammina
d Propylamin

6756 propyl butyrate
f butyrate de propyle
e butirato de propilo
i butirrato di propile
d Propylbutyrat

6757 propyl chlorosulphonate
f chlorosulfonate de propyle
e clorosulfonato de propilo

i clorosolfonato di propile
d Propylchlorsulfonat

6758 propylene
  f propylène
  e propileno
  i propilene
  d Propylen

6759 propylene carbonate
  f carbonate de propylène
  e carbonato de propileno
  i carbonato di propilene
  d Propylencarbonat

6760 propylene chlorohydrin
  f chlorhydrine propylénique
  e clorhidrina propilénica
  i cloroidrina di propilene
  d Propylenchlorhydrin

6761 propylenediamine
  f propylènediamine
  e propilendiamina
  i propilenediammina
  d Propylendiamin

6762 propylene dichloride
  f bichlorure de propylène
  e dicloruro de propileno
  i bicloruro di propilene
  d Propylendichlorid

6763 propylene glycol
  f propylèneglycol
  e propilenglicol
  i glicolo di propilene
  d Propylenglykol

6764 propylene glycol monoricinoleate
  f monoricinoléate de propylèneglycol
  e monorricinoleato de propilenglicol
  i monoricinoleato di propileneglicolo
  d Propylenglykolmonoricinoleat

6765 propylene glycol phenyl ether
  f éther phénylique de propylèneglycol
  e éter fenílico del propilenglicol
  i etere glicolfenilpropilenico
  d Propylenglykolphenyläther

6766 propylene oxide
  f oxyde de propylène
  e óxido de propileno
  i ossido di propilene
  d Propylenoxyd

6767 propylhexedrine
  f propylhexédrine

e propilhexedrina
i propilessedrina
d Propylhexedrin

6768 propyl hydroxybenzoate; propylparaben
  f hydroxybenzoate de propyle
  e hidroxibenzoato de propilo
  i idrossibenzoato di propile
  d Propylhydroxybenzoat

6769 propyl mercaptan
  f propylmercaptan
  e propilmercaptano
  i propilmercaptano
  d Propylmercaptan

* **propylparaben** → **6768**

6770 propyl propionate
  f propionate de propyle
  e propionato de propilo
  i propionato di propile
  d Propylpropionat

6771 propylthiouracil
  f propylthiouracil
  e propiltiouracilo
  i propiltiouracile
  d Propylthiouracil

6772 protactinium
  f protactinium
  e protactinio
  i protattinio
  d Protaktinium

6773 protease
  f protéase
  e proteasa
  i proteasi
  d Protease

6774 proteins
  f protéines
  e proteínas
  i proteine
  d Proteine

6775 proteolytic
  f protéolytique
  e proteolítico
  i proteolitico
  d eiweißspaltend

6776 protoblast
  f protoblaste
  e protoblasto
  i protoblasto
  d membranöse Zelle

6777 protoplasm
f protoplasme
e protoplasma
i protoplasma
d Protoplasma

6778 protoplasmatic
f protoplasmique
e protoplásmico
i protoplasmico
d protoplasmatisch

6779 prototrophic
f prototrophique
e protótrofo
i prototrofico
d prototroph

6780 protoveratrine
f protovératrine
e protoveratrina
i protoveratrina
d Protoveratrin

6781 proustite
f proustite
e proustita
i proustite
d Proustit

6782 Prussian blue; Parisian blue
f bleu de Prusse; bleu de Paris
e azul de Prusia; azul de París
i blu di Prussia
d Preußischblau

6783 Prussian brown
f brun de Prusse
e marrón de Prusia
i marrone di Prussia
d Preußischbraun

6784 pseudoacid
f pseudoacide
e pseudoácido
i pseudoacido
d Pseudosäure

6785 pseudobase
f pseudobase
e pseudobase
i pseudobase
d Pseudobase

6786 pseudocumene
f pseudocumène
e pseudocumeno
i pseudocumene
d Pseudocumol

6787 pseudomorph
f pseudomorphe
e pseudomorfo
i pseudomorfo
d Afterkristall

6788 pseudomorphic
f pseudomorphe
e pseudomorfo
i pseudomorfo
d pseudomorph

6789 pseudoracemic
f pseudoracémique
e pseudoracémico
i pseudoracemico
d pseudoracemisch

6790 psilomelane
f psilomélane
e psilomelano
i psilomelano
d Psilomelan

6791 psychochemistry
f psychochimie
e sicoquímica
i psicochimica
d Psychochemie

6792 psychrometer
f psychromètre
e sicrómetro
i psicrometro
d Feuchtigkeitsmesser; Psychrometer

6793 ptomaines
f ptomaïnes
e ptomaínas
i ptomaine
d Ptomaine

6794 pug mill
f malaxeur; pétrin
e molino de amasar
i impastatrice
d Tonmühle

6795 pulverized chalk
f craie pulvérisée
e creta en polvo
i creta in polvere
d feingemahlene Kreide

6796 pulverizing
f pulvérisation
e pulverización
i pulverizzazione
d Pulverisieren

6797 pumice
  f pierre ponce; pumite
  e piedra pómez
  i pomice
  d Bims; Bimstein

6798 pump mixer settler
  f mélangeur-clarificateur à pompes
  e mezclador-asentador de bombas
  i mescolatore-chiarificatore a pompe
  d Misch- und Klärgerät mit Umlaufpumpen

6799 punty
  f pontil; pontis
  e puntero; puntel; pontil
  i puntello
  d Hefteisen

6800 pure substance
  f substance pure
  e substancia pura
  i sostanza pura
  d Reinstoff

6801 purine
  f purine
  e purina
  i purina
  d Purin

6802 puromycin
  f puromycine
  e puromicina
  i puromicina
  d Puromycin

6803 purpurin
  f purpurine
  e purpurina
  i purpurina
  d Purpurin

6804 putrefaction
  f putréfaction
  e putrefacción
  i putrefazione
  d Fäulnis

6805 putrefy *v*
  f se putréfier
  e putrificar
  i putrificarsi
  d verfaulen; vermodern

6806 putrescence
  f putrescence
  e putrescencia
  i putrescenza
  d Putreszenz

6807 putrid
  f putride
  e pútrido; podrido
  i putrido
  d verfault

6808 putty
  f mastic
  e masilla; mastique
  i mastice
  d Kitt

* puzzolana → 6668

6809 pycnometer
  f picnomètre; flacon à densité
  e picnómetro
  i picnometro
  d Pyknometer

6810 pycnosis
  f pycnose
  e picnosis
  i picnosi
  d Pyknose

6811 pycnotic
  f pycnotique
  e picnótico
  i picnotico
  d pyknotisch

6812 pyrargyrite
  f pyrargyrite; argent rouge antimonial
  e pirargirita
  i pirargirite
  d Pyrargyrit

6813 pyrathiazine hydrochloride
  f chlorhydrate de pyrathiazine
  e hidrocloruro de piratiacina
  i cloridrato di piratiazina
  d Pyrathiazinhydrochlorid

6814 pyrazinamide
  f pyrazinamide
  e piracinamida
  i pirazinammide
  d Pyrazinamid

6815 pyrazolone
  f pyrazolone
  e pirazolona
  i pirazolone
  d Pyrazolon

6816 pyrethrin
  f pyréthrine
  e piretrina

i piretrina
d Pyrethrin

6817 pyridine
f pyridine
e piridina
i piridina
d Pyridin

6818 pyridinium bromide perbromide
f perbromure de bromure de pyridinium
e perbromuro bromuro de piridinio
i perbromuro di bromuro di piridinio
d Pyridiniumbromidperbromid

6819 pyridostigmine bromide
f bromure de pyridostigmine
e bromuro de piridostigmina
i bromuro di piridostigmina
d Pyridostigminbromid

6820 pyridoxine; vitamin B₆
f pyridoxine
e piridoxina
i piridossina
d Pyridoxin

6821 pyrilamine maleate
f maléate de pyrilamine
e maleato de pirilamina
i maleato di pirilammina
d Pyrilaminmaleat

6822 pyrimethamine
f pyriméthamine
e pirimetamina
i pirimetammina
d Pyrimethamin

6823 pyrimidines
f pyrimidines
e pirimidinas
i pirimidine
d Pyrimidine

6824 pyrite
f pyrite
e pirita
i pirite
d Pyrit

6825 pyrites burner
f four à griller les pyrites
e quemador de piritas; horno de piritas
i forno per piriti
d Pyritbrenner

6826 pyro-
Prefix denoting a substance formed or

obtained by heat.
f pyro-
e piro-
i piro-
d Pyro-

6827 pyrobutamine phosphate
f phosphate de pyrobutamine
e fosfato de pirobutamina
i fosfato di pirobutammina
d Pyrobutaminphosphat

6828 pyrocatechol
f pyrocatéchine
e pirocatecol
i pirocatecolo
d Pyrokatechin

6829 pyrochemistry
f pyrochimie
e piroquímica
i pirochimica
d Pyrochemie

6830 pyrogallic acid
f acide pyrogallique
e ácido pirogálico
i acido pirogallico
d Pyrogallussäure

6831 pyroligneous acid
f acide pyroligneux
e ácido piroleñoso
i acido piroligneo
d Holzessig

6832 pyrolusite
f pyrolusite
e pirolusita
i pirolusite
d Pyrolusit

6833 pyrolysis
f pyrolyse
e pirólisis
i pirolisi
d Pyrolyse

6834 pyromellitic acid
f acide pyromellitique
e ácido piromelítico
i acido piromellitico
d Pyromellithsäure

6835 pyrometer
f pyromètre
e pirómetro
i pirometro
d Pyrometer

6836 pyrones
 f pyrones
 e pironas
 i pironi
 d Pyrone

6837 pyrophoric
 f pyrophorique
 e pirofórico
 i piroforico
 d pyrophor; luftentzündlich

6838 pyrophosphoric acid
 f acide pyrophosphorique
 e ácido pirofosfórico
 i acido pirofosforico
 d Pyrophosphorsäure

6839 pyrophyllite
 f pyrophyllite
 e pirofilita
 i pirofillite
 d Pyrophyllit

6840 pyrotartaric acid; methylsuccinic acid
 f acide pyrotartarique
 e ácido pirotartárico
 i acido pirotartarico
 d Pyroweinsteinsäure

6841 pyroxilin
 f pyroxyline; pyroxyle
 e piroxilina
 i pirossilina
 d Pyroxylin

6842 pyrrhotite
 f pyrrhotine
 e pirrotita
 i pirrotite
 d Pyrrothin; Magnetkies

6843 pyrrole
 f pyrrole
 e pirrol
 i pirrolo
 d Pyrrol

6844 pyrrolidone
 f pyrrolidone
 e pirrolidona
 i pirrolidone
 d Pyrrolidon

6845 pyruvic aldehyde
 f aldéhyde pyruvique
 e aldehído pirúvico
 i aldeide piruvica
 d Methylglyoxal

6846 pyrvinium chloride
 f chlorure de pyrvinium
 e cloruro de pirvinio
 i cloruro di pirvinio
 d Pyrviniumchlorid

Q

* **quadrivalent** → 8175

6847 qualitative analysis
  f analyse qualitative
  e análisis cualitativo
  i analisi qualitativa
  d qualitative Analyse

6848 quantitative analysis
  f analyse quantitative
  e análisis cuantitativo
  i analisi quantitativa
  d quantitative Analyse

6849 quantum
  f quantum
  e cuanto
  i quanto
  d Quant

6850 quarter v
  f diviser en quatre parties
  e cuartear
  i dividere in quattro parti
  d vierteln

6851 quartz
  f quartz; cristal de roche
  e cuarzo
  i quarzo
  d Quartz; Bergkristall

6852 quassia
  f quassie; bois de quassia
  e cuasia
  i quassia
  d Bitterholz; Quassia

6853 quaternary ammonium compounds
  f composés d'ammonium quaternaires
  e compuestos amónicos cuaternarios
  i composti di ammonio quaternario
  d quartäre Ammoniumverbindungen

* **quebrachine** → 8902

6854 quebrachitol
  f québrachitol
  e quebrachitol
  i quebracitolo
  d Quebrachitol

6855 quebracho
  f québracho
  e quebracho

  i quebracho
  d Quebracho

6856 quench v
  f éteindre
  e extinguir
  i spegnere
  d löschen

6857 quenching
  f trempe; extinction
  e enfriamiento rápido
  i raffreddamento rapido
  d Abschrecken; Löschung

6858 quenching gas
  f gaz de coupage
  e gas de extinción
  i gas d'estinzione
  d Löschgas

6859 quercetin; tetrahydroxyflavonol
  f quercétine
  e quercetina
  i quercetina
  d Quercetin

6860 quercitol
  f quercitol; quercite
  e quercitol
  i quercitolo
  d Quercit

6861 quercitron
  f quercitron
  e quercitrón
  i quercitrone
  d Quercitron

6862 quicking
  f placage au mercure
  e plateado preliminar con mercurio
  i galvanizzazione preliminare con mercurio
  d Verquickung

* **quicklime** → 1369

* **quicksilver** → 5248

6863 quinaldine
  f quinaldine
  e quinaldina
  i chinaldina
  d Chinaldin

6864 quinhydrone
  f quinhydrone
  e quinhidrona

i chinidrone
d Chinhydron

6865 quinhydrone electrode
f électrode à la quinhydrone
e electrodo de quinhidrona
i elettrodo al chinidrone
d Chinhydronelektrode

6866 quinhydrone half-cell
f demi-cellule à la quinhydrone
e semicelda de quinhidrona
i semicella di chinidrone
d Chinhydronhalbzelle

6867 quinic acid
f acide quinique
e ácido quínico
i acido chinico
d Chinasäure

6868 quinidine
f quinidine
e quinidina
i chinidina
d Chinidin; Conchinin

6869 quinidine sulphate
f sulfate de quinidine
e sulfato de quinidina
i solfato di chinidina
d Chinidinsulfat

6870 quinine
f quinine
e quinina
i chinino
d Chinin

6871 quinine carbacrylic resin
f résine carbacrylique de quinine
e resina quininacarbacrílica
i resina chininocarbacrilica
d Chinincarbacrylsäureharz

6872 quinine ethylcarbonate
f éthylcarbonate de quinine
e etilcarbonato de quinina
i etilcarbonato di chinino
d Chininäthylcarbonat

6873 quinine hydrochloride
f chlorhydrate de quinine
e hidrocloruro de quinina
i idrocloruro di chinino
d Chininhydrochlorid

6874 quinine sulphate
f sulfate de quinine

c sulfato de quinina
i solfato di chinino
d Chininsulfat

6875 quinoline; leucoline
f quinoléine
e quinolina; leucolina
i chinolina
d Chinolin

6876 quinoline yellow
f jaune de quinoléine
e amarillo de quinolina
i giallo di chinolina
d Chinolingelb

6877 quinone
f quinone
e quinona
i chinone
d Chinon

6878 quinoxalines
f quinoxalines
e quinoxalinas
i chinossaline
d Chinoxaline

* **quinquivalent** → 6181

6879 quintessence
f quintessence
e quintaesencia
i quintessenza
d Quintessenz

R

* racemation → 6882

6880 racemic acid
f acide racémique
e ácido racémico
i acido racemico
d Racemsäure

6881 racemic compounds
f composés racémiques
e compuestos racémicos
i composti racemici
d Racemverbindungen

6882 racemization; racemation
f racémisation
e racemización
i racemizzazione
d Racemisierung; optische Inaktivierung

6883 racephedrine hydrochloride
f chlorhydrate de racéphédrine
e hidrocloruro de racefedrina
i cloridrato di racefedrina
d Racephedrinhydrochlorid

6884 rack v
f soutirer
e transvasar
i riempire
d abfüllen

6885 racking cock
f canule
e espitón de trasvase
i spinone dell'empifusti
d Abfüllhahn

6886 racking hose
f tuyau de soutirage
e tubo para transvasar
i tubo per travaso
d Abfüllschlauch

6887 racking room
f local de soutirage
e sala de llenado; sala de trasvase
i locale d'infustamento; locale
  d'imbottigliamento
d Abfüllraum

6888 racking square
f cuve de soutirage
e cuba de envasado
i tino per infustamento
d Abfüllbütte

6889 radiation
f rayonnement
e radiación
i radiazione
d Strahlung

6890 radiation catalysis
f catalyse par rayonnement
e catálisis por radiación
i catalisi a radiazioni
d Strahlungskatalyse

6891 radiation chemistry
f chimie du rayonnement
e química de radiación
i chimica di radiazione
d Strahlungschemie

6892 radiation decomposition
f décomposition par rayonnement
e decomposición por radiación
i decomposizione per radiazione
d Radiolyse

6893 radiation loss
f perte par rayonnement
e pérdida por radiación
i perdità per radiazione
d Strahlungsverlust

6894 radiation polymerization
f polymérisation par rayonnement
e polimerización por radiación
i polimerizzazione per radiazione
d Strahlungspolymerisation

6895 radiator
f radiateur
e radiador
i radiatore
d Heizkörper

6896 radical
f radical
e radical
i radicale
d Radikal

6897 radio-
Prefix denoting radioactivity.
f radio-
e radio-
i radio-
d Radio-

6898 radioactinium
f radioactinium
e radiactinio

i radioattinio
d Radioaktinium

6899 radioactive
f radioactif
e radioactivo
i radioattivo
d radioaktiv

6900 radioactive paint
f peinture radioactive
e pintura radioactiva
i vernice radioattiva
d radioaktive Farbe

6901 radioactivity
f radioactivité
e radioactividad
i radioattività
d Radioaktivität

6902 radiochemical purity
f pureté radiochimique
e pureza radioquímica
i purità radiochimica
d radiochemische Reinheit

6903 radiochemistry
f radiochimie
e radioquímica
i radiochimica
d Radiochemie

6904 radioelement
f radioélément
e radioelemento
i elemento radioattivo
d Radioelement

6905 radiofrequency welding; radio welding
f soudage à haute fréquence
e soldadura por calor de corrientes de hiperfrecuencia
i saldatura a radiofrequenza
d Hochfrequenzschweißen

6906 radiolysis of solvents
f décomposition radiolytique des solvants
e descomposición radiolítica de los solventes
i decomposizione radiolitica dei solventi
d Radiolyse-Zerlegung von Lösungsmitteln

6907 radiometric analysis
f analyse radiométrique
e análisis radiométrico
i analisi radiometrica
d radiometrische Analyse

6908 radiopaque
f opaque aux radiations
e radioopaco
i opaco alla radiazione
d strahlenundurchlässig

6909 radiothorium
f radiothorium
e radiotorio
i radiotorio
d Radiothorium

* **radio welding** → 6905

6910 radium
f radium
e radio
i radio
d Radium

6911 radium bromide
f bromure de radium
e bromuro de radio
i bromuro di radio
d Radiumbromid

6912 radium carbonate
f carbonate de radium
e carbonato de radio
i carbonato di radio
d Radiumcarbonat

6913 radium chloride
f chlorure de radium
e cloruro de radio
i cloruro di radio
d Radiumchlorid

* **radium emanation** → 6915

6914 radium sulphate
f sulfate de radium
e sulfato de radio
i solfato di radio
d Radiumsulfat

6915 radon; emanon; radium emanation
f radon; émanation du radium
e radón
i radon
d Radon; Radiumemanation

6916 rafaelite
f rafaélite
e rafaelita
i rafaelite
d Rafaelit

6917 raffinate
f raffinat
e residuos de refinado
i raffinato
d Raffinat

6918 raffinate layer
f couche de raffinant
e capa de refinado
i strato di raffinato
d Raffinatschicht

6919 raffinose
f raffinose
e rafinosa
i raffinosio
d Raffinose

6920 rag pulp
f pulpe de chiffons
e pasta de trapos
i polpa di stracci
d Lumpenbrei

6921 raise *v* steam
f produire de la vapeur
e producir vapor
i produrre vapore
d Dampf erzeugen

6922 raising
f récupération de l'huile
e recuperación de aceite de aguas de la gamuza
i estrazione d'olio
d Heben in der Kleienbeize

6923 ramie
f ramie
e ramio
i ramie
d Ramiefaser

6924 rancid
f rance
e ráncido
i rancido
d ranzig

6925 rancidity
f rancidité
e rancidez
i rancidità
d Ranzigkeit

6926 random
f par hazard
e fortuito; casual; al azar

i a casaccio
d zufällig

6927 randomization
f randomisation
e randomización
i randomizzazione
d Randomisation

6928 Raoult's law
f loi de Raoult
e ley de Raoult
i legge di Raoult
d Raoultsches Gesetz

6929 rapid hardening Portland cement
f ciment Portland prompt
e cemento Portland de endurecimiento rápido
i cemento Portland a presa rapida
d schnellhärtender Portlandzement

6930 rare earth elements
f éléments des terres rares
e elementos de tierras raras
i elementi di terre rare
d seltene Erdelemente

6931 rarefaction
f raréfaction
e rarefacción; enrarecimiento
i rarefazione
d Verdünnung

6932 rarefaction of air
f raréfaction de l'air
e enrarecimiento del aire
i rarefazione dell'aria
d Luftverdünnung

6933 rarefy *v*
f raréfier
e enrarecer; rarificar
i rarefare
d verdünnen

6934 Raschig process
f procédé Raschig
e proceso Raschig
i processo Raschig
d Raschig-Phenol-Verfahren

6935 Raschig rings
f anneaux Raschig
e anillos Raschig
i anelli Raschig
d Raschig-Ringe

6936 rate *(storage battery)*
f régime
e régimen
i corrente di regime
d Nennentladestrom

6937 rated power
f puissance nominale
e potencia de régimen; potencia normal
i potenza nominale
d Nennleistung

6938 rating *(storage battery)*
f capacité nominale
e capacidad nominal
i capacità nominale
d Nennkapazität; Nennleistung

6939 rauwolfia
f rauwolfia
e rauwolfia
i rauwolfia
d Rauwolfia

6940 raw linseed oil
f huile de lin brute
e aceite de linaza crudo
i olio di lino grezzo
d rohes Leinöl

6941 raw material
f matière primaire
e materia prima
i materia prima
d Rohstoff

6942 raw oil
f huile brute
e aceite crudo
i olio di lino grezzo
d Rohöl

6943 raw sienna
f terre de Sienne brute
e pigmento de Siena
i terra di Siena grezza
d rohe Sienna

6944 Raymond mill
f broyeur Raymond
e molino de Raymond
i frantoio Raymond
d Pendelmühle; Raymond-Mühle

6945 rayon
f rayonne
e rayón
i rayon
d Kunstseide; Reyon

6946 reaction
f réaction
e reacción
i reazione
d Reaktion

6947 reaction time
f temps de réaction
e tiempo de reacción
i tempo di reazione
d Reaktionszeit

6948 reaction velocity
f vitesse de réaction
e velocidad de reacción
i velocità di reazione
d Reaktionsgeschwindigkeit

6949 reactivation
f réactivation
e reactivación
i riattivazione; rigenerazione
d Reaktivierung

6950 reactive
f réactif
e reactivo
i reattivo; reagente
d reagierend

6951 reactivity
f réactivité
e reactividad
i reattività
d Reaktivität

6952 reactor
f réacteur
e reactor
i reattore
d Reaktor

6953 reagent
f réactif
e reactivo
i reagente
d Reagenz

6954 reagent bottle
f flacon à réactifs
e frasco para reactivos
i bottiglia per reagenti
d Reagenzienflasche

6955 realgar
f réalgar
e rejalgar; arsénico sulfurado rojo
i realgar; sandracca
d Realgar; rotes Arsensulfid

6956 reboiler
f bouilleur
e evaporador
i bollitore
d Verdampfer

6957 receiver
f réservoir; récipient
e receptor; recipiente
i ricevitore
d Sammelbehälter

6958 recessive
f récessif
e recesivo
i recessivo
d rezessiv

6959 recessiveness
f récessivité
e recesividad
i recessività
d Rezessivität

6960 recipient bacterium
f bactérie réceptrice
e bacteria recipiente
i batterio ricettore
d Rezeptor-Bakterium

6961 recirculation
f recyclage
e recirculación
i ricircolazione; circolazione continua
d Zurückführung

6962 reclaim v
f récupérer
e recuperar
i ricuperare
d wiedergewinnen

6963 reclaimed rubber
f caoutchouc régénéré
e goma regenerada
i gomma rigenerata
d Regenerat

6964 reclaiming digester
f digesteur à régénération
e digestor de regeneración
i digestore per rigenerare
d Regenerier-Digestor

6965 reclaiming oils
f huiles pour régénérer le caoutchouc
e aceites para regenerar la goma
i oli per rigenerazione
d Ölsorten für Regenerataufbereitung

6966 recombination
f recombinaison
e recombinación
i ricombinazione
d Rekombination

6967 recombine v
f recombiner
e recombinar
i ricombinare
d wiedervereinigen

6968 recooler
f réfrigérant
e reenfriador
i refrigeratore a circolazione
d Nachkühler

6969 recorder
f enregistreur
e registrador; contador
i registratore
d Registrierapparat; Schreiber

6970 recording chart
f bande enregistreuse
e carta de registro
i scheda di registrazione
d Registrierband; Registrierstreifen

6971 recording manometer
f manomètre enregistreur
e manómetro registrador
i manometro registratore
d Druckschreiber

6972 recovery
f récupération
e recuperación
i ricupero
d Wiedergewinnung; Regenerierung

6973 recrystallization
f recristallisation
e recristalización
i ricristallizzazione
d Umkristallisieren

6974 recrystallize v
f recristalliser
e recristalizar
i ricristallizzare
d wiederkristallisieren

6975 rectification
f rectification
e rectificación
i rettificazione
d Rektifikation

6976 rectifier
  f redresseur
  e rectificador
  i rettificatore
  d Rektifikator

6977 rectify *v*
  f rectifier
  e rectificar
  i rettificare
  d rektifizieren

6978 rectifying column
  f colonne à rectifier
  e columna rectificadora
  i colonna rettificatrice
  d Rektifiziersäule

6979 rectifying section
  f section de rectification
  e sección de rectificación
  i sezione di rettifica
  d Rektifikationsteil

6980 reculture *v*
  f inoculer
  e inocular
  i trapiantare
  d abimpfen

  * **red antimony** → 4617

6981 red copper oxide
  f oxyde de cuivre rouge
  e óxido rojo de cobre
  i ossido di rame rosso
  d rotes Kupferoxyd

6982 red iron oxide
  f oxyde rouge de fer
  e óxido rojo de hierro
  i ossido ferrico
  d Eisenoxydrot

  * **red lead** → 5506

6983 red mercuric oxide
  f oxyde de mercure rouge
  e óxido mercúrico rojo
  i ossido mercurico rosso
  d rotes Quecksilberoxyd

6984 red mercuric sulphide
  f sulfure de mercure rouge
  e sulfuro mercúrico rojo
  i solfuro rosso di mercurio
  d rotes Quecksilbersulfid

6985 Redonda phosphate
  f phosphate de Redonda
  e fosfato de aluminio
  i fosfato Redonda
  d Aluminiumphosphat

6986 redox processes
  f procédés d'oxydo-réduction
  e procedimientos redox
  i processi redox
  d Redoxverfahren

6987 redruthite
  f rédruthite
  e redrutita
  i calcocite
  d Kupferglanz

6988 reduced resin
  f résine modifiée
  e resina modificada
  i resina modificata
  d verschnittenes Harz

6989 reducer
  f cuve de réduction
  e cuba de reducción
  i riduttore
  d Reduktionsgefäß

6990 reducing agent
  f réducteur; agent réducteur
  e agente de reducción; reductor
  i agente riducente; agente di riduzione
  d Reduktionsmittel

6991 reducing flame
  f flamme réductrice
  e llama reductora; llama desoxidante
  i fiamma riduttrice
  d reduzierende Flamme

6992 reductase
  f réductase
  e reductasa
  i riduttasi
  d Reduktase

6993 reduction
  f réduction
  e reducción
  i riduzione
  d Reduktion

6994 reductional mitosis
  f mitose réductionnelle
  e mitosis reductora
  i mitosi eterotipica
  d Reduktionsmitose

6995 reduction of area
f diminution de la section transversale
e reducción de la sección transversal
i diminuzione della sezione trasversale
d Querschnittsverminderung

6996 reductive infection
f infection réductive
e infección reductiva
i infezione riduttiva
d Reduktiv-Infektion

6997 reference electrode
f électrode de référence
e electrodo de referencia
i elettrodo di riferimento
d Bezugselektrode

6998 refinery
f raffinerie
e refinería
i raffineria
d Raffinerie

6999 refinery gas
f gaz de raffinerie
e gas de refinería
i gas di raffinazione
d Raffineriegas

7000 reflux
f reflux
e reflujo
i riflusso
d Rückfluß; Rückstrom

7001 reflux boiling
f ébullition au reflux
e ebullición del reflujo
i bollitura a riflusso
d Rückflußkochen

7002 reflux ratio
f taux de reflux
e relación de reflujo
i rapporto di riflusso
d Rücklaufverhältnis

7003 reforming
f réformation; reforming
e "reforming"
i operazione di "reforming"
d "Reforming"

7004 refractories
f réfractaires
e productos refractarios
i materiale refrattario
d feuerfeste Materialien

7005 refractory materials
f matériaux réfractaires
e materiales refractarios
i materiali refrattari
d feuerfeste Materialien

7006 refrigerant
f réfrigérant
e refrigerante
i refrigerante
d Kühlmittel

7007 regain
f teneur en humidité
e contenido de humedad
i tolleranza di umidità
d Feuchtigkeitsgehalt

7008 regenerate v
f régénérer
e regenerar
i riattivare
d wiedergewinnen

7009 regenerated cellulose
f cellulose régénérée
e celulosa regenerada
i cellulosa rigenerata
d regenerierte Cellulose

7010 regenerated fibre
f fibre régénérée
e fibra regenerada
i fibra rigenerata
d regenerierte Faser

7011 regenerated leach liquor
f liqueur de lessive régénérée
e líquido de solución regenerada
i liscivia rigenerata
d regenerierte Lauge

7012 regeneration
f régénération
e regeneración
i rigenerazione
d Aufarbeitung

7013 regenerative furnace
f four régénérateur
e horno de recuperación
i forno a ricupero di calore
d Regenerativofen

7014 regenerator
f régénérateur; récupérateur
e regenerador; termocambiador
i ricuperatore (di calore)
d Regenerator

7015 regulator box
  f cuve régulatrice
  e cuba de regulación
  i scatola di distribuzione
  d Reglerkasten

7016 reinforced concrete
  f béton armé
  e hormigón armado
  i cemento armato
  d armierter Beton; Eisenbeton

7017 reinforced plastics
  f plastiques renforcés
  e plásticos reforzados
  i resine sintetiche armate
  d verstärkte Kunststoffe

7018 reinforcing filler
  f charge renforçante
  e carga reforzadora
  i carica rinforzante
  d verstärkendes Füllmittel

7019 relational coiling
  f enroulement réciproque
  e arrollamiento recíproco
  i avvolgimento reciproco
  d Relationsspirale

7020 relative chalk rating
  f coefficient de résistance au farinage
  e coeficiente de resistencia a la
     desintegración
  i coefficiente di resistenza allo
     sfarinamento
  d Kalkwert

7021 relative viscosity
  f viscosité relative
  e viscosidad relativa
  i viscosità relativa
  d relative Viskosität

7022 relative volatility
  f volatilité relative
  e volatilidad relativa
  i volatilità relativa
  d relative Flüchtigkeit

7023 relief valve
  f vanne de détente; détendeur
  e válvula de seguridad
  i valvola di sicurezza
  d Überdruckventil

7024 relieving
  f dégagement
  e liberación

  i eliminazione
  d Reliefieren

7025 remote-reading instrument
  f instrument de lecture à distance
  e instrumento medidor teleaccionado
  i strumento a telelettura
  d Meßgerät mit Fernablesung

  * **rennase** → 1841

7026 rennet casein
  f caséine caillée
  e caseína al cuajo
  i caseina cagliata
  d Labkasein

  * **rennin** → 1841

7027 reparable mutant
  f mutant réparable
  e mutante reparable
  i mutante restituibile
  d reparable Mutante

7028 Reppe process
  f procédé de Reppe
  e proceso de Reppe
  i processo Reppe
  d Reppe-Verfahren

7029 repressor
  f répresseur
  e represor
  i repressore
  d Repressor

7030 reprocess *v*
  f récupérer
  e regenerar
  i rilavorare
  d wiederaufarbeiten

7031 rereeling machine
  f rebobineuse
  e máquina rebobinadora
  i riaspatrice; riincannatrice
  d Maschine zum Wiederaufwickeln

7032 rescinnamine
  f rescinnamine
  e rescinamina
  i rescinnammina
  d Rescinnamin

7033 reserpine
  f réserpine
  e reserpina

i reserpina
d Reserpin

7034 residual affinity
f affinité résiduelle
e afinidad remanente
i affinità residua
d Affinitätsrest

7035 residual chromosome
f chromosome résiduel
e cromosoma residual
i cromosoma residuale
d Residualchromosom

7036 residual current
f courant résiduel
e corriente remanente
i corrente residua
d Reststrom

7037 residual heterozygote
f hétérozygote résiduel
e heterocigoto residual
i eterozigote residuale
d Residual-Heterozygote

7038 residual tack
f adhésivité résiduelle
e pegajosidad remanente
i adesività residua
d Restklebrigkeit

7039 resilience
f résilience
e resiliencia
i resilienza
d Rückprallelastizität

7040 resin
f résine
e resina
i resina
d Harz

7041 resinates
f résinates
e resinatos
i resinati
d Resinate

7042 resin bonded
f lié par la résine
e unido por resina
i unito dalla resina
d harzgebunden

7043 resin bonded plywood
f contre-plaqué à la résine
e madera contrachapada encolada con resina sintética
i legno compensato con adesivo in resina sintetica
d kunstharzverleimtes Sperrholz

7044 resin impregnated
f imprégné de résine
e impregnado por resina
i impregnato con resina
d harzgetränkt

7045 resinols
f résinols
e resinoles
i resinoli
d Resinole

7046 resin-rich area
f zone riche en résine
e zona alta en resina
i zona ricca di resine
d harzreiche Stelle

7047 resin-starved area
f zone pauvre en résine
e zona baja en resina
i zona d'insufficienza di resine
d harzarme Stelle

7048 resist
f matière de protection
e materia aislante; revestimiento protector
i rivestimento isolante
d Abdeckmittel; Schutzmasse

7049 resol resins
f résines de résol
e resinas de resol
i resine di resolo
d Resolharze

7050 re-solution
f redissolution
e redisolución
i ridissoluzione
d Wiederauflösung

7051 resonance
f résonance
e resonancia
i risonanza
d Resonanz

7052 resorcinol; resorcin
f résorcine
e resorcina
i resorcinolo
d Resorcin

7053 resorcinol diglycidyl ether
  f éther diglycidylique de résorcine
  e éter diglicidílico de resorcina
  i etere diglicidilico di resorcinolo
  d Resorcindiglycidyläther

7054 resorcinol monoacetate
  f monoacétate de résorcine
  e monoacetato de resorcina
  i monoacetato di resorcinolo
  d Resorcinmonoacetat

7055 resorcinol monobenzoate
  f monobenzoate de résorcine
  e monobenzoato de resorcina
  i monobenzoato di resorcinolo
  d Resorcinmonobenzoat

7056 resorcinol test
  f essai à la résorcine
  e prueba de resorcina
  i prova di resorcinolo
  d Resorcinprobe

7057 resorcylic acid
  f acide résorcylique
  e ácido resorcílico
  i acido resorcilico
  d Resorcylsäure

7058 restriction gene
  f gène restrictif
  e gen restrictivo
  i gene restrittivo
  d Restriktionsgen

7059 retarder
  f agent retardant
  e retardador
  i ritardante; rallentatore
  d Verzögerer

7060 retinol
  f rétinol
  e retinol
  i retinolo
  d Retinin

7061 retort
  f cornue
  e retorta
  i storta
  d Retorte; Destillationskolben

7062 retort stand
  f support de cornue
  e portarretorta
  i sostegno per la storta
  d Retortenhalter

7063 reversal heteropycnosis
  f hétéropycnose reverse
  e heteropicnosis de inversión
  i eteropicnosi invertita
  d Heteropyknoseumkehr

7064 reverse-current cleaning
  f dégraissage anodique
  e desengrase anódico
  i sgrassaggio anodico
  d anodische Reinigung

7065 reverse-roll coater
  f machine à rouleaux inversés
  e revestidora de rodillos de rotación
    inversa
  i verniciatrice a rullo a rotazione inversa
  d gleichsinnig laufende
    Walzenauftragmaschine

7066 reversible cell
  f élément réversible
  e pila reversible; elemento reversible
  i pila reversibile
  d umkehrbares Element

7067 reversible gel
  f gel réversible
  e gel reversible
  i gel reversibile
  d reversibles Gel

7068 reversible process
  f réaction réversible
  e reacción reversible
  i reazione reversibile
  d reversibler Vorgang

7069 reversible reaction
  f réaction réversible
  e reacción reversible
  i reazione reversibile
  d umkehrbare Reaktion

7070 revivification
  f réduction
  e reactivación
  i riattivazione
  d Wiederbelebung

7071 revulsive
  f révulsif
  e revulsivo
  i rivulsivo
  d Ableitungsmittel

7072 Reynolds number
  f nombre de Reynolds
  e número de Reynolds

i numero di Reynolds
d Reynoldsche Zahl

7073 rH value
f valeur rH
e valor rH
i valore rH
d rH-Wert

7074 rhenium
f rhénium
e renio
i renio
d Rhenium

7075 rheological instrumentation
f instrumentation rhéologique
e instrumentación reológica
i strumenti reologici
d rheologische Instrumentierung

7076 rheology
f rhéologie
e reología
i reologia
d Fließkunde; Fließlehre

7077 rheometer
f consistomètre
e consistómetro
i consistometro
d Konsistenzmesser

7078 rheopectic
f rhéopectique
e reopéctico
i reopectico
d rheopektisch

7079 rheopexy
f rhéopexie
e reopexia
i reopessia
d Rheopexie

7080 rhodamines
f rhodamines
e rodaminas
i rodammine
d Rhodamine

7081 rhodanine
f rhodanine
e rodanina
i rodanina
d Rhodanin

7082 rhodanizing
f galvanoplastie au rhodium

e galvanoplastiado con rodio; rodiado
i rodanizzazione
d Rhodanisieren

7083 rhodinol
f rhodinol
e rodinol
i rodinolo
d Rhodinol

7084 rhodinyl acetate
f acétate de rhodinyle
e acetato de rodinilo
i acetato di rodinile
d Rhodinylacetat

7085 rhodium
f rhodium
e rodio
i rodio
d Rhodium

7086 rhodochrosite
f rhodochrosite
e rodocrosita
i rodocrosite
d Manganspat

7087 rhodonite
f rhodonite
e rodonita
i rodonite
d Rhodonit; Kieselmangan

7088 rhodopsin
f rhodopsine
e rodopsina
i rodopsina
d Rhodopsin

7089 rhubarb
f rhubarbe
e ruibarbo
i rabarbaro
d Rhabarber

7090 riboflavine; vitamin B$_2$
f riboflavine
e riboflavina
i riboflavina
d Riboflavin

7091 ribonuclease
f ribonucléase
e ribonucleasa
i ribonucleasi
d Ribonuklease

7092 ribonucleic acid
  f acide ribonucléique
  e ácido ribonucleico
  i acido ribonucleico
  d Ribonukleinsäure

7093 ribose
  f ribose
  e ribosa
  i ribosio
  d Ribose

7094 ribosomal
  f ribosomal
  e ribosomal
  i ribosomale
  d ribosomal

7095 ribosome
  f ribosome
  e ribosoma
  i ribosoma
  d Ribosom

7096 rich gas
  f gaz riche
  e gas rico
  i gas ricco
  d hochwertiges Gas; Reichgas

7097 rich solvent
  f solvant riche
  e solvente rico
  i solvente ricco
  d angereichertes Lösungsmittel

7098 ricinoleic acid
  f acide ricinoléique
  e ácido ricinoleico
  i acido ricinoleico
  d Ricinolsäure

7099 ricinoleyl alcohol
  f alcool ricinoléylique
  e alcohol ricinoleílico
  i alcool ricinoleilico
  d Ricinoleylalkohol

  * **Ricinus communis** → 1503

7100 rider *(of a balance)*
  f cavalier
  e jinete
  i cavaliere
  d Reiter

7101 riffler; sand trap
  f sablier
  e cuba separadora

  i traversa
  d Sandfang

7102 rigid PVC
  f PVC rigide
  e plástico rígido de cloruro de polivinilo
  i cloruro di polivinile rigido
  d Hart-PVC

7103 ring chromatid
  f chromatide en anneau
  e cromatidio anular
  i cromatido anulare
  d Ringchromatide

7104 ring chromosome
  f chromosome en anneau
  e cromosoma anular
  i cromosoma anulare
  d Ringchromosom

  * **Rinmann's green** → 1901

7105 rinser
  f rinceuse
  e enjuagadora
  i spruzzatrice
  d Ausspritzapparat

7106 ripple finish
  f finissage ondulé
  e acabado ondulado
  i rifinizione a ondulazione
  d Kräusellackierung

7107 ripple tray
  f plateau ondulé
  e bandeja ondulada de destilación
  i bacinella increspata
  d Riffelplatte

7108 ristocetin
  f ristocétine
  e ristocetina
  i ristocetina
  d Ristocetin

7109 rivelling
  f formation de rides
  e formación de pequeñas arrugas
  i raggrinzatura
  d Kräuselung

7110 RNA messenger
  f ARN messager
  e mensajero ARN
  i messaggero RNA
  d Messenger RNS

7111 Rochelle salt
 f sel de Seignette
 e sal de Rochelle
 i sale di Rochelle
 d Rochellesalz; Kaliumnatriumtartrat

7112 Rockwell hardness
 f dureté Rockwell
 e dureza Rockwell
 i grado di durezza alla prova Rockwell
 d Rockwell-Härte

7113 rod mill
 f train de serpentage
 e tren laminador de redondos; trituradora
   de varillas
 i molino a barre
 d Stabmühle

7114 roller-coating enamel
 f émail pour application par machine
 e barniz para revestidora de rodillos
 i smalto per verniciatrice a rulli
 d Emaillelack für Walzenauftrag

7115 roller mill
 f malaxeur; mélangeur à cylindres
 e mezcladora de cilindros
 i mescolatore a cilindri
 d Walzenmühle; Mischwalzwerk

7116 roller process
 f procédé à rouleaux
 e proceso de rodillos
 i processo a rulli
 d Walzverfahren

7117 roll sulphur
 f soufre en canon
 e azufre moldeado
 i zolfo in cannelli
 d Stangenschwefel

7118 room temperature
 f température ambiante
 e temperatura ambiente
 i temperatura ambiente
 d Zimmertemperatur

7119 ropiness
 f tendance à être filandreux
 e viscosidad
 i cordonatura
 d Streifenbildung

7120 roscoelite
 f roscoélithe
 e roscoelita

 i roscolite
 d Roscoelit

7121 Rose crucible
 f creuset de Rose
 e crisol de Rose
 i crogiuolo Rose
 d Rosetiegel

7122 rosemary oil
 f essence de romarin
 e aceite de romero
 i essenza di rosmarino
 d Rosmarinöl

7123 rose oil
 f essence de rose
 e esencia de rosa
 i essenza di rosa
 d Rosenöl

7124 rosin; colophony
 f colophane
 e colofonia
 i colofonia
 d Colophonium

7125 rosin oil
 f huile de résine
 e aceite de colofonia
 i olio di resina
 d Harzöl

7126 rosin pitch
 f poix de colophane
 e pez rubia
 i pece di resina
 d Harzpech

7127 rosin spirit
 f essence de colophane
 e espíritu de colofonia
 i spirito di resina
 d Harzspiritus

7128 rotary column
 f colonne à rotation
 e columna de rotación
 i colonna a rotazione
 d Drehungskolonne

7129 rotary crusher
 f broyeur rotatif; concasseur rotatif
 e mañacadora giratoria
 i frantoio rotativo
 d Kreiselbrecher

7130 rotary distillation
 f distillation rotative

e destilación rotatoria
i distillazione a rotazione
d Rotationsdestillation

7131 rotary dryer
f séchoir rotatif
e secador rotatorio; secadora giratoria
i essicatoio rotante
d Trommeltrockner

7132 rotary filter
f filtre rotatif
e filtro rotatorio
i filtro rotante
d Drehfilter; Trommelfilter

7133 rotary kiln
f four rotatif
e horno rotatorio
i fornace rotante
d Drehofen

7134 rotating-disk contactor
f contacteur à disques rotatifs
e contactor de discos giratorios
i contattore a dischi rotanti
d Drehscheibenextraktor

7135 rotating-disk process
f procédé à disque rotatif
e proceso del disco giratorio
i processo a disco rotante
d Drehscheibenverfahren

7136 rotenone
f roténone
e rotenona
i rotenone
d Rotenon

7137 roughage
f matière cellulosique
e substancia celulósica
i sostanza cellulosica
d Rohfasern

7138 roughing
f vide préalabe
e vacío preliminar
i prevuoto
d Herstellung eines Vorvakuums

7139 round bottomed flask
f ballon rond
e matraz redondo
i matraccio a fondo rotondo
d Rundkolben

7140 rouse *v*
f activer la fermentation
e activar la fermentación
i ravvivare
d Gärung beleben

7141 Roussin's black salt
f sel noir de Roussin
e sal negra de Roussin
i sale nero di Roussin
d schwarzes Roussinsalz

7142 Roussin's red salt
f sel rouge de Roussin
e sal roja de Roussin
i sale rosso di Roussin
d rotes Roussinsalz

7143 Roussin's salts
f sels de Roussin
e sales de Roussin
i sali di Roussin
d Roussinsalze

7144 rubber
f gomme; caoutchouc
e caucho; goma
i gomma
d Gummi; Kautschuk

7145 rubber accelerators
f accélérateurs pour le caoutchouc
e aceleradores del caucho
i acceleratori della gomma
d Gummibeschleuniger

7146 rubber hydrochloride
f chlorhydrate de caoutchouc
e hidrocloruro de caucho
i gomma idroclorurata
d Gummihydrochlorid

7147 rubberizing
f caoutchoutage
e engomado
i gommatura
d Gummieren

7148 rubberseed oil
f huile de graines de caoutchouc
e aceite de semillas de caucho
i olio semi di gomma
d Kautschuksamenöl

7149 rubber solvent
f solvant à base de caoutchouc
e disolvente de caucho
i solvente della gomma
d Gummilösemittel

7150 rubber sponge
 f caoutchouc spongieux
 e esponja de caucho
 i gomma spugnosa
 d Schaumgummi

 * rubbing varnish → 6498

7151 rubeanic acid
 f acide rubéanique
 e ácido rubeánico
 i acido rubianico
 d Rubeanwasserstoff

7152 rubidium
 f rubidium
 e rubidio
 i rubidio
 d Rubidium

7153 rubidium chloride
 f chlorure de rubidium
 e cloruro de rubidio
 i cloruro di rubidio
 d Rubidiumchlorid

7154 rue oil
 f essence de rue
 e aceite de ruda
 i olio essenziale di ruta
 d Rautenöl

7155 rumbling; tumbling
 f dessablage au tonneau
 e tratamiento en tambor giratorio
 i trattamento a botte rotante
 d Trommeln

7156 run down
 f déchargé
 e descargado
 i scarica
 d entladen

7157 run-of-retort coke
 f coke brut
 e coque bruto
 i coke greggio
 d Rohkoks

7158 rust
 f rouille
 e rust
 i basidiomiceto della ruggine delle piante
 d Rost

7159 ruthenium
 f ruthénium
 e rutenio
 i rutenio
 d Ruthenium

7160 ruthenium chloride
 f chlorure de ruthénium
 e cloruro de rutenio
 i cloruro di rutenio
 d Rutheniumchlorid

7161 rutile
 f rutile
 e rutilo
 i rutilo
 d Rutil

7162 rutin; rutoside
 f rutine
 e rutina
 i rutina
 d Rutin

S

7163 sabadilla
  f sabadiline
  e sabadilla; cebadilla
  i sabadilla
  d Sabadilla

7164 saccharimeter
  f saccharimètre
  e sacarímetro
  i saccarimetro
  d Saccharimeter; Polarisationsapparat

7165 saccharimetry
  f saccharimétrie
  e sacarimetría
  i saccarimetria
  d Zuckermessung

7166 saccharin
  f saccharine
  e sacarina
  i saccarina
  d Saccharin

7167 saccharometer
  f saccharimètre
  e sacarómetro
  i saccarometro
  d Saccharometer

7168 saccharose; sucrose
  f saccharose
  e sacarosa; sucrosa
  i saccarosio
  d Saccharose; Sucrose

7169 safety funnel
  f tube de sûreté
  e tubo de seguridad
  i tubo di sicurezza
  d Sicherheitsröhre

7170 safety valve
  f soupape de sûreté
  e válvula de seguridad
  i valvola di sicurezza
  d Ablaßventil; Sicherheitsventil

7171 safflower oil
  f huile de carthame
  e aceite de cártamo
  i olio di cartamo
  d Saffloröl

7172 saffron
  f safran

  e azafràn
  i zafferano
  d Safran

7173 safranines
  f safranines
  e safraninas
  i safranine
  d Safranine

7174 safrole
  f safrol
  e safrol
  i safrolo
  d Safrol

7175 sagger; saggar
  f cassette réfractaire
  e caja refractaria de loza
  i cassetta refrattaria
  d Kapsel

* **sal ammoniac → 462**

7176 salicin
  f salicine
  e salicina
  i salicina
  d Salicin

7177 salicyl alcohol
  f alcool salicylique
  e alcohol salicílico
  i alcool salicilico
  d Salicylalkohol

7178 salicylaldehyde
  f salicylaldéhyde
  e salicilaldehído
  i aldeide salicilica
  d Salicylaldehyd

7179 salicylamide
  f salicylamide
  e salicilamida
  i salicilammide
  d Salicylamid

7180 salicylanilide
  f salicylanilide
  e salicilanilida
  i salicilanilide
  d Salicylanilid

7181 salicylazosulphapyridine
  f salicylazosulfapyridine
  e salicilazosulfapiridina
  i salicilazosolfapiridina
  d Salicylazosulfapyridin

7182 salicylic acid
 f acide salicylique
 e ácido salicílico
 i acido salicilico
 d Salicylsäure

7183 saline
 f salin
 e salino
 i salino
 d salzig; salzhaltig

7184 salinity
 f salinité; salure
 e salinidad; salsedumbre
 i salinità
 d Salzgehalt

7185 salinometer
 f salinomètre
 e salinómetro
 i salinometro
 d Salzmesser

7186 salivin
 f ptyaline
 e salivina
 i ptialina
 d Ptyalin

7187 salol
 f salol
 e salol
 i salolo
 d Salol

7188 salt
 f sel
 e sal
 i sale
 d Salz

7189 salt bath furnace
 f four à bain de sel
 e horno de baño de sales
 i forno a bagno di sale
 d Salzbadofen

7190 salt cake
 f sulfate de sodium brut
 e torta de sal
 i solfato di sodio commerciale
 d Natriumsulfatkuchen; Salzkuchen

7191 saltern
 f saunerie
 e salina
 i salina
 d Saline

7192 salt glaze
 f vernissage par salage
 e vidriado a la sal
 i smaltatura a sale
 d Salzglasur

7193 salt grainer
 f évaporateur de sel
 e evaporadora de sal
 i concentratore a serpentino
 d Salzkocher

7194 salting out
 f précipitation par un sel
 e precipitación de un electrodo
 i precipitazione di un elettrolito
 d Aussalzung

7195 salting-out agent
 f agent relargant
 e agente de precipitación por adición de sal
 i agente di precipitazione per addizione di sale
 d Aussalzungsstoff

* **saltpetre → 6630**

* **salts of Sorrel → 6643**

7196 salt spray test
 f essai à eau saline
 e ensayo de niebla salina
 i prova al sale
 d Salzsprühversuch

7197 samarium
 f samarium
 e samario
 i samario
 d Samarium

7198 samarskite
 f samarskite
 e samarskita; samarsquita
 i samarschite
 d Samarskit

7199 sample
 f échantillon; spécimen
 e muestra; prueba
 i campione; provino; esemplare
 d Muster; Probe; Exemplar

7200 sampler
 f échantillonneur
 e sacamuestras
 i sonda per campioni
 d Musternehmer; Probenehmer

7201 sample reducer
 f découpeuse d'échantillons
 e divisor de muestras
 i frazionatore di campioni
 d Musterzerkleinerer

7202 sampling
 f prélèvement d'échantillons
 e extracción de una muestra
 i campionatura
 d Probeentnahme

7203 sampling cock
 f robinet de prise
 e grifo de muestreo
 i rubinetto di presa
 d Probehahn; Probierhahn

7204 sand
 f sable
 e arena
 i sabbia
 d Sand

7205 sandalwood oil
 f essence de bois de santal
 e aceite de madera de sándalo
 i essenza di legno di sandalo
 d Sandelholzöl

7206 sandarac gum
 f gomme de genévrier; sandaraque
 e goma sandáraça
 i sandracca
 d Sandarakgummi

7207 sand bath
 f bain de sable
 e baño de arena
 i bagno di sabbia
 d Sandbad

7208 sand filter
 f filtre à sable
 e filtro de arena
 i filtro a sabbia
 d Sandfilter

7209 sandstone
 f grès
 e piedra arenisca
 i arenaria; gres
 d Sandstein

 * **sand trap** → **7101**

7210 santalol
 f santalol
 e santalol

 i santalolo
 d Santalol

7211 santalyl acetate
 f acétate de santalyle
 e acetato de santalilo
 i acetato di santalile
 d Santalylacetat

7212 santonin; santolactone
 f santonine
 e santonina
 i santonina
 d Santonin

7213 sap
 f sève
 e savia
 i sugo
 d Saft

7214 saponaceous
 f saponacé
 e saponáceo
 i saponaceo
 d seifig

7215 saponification
 f saponification
 e saponificación
 i saponificazione
 d Verseifung

7216 saponification number
 f nombre de saponification
 e número de saponificación
 i numero di saponificazione
 d Verseifungszahl

7217 saponify *v*
 f saponifier
 e saponificar
 i saponificare
 d verseifen

7218 saponifying agent
 f agent de saponification
 e agente de saponificación
 i agente di saponificazione
 d Verseifungsmittel

7219 saponin
 f saponine
 e saponina
 i saponina
 d Saponin

7220 saran
 f saran

e nombre genérico de polímeros y
copolímeros de cloruro de vinilideno
i saran
d Saran

7221 sarcosine
f sarcosine
e sarcosina
i sarcosina
d Sarkosin

7222 sardonyx
f sardonyx
e sardónice
i sardonica
d Sardonyx

7223 sarsaparilla
f salsepareille
e zarzaparrila
i sarsaparilla
d Sassaparilwurzel

7224 sassafras oil
f essence de sassafras
e aceite de sasafrás
i essenza di sassafrasso
d Sassafrasöl

7225 sassolite
f sassoline
e sasolita
i sassolite
d Sassolin

7226 SAT-chromosome
f SAT-chromosome
e cromosoma SAT
i SAT-cromosoma
d SAT-Chromosom

7227 satellite nucleolus
f nucléole satélite
e nucléolo satélite
i nucleolo satellite
d Satellitennukleolus

7228 satin white
f blanc satin
e blanco satín
i bianco di calcite
d Atlasweiß; Satinweiß

7229 saturant
f agent d'imprégnation
e saturante
i saturante
d Imprägniermittel

7230 saturated compound
f composé saturé
e compuesto saturado
i composto saturo
d gesättigte Verbindung

7231 saturated hydrocarbons
f hydrocarbures saturés
e hidrocarburos saturados
i idrocarburi saturi
d gesättigte Kohlenwasserstoffe

7232 saturated solution
f solution saturée
e solución saturada
i soluzione satura
d gesättigte Lösung

7233 saturated steam
f vapeur saturée
e vapor saturado
i vapore saturo
d gesättigter Dampf

7234 saturated vapour pressure
f pression de vapeur saturée
e presión de vapor saturado
i pressione di vapore saturo
d Sattdampfdruck

7235 saturation
f saturation
e saturación
i saturazione
d Sättigung

7236 scald *v*
f échauder
e hervir; escaldar
i sterilizzare con ebollizione
d brühen

7237 scald *v* **out**
f échauder; ébouillanter
e escaldar; limpiar con agua hirviente
i pulire con acqua bollente
d ausbrühen

7238 scale
f incrustation; calcin
e incrustación
i incrostazione
d Kesselstein; Kruste

7239 scale-down *v*
f réduire l'échelle
e reducir a escala
i riprodurre a scala ridotta
d maßstäblich verkleinern

7240 scale-up *v*
f augmenter l'échelle
e aumentar a escala
i riprodurre a scala maggiore
d maßstäblich vergrößern

7241 scandium
f scandium
e escandio
i scandio
d Scandium

* **scarlet chrome** → 1799

7242 scarlet lake
f laque écarlate
e laca escarlata
i lacca scarlatta
d Scharlachlack

7243 scavenger
f épurateur
e depurador
i antipiombo; depuratore
d Spülmittel

7244 scavenging
f coprécipitation
e depuración
i coprecipitazione
d Reinigungsfällung

7245 Schäffer's salt
f sel de Schäffer
e sal de Schäffer
i sale di Schäffer
d Schäffersches Salz

7246 scheelite
f schellin; scheelite
e scheelita; esquelita
i sceelite
d Scheelspat; Scheelerz; Scheelit

7247 Schiff bases
f bases de Schiff
e bases de Schiff
i basi di Schiff
d Schiffsche Basen

7248 schoepite
f schoepite
e schoepita
i schoepite
d Schöpit

7249 schroeckingerite
f schroeckingérite
e schroeckingerita

i schroeckingerite
d Schröckingerit; Dakeit

7250 Schweinfurt green
f vert de Schweinfurt
e verde de Schweinfurt
i verde di Schweinfurt
d Schweinfurtergrün

7251 scintillation
f scintellement
e centelleo; escintillación
i scintillamento
d Flimmern

7252 scission
f scission
e escisión
i scissione
d Spaltung

7253 sclerometer
f schléromètre
e esclerómetro
i sclerometro
d Ritzhärteprüfer; Sklerometer

* **scleroproteins** → 274

7254 scoop
f cuiller
e cuchara
i cucchiaio
d Löffel

7255 scoparium
f scoparium
e scoparius
i scoparia
d Besenkraut; Scoparium

7256 scopolamine; hyoscine
f scopolamine
e escopolamina
i scopolamina
d Scopolamin; Hyoscin

7257 scopoline; oscine
f scopoline
e escopolina
i scopolina
d Scopolin

7258 scorch *v*
f brûler superficiellement
e chamuscar
i scottare
d versengen

7259 scorification
f scorification
e escorificación
i scorificazione
d Verschlackung

7260 scorifier
f scorificateur
e escorificador
i scorificatoio
d Ansiedescherben

7261 scorodite
f scorodite
e escorodita
i scorodite
d Arseniksinter

7262 scour *v*
f nettoyer
e pulir
i pulire
d abreiben

7263 screen
f tamis
e tamiz; criba
i staccio
d Sieb

7264 screen classifier
f trieur à tamis
e clasificador de tamiz
i vaglio classificatore
d Siebplansichter

7265 scrubber tower
f tour de lavage
e torre lavadora; torre depuradora
i torre di depurazione
d Skrubber; Rieselturm

7266 scrubbing
f épuration; lavage
e lavado; depuración
i depurazione; lavaggio
d Berieselung; Wachsen

7267 scum *v*
f écumer
e sacar la espuma
i schiumare
d abschäumen

7268 scum
f écume
e espuma
i schiuma; spuma
d Abschaum

7269 scump
f coagulum de latex écumé
e coágulo de espuma de látex
i coagulo di gomma piuma
d Koagulat aus Latexschaum

7270 sealer
f peinture isolante
e pintura hermética
i fissativo
d Isoliergrund

7271 sebacic acid
f acide sébacique
e ácido sebácico
i acido sebacico
d Sebacinsäure

7272 sebaconitrile
f sébaconitrile
e sebaconitrilo
i sebaconitrile
d Sebaconitril

7273 sec-
Prefix denoting secondary.
f sec-
e sec-
i sec-
d Sek-

7274 secobarbital
f sécobarbital
e secobarbital
i secobarbitale
d Secobarbital

7275 secobarbital sodium
f sécobarbital sodium
e secobarbital sódico
i secobarbitale sodico
d Secobarbitalnatrium

7276 secondary alcohols
f alcools secondaires
e alcoholes secundarios
i alcooli secondari
d sekundäre Alkohole

7277 secondary amide
f amide secondaire
e amida secundaria
i ammide secondaria
d sekundäres Amid

7278 secondary cell
f élément secondaire
e elemento secundario

i elemento secondario
d Sammlerbatterie

7279 secondary polyploidy
f polyploïdie secondaire
e poliploidia secundaria
i poliploidismo secondario
d sekundäre Polyploidie

7280 secondary reaction
f réaction secondaire
e reacción secundaria
i reazione secondaria
d Sekundärreaktion

7281 sedative
f sédatif
e sedativo
i sedativo
d Beruhigungsmittel

7282 sediment
f sédiment
e sedimento
i sedimento
d Absatz

7283 sedimentation
f sédimentation
e sedimentación
i sedimentazione
d Absetzung

7284 sedimentation potential
f potentiel de sédimentation
e potencial de sedimentación
i potenziale di sedimentazione
d Sedimentationspotential

7285 seediness
f rugosité
e aspecto granuloso
i puntinatura
d griesiges Aussehen

7286 seep *v*
f suinter
e rezumar; infiltrarse
i infiltrare
d durchsickern

7287 Seger cone
f montre de Seger
e cono de Seger
i cono di Seger
d Seger-Kegel

7288 segmental allopolyploid
f allopolyploïde segmentaire

e alopoliploide segmental
i allopoliploide segmentale
d Segmentallopolyploid

* **segmentation cavity** → 1030

7289 segregant
f ségrégant
e segregante
i segregante
d Spalter

7290 segregated oils
f huiles séparées
e aceites segregados
i oli segregati
d abgespaltene Öle

7291 segregator
f séparateur d'urines
e segregador
i separatore
d Trennapparat

7292 selective factor
f facteur sélectif
e factor selectivo
i fattore selettivo
d Selektionsfaktor

7293 selenium
f sélénium
e selenio
i selenio
d Selen

7294 selenium diethyldithiocarbamate
f diéthyldithiocarbamate de sélénium
e dietilditiocarbamato de selenio
i dietilditiocarbammato di selenio
d Selendiäthyldithiocarbamat

7295 selenium dioxide
f bioxyde de sélénium
e dióxido de selenio
i biossido di selenio
d Selendioxyd

7296 selenium red
f rouge de sélénium
e rojo de selenio
i rosso di selenio
d Selenrot

7297 selenium sulphide
f sulfure de sélénium
e sulfuro de selenio
i solfuro di selenio
d Selensulfid

7298 **selenous acid**
  f acide sélénieux
  e ácido selenioso
  i acido selenioso
  d selenige Säure

7299 **self-discharge**
  f décharge spontanée
  e descarga espontánea
  i scarica spontanea
  d Selbstentladung

7300 **self-fertile**
  f autofertile
  e autofértil
  i autofertile
  d selbstfertil

7301 **self-fertility**
  f autofertilité
  e autofertilidad
  i autofertilità
  d Selbstfertilität

7302 **self-ignition temperature**
  f température d'auto-inflammation
  e temperatura de autoinflamación
  i temperatura di autoaccensione
  d Selbstentzündungstemperatur

7303 **semen**
  f sperme
  e semen
  i sperma
  d Samen

7304 **semi-allele**
  f semi-allèle
  e semialelo
  i semiallelo
  d Semiallel

7305 **semi-automatic cycle**
  f cycle semi-automatique
  e ciclo semiautomático
  i ciclo semiautomatico
  d halbautomatischer Preßzyklus

7306 **semi-automatic electroplating**
  f galvanoplastie semi-automatique
  e galvanoplastia semiautomática
  i galvanoplastica semiautomatica
  d halbautomatische Galvanisierung

7307 **semicarbazide hydrochloride**
  f chlorhydrate de semi-carbazide
  e hidrocloruro de semicarbacida
  i cloridrato di semicarbazide
  d Semicarbazidhydrochlorid

7308 **semichemical pulp**
  f pâte mi-chimique
  e pasta semiquímica
  i polpa semichimica
  d halbchemischer Zellstoff

7309 **semichiasma**
  f semi-chiasma
  e semiquiasma
  i semichiasma
  d Semichiasma

7310 **semicoke**
  f coke de distillation à basse température
  e semicoque; coque de destilación a baja temperatura
  i coke di distillazione a bassa temperatura
  d Halbkoks; Schwelkoks

7311 **semiheterotypic**
  f semi-hétérotypique
  e semiheterotípico
  i semieterotipico
  d semiheterotypisch

7312 **semikaryotype**
  f semi-caryotype
  e semicariótipo
  i semicariotipo
  d Semikaryotyp

7313 **semilethal**
  f semi-léthal
  e semiletal
  i semiletale
  d semiletal

7314 **semipermeable membrane**
  f membrane semi-perméable
  e membrana semipermeable
  i membrana semipermeabile
  d halbdurchlässige Wand

7315 **semiunivalent**
  f semi-univalent
  e semiunivalente
  i semiunivalente
  d semiunivalent

7316 **senarmontite**
  f sénarmontite
  e senarmontita
  i senarmontite
  d Senarmontit

7317 **senega**
  f polygala de Virginie
  e senega

i poligala
d Senegawurzel

7318 senna
f séné
e sen
i senna
d Sennesblätter

7319 sensitivity to light
f sensibilité à la lumière
e sensibilidad a la luz
i sensibilità alla luce
d Lichtempfindlichkeit

7320 separate *v*
f séparer; isoler
e separar; aislar
i separare; isolare
d trennen; ausscheiden

7321 separating unit
f groupe de séparation
e grupo de separación
i gruppo di separazione
d Trenngruppe

7322 separation
f séparation
e separación
i separazione
d Trennung

7323 separation column
f colonne de séparation
e columna de separación
i colonna di separazione
d Trennsäule

7324 separation potential
f potentiel de séparation
e potencial de separación
i potenziale di separazione
d Trennpotential

7325 sepia
f sépia
e sepia
i seppia
d Sepia; Sepiabraun

7326 septisomic
f septisomique
e septisómico
i settisomico
d septisom

7327 sequestration
f séquestration

e secuestración
i sequestrazione
d Absonderung; Sequestration

7328 series system
f système série
e sistema en serie
i sistema in serie
d bipolare Schaltung

7329 serine
f sérine
e serina
i serina
d Serin

7330 seroculture
f séroculture
e serocultivo
i sierocoltura
d Serokultur

7331 serosity
f sérosité
e serosidad
i sierosità
d Serosität

7332 serotonin
f sérotonine
e serotonina
i serotonina
d Serotonin

7333 serpentine
f serpentine
e serpentina
i serpentina
d Serpentin; Schlangenstein

7334 serum
f sérum
e suero
i siero
d Serum

7335 sesame oil
f huile de sésame
e aceite de sésamo
i olio di sesamo
d Sesamöl

7336 sesamolin
f sésamoline
e sesamolina
i sesamolina
d Sesamolin

7337 sesquidiploidy
f sesquidiploïdie
e sesquidiploidia
i sesquidiploidismo
d Sesquidiploidie

7338 setting temperature
f température de prise
e temperatura de curación
i temperatura di presa
d Härtetemperatur

7339 settle *v*
f déposer; sédimenter
e depositar; sedimentar
i sedimentarsi; depositarsi
d sich ablagern

7340 settling
f décantation
e decantación
i decantazione
d Abklären

7341 settling vat
f cuve à défécation
e cuba de clarificación
i tino di chiarificazione
d Klärbottich

7342 set-up cure
f prévulcanisation
e prevulcanización
i prevulcanizzazione
d Vorvulkanisation

7343 sewage
f eaux résiduelles
e aguas cloacales
i acque luride
d Abwasser

7344 sewage sludge
f boue d'égouts
e fango cloacal
i fango d'acque luride
d Abwasserschlamm

7345 sex chromatin
f chromatine sexuelle
e cromatina sexual
i cromatina sessuale
d Geschlechtschromatin

7346 sex chromosome
f chromosome sexuel
e cromosoma sexual
i cromosoma sessuale
d Geschlechtschromosom

7347 sex dimorphism
f dimorphisme sexuel
e dimorfismo sexual
i dimorfismo sessuale
d Geschlechtsdimorphismus

7348 sexivalent
f hexavalent
e hexavalente
i esavalente
d sechswertig

7349 shadowing
f effet d'écran
e efecto de pantalla
i effetto di schermo
d Abschirmung

7350 shake
f secouage
e vibración transversal
i scuotitura
d Schütteln

7351 shake-up flask
f flacon d'agitation
e frasco de agitación
i recipiente per agitare
d Schüttelflasche

7352 shale oil
f huile de schiste
e aceite de esquistosa
i olio di schisto
d Schieferöl

7353 shark liver oil
f huile de foie de requin
e aceite de hígado de tiburón
i olio di fegato di pescecane
d Haifischtran

7354 sharp paint
f peinture à séchage rapide
e pintura magra
i pittura a essicamento rapido
d schnelltrocknende Farbe

7355 shattering
f fragmentation
e fragmentación
i frammentazione
d Fragmentation

7356 sheariness
f effet gras
e efecto opalescente
i lattescenza
d Fettigkeit

7357 sheathed pyrometer
 f canne pyrométrique
 e caña pirométrica
 i canna pirometrica
 d Pyrometerstab; Pyrometerrohr

7358 shellac
 f gomme-laque
 e goma laca
 i gomma lacca
 d Schellack

7359 shell-and-tube exchanger
 f échangeur à faisceaux
 e termointercambiador de carcase y tubos
 i scambiatore di calore a involucro e tubi
 d geschlossener Wärmeaustauscher

7360 shell-flour filler
 f coquille de charge
 e material rellenador (fabricado de polvo
    de cáscaras)
 i carica inerte di farina di gusci
 d Füller aus zermahlenen Schalen

7361 shield
 f écran
 e pantalla
 i schermo
 d Abschirmung

7362 shoot
 f rejet
 e brote
 i germoglio
 d Schößling

7363 short oil varnish
 f vernis léger
 e barniz con poco aceite; barniz de secado
    rápido
 i vernice a basso tenore di olio
 d Lack mit niedrigem Ölgehalt

7364 shot blasting
 f grenaillement
 e chorreo con granalla
 i granigliatura
 d Kugelstrahlen

7365 shredder
 f délisseuse
 e trituradora; desmenuzadora
 i trinciatrice
 d Zerkleinerungsmaschine; Zerfaserer

7366 shrinkable film
 f pellicule rétrécissable
 e film de politeno encogible

 i pellicola restringibile
 d einlaufende Folie

7367 shrinkage water
 f eau colloïdale
 e agua coloidal
 i acqua di ritiro
 d Schwindwasser

 * **siccative** → **2866**

7368 side chain
 f chaîne latérale
 e cadena lateral
 i catena laterale
 d Seitenkette

7369 siderite; chalybdite; spathic iron ore
 f sidérite; fer spathique
 e siderita; calibdita; hierro espático
 i siderite; carbonato di ferro
 d Siderit; Eisenspat

7370 sienna
 f terre de Sienne
 e siena
 i terra di Siena
 d Siennaerde

7371 sieve *v*
 f cribler; tamiser
 e tamizar; cribar
 i stacciare
 d sieben

7372 sieve analysis
 f analyse au tamis
 e análisis granulométrico
 i analisi al setaccio
 d Siebanalyse

7373 sieve residue
 f résidu de tamis
 e residuo de tamiz
 i residuo di stacciatura
 d Siebrückstand

7374 sieve-shaker
 f vibreur de filtre
 e vibradora de tamices
 i scuotitore per vaglio
 d Siebschüttelvorrichtung

7375 sieve tray
 f plateau perforé
 e bandeja de tamices
 i vassoio di staccio
 d Siebplatte

7376 sight hole
 f regard
 e mirilla
 i spia; traguardo
 d Schauluke

7377 silane
 f silane
 e silano
 i silano
 d Silan

7378 silica
 f terre siliceuse; silice
 e sílice
 i terra silicea
 d Kiesel; Kieselerde

7379 silica gel
 f gel de silice
 e gel de sílice
 i gel di silice
 d Silikagel

7380 silicate
 f silicate
 e silicato
 i silicato
 d Silikat

7381 silicate paints
 f peintures au silicate
 e pinturas de silicato
 i vernici al silicato
 d Silikatfarben

7382 silicic acid
 f acide silicique
 e ácido silícico
 i acido silicico
 d Kieselsäure

7383 silicon
 f silicium
 e silicio
 i silicio
 d Silizium

7384 silicon carbide
 f carbure de silicium
 e carburo de silicio
 i carburo di silicio
 d Siliziumcarbid; Karborund

7385 silicone fluids
 f fluides silicones
 e fluidos silicónicos
 i liquidi siliconici
 d Silikonlösungen

7386 silicone resins
 f résines silicones
 e resinas silicónicas
 i resine siliconiche
 d Silikonharze

7387 silicone rubber
 f caoutchouc silicone
 e goma silicónica
 i gomma siliconica
 d Silikonkautschuk

7388 silicones
 f silicones
 e siliconas
 i siliconi
 d Silikone

7389 silicon nitride
 f nitrure de silicium
 e nitruro de silicio
 i nitruro di silicio
 d Siliziumnitrat

7390 silicon tetrachloride
 f tétrachlorure de silicium
 e tetracloruro de silicio
 i tetracloruro di silicio
 d Siliziumtetrachlorid; Chlorsilizium

7391 silicon tetrafluoride
 f tétrafluorure de silicium
 e tetrafluoruro de silicio
 i tetrafluoruro di silicio
 d Siliziumtetrafluorid

7392 silicotungstic acid
 f acide silicotungstique
 e ácido silicotúngstico
 i acido silicotungstico
 d Kieselwolframsäure

7393 silking
 f éclat soyeux
 e aspecto sedoso
 i setosità
 d seidenartiges Ansehen

7394 sillimanite
 f sillimanite
 e silimanita
 i sillimanite
 d Sillimanit

7395 siloxane
 f siloxane
 e siloxano
 i silossano
 d Siloxan

7396 silver
 f argent
 e plata
 i argento
 d Silber

7397 silver arsphenamine
 f arsphénamine d'argent
 e arsfenamina de plata
 i arsfenammina d'argento
 d Silberarsphenamin

7398 silver bromide
 f bromure d'argent
 e bromuro de plata
 i bromuro d'argento
 d Silberbromid; Bromsilber

7399 silver chloride
 f chlorure d'argent
 e cloruro de plata
 i cloruro d'argento
 d Silberchlorid; Chlorsilber

7400 silver chloride cell
 f pile au chlorure d'argent
 e pila de cloruro de plata
 i pila a cloruro d'argento
 d Silberchloridelement

7401 silver chromate
 f chromate d'argent
 e cromato de plata
 i cromato d'argento
 d Silberchromat

 * silver, colloidal ~ → 1970

7402 silver cyanide
 f cyanure d'argent
 e cianuro de plata
 i cianuro d'argento
 d Silbercyanid; Cyansilber

7403 silver halides
 f halogénures d'argent
 e haluros de plata
 i alogenuri d'argento
 d Silberhalogenide

7404 silver iodide
 f iodure d'argent
 e yoduro de plata
 i ioduro d'argento
 d Silberjodid; Jodsilber

7405 silver nitrate
 f nitrate d'argent
 e nitrato de plata

 i nitrato d'argento
 d Silbernitrat

7406 silver nitrite
 f nitrite d'argent
 e nitrito de plata
 i nitrito d'argento
 d Silbernitrit

7407 silver oxide
 f oxyde d'argent
 e óxido de plata
 i ossido d'argento
 d Silberoxyd

7408 silver oxide cell
 f pile à oxyde d'argent
 e pila de óxido de plata
 i pila a ossido d'argento
 d Silberoxydelement

7409 silver phosphate
 f phosphate d'argent
 e fosfato de plata
 i fosfato d'argento
 d Silberphosphat

7410 silver picrate
 f picrate d'argent
 e picrato de plata
 i picrato d'argento
 d Silberpikrat

7411 silver potassium cyanide
 f cyanure d'argent et potassium
 e cianuro de plata y potasio
 i cianuro di potassio e argento
 d Silberkaliumcyanid

7412 silver protein
 f protéine argentée
 e proteína argéntica
 i proteina d'argento
 d Proteinsilber

7413 silver screen
 f crible; tamis
 e tamiz de pulpa basta
 i crivello d'acciaio forato
 d rotierender Astfänger

7414 silver storage battery
 f accumulateur à l'argent
 e acumulador de plata
 i accumulatore ad argento
 d Silber-Zink-Akkumulator

7415 silver sulphate
 f sulfate d'argent

e sulfato de plata
i solfato d'argento
d Silbersulfat

7416 simaruba
f simaruba
e simaruba
i simaruba
d Simaruba

7417 Simons process
f procédé de Simons
e proceso de Simons; método Simons
i processo Simons
d Simonsches Verfahren

7418 single chromosome compensating substitution line
f lignée de substitution compensatoire pour un chromosome
e línea de substitución compensatoria a cromosoma simple
i linea di sostituzione di compensazione a cromosoma unico
d Linie mit einem substituierten Chromosom

7419 single-effect evaporator
f évaporateur à simple effet
e evaporador de efecto simple
i evaporatore ad effetto singolo
d Einstufenverdampfer

7420 singlet
f singlet
e unión de un solo electrón
i singoletto
d Einelektronverbindung

7421 sinkers *(brewing)*
f grains plongeurs
e granos pesados
i grani pesanti
d Sinker

7422 sinter *v*
f fritter
e sinterizar; fritar
i sinterizzare
d sintern

7423 sintered plate
f plaque frittée
e placa sinterizada
i piastra sinterizzata
d Sinterplatte

7424 sintering
f frittage

e sinterización
i sinterizzazione
d Sintern

7425 sintering furnace
f four de frittage
e horno de sinterizar
i forno di sinterizzazione
d Sinterofen

* **siphon → 8004**

7426 siphon *v* **off**
f siphonner
e sifonar
i sifonare
d abhebern

7427 sisal
f sisal; agave d'Amérique
e sisal
i canapa sisaliana
d Sisalhanf

7428 sister chromatids
f chromatides sœurs
e cromatidios hermanos
i cromatidi fratelli
d Schwesterchromatiden

7429 sitosterol
f sitostérol
e sitosterol
i sitosterolo
d Sitosterin; Sitosterol

7430 size
f apprêt
e apresto
i colla; appretto
d Leim; Schlichte

7431 sizing
Separating material in different sizes.
f classement par grosseur
e clasificación volumétrica
i classifica granulometrica
d Sortieren nach der Größe

7432 sizing
Giving paper a degree of water resistance.
f encollage
e encolado impermeable del papel
i collatura
d Leimung; Leimen

7433 skatole
f skatol

e escatol
i scatolo
d Skatol

7434 skim *v*
f écumer
e quitar la espuma
i schiumare
d abheben; abschäumen

7435 skimmings
f mélange de bière et levure
e mezcla de levadura y cerveza
i scrematura
d Abgeschäumtes

7436 skinning
f formation de peau
e formación de una capa
i formazione di pelli
d Hautbildung

7437 skutterudite
f skuttérudite
e skuterudita
i minerale di nichel e cobalto
d Skutterudit

7438 slab dissolver
f récipient de solution pour plaques
  enrichies
e recipiente de solución para placas
  enriquecidas
i recipiente di soluzione per piastre
  arricchite
d Lösungsgefäß für angereicherte Platten

* **slack lime** → **4881**

7439 slack melt copal
f copal à basse température
e copal fundido a baja temperatura
i copale parzialmente fusa a bassa
  temperatura
d Tieftemperaturkopal

7440 slack wax
f paraffine brute
e parafina cruda
i cera greggia di petrolio
d Rohparaffin

7441 slag
f crasses
e escoria
i scoria
d Schlacke

7442 slag brick
f brique de laitier
e ladrillo de escoria
i mattone di loppa
d Schlackenziegel

7443 slag cement
f ciment de laitier
e cemento de escoria
i cemento di scorie
d Schlackenzement

7444 slake *v*
f éteindre
e apagar (la cal)
i spegnere
d ablöschen

* **slaked lime** → **4881**

7445 slate
f ardoise
e pizarra
i lastra di ardesia; lavagna
d Schiefer

7446 slate flour
f poudre d'ardoise
e polvo de pizarra
i farina d'ardesia
d Schiefermehl

7447 sleepiness
f manque de lustre
e aspecto deslustrado
i opacità
d Weichheit; Teigigkeit

7448 slime
  Fine insoluble material formed in
  electrolysis or in electrodeposition.
f boue
e limo
i fanghi di deposizione
d Rückstand; Schlamm

7449 slimes
  Material of less than 200 mesh which
  settles very slowly and must be leached
  by agitation.
f boue
e finos; fangos
i melma
d Schlamm

7450 slip
  Clay mixed with water.
f pâte
e pasta de acrilla acuosa

   i argilla liquida
   d Schlamm

7451 slip
Condition of a coating in which surface tackiness is virtually absent and the film seems as if lubricated.
   f poli
   e efecto resbaladizo (pinturas)
   i levigatezza
   d Glätte

7452 slip
Resistance of a plastics film to sliding movement.
   f résistance au glissement
   e resistencia de deslizamiento
   i resistenza allo scivolamento
   d Gleitsicherheit

7453 slip additive
   f additif de lubrification
   e aditivo de lubrificación
   i agente di lubrificazione
   d Gleitmittel

7454 sludge
   f boue
   e fango
   i fango
   d Schlamm

7455 slurry
Mixture of finely-divided solids with water.
   f boue; bouillie
   e lechada; mezcla semifluida
   i fango; fanghiglia
   d Brei; Schlamm

7456 slurry
Wet mixed raw materials ready for burning Portland cement.
   f coulis; lait
   e mezcla semifluida de crudos
   i impasto di cemento Portland e acqua
   d Zementbrühe; dünner Mörtel

7457 slurry pump
   f pompe à boue
   e bomba de lechada
   i pompa per fanghiglia
   d Schlammpumpe

7458 slush casting
   f coulée de surface
   e fundición en hueco
   i fusione in conchiglia a rovesciamento
   d Gießen von Hohlkörpern

7459 smalt
   f smalt
   e esmaltín; esmalte
   i smaltino
   d Smalte; Kobaltblau

7460 smaltite
   f smaltine
   e esmaltita; esmaltina
   i smaltina
   d Glanzkobalt; Smaltin

7461 smelting furnace
   f four de fusion
   e cámara de fundición; horno de fundición
   i forno fusorio
   d Schmelzofen

7462 smithsonite
   f smithsonite
   e esmitsonita
   i smitsonite
   d Smithsonit

7463 smoke
   f fumée
   e humo
   i fumo
   d Rauch

7464 smokeless powder
   f poudre sans fumée
   e pólvora sin humo
   i polvere senza fumo
   d rauchloses Pulver

7465 smut
   f rouille
   e ustílago; tizón
   i carie
   d Faulbrand

7466 soaking period
   f trempage
   e tiempo necesario para la vitrificación
   i periodo di imbibizione
   d Ausgleichzeit

7467 soap
   f savon
   e jabón
   i sapone
   d Seife

7468 soap-consuming power
   f capacité d'absorption du savon
   e cantidad de jabón precipitado
   i potenza di consumo del sapone
   d Seifeaufnahmevermögen

* soapstone → 7786

* soda ash → 553

7469 **soda cellulose**
f cellulose à soude
e celulosa sódica
i cellulosa di sodio
d Natroncellulose

7470 **soda lime**
f chaux sodée
e cal sodada
i calce sodata
d Natronkalk

7471 **soda-lime glass**
f verre de chaux sodée
e vidrio de sosa y cal
i vetro di calce sodata
d Natronkalkglas; Solinglas

7472 **Soderberg electrode**
f électrode de Soderberg
e electrodo de Soderberg
i elettrodo di Soderberg
d Soderberg-Elektrode

7473 **sodium**
f sodium
e sodio
i sodio
d Natrium

7474 **sodium abietate**
f abiétate de soude
e abietato sódico
i abietato di sodio
d Natriumabietat

7475 **sodium acetate**
f acétate de sodium
e acetato sódico
i acetato di sodio
d Natriumacetat

7476 **sodium acetrizoate**
f acétrizoate de sodium
e acetrizoato sódico
i acetrizoato di sodio
d Natriumacetrizoat

7477 **sodium acetylarsanilate**
f acétylarsanilate de sodium
e acetilarsanilato sódico
i acetilarsanilato di sodio
d Natriumacetylarsanilat

7478 **sodium alginate**
f alginate de sodium
e alginato sódico
i alginato di sodio
d Natriumalginat

7479 **sodium aluminate**
f aluminate de sodium
e aluminato sódico
i alluminato di sodio
d Natriumaluminat

7480 **sodium aluminium hydride**
f hydrure de sodium et d'aluminium
e hidruro sódico-alumínico
i idruro di sodio e alluminio
d Natriumaluminiumhydrid

7481 **sodium aluminium silicofluoride**
f silicofluorure de sodium et d'aluminium
e silicofluoruro sódico-alumínico
i silicofluoruro di sodio e alluminio
d Natriumaluminiumsilicofluorid

7482 **sodium amalgam**
f amalgame de sodium
e amalgama de sodio
i amalgama di sodio
d Natriumamalgam

7483 **sodium amide**
f amide de sodium
e amida sódica
i ammide sodica
d Natriumamid

* **sodium aminoarsonate** → 7489

7484 **sodium para-aminobenzoate**
f para-aminobenzoate de sodium
e para-aminobenzoato sódico
i para-aminobenzoato di sodio
d Natrium-para-aminobenzoat

7485 **sodium para-aminohippurate**
f para-aminohippurate de sodium
e para-aminohipurato sódico
i para-aminoippurato di sodio
d Natrium-para-aminohippurat

7486 **sodium para-aminosalicylate**
f para-aminosalicylate de sodium
e para-aminosalicilato sódico
i para-aminosalicilato di sodio
d Natrium-para-aminosalicylat

7487 **sodium ammonium phosphate**
f phosphate de sodium et d'ammonium
e fosfato sódico-amónico

i fosfato di sodio e ammonio
d Natriumammoniumphosphat

7488 sodium antimonate
f antimoniate de sodium
e antimoniato sódico
i antimoniato sodico
d Natriumantimonat

7489 sodium arsanilate; sodium aminoarsonate
f arsanilate de sodium
e arsanilato sódico
i arsanilato sodico
d Natriumarsanilat

7490 sodium arsenate
f arséniate de sodium
e arseniato sódico
i arseniato sodico
d Natriumarsenat

7491 sodium arsenite
f arsénite de sodium
e arsenito sódico
i arsenito sodico
d Natriumarsenit

7492 sodium ascorbate
f ascorbate de sodium
e ascorbato sódico
i ascorbato sodico
d Natriumascorbat

7493 sodium azide
f azide de sodium; azothydrure de sodium
e azida sódica
i azide di sodio
d Natriumazid

7494 sodium benzoate
f benzoate de sodium
e benzoato sódico
i benzoato sodico
d Natriumbenzoat

7495 sodium bicarbonate; baking soda
f bicarbonate de sodium
e bicarbonato sódico
i bicarbonato di sodio
d Natriumbicarbonat

7496 sodium bifluoride
f bifluorure de sodium
e bifluoruro sódico
i bifluoruro di sodio
d Natriumbifluorid

7497 sodium biphosphate
f biphosphate de sodium
e bifosfato sódico
i bifosfato di sodio
d Natriumbiphosphat

7498 sodium bisulphate
f bisulfate de sodium
e bisulfato sódico
i bisolfato di sodio
d Natriumbisulfat

7499 sodium bisulphite
f bisulfite de sodium
e bisulfito sódico
i bisolfito di sodio
d Natriumbisulfit

7500 sodium bitartrate
f bitartrate de sodium
e bitartrato sódico
i bitartrato di sodio
d Natriumbitartrat

7501 sodium borate
f borate de sodium
e borato sódico
i borato di sodio
d Natriumborat

7502 sodium borate perhydrate
f borate-perhydrate de sodium
e borato sódico perhidratado
i borato sodico peridratato
d Natriumboratperhydrat

7503 sodium borohydride
f borohydrure de sodium
e borohidruro sódico
i boroidruro di sodio
d Natriumborohydrid

7504 sodium bromate
f bromate de sodium
e bromato sódico
i bromato di sodio
d Natriumbromat

7505 sodium bromide
f bromure de sodium
e bromuro sódico
i bromuro di sodio
d Natriumbromid

7506 sodium cacodylate
f cacodylate de sodium
e cacodilato sódico
i cacodilato di sodio
d Natriumcacodylat

7507 sodium caprylate; sodium octoate
  f caprylate de sodium
  e caprilato sódico
  i caprilato di sodio
  d Natriumcaprylat

7508 sodium carbonate
  f carbonate de sodium
  e carbonato sódico
  i carbonato di sodio
  d Natriumcarbonat; Soda; Natron

7509 sodium carbonate peroxide
  f peroxyde du carbonate de sodium
  e peróxido del carbonato sódico
  i perossido carbonato di sodio
  d Natriumcarbonatperoxyd

7510 sodium carboxymethylcellulose
  f carboxyméthylcellulose de sodium
  e carboximetilcelulosa sódica
  i carbossimetilcellulosa di sodio
  d Natriumcarboxymethylcellulose

7511 sodium chlorate
  f chlorate de sodium
  e clorato sódico
  i clorato di sodio
  d Natriumchlorat

7512 sodium chloride
  f chlorure de sodium
  e cloruro sódico
  i cloruro di sodio
  d Natriumchlorid; Salz

7513 sodium chlorite
  f chlorite de sodium
  e clorito sódico
  i clorito di sodio
  d Natriumchlorit

7514 sodium chloroacetate
  f chloroacétate de sodium
  e cloroacetato sódico
  i cloroacetato di sodio
  d Natriumchloracetat

7515 sodium chlorotoluene sulphonate
  f chlorotoluène-sulfonate de sodium
  e clorotoluensulfonato sódico
  i clorotoluenesolfonato di sodio
  d Natriumchlortoluolsulfonat

7516 sodium chromate
  f chromate de sodium
  e cromato sódico
  i cromato di sodio
  d Natriumchromat

7517 sodium chromate tetrahydrate
  f chromate de sodium tétrahydraté
  e cromato sódico tetrahidratado
  i tetraidrato cromato di sodio
  d Natriumchromattetrahydrat

7518 sodium citrate
  f citrate de sodium
  e citrato sódico
  i citrato di sodio
  d Natriumcitrat

7519 sodium cyanate
  f cyanate de sodium
  e cianato sódico
  i cianato di sodio
  d Natriumcyanat

7520 sodium cyanide
  f cyanure de sodium
  e cianuro sódico
  i cianuro di sodio
  d Natriumcyanid

7521 sodium cyclamate
  f cyclamate de sodium
  e ciclamato sódico
  i ciclamato di sodio
  d Natriumcyclamat

7522 sodium dehydroacetate
  f déhydroacétate de sodium
  e dehidroacetato sódico
  i deidroacetato di sodio
  d Natriumdehydroacetat

7523 sodium diacetate
  f diacétate de sodium
  e diacetato sódico
  i diacetato di sodio
  d Natriumdiacetat

7524 sodium diatrizoate
  f diatrizoate de sodium
  e diatrizoato sódico
  i diatrizoato di sodio
  d Natriumdiatrizoat

7525 sodium dichloroisocyanurate
  f dichloroisocyanurate de sodium
  e dicloroisocianurato sódico
  i dicloroisocianurato di sodio
  d Natriumdichlorisocyanurat

7526 sodium dichlorophenoxyacetate
  f dichlorophénoxyacétate de sodium
  e diclorofenoxiacetato sódico
  i diclorofenossiacetato di sodio
  d Natriumdichlorphenoxyacetat

7527 sodium dichromate
 f bichromate de sodium
 e dicromato sódico
 i bicromato di sodio
 d Natriumdichromat

7528 sodium dimethyldithiocarbamate
 f diméthyldithiocarbamate de sodium
 e dimetilditiocarbamato sódico
 i dimetilditiocarbammato di sodio
 d Natriumdimethyldithiocarbamat

7529 sodium dinitro-orthocresylate
 f dinitroorthocrésylate de sodium
 e dinitroortocresilato sódico
 i dinitroortocresilato di sodio
 d Natriumdinitro-ortho-cresylat

7530 sodium diuranate
 f diuranate de sodium
 e diuranato sódico
 i diuranato di sodio
 d Natriumdiuranat

7531 sodium dodecylbenzene sulphonate
 f dodécylbenzène-sulfonate de sodium
 e dodecilbencenosulfonato sódico
 i solfonato dodecilbenzenico di sodio
 d Natriumdodecylbenzolsulfonat

7532 sodium ethylate
 f éthylate de sodium
 e etilato sódico
 i etilato di sodio
 d Natriumäthylat

7533 sodium ferricyanide
 f ferricyanure de sodium
 e ferricianuro sódico
 i ferricianuro di sodio
 d Natriumferricyanid

7534 sodium ferrocyanide
 f ferrocyanure de sodium
 e ferrocianuro sódico
 i ferrocianuro di sodio
 d Natriumferrocyanid

7535 sodium fluoride
 f fluorure de sodium
 e fluoruro sódico
 i fluoruro di sodio
 d Natriumfluorid

7536 sodium fluoroacetate
 f fluoroacétate de sodium
 e fluoacetato sódico
 i fluoroacetato di sodio
 d Natriumfluoracetat

**7537 sodium fluorosilicate; sodium
    silicofluoride**
 f fluosilicate de sodium; silicofluorure de
    sodium
 e fluosilicato sódico
 i fluosilicato di sodio; silicofluoruro di
    sodio
 d Natriumkieselfluorid;
    Natriumsilikofluorid; Kieselfluornatrium

7538 sodium formaldehyde sulphoxylate
 f formaldéhyde-sulfoxylate de sodium
 e formaldehído-sulfoxilato sódico
 i idrosolfito di formaldeide
 d Natriumformaldehydsulfoxylat

7539 sodium formate
 f formiate de sodium
 e formiato sódico
 i formiato di sodio
 d Natriumformiat

7540 sodium gentisate
 f gentisate de sodium
 e gentisato sódico
 i gentisato di sodio
 d Natriumgentisat

7541 sodium glucoheptinate
 f glucoheptinate de sodium
 e glucoheptinato sódico
 i glucoeptinato di sodio
 d Natriumglucoheptinat

7542 sodium gluconate
 f gluconate de sodium
 e gluconato sódico
 i gluconato di sodio
 d Natriumgluconat

7543 sodium glutamate
 f glutamate de sodium
 e glutamato sódico
 i glutamato di sodio
 d Natriumglutamat

7544 sodium glycerophosphate
 f glycérophosphate de sodium
 e glicerofosfato sódico
 i glicerofosfato di sodio
 d Natriumglycerophosphat

**7545 sodium gynocardate; sodium
    hydnocarpate**
 f gynocardate de sodium
 e ginocardato sódico
 i ginocardato di sodio
 d Natriumgynocardat

7546 sodium hexylene glycol monoborate
  f hexylèneglycolmonoborate de sodium
  e hexilenglicolmonoborato sódico
  i monoborato glicolesilenico di sodio
  d Natriumhexylenglykolmonoborat

  * **sodium hydnocarpate** → 7545

7547 sodium hydride
  f hydrure de sodium
  e hidruro sódico
  i idruro di sodio
  d Natriumwasserstoff

7548 sodium hydrosulphide
  f hydrosulfure de sodium
  e hidrosulfuro sódico
  i idrosolfuro di sodio
  d Natriumhydrosulfid

7549 sodium hydrosulphite
  f hydrosulfite de sodium
  e hidrosulfito sódico
  i idrosolfito di sodio
  d Natriumhydrosulfit

  * **sodium hydroxide** → 1535

7550 sodium hypochlorite
  f hypochlorite de sodium
  e hipoclorito sódico
  i ipoclorito di sodio
  d Natriumhypochlorit

7551 sodium hypophosphite
  f hypophosphite de sodium
  e hipofosfito sódico
  i ipofosfito di sodio
  d Natriumhypophosphit

7552 sodium iodate
  f iodate de sodium
  e yodato sódico
  i iodato di sodio
  d Natriumjodat

7553 sodium iodide
  f iodure de sodium
  e yoduro sódico
  i ioduro di sodio
  d Natriumjodid

7554 sodium iron pyrophosphate
  f pyrophosphate sodique de fer
  e pirofosfato sódico de hierro
  i pirofosfato sodico di ferro
  d Natriumeisenpyrophosphat

7555 sodium isopropyl xanthate
  f xanthate isopropylique de sodium
  e isopropilxantato sódico
  i isopropilxantato di sodio
  d Natriumisopropylxanthat

7556 sodium lactate
  f lactate de sodium
  e lactato sódico
  i lattato di sodio
  d Natriumlactat

7557 sodium lauryl sulphate
  f laurylsulfate de sodium
  e laurilsulfato sódico
  i solfato di laurile e sodio
  d Natriumlaurylsulfat

7558 sodium lignosulphate
  f lignosulfate de sodium
  e lignosulfato sódico
  i lignosolfato di sodio
  d Natriumlignosulfat

7559 sodium metaborate
  f métaborate de sodium
  e metaborato sódico
  i metaborato di sodio
  d Natriummetaborat

7560 sodium metanilate
  f métanilate de sodium
  e metanilato sódico
  i metanilato di sodio
  d Natriummetanilat

7561 sodium metaphosphate
  f métaphosphate de sodium
  e metafosfato sódico
  i metafosfato di sodio
  d Natriummetaphosphat

7562 sodium metasilicate
  f métasilicate de sodium
  e metasilicato sódico
  i metasilicato di sodio
  d Natriummetasilikat

7563 sodium metavanadate
  f métavanadate de sodium
  e metavanadato sódico
  i metavanadato di sodio
  d Natriummetavanadat

7564 sodium methylate
  f méthylate de sodium
  e metilato sódico
  i metilato di sodio
  d Natriummethylat

7565 sodium methyl oleoyl taurate
 f méthyloléoyltaurate de sodium
 e metiloleoiltaurato sódico
 i metiloleoiltaurato di sodio
 d Natriummethyloleoyltaurat

7566 sodium molybdate
 f molybdate de sodium
 e molibdato sódico
 i molibdato di sodio
 d Natriummolybdat

7567 sodium monoxide
 f monoxyde de sodium
 e monóxido sódico
 i monossido di sodio
 d Natriummonoxyd

7568 sodium naphthalenesulphonate
 f naphtalène-sulfonate de sodium
 e naftalenosulfonato sódico
 i naftalenesolfonato di sodio
 d Natriumnaphthalinsulfonat

7569 sodium naphthenate
 f naphténate de sodium
 e naftenato sódico
 i naftenato di sodio
 d Natriumnaphthenat

7570 sodium naphthionate
 f naphtionate de sodium
 e naftionato sódico
 i naftionato di sodio
 d Natriumnaphthionat

7571 sodium nitrate; Chile salpetre
 f nitrate de sodium
 e nitrato sódico
 i nitrato di sodio
 d Natriumnitrat; Chilesalpeter

7572 sodium nitrite
 f nitrite de sodium
 e nitrito sódico
 i nitrito di sodio
 d Natriumnitrit

7573 sodium nitroferricyanide
 f nitroferricyanure de sodium
 e nitroferricianuro sódico
 i nitroferricianuro di sodio
 d Natriumnitroferricyanid

7574 sodium nitroprussiate
 f nitroprussiate de sodium
 e nitroprusiato sódico
 i nitroprussiato di sodio
 d Natriumnitroprussiat

* **sodium octoate** → 7507

7575 sodium oleate
 f oléate de sodium
 e oleato sódico
 i oleato di sodio
 d Natriumoleat

7576 sodium oxalate
 f oxalate de sodium
 e oxalato sódico
 i ossalato di sodio
 d Natriumoxalat

7577 sodium pentaborate
 f pentaborate de sodium
 e pentaborato sódico
 i pentaborato di sodio
 d Natriumpentaborat

7578 sodium pentachlorophenate
 f pentachlorophénate de sodium
 e pentaclorofenato sódico
 i pentaclorofenato di sodio
 d Natriumpentachlorphenat

7579 sodium perborate
 f perborate de sodium
 e perborato sódico
 i perborato di sodio
 d Natriumperborat

7580 sodium perchlorate
 f perchlorate de sodium
 e perclorato sódico
 i perclorato di sodio
 d Natriumperchlorat

7581 sodium periodate
 f periodate de sodium
 e peryodato sódico
 i periodato di sodio
 d Natriumperjodat

7582 sodium permanganate
 f permanganate de sodium
 e permanganato sódico
 i permanganato di sodio
 d Natriumpermanganat; Z-Stoff

7583 sodium peroxide
 f peroxyde de sodium
 e peróxido sódico
 i perossido di sodio
 d Natriumperoxyd

7584 sodium persulphate
 f persulfate de sodium
 e persulfato sódico

i persolfato di sodio
d Natriumpersulfat

7585 sodium phenate
f phénate de sodium
e fenato sódico
i fenato di sodio
d Natriumphenat

7586 sodium ortho-phenylphenate
f orthophénylphénate de sodium
e ortofenilfenato sódico
i ortofenilfenato di sodio
d Natrium-ortho-phenylphenat

7587 sodium phenylphosphinate
f phénylphosphinate de sodium
e fenilfosfinato sódico
i fenilfosfinato di sodio
d Natriumphenylphosphinat

* **sodium phosphate, dibasic ~**
   → 2501

* **sodium phosphate, monobasic ~**
   → 5569

* **sodium phosphate, tribasic ~**
   → 8397

7588 sodium phosphomolybdate
f phosphomolybdate de sodium
e fosfomolibdato sódico
i fosfomolibdato di sodio
d Natriumphosphomolybdat

7589 sodium phosphotungstate
f phosphotungstate de sodium
e fosfotungstato sódico
i fosfotungstato di sodio
d Natriumphosphowolframat

7590 sodium picramate
f picramate de sodium
e picramato sódico
i picramato di sodio
d Natriumpicramat

7591 sodium polysulphide
f polysulfure de sodium
e polisulfuro sódico
i polisolfuro di sodio
d Natriumpolysulfid

7592 sodium potassium carbonate
f carbonate de sodium et de potassium
e carbonato sódico-potásico
i carbonato di sodio e potassio
d Natriumkaliumcarbonat

7593 sodium propionate
f propionate de sodium
e propionato sódico
i propionato di sodio
d Natriumpropionat

7594 sodium pyrophosphate
f pyrophosphate de sodium
e pirofosfato sódico
i pirofosfato di sodio
d Natriumpyrophosphat

7595 sodium ricinoleate
f ricinoléate de sodium
e ricinoleato sódico
i ricinoleato di sodio
d Natriumricinoleat

7596 sodium salicylate
f salicylate de sodium
e salicilato sódico
i salicilato di sodio
d Natriumsalicylat

7597 sodium sarcosinate
f sarcosinate de sodium
e sarcosinato sódico
i sarcosinato di sodio
d Natriumsarcosinat

7598 sodium selenate
f séléniate de sodium
e seleniato sódico
i seleniato di sodio
d Natriumselenat

7599 sodium selenite
f sélénite de sodium
e selenito sódico
i selenite di sodio
d Natriumselenit

7600 sodium sesquicarbonate
f sesquicarbonate de sodium
e sesquicarbonato sódico
i sesquicarbonato di sodio
d Natriumsesquicarbonat

7601 sodium sesquisilicate
f sesquisilicate de sodium
e sesquisilicato sódico
i sesquisilicato di sodio
d Natriumsesquisilikat

7602 sodium silicate; water glass
f silicate de sodium; verre soluble
e silicato sódico; vidrio soluble
i silicato di sodio

d Natronwasserglas; Natriumsilikat;
   Wasserglas

* **sodium silicofluoride** → 7537

7603 sodium sodioacetate
  f sodioacétate de sodium
  e sodioacetato sódico
  i sodioacetato di sodio
  d Natriumsodioacetat

7604 sodium stannate
  f stannate de sodium
  e estannato sódico
  i stannato di sodio
  d Natriumstannat

7605 sodium stearate
  f stéarate de sodium
  e estearato sódico
  i stearato sodico
  d Natriumstearat

7606 sodium succinate
  f succinate de sodium
  e succinato sódico
  i succinato sodico
  d Natriumsuccinat

7607 sodium sulphanilate
  f sulfanilate de sodium
  e sulfanilato sódico
  i solfanilato sodico
  d Natriumsulfanilat

7608 sodium sulphate
  f sulfate de sodium
  e sulfato sódico
  i solfato di sodio
  d Natriumsulfat

7609 sodium sulphide
  f sulfure de sodium
  e sulfuro sódico
  i solfuro di sodio
  d Natriumsulfid; Schwefelnatrium;
     Schwefelnatron

7610 sodium sulphite
  f sulfite de sodium
  e sulfito sódico
  i solfito di sodio
  d Natriumsulfit

7611 sodium sulphoricinoleate
  f sulforicinoléate de sodium
  e sulforricinoleato sódico
  i solforicinoleato di sodio
  d Natriumsulforicinoleat

7612 sodium suramin
  f suramine de sodium
  e suramina sódica
  i suramina di sodio
  d Natriumsuramin

7613 sodium tartrate
  f tartrate de sodium
  e tartrato sódico
  i tartrato sodico
  d Natriumtartrat; Natronweinstein

7614 sodium tetrachlorophenate
  f tétrachlorophénate de sodium
  e tetraclorofenato sódico
  i tetraclorofenato di sodio
  d Natriumtetrachlorphenat

7615 sodium tetradecyl sulphate
  f tétradécylsulfate de sodium
  e tetradecilsulfato sódico
  i solfato tetradecilsodico
  d Natriumtetradecylsulfat

7616 sodium tetrasulphide
  f tétrasulfure de sodium
  e tetrasulfuro sódico
  i tetrasolfuro di sodio
  d Natriumtetrasulfid

7617 sodium thiocyanate
  f thiocyanate de sodium
  e tiocianato sódico
  i tiocianato di sodio
  d Natriumthiocyanat

7618 sodium thioglycolate
  f thioglycolate de sodium
  e tioglicolato sódico
  i tioglicolato sodico
  d Natriumthioglykolat

7619 sodium thiosulphate
  f thiosulfate de sodium
  e tiosulfato sódico
  i tiosolfato di sodio
  d Natriumthiosulfat; Fixiernatron

7620 sodium toluenesulphonate
  f toluène-sulfonate de sodium
  e toluensulfonato sódico
  i toluenesolfonato di sodio
  d Natriumtoluolsulfonat

7621 sodium trichloroacetate
  f trichloroacétate de sodium
  e tricloroacetato sódico
  i tricloroacetato di sodio
  d Natriumtrichloracetat

7622 sodium tripolyphosphate
f tripolyphosphate de sodium
e tripolifosfato sódico
i tripolifosfato di sodio
d Natriumtripolyphosphat

7623 sodium tungstate
f tungstate de sodium
e tungstato sódico
i tungstato di sodio
d Natriumwolframat

7624 sodium undecylenate
f undécylénate de sodium
e undecilenato sódico
i undecilenato di sodio
d Natriumundecylenat

7625 sodium uranate
f uranate de sodium
e uranato de sodio
i uranato di sodio
d Natriumuranat

7626 sodium valerate
f valérianate de sodium
e valerianato sódico
i valerato di sodio
d Natriumvalerianat

7627 sodium xylene sulphonate
f xylène-sulfonate de sodium
e xilensulfonato sódico
i xilenesolfonato di sodio
d Natriumxylolsulfonat

7628 sodium zirconium glycolate
f glycolate de sodium et de zirconium
e glicolato de sodio y zirconio
i glicolato di sodio e zirconio
d Natriumzirkoniumglykolat

7629 sodium zirconium lactate
f lactate de sodium et de zirconium
e lactato de sodio y zirconio
i lattato di sódio e zirconio
d Natriumzirkoniumlactat

7630 "softening"
f adoucissement
e ablandamiento; depuración del plomo
i purificazione del piombo in fornace a riverbero
d Bleiraffination

7631 softening agents
f plastifiants
e agentes ablandantes; agentes de reblandecimiento

i agenti emollienti
d Weichmacher

7632 softening point
f point de ramollissement
e temperatura de reblandecimiento
i punto di rammollimento
d Erweichungspunkt

7633 softening range
f plage de ramollissement
e serie de temperaturas de reblandecimiento
i intervallo di rammollimento
d Erweichungsbereich

7634 soft flow; easy flow; free flowing
f haute fluidité
e alta fluidez
i alta fluidità
d gute Fließfähigkeit

7635 softness index
f indice de fluidité
e índice de plasticidad
i indice di consistenza
d Weichheitszahl

7636 soft soaps
f savons mous
e jabones blandos
i saponi molli
d Schmierseifen

7637 sol
f sol
e sol
i sol
d Sol; kolloide Lösung

7638 solar salt
f sel marin
e sal marina
i sale marino
d Sonnensalz

7639 solation
f gélatinisation
e solación; gelatinización
i gelatinizzazione
d Gelatineprozeß

7640 solid solution
f solution solide
e solución cristalizada
i soluzione solida
d feste Lösung

7641 solid state
f état solide
e estado sólido
i stato solido
d fester Aggregatzustand

7642 solubility curve
f courbe de solubilité
e curva de solubilidad
i curva di solubilità
d Löslichkeitskurve

7643 soluble colourant
f colorant soluble
e colorante soluble
i colorante solubile
d löslicher Farbstoff

7644 soluble nylon resin
f nylon sous forme de résin soluble
e resina soluble de nylon
i resina di nylon solubile
d lösliches Nylonharz

7645 solute
f substance dissouse
e soluto
i soluto
d aufgelöster Stoff

7646 solution
f solution
e solución
i soluzione
d Lösung

7647 solution adhesive
f solution adhésive
e solución adhesiva
i soluzione adesiva
d flüssiger Klebstoff

7648 solutrope
f solutrope
e solutropo
i miscela solutropica
d solutropische Mischung

7649 solvate
f solvate
e solvato
i solvato
d Solvat

7650 solvation
f solvation
e solvación; solvatación
i solvatazione
d Solvatisierung

7651 Solvay process; ammonia-soda process
f procédé Solvay
e proceso Solvay; proceso de la soda al amoníaco
i processo Solvay
d Solvay-Verfahren

7652 solvent
f solvant; dissolvant
e solvente
i solvente
d Lösungsmittel

7653 solvent blushing
f turbidité
e aspecto lechoso
i torbidità
d Trübung

7654 solvent cleaning
f dégraissage au solvent
e desengrase por disolvente
i sgrassaggio con solventi
d Reinigung mit organischen Lösungsmitteln

7655 solvent emulsion degreasing
f dégraissage par émulsion
e desengrase por emulsión disolvente
i sgrassatura con emulsione
d Emulsionsentfettung

7656 solvent extraction
f extraction par dissolvant
e extracción por solvente
i estrazione per solvente
d Flüssig-Flüssig-Extraktion; Flüssigkeitsextraktion

7657 solvent naphtha
f solvant naphta
e nafta disolvente
i idrocarboni del benzene
d Solventnaphtha

7658 solvent recovery
f récupération du solvant
e recobrado de disolvente
i ricupero solvente
d Lösemittelwiedergewinnung

7659 solvent resistance
f résistance à un solvant
e resistencia a un solvente
i resistenza ai solventi
d Lösemittelbeständigkeit

7660 solvent tolerance
f tolérance aux solvants

e tolerancia de solución
i tolleranza di soluzione
d Lösemittelaufnahmefähigkeit

7661 solvent-type adhesive
f colle au solvant
e cola en solución
i collante a solvente
d flüssiger Klebstoff

7662 somatic
f somatique
e somático
i somatico
d somatisch

7663 somatic mutation
f mutation somatique
e mutación somática
i mutazione somatica
d somatische Mutation

7664 somatic reduction
f réduction somatique
e reducción somática
i riduzione somatica
d somatische Reduktion

7665 somatoplastic
f somatoplastique
e somatoplástico
i somatoplastico
d somatoplastisch

7666 soot
f suie
e hollín
i fuliggine
d Ruß

7667 sorbic acid
f acide sorbique
e ácido sórbico
i acido sorbico
d Sorbinsäure

7668 sorbitol; hexahydric alcohol
f sorbitol
e sorbitol; sorbita
i sorbitolo
d Sorbit; Zuckeralkohol

7669 sorbose
f sorbose
e sorbosa
i sorbosio
d sorbose

7670 Sorel's cement
f ciment de magnésie
e cemento Sorel
i cemento Sorel
d Sorelzement

7671 sorption
f sorption
e sorción
i adsorbimento; assorbimento
d Sorption

7672 soured wort
f moût acide
e mosto acidulado
i mosto acidificato
d saure Würze

7673 "sour gas"
f gaz naturel acide
e gas natural ácido
i gas naturale acido
d saures Gas; schwefelwasserstoffhaltiges Gas

7674 sour gasoline *(US)*
f essence acide
e gasolina mezclada con mercaptanos
i benzina acida
d Benzin mit Mercaptanzusatz

7675 soya bean oil
f huile de soya
e aceite de semilla de soja
i olio di soia
d Sojabohnenöl

7676 space velocity
f vitesse spatiale
e velocidad espacial
i velocità spazio
d Raumgeschwindigkeit

7677 Spanish red oxide
f oxyde rouge d'Espagne
e óxido rojo español
i ossido rosso di Spagna
d Spanischrot

7678 spar
f spath
e espato
i spato
d Spat

7679 sparge pipe; sparger
f tuyau perforé
e tubo rociador

    i tubo di dispersione
    d Zerstäuber

7680 sparge water
    f eau d'aspersion
    e agua para el rociado
    i acqua di dispersione
    d Zerstäubungswasser

7681 sparteine sulphate
    f sulfate de spartéine
    e sulfato de esparteína
    i solfato di sparteina
    d Sparteinsulfat

7682 spar varnish
    f vernis imperméable
    e barniz para usos marinos
    i vernice impermeabile
    d Spatlack

    * **spathic iron ore** → **7369**

7683 spatula
    f spatule
    e espátula
    i spatola
    d Spatel

7684 spearmint oil
    f essence
    e esencia de menta romana
    i essenza di menta
    d Spearmintöl

7685 speciation
    f spéciation
    e especiación
    i speciazione
    d Speciation

7686 specific capacity
    f capacité spécifique
    e capacidad específica
    i capacità specifica
    d spezifische Kapazität

7687 specific gravity
    f poids spécifique
    e peso específico
    i peso specifico
    d spezifisches Gewicht

7688 specific gravity flask
    f flacon à densité
    e vaso de medir las densidades
    i picnometro
    d Dichtigkeitsmesser

7689 specific heat
    f chaleur spécifique
    e calor específico
    i calore specifico
    d spezifische Wärme

7690 specific viscosity
    f viscosité spécifique
    e viscosidad específica
    i viscosità specifica
    d spezifische Viskosität

7691 specific volume
    f volume spécifique
    e volumen específico
    i volume specifico
    d Eigenvolumen

7692 spectroscope
    f spectroscope
    e espectroscopio
    i spettroscopio
    d Spektroskop

7693 spectrum
    f spectre
    e espectro
    i spettro
    d Spektrum

7694 speed of water absorption
    f vitesse d'absorption d'eau
    e velocidad de absorción de agua
    i velocità di assorbimento dell'acqua
    d Wasseraufnahmegeschwindigkeit

7695 spelter
    f zinc
    e cinc
    i zinco
    d Zink

7696 spent oxide
    f oxyde de fer usagé
    e óxido de hierro usado
    i ossido di ferro sfruttato
    d verbrauchtes Eisenoxyd

7697 spent wash
    f lessive résiduaire
    e vinazas
    i residuo di distilleria
    d Ablauge

7698 spermaceti
    f blanc de baleine; spermaceti
    e espermaceti; esperma de ballena
    i spermaceti
    d Spermazet; Walrat

7699 spermatid
f spermatide
e espermatida
i spermatidio
d Spermatide

7700 spermatocyte
f spermatocyte
e espermatócito
i spermatocita
d Spermatozyt

7701 spermatozoid
f spermatozoïde
e espermatozoide
i spermatozoide
d Spermatozoid

7702 spermin(e)
f spermine
e espermina
i spermina
d Spermin

7703 sperm oil
f huile de baleine; huile de spermaceti
e aceite de esperma
i olio di spermaceti
d Spermwalöl; Spermazetöl

7704 Sperry process
f procédé Sperry
e método Sperry
i processo Sperry
d Sperry-Verfahren

* **sphalerite** → 1041

7705 spherome
f sphérome
e esferoma
i sferoma
d Sphärom

7706 spheroplast
f sphéroplaste
e esferoplasto
i sferoplasto
d Sphäroplast

7707 spherosome
f sphérosome
e esferosoma
i sferosoma
d Sphärosom

7708 spicular
f aciculaire
e acicular

i aciculare
d nadelförmig

* **spider cell** → 699

7709 spiegeleisen
f fonte spéculaire; fonte miroitante
e hierro especular
i ghisa speculare
d Spiegeleisen; Hartfloß

7710 spike oil; lavender spike oil
f huile d'aspic
e aceite de espliego
i olio di spigo di lavanda
d Spiköl

7711 spinel
f candite; spinelle
e espinela
i spinello
d Spinell

7712 spin v off *(sugar)*
f centrifuger
e centrifugar
i centrifugare
d zentrifugieren

7713 spirem
f spirème
e espirema
i spirema
d Spirem

7714 spirit
f esprit
e espíritu
i spirito
d Spiritus

7715 spirits of hartshorn
f carbonate d'ammoniaque
e espíritu de cuerno de ciervo
i ammoniaca in soluzione acquosa
d Hirschhorngeist

7716 spirits of salt
f esprit de sel
e espíritu de sal
i spirito di sale
d Salzsäure

* **spirits of wine** → 3138

7717 splash arms
f agitateurs
e brazos de chapoteo

i sbattitori
d Spritzarme

7718 splenocyte
f splénocyte
e esplenócito
i splenocita
d Milzzelle

* **spodium** → 556

7719 spodumene
f spodumène; triphane
e espodúmeno
i silicato di litio e alluminio
d Spodumen; Triphan

7720 sponge
f éponge
e esponja
i deposizione spugnosa
d Schwamm

7721 sponging agent
f agent gonflant
e agente esponjante
i agente soffiante
d Treibmittel

7722 spongy platinum
f mousse de platine
e negro de platino
i spugna di platino
d Platinschwamm

7723 spontaneous combustion
f combustion spontanée
e combustión espontánea
i combustione spontanea
d Selbstverbrennung

7724 spore
f spore
e espora
i spora
d Spore; Keimzelle

7725 sporocyte
f sporocyte
e esporócito
i sporocito
d Sporocyte

7726 sporophyte
f sporophyte
e esporófito
i sporofito
d Sporophyt

7727 spotting
f tachetures
e pecas
i macchiettature
d Fleckigwerden

7728 spray dryer
f sécheur pulvérisateur
e secador de pulverización
i essiccatoio da polverizzazione
d Zerstäubungstrockner

7729 spray tower
f tour d'arrosage
e torre de pulverización
i torre di spruzzatura
d Rieselturm

7730 spreading power
f pouvoir courant
e poder de cubrición
i distendibilità
d Ausgiebigkeit

7731 sprout *v*
f germer
e germinar
i germogliare
d keimen

7732 squalane
f squalène
e triterpeno con seis enlaces dobles
i squalano
d Squalen

7733 stabilizer
f stabilisant
e estabilizador
i stabilizzatore
d Stabilisator

7734 stage compressor
f compresseur étagé
e compresor por etapas
i compressore a stadi
d Stufenkompressor

7735 stage-wise contactor
f appareil de contact par étage
e aparato de contacto por etapa
i apparecchio di contatto per stadio
d Kontaktapparat je Stufe

7736 stagnation point
f point de stagnation
e punto neutro
i punto di stagnamento
d Staupunkt

7737 stain
f mordant
e tinte; solución colorante
i mordente
d Beize

7738 stain spots
f souillures
e mohos
i macchie da essudazione
d Ausblühungen

7739 standard deviation
f déviation normale
e desviación normal
i deviazione normale
d normale Abweichung

7740 standard electrode potential
f tension standard d'une électrode
e tensión normal de un electrodo
i potenziale normale di elettrodo
d Normalpotential einer Elektrode

7741 standard hydrogen electrode
f électrode standard à hydrogène
e electrodo de hidrógeno
i elettrodo a idrogeno
d Normalwasserstoffelektrode

7742 standard solution
f solution normale
e solución normal
i soluzione titolata
d Normallösung

7743 stand oil
f huile siccative
e aceite secante; aceite contraido
i standolio
d Standöl

7744 stannane
f hydrure d'étain
e hidruro de estaño
i idruro di stagno
d Zinnhydrid

7745 stannates
f stannates
e estannatos
i stannati
d Stannate; zinnsaure Salze

7746 stannic acids
f acides stanniques
e ácidos estánnicos
i acidi stannici
d Zinnsäuren

7747 stannic chloride
f chlorure stannique
e cloruro estánnico
i cloruro stannico
d Stannichlorid; Zinnchlorid

7748 stannic chromate
f chromate stannique
e cromato estánnico
i cromato stannico
d Stannichromat

7749 stannic oxide
f oxyde stannique
e óxido estánnico
i ossido stannico
d Zinnasche; Zinnkalk; Zinnoxyd

7750 stannite; stannine; bell metal ore; tin pyrites
f stannite; stannine; étain pyriteux
e estannita; pirita de estaño
i stannite
d Zinnkies

7751 stannous chloride; tin bichloride
f chlorure stanneux; bichlorure d'étain
e cloruro estannoso; dicloruro de estaño
i cloruro stannoso; bicloruro di stagno
d Stannochlorid; Zinnchlorür; Zinndichlorid

7752 stannous chromate
f chromate stanneux
e cromato estannoso
i cromato stannoso
d Stannochromat

7753 stannous ethylhexoate
f éthylhexoate stanneux
e etilhexoato estannoso
i etilesoato stannoso
d Stannoäthylhexoat

7754 stannous oleate
f oléate stanneux
e oleato estannoso
i oleato stannoso
d Stannooleat

7755 stannous oxalate
f oxalate stanneux
e oxalato estannoso
i ossalato stannoso
d Stannooxalat; Zinnoxalat

7756 stannous oxide
f oxyde stanneux
e óxido estannoso

i ossido stannoso
d Stannooxyd

7757 stannous sulphate
f sulfate stanneux
e sulfato estannoso
i solfato stannoso
d Stannosulfat

7758 stannous sulphide
f sulfure stanneux
e sulfuro estannoso
i solfuro stannoso
d Stannosulfid; Zinnsulfür

7759 stannous tartrate
f tartrate stanneux
e tartrato estannoso
i tartrato stannoso
d Stannotartrat

7760 starch
f amidon
e almidón
i amido
d Stärke

7761 starch content
f teneur en amidon
e contenido en almidón
i tenore di amido
d Stärkegehalt

7762 starter battery
f batterie de démarrage
e batería de arranque
i batteria di avviamento
d Anlaßbatterie

7763 starvation
f carence
e insuficiencia en ácido nucleico
i insufficienza in acido nucleico
d Starvation

7764 static charge
f charge électrostatique
e carga estática
i carica elettrostatica
d statische Aufladung

7765 static detector
f détecteur statique
e detector estático
i rilevatore di carica statica
d statischer Detektor

7766 static electrode potential
f tension statique d'une électrode

e tensión estática de un electrodo
i potenziale statico di elettrodo
d statisches Elektrodenpotential

7767 station
f position
e posición; emplazamiento
i posizione
d Position; Stellung

7768 statistical test
f test statistique
e prueba estadística
i prova statistica
d statistisches Prüfverfahren

7769 steam distillation
f distillation à la vapeur
e destilación por vapor
i distillazione a vapore
d Wasserdampfdestillation

7770 steam ejector
f éjecteur à vapeur
e aspirador accionado por vapor
i eiettore di vapore
d Dampfstrahlapparat

7771 steam funnel
f entonnoir à tuyau de vapeur
e embudo con tubo de vapor
i imbuto con tubo di vapore
d Dampftrichter

7772 steam injector
f injecteur à vapeur
e vapoinyector
i iniettore di vapore
d Dampfinjektor

7773 steam jacket
f chemise de vapeur
e camisa de vapor
i camicia di vapore
d Dampfmantel

7774 steam jacketed vulcanizing pan
f chaudière à vulcaniser chemisée de vapeur
e caldera de vulcanizar con camisa de vapor
i caldaia di vulcanizzazione con camicia di vapore
d Vulkanisierkessel mit Dampfmantel

7775 steam pan
f chaudière pour vulcanisation en vapeur
e caldera calentada por vapor

i vulcanizzatore a vapore
d Dampfheizkessel

* **steam pressure** → 8715

7776 steam seal
f joint de vapeur
e estanqueidad por vapor
i tenuta a vapore
d Dampfstopfbüchse

7777 steam sterilizer
f étuve
e esterilizador de vapor
i pentola di Koch
d Dampftropf

7778 steam trap
f séparateur d'eau
e separador de agua
i essiccatore di vapore
d Dampfentwässerer

7779 steam vulcanization
f vulcanisation à la vapeur
e vulcanización por vapor
i vulcanizzazione in vapore
d Dampfvulkanisation

7780 stearate
f stéarate
e estearato
i stearato
d Stearat

7781 stearic acid
f acide stéarique
e ácido esteárico
i acido stearico
d Stearinsäure; Talgsäure

7782 stearin
f stéarine
e estearina
i stearina
d Stearin

7783 stearyl alcohol
f alcool stéarylique
e alcohol estearílico
i alcool stearilico
d Stearylalkohol

7784 stearyl mercaptan
f stéarylmercaptan
e estearilmercaptano
i mercaptano di stearile
d Stearylmercaptan

7785 stearyl methacrylate
f méthacrylate de stéaryle
e metacrilato estearílico
i metacrilato di stearile
d Stearylmethacrylat

7786 steatite; soapstone
f stéatite
e esteatita
i steatite
d Steatit; Speckstein

7787 steep *v*
f tremper; macérer
e enriar; empapar; sumergir
i macerare
d eintauchen

7788 steeping
f trempage
e maceración; impregnación
i macero
d Eintauchen

7789 Steffens process
f procédé Steffens
e proceso Steffens
i processo Steffens
d Steffensches Verfahren

7790 step allelomorphism
f allélomorphisme graduel
e alelomorfismo gradual
i allelomorfismo graduale
d Treppenallelomorphismus

7791 stephanite; brittle silver ore
f stéphanite
e estefanita
i stefanite
d Stephanit; Röscherz; Schwarzsilberglanz

7792 stereoblock polymer
f polymère en blocs isotactiques
e polímero cuya estructura estereo-
específica está separada por secciones de
diferente estructura
i polimero in stereoblocchi
d Stereoblockpolymer

7793 stereochemistry
f stéréochimie
e estereoquímica
i stereochimica
d Stereochemie; Raumchemie

7794 stereoisomers
f stéréoïsomères
e estereoisómeros

i stereoisomeri
d Stereoisomere

7795 stereoregular polymers
f polymères stéréospécifiques
e polímeros estereorregulares
i polimeri stereospecifici
d regelmäßige Stereopolymere

7796 stereospecific catalysts
f catalyseurs stéréospécifiques
e catalizadores estereoespecíficos
i catalizzatori stereospecifici
d stereospezifische Katalysatoren

7797 stereospecific polymers
f polymères stéréospécifiques
e polímeros estereoespecíficos
i polimeri stereospecifici
d stereospezifische Polymere

7798 steric hindrance
f empêchement stérique
e impedimento estérico
i ostacolo sterico
d räumliche Behinderung

7799 sterile
f stérile
e estéril; esterilizado
i sterile
d steril; keimfrei

7800 sterile air
f air stérile
e aire esterilizado
i aria sterile
d sterile Luft

7801 steroids
f stéroïdes
e esteroides
i steroidi
d Stereoide

7802 sterols
f stérols
e esteroles
i steroli
d Sterole

7803 stibine
f stibine
e estibina
i stibina
d Stibin

* stibnite → 614

7804 stibophen
f stibophène
e estibofeno
i stibofene
d Stibophen

7805 stick lac
f laque en baton
e goma laca natural; laca en barras
i lacca in bastoni
d Rohlack

7806 sticky association
f association par agglutination
e asociación por aglutinación
i associazione per agglutinazione
d durch Verklebung entstandener
Chromosomenverband

7807 sticky effect
f effet d'agglutination
e efecto de aglutinación
i effetto d'agglutinazione
d Verklebungseffekt

7808 stiff flow
f grande dureté
e baja fluidez
i fluidità bassa
d schlechter Fluß

7809 stigmasterol
f stigmastérol; stigmastérine
e estigmasterol
i stigmasterolo
d Stigmasterol

7810 stilbamidine isethionate
f iséthionate de stilbamidine
e isetionato de estilbamidina
i isetionato di stilbamidina
d Stilbamidinisäthionat

7811 stilbene
f stilbène
e estilbeno
i stilbene
d Stilben

7812 stilbite
f stilbite
e estilbita
i stilbite
d Stilbit; Desmin

7813 still
f appareil de distillation
e aparato de destilación; alambique

i distilleria
d Destillierapparat

7814 still body; still pot
f corps d'alambic
e vientre del alambique
i ventre dell'alambicco
d Destillierkessel

7815 stillingia oil
f huile de stillingie
e aceite de estilingia
i olio di stilingia
d Stillingiaöl

7816 stillion
f bock de soutirage
e estante
i riempifusti
d Abfüllbock

* **still pot → 7814**

7817 stirring vane column
f colonne à ailettes mélangeuses
e solumna de aletas mezcladoras
i colonna ad alette mescolatrici
d Kolonne mit Mischflügeln

7818 stirrups *(of a balance)*
f étriers
e estribos
i staffe
d Bügel

7819 stock
f pâte
e pasta de papel húmeda
i mescola
d Zeug

7820 stoichiometry
f stoechiométrie
e estequiometría
i stoichiometria
d Elementenmessung

7821 stone mill
f concasseur
e molino de muelas; machacadora
i macina di pietra
d Schlagmühle; Brechmaschine

7822 stoneware
f grès
e gres
i gres
d Steingut

7823 stop-off lacquer
f vernis isolant
e laca aislante
i vernice isolante
d Isolierlack

7824 stopper
f mastice
e empaste; masilla
i stucco
d Kitt

7825 stopping off
f isolation
e aislamiento
i isolante
d Abdecken

7826 storage cell
f élément secondaire; accumulateur
e acumulador; elemento secundario
i accumulatore; elemento secondario
d Akkumulatorenbatterie; Sammlerbatterie

7827 stove-kiln
f four à briques
e horno de cochura
i forno
d Backsteinofen

7828 stoving finish
f émail à four
e esmaltado a la estufa
i smalto a fuoco
d Einbrennlack

7829 stramonium
f stramonine
e estramonio
i stramonio
d Stechapfel; Stramonium

7830 straw pulp
f pulpe de paille
e pulpa de paja
i polpa di paglia
d Strohzellstoff

7831 streamline flow
f flux à vitesse uniforme
e flujo de velocidad uniforme
i flusso di velocità uniforme
d gleichmäßige Strömung

* **stream, on ~ → 5971**

7832 strength
f puissance
e fuerza; potencia

    i potenza
    d Stärke

7833 strepsineme
    f strepsinème
    e estrepsinema
    i strepsinema
    d Strepsinema

7834 streptodornase
    f streptodornase
    e estreptodornasa
    i streptodornasi
    d Streptodornase

7835 streptoduocin
    f streptoduocine
    e estreptoduocina
    i streptoduocina
    d Streptoduocin

7836 streptokinase
    f streptokinase
    e estreptoquinasa
    i streptocinasi
    d Streptokinase

7837 streptolin
    f streptoline
    e estreptolina
    i streptolina
    d Streptolin

7838 streptomycin
    f streptomycine
    e estreptomicina
    i streptomicina
    d Streptomycin

    * **streptonivicin** → 5852

7839 streptothricin
    f streptothricine
    e estreptotricina
    i streptotricina
    d Streptothricin

7840 strike
    f dépôt amorce
    e depósito primario
    i deposito attivante
    d Untergalvanisierung; Vorgalvanisierung

7841 strike bath
    f bain d'amorçarge
    e baño primario
    i bagno di attivazione
    d Vorgalvanisierbad

7842 striking
    f amorçage
    e iniciación de baño primario
    i attivazione
    d Vorgalvanisieren

7843 strip
    f bain dépouillage
    e baño de eliminación
    i bagno di degalvanizzazione
    d Entplattierungsbad; Abbeizbad

7844 stripper
    f décapant
    e quitapinturas
    i sverniciatore
    d Abbeizmittel

7845 stripper tank
    f cuve de dépouillage
    e tanque de despojamiento
    i cella per strappamento
    d Mutterblechbad

7846 stripping compound
    f composé de dépouillage
    e compuesto de despojamiento
    i materiale da strappamento
    d Trennsubstanz

7847 stroboscope
    f stroboscope
    e estroboscopio
    i stroboscopio
    d Stroboskop

7848 strontianite
    f strontianite
    e estroncianita
    i stronzianite
    d Strontianit

7849 strontium
    f strontium
    e estroncio
    i stronzio
    d Strontium

7850 strontium bromide
    f bromure de strontium
    e bromuro de estroncio
    i bromuro di stronzio
    d Strontiumbromide

7851 strontium carbonate
    f carbonate de strontium
    e carbonato de estroncio
    i carbonato di stronzio
    d Strontiumcarbonat

7852 strontium chlorate
f chlorate de strontium
e clorato de estroncio
i clorato di stronzio
d Strontiumchlorat

7853 strontium chromate; strontium yellow
f chromate de strontium; jaune de strontium
e cromato de estroncio; amarillo de estroncio
i cromato di stronzio; giallo di stronzio
d Strontiumchromat; Strontiumgelb

7854 strontium hydroxide
f hydroxyde de strontium
e hidróxido de estroncio
i idrossido di stronzio
d Strontiumhydroxyd

7855 strontium nitrate
f nitrate de strontium
e nitrato de estroncio
i nitrato di stronzio
d Strontiumnitrat

7856 strontium oxide
f oxyde de strontium
e óxido de estroncio
i ossido di stronzio
d Strontiumoxyd

7857 strontium peroxide
f peroxyde de strontium
e peróxido de estroncio
i perossido di stronzio
d Strontiumperoxyd; Strontiumsuperoxyd

7858 strontium sulphate
f sulfate de strontium
e sulfato de estroncio
i solfato di stronzio
d Strontiumsulfat

7859 strontium sulphide
f sulfure de strontium
e sulfuro de estroncio
i solfuro di stronzio
d Strontiumsulfid

7860 strontium titanate
f titanate de strontium
e titanato de estroncio
i titanato di stronzio
d Strontiumtitanat

7861 strontium unit
f unité de strontium
e unidad de estroncio

i unità di stronzio
d Strontium-Einheit

7862 strontium white
f blanc de strontium
e blanco de estroncio
i bianco di stronzio
d Strontiumsulfat; Strontiumweiß

* **strontium yellow** → 7853

7863 strophanthin
f strophantine
e estrofantina
i strofantina
d Strophanthin

7864 structural formula
f formule rationnelle; formule de constitution
e fórmula estructural
i formula di struttura
d Wertigkeitsformel

7865 strychnine
f strychnine
e estricnina
i stricnina
d Strychnin

7866 studtite
f studtite
e estudtita
i studtita
d Studtit

7867 styphnic acid
f acide styphnique
e ácido estífnico
i acido stifnico
d Styphninsäure

7868 styrallyl acetate
f acétate de styrallyle
e acetato estiralílico
i acetato di stirallile
d Styrallylacetat

7869 styrallyl alcohol
f alcool styrallylique
e alcohol estiralílico
i alcool stirallilico
d Styrallylalkohol

7870 styrax
f styrax; aliboufier
e estoraque
i storace
d Storax; Styrax

7871 styrene; phenylethylene
 f styrolène
 e estireno
 i stirolo
 d Styrol

7872 styrene-butadiene rubber
 f caoutchouc de styrolène-butadiène
 e goma de estireno-butadieno
 i gomma al butadiene-stirolo
 d Styrolbutadienkautschuk

7873 styrene nitrosite
 f nitrosite de styrène
 e nitrosita de estireno
 i nitrostirolo
 d Styrolnitrosit

7874 styrene oxide
 f oxyde de styrolène
 e óxido de estireno
 i ossido di stirolo
 d Styroloxyd

7875 styrene resins
 f résines de styrolène
 e resinas de estireno
 i resine allo stirolo
 d Styrolharze

7876 styrolene bromide
 f bromostyrol
 e bromuro de estiroleno
 i bromostirolo
 d Bromstyrol

7877 subatomics
 f subatomique
 e ciencia subatómica
 i subatomica
 d Subatomik

7878 subchromatid
 f sous-chromatide
 e subcromátida
 i subcromatidio
 d Subchromatide

7879 subculture
 f sous-culture
 e subcultivo
 i subcoltura
 d Zweitkultur

7880 suberic acid
 f acide subérique
 e ácido subérico
 i acido suberico
 d Suberinsäure; Kork

7881 subgene
 f sous-gène
 e subgen
 i subgene
 d Subgen

7882 sublimate
 f sublimé
 e sublimado
 i sublimato
 d Sublimat

7883 sublimation
 f sublimation
 e sublimación
 i sublimazione
 d Sublimation

7884 sublimed white lead
 f plomb blanc sublimé
 e plomo blanco sublimado
 i bianco sublimato di piombo
 d Bleiweiß

7885 submarine blasting gelatine
 f gélatine explosive sous-marine
 e gelatina para explosiones submarinas
 i gelatina di nitroglicerina per uso
   subacqueo
 d Unterwassersprenggelatine

7886 submicron
 f submicron
 e submicrón
 i submicron
 d Submikron

7887 submicrosome
 f sous-microsome
 e submicrosoma
 i submicrosoma
 d Submikrosom

7888 submicrostructure
 f structure submicroscopique
 e submicroestructura
 i submicrostruttura
 d Submikrostruktur

7889 substitution
 f substitution
 e substitución
 i sostituzione
 d Substitution

7890 subtilin
 f subtiline
 e subtilina

i subtilina
d Subtilin

7891 succinic acid
f acide succinique
e ácido succínico
i acido succinico
d Bernsteinsäure

7892 succinic anhydride
f anhydride succinique
e anhidrido succínico
i anidride succinica
d Bernsteinsäureanhydrid

7893 succinimide
f succinimide
e succinimida
i succinimmide
d Succinimid

**7894 succinylcholine bromide;
suxamethonium bromide**
f bromure de succinylcholine
e bromuro de succinilcolina
i bromuro di succinilcolina
d Succinylcholinbromid

7895 succinylsulphathiazole
f succinylsulfathiazole
e succinilsulfatiazol
i succinilsolfatiazolo
d Succinylsulfathiazol

7896 suck *v* out
f aspirer
e aspirar
i estrarre per aspirazione
d absaugen

* **sucrol** → **2893**

* **sucrose** → **7168**

7897 sucrose monostearate
f monostéarate de saccharose
e monoesterato de sacarosa
i monostearato di saccarosio
d Sucrosemonostearat

7898 sucrose octo-acetate
f octa-acétate de saccharose
e octaacetato de sacarosa
i ottacetato di saccarosio
d Sucrose-octa-acetat

7899 suction boxes
f caisses aspirantes
e canales de desagüe

i cassette aspiranti
d Saugkästen; Sauger

7900 suction gas producer
f gazogène à aspiration
e gasógeno
i gassogeno ad aspirazione
d Unterdruckgaserzeuger

7901 suction-type hydrometer
f hydromètre à suction
e hidrómetro de succión
i idrometro ad aspirazione
d Saughydrometer

7902 suede finish
t tını daım
e acabado de aspecto aterciopelado
i fioccosità
d Feinkräuselanstrich

7903 sugar crusher
f broyeur de cannes de sucre
e machacadora de azúcar
i frantoio per canna da zucchero
d Zuckerrohrquetsche

* **sugar of lead** → **4758**

7904 Suida process
f procédé Suida
e proceso de Suida
i processo Suida
d Suida-Verfahren

7905 sulphacetamide
f sulfacétamide
e sulfacetamida
i solfacetammide
d Sulfacetamid

7906 sulphadiazine
f sulfadiazine
e sulfadiacina; sulfadiazina
i solfadiazina
d Sulfadiazin

7907 sulphadimethoxine
f sulfadiméthoxine
e sulfadimetoxina
i solfadimetossina
d Sulfadimethoxin

* **sulphadimidine** → **7910**

7908 sulphaguanidine
f sulfaguanidine
e sulfaguanidina

i solfaguanidina
d Sulfaguanidin

7909 sulphamerazine; sulphamethyldiazine
f sulfamérazine
e sulfameracina; sulfamerazina
i solfamerazina
d Sulfamerazin

7910 sulphamethazine; sulphadimidine
f sulfaméthazine
e sulfametacina
i solfametazina
d Sulfamethazin

7911 sulphamethiazole
f sulfaméthiazole
e sulfametiazol
i solfametiazolo
d Sulfamethiazol

7912 sulphamethoxypyridazine
f sulfaméthoxypyridazine
e sulfametoxipiridacina
i solfametossipiridazina
d Sulfamethoxypyridazin

* sulphamethyldiazine → 7909

7913 sulphamic acid
f acide sulfamique
e ácido sulfámico
i acido solfamico
d Sulfaminsäure

7914 sulphanilamide
f sulfanilamide
e sulfanilamida
i solfanilammide
d Sulfanilamid

7915 sulphanilic acid
f acide sulfanilique
e ácido sulfanílico
i acido solfanilico
d Sulfanilsäure

7916 sulphapyridine
f sulfapyridine
e sulfapiridina
i solfapiridina
d Sulfapyridin

7917 sulphaquinoxaline
f sulfaquinoxaline
e sulfaquinoxalina
i solfachinossalina
d Sulfachinoxalin

7918 sulpharsphenamine
f sulfarsphénamine
e sulfarsfenamina
i solfarsfenammina
d Sulfarsphenamin

7919 sulphate process
f procédé au sulfate
e método del sulfato
i processo al solfato
d Sulfatverfahren

7920 sulphates
f sulfates
e sulfatos
i solfati
d Sulfate; Schwefelsalze

7921 sulphathiazole
f sulfathiazole
e sulfatiazol
i solfatiazolo
d Sulfathiazol

7922 sulphating
f sulfatage
e sulfatación
i solfatazione
d Sulfatation

7923 sulphides
f sulfures
e sulfuros
i solfuri
d Sulfide

**7924 sulphinpyrazone; sulphoxyphenyl-
pyrazolidone**
f sulfinylpyrazone
e sulfinpirazona
i solfinpirazone
d Sulfinpyrazon

7925 sulphite liquor
f lessive sulfitée
e licor de sulfitos
i brodo al solfito
d Sulfitlauge

7926 sulphite process
f procédé au sulfite
e método del sulfito
i processo al solfito
d Sulfitverfahren

7927 sulphobenzoic acid
f acide sulfobenzoïque
e ácido sulfobenzoico

i acido solfobenzoico
d Sulfobenzoesäure

7928 sulphobromophthalein sodium
f sulfobromophtaléine-sodium
e sulfobromoftaleína sódica
i solfobromoftaleina sodica
d Sulfobromphthaleinnatrium

7929 sulphocyanides; thiocyanates
f sulfocyanures
e sulfocianuros
i solfocianuri
d Thiocyanide

* **sulphonal** → **7934**

* **sulphonated castor oil** → **8581**

7930 sulphonated oil
f huile sulfonée
e aceite sulfonado
i olio solfonato
d geschwefeltes Öl

7931 sulphonation
f sulfonation
e sulfonación
i solfonazione
d Sulfierung

7932 sulphonator
f sulfonateur
e vasija de sulfonación
i solfonatore
d Sulfoniergefäß

7933 sulphonethylmethane; methylsulphonal
f sulfonéthylméthane
e sulfonetilmetano
i solfonetilmetano
d Sulfonäthylmethan

7934 sulphonmethane; sulphonal
f sulfonméthane
e sulfonmetano
i solfonmetano
d Sulfonal

7935 sulphonyldianiline
f sulfonyldianiline
e sulfonildianilina
i solfonildianilina
d Sulfonyldianilin

7936 sulphophthalic acid
f acide sulfophtalique
e ácido sulfoftálico

i acido solfoftalico
d Sulfophthalsäure

7937 sulphosalicylic acid
f acide sulfosalicylique
e ácido sulfosalicílico
i acido solfasalicilico
d Sulfosalicylsäure

* **sulphoxyphenylpyrazolidone** → **7924**

7938 sulphur
f soufre
e azufre
i zolfo
d Schwefel

7939 sulphurated lime
f chaux sulfurée
e cal sulfurada
i calce solforata
d sulfurierter Kalk

7940 sulphur blooming
f efflorescence du soufre
e eflorescencia de azufre
i efflorescenza di zolfo
d Ausschwefelung

7941 sulphur chloride
f chlorure de soufre
e cloruro de azufre
i cloruro di zolfo
d Schwefelchlorid; Chlorschwefel

7942 sulphur dichloride
f bichlorure de soufre
e dicloruro de azufre
i bicloruro di zolfo
d Schwefeldichlorid

7943 sulphur dioxide
f bioxyde de soufre
e dióxido de azufre
i biossido di zolfo
d Schwefeldioxyd

* **sulphuretted hydrogen** → **4203**

7944 sulphur hexafluoride
f hexafluorure de soufre
e hexafluoruro de azufre
i esafluoruro di zolfo
d Schwefelhexafluorid

7945 sulphuric acid; oil of vitriol
f acide sulfurique; huile de vitriol
e ácido sulfúrico; aceite de vitriolo ·

i acido solforico; olio di vetriolo
d Schwefelsäure

* **sulphuric acid, fuming ~ → 3624**

7946 sulphurous acid
f acide sulfureux
e ácido sulfuroso
i acido solforoso
d schweflige Säure

7947 sulphur trioxide
f trioxyde sulfurique; anhydride sulfurique
e trióxido de azufre
i triossido di zolfo; anidride solforica
d Schwefeltrioxyd

7948 sulphuryl chloride
f chlorure de sulphuryle
e cloruro de sulfurilo
i cloruro di solforile
d Sulfurylchlorid

7949 sumbul; musk root
f sumbul
e sumbul
i euriangio
d Sumbul; Moschuswurzel

7950 sunflower oil
f huile de tournesol
e aceite de girasol
i olio di girasole
d Sonnenblumenöl

7951 superactivity
f suractivité
e superactividad
i soprattività
d Übertätigkeit

7952 supercalendering
f supercalandrage
e supercalandrdo; supersatinado
i supersatinazione
d Superkalandrieren

7953 supercooling
f surfusion
e sobrefusión
i surraffreddamento
d Unterkühlung

7954 superheated steam
f vapeur surchauffée
e vapor sobrecalentado
i vapore surriscaldato
d Heißdampf

7955 superheater
f surchauffeur
e recalentador
i surriscaldatore
d Überhitzer

7956 superheater coils
f serpentin surchauffeur
e tubería helicoidal del recalentador
i serpentine per surriscaldatore
d Überhitzerschlangen

7957 supernate
f couche surnageante
e capa sobrenadante
i liquido sovrastante
d Überschicht

7958 superphosphate
f superphosphate
e superfosfato
i superfosfato
d Superphosphat

7959 supersaturation
f sursaturation
e supersaturación
i supersaturazione
d Übersättigung

7960 suppressible mutant
f mutant supprimable
e mutante supresible
i mutante sopprimibile
d suppressible Mutante

7961 surface active agents
f agents tensio-actifs
e compuestos variantes de tensión interfacial
i agenti tensioattivi
d Oberflächenaktivstoffe

7962 surface combustion
f combustion de surface
e combustión superficial
i combustione superficiale
d Oberflächenverbrennung

7963 surface haze
f mat de surface
e aspecto nuboso superficial
i opacità della superficie
d Glanzlosigkeit (einer Oberfläche)

7964 surface tension
f tension superficielle
e tensión superficial

i tensione superficiale
d Oberflächenspannung

7965 surface waviness
f surface ondulée
e ondulación superficial
i ondulazione della superficie
d Oberflächenwelligkeit

7966 surfactant
f agent tensio-actif
e agente activo de superficie
i agente tensio-attivo
d Netzmittel

7967 susceptibility
f susceptibilité
e susceptibilidad
i suscettibilità
d Empfänglichkeit

7968 suspension
f suspension
e suspensión
i sospensione
d Suspension

7969 suspensoid
f suspensoïde
e suspensoide; substancia en suspenso
i sospensoide
d Suspensoid; disperse Phase

* **suxamethonium bromide** → **7894**

7970 sweat *v*
f transuder
e exudar
i trasudare
d durchschwitzen

7971 sweating
f exsudation; suintement
e exudación
i trasudamento
d Schwitzen; Ausschwitzen

7972 Swedish black
f noir de Suède
e negro sueco
i nero svedese
d schwedisches Schwarz

7973 "sweetened" gasoline *(US)*
f essence adoucie
e gasolina desmercaptanizada
i benzina desolforata
d ausgesüßtes Benzin

7974 sweetening
f adoucissement
e desmercaptanización
i desolforazione
d Entschwefelung

7975 sweet spirits of nitre
f esprit de nitre dulcifié
e espíritu dulce de nitro
i spirito di etere nitroso
d versüßter Salpetergeist

7976 sweet wort
f moût adouci
e mosto dulce
i mosto dolce
d süße Würze

* **swelling** → **4394**

7977 swimmers
f grains flottants; flotteurs
e granos flotantes de cebada
i grani galleggianti
d Schwimmer

7978 swing pipe
f tuyau articulé
e tubo oscilante
i tubo oscillante
d Schwenkrohr

7979 sylvanite
f sylvanite
e silvanita
i silvanite
d Silvanit; Tellursilberblei

7980 sylvinite
f sylvinite
e silvinita
i silvinite
d Sylvinit

7981 sylvite
f sylvite
e silvita
i silvite
d Hartsalz; Silvin

7982 sym-
Prefix denoting organic compounds having a symmetrical structure.
f sym-
e sim-
i sim-
d Sym-

7983 symbiosis
 f symbiose
 e simbiosis
 i simbiosi
 d Symbiose

7984 symmetry
 f symétrie
 e simetría
 i simmetria
 d Symmetrie

7985 sympatholytic
 f sympatholytique
 e simpatolítico
 i simpatolitico
 d sympatholytisch

7986 synapsis
 f synapsis
 e sinapsis
 i sinapsi
 d Synapsis

7987 synaptic
 f synaptique
 e sináptico
 i sinaptico
 d synaptisch

7988 syncaryon; synkarion
 f synkaryon
 e sincarión
 i sincarion
 d Synkaryon

7989 syncyte
 f syncyte
 e síncito
 i sincito
 d Syncyte

 * syndet → 7998

7990 syndiotactic
 f syndiotactique
 e sindiotáctico
 i sindiotattico
 d syndiotaktisch

7991 syneresis
 f synérèse
 e sinéresis
 i sineresi
 d Synärese; Ausschwitzen

7992 synergist
 f synergiste
 e producto sinergético

 i sostanza sinergetica
 d Synergist; synergisches Präparat

7993 synergistic additives
 f additifs synergiques
 e aditivos sinérgicos
 i additivi sinergici
 d synergische Zusatzmittel

 * synkarion → 7988

7994 syntactic foams
 f mousses syntactiques
 e plásticos alveolares sintácticos
 i spugne plastiche sintattiche
 d syntaktische Schaumstoffe

7995 synthesis
 f synthèse
 e síntesis
 i sintesi
 d Synthese

7996 synthesis gas
 f gaz synthèse
 e gas de síntesis
 i gas di sintesi
 d Synthesegas

7997 synthetic
 f synthétique
 e sintético
 i sintetico
 d synthetisch

7998 synthetic detergent; syndet
 f détergent synthétique
 e detergente sintético
 i detergente sintetico
 d synthetisches Reinigungsmittel

7999 synthetic resin adhesive
 f colle synthétique
 e adhesivo de resina sintética
 i adesivo di resina sintetica
 d Harzklebstoff

8000 synthetic resins
 f résines synthétiques
 e resinas sintéticas
 i resine sintetiche
 d Kunstharze

8001 synthetic rubber
 f caoutchouc synthétique
 e caucho sintético
 i gomma sintetica
 d synthetischer Gummi

8002 synthetics
  f produits synthétiques
  e productos sintéticos
  i sintetici
  d Kunststoffe

8003 syntroph
  f syntrophe
  e síntrofo
  i sintrofo
  d syntroph

8004 syphon; siphon
  f siphon
  e sifón
  i sifone
  d Syphon

8005 syrosingopine
  f syrosingopine
  e sirosingopina
  i sirosingopina
  d Syrosingopin

8006 syrup
  f sirop
  e jarabe
  i sciroppo
  d Syrup

8007 systemic
  f général
  e general
  i sistemico
  d allgemein

T

8008 table rolls
f rouleaux de support
e rodillos soporte
i rulli di supporto
d Rollgang

8009 tablet
f comprimé
e comprimido; tableta
i pastiglia
d Tablette

8010 tacciometer
f viscosimètre
e medidor de pegajosidad
i misuratore di appiccicosità
d Klebrigkeitsmesser

8011 tachysterol
f tachystérine
e taquisterol
i tachisterolo
d Tachysterin

8012 tack
f état collant
e pegajosidad
i adesività
d Klebrigkeit

8013 tack-free
f sec hors-poisse
e no pegajoso
i non appiccicoso
d nicht klebrig

8014 tackifier
f agent poisseux
e agente de pegajosidad
i agente di appiccicosità
d Klebrigmacher

8015 tacky
f gluant
e pegajoso
i appiccicoso
d klebrig

8016 tactic
f tactique
e táctico
i tattico
d taktisch

8017 tag v
f marquer

e marbetear; identificar
i contrassegnare con un cartellino
d markieren

8018 tagged atom
f atome marqué
e átomo marcado
i atomo marcato
d markiertes Atom

8019 tailings
Residue of low metal content.
f résidus
e residuos
i residui
d Rückstand

8020 tailings
Refuse from barley.
f déchets d'orge
e desperdicios
i scarto dell'orzo
d Gersteabfälle

8021 take v **fire**
f s'enflammer
e inflamarse
i infiammarsi
d sich entzünden

8022 talbutal
f talbutal
e talbutal
i talbutal
d Talbutal

8023 talc; talcum
f talc
e talco
i talco
d Talk; Talkum

8024 tall oil
f résine liquide
e aceite de resina
i tallolio
d Tallöl

8025 tallow
f suif
e sebo
i sego
d Talg

8026 tallow seed oil
f huile de suif végétale
e aceite de sebo vegetal
i olio di semi di sego
d Stillingiaöl

8027 tan *v*
f tanner
e curtir
i conciare
d gerben

8028 tandem association
f association en tandem
e asociación tándem
i associazione in tandem
d Tandemassoziation

8029 tandem duplication
f duplication en tandem
e duplicación tándem
i duplicazione in tandem
d Tandemduplikation

8030 tandem fusions
f fusions en tandem
e fusiones tándem
i fusioni in tandem
d Tandemfusionen

8031 tandem ring
f anneau-tandem
e anillo tándem
i anello tandem
d Tandemring

8032 tanker
f bateau-citerne; camion-citerne
e bugue cisterna; camión cisterna; vagón cisterna
i petroliera nave cisterna; autobotte
d Tanker; Tankwagen; Tankschiff

8033 tank voltage
f tension de cuve
e tensión de cuba
i tensione di cella
d Badspannung

8034 tannic acid
f acide tannique
e ácido tánico
i acido tannico
d Gerbsäure; Tanninsäure

8035 tannin
f tanin
e tanino
i tannino
d Tannin

8036 tanning
f tannage
e curtimiento; curtido

i conciatura
d Gerben

8037 tannoform
f tannoforme
e tanoformo
i tannoformio
d Tannoform

8038 tantalite
f tantalite
e tantalita
i tantalite
d Tantalite; Kolumbit

8039 tantalum
f tantale
e tántalo
i tantalio
d Tantal

8040 tantalum carbide
f carbure de tantale
e carburo de tántalo
i carburo di tantalio
d Tantalcarbid

8041 tantalum chloride
f chlorure de tantale
e cloruro de tántalo
i cloruro di tantalio
d Tantalchlorid

8042 tantalum oxide
f oxyde de tantale
e óxido de tántalo
i ossido di tantalio
d Tantaloxyd

8043 tar
f goudron
e alquitrán
i catrame; bitume
d Teer

8044 tar acids
f acides du goudron
e ácidos del alquitrán
i acidi di catrame
d Teersäuren

8045 taraxacum
f pissenlit; taraxacum
e taraxacón
i taraxacum
d Taraxacum

8046 tare
f tare

e tara
i tara
d Tara

8047 tared filter
f filtre taré
e filtro tarado
i filtro tarato
d tarierter Filter

8048 target
f plaque de déchiquetage
e meta; placa de paro
i piastra d'arresto
d Auftreffplatte

8049 tarnish
f ternissement
e descoloración superficial
i appannamento
d Anlaufen

8050 tartar
f tartre
e tártaro
i tartaro
d Weinstein

8051 tartar emetic
f tartre émétique
e tártaro emético
i tartaro emetico
d Brechweinstein

8052 tartaric acid
f acide tartarique
e ácido tartárico
i acido tartarico
d Weinsteinsäure

8053 taurine
f taurine
e taurina
i taurina
d Taurin

8054 tautomerism; dynamic isomerism
f tautomérie; isomérisme dynamique
e tautomerismo; isomerismo dinámico
i tautomerismo; isomerismo dinamico
d Tautomerie

8055 tear test
f essai de déchirement; essai de déchirure
e ensayo de desgarramiento
i prova di lacerazione
d Reißversuch; Einreißprobe

8056 technetium
f technécium
e tecnecio
i tecnezio
d Technetium

8057 telechromomere
f téléchromomère
e telecromómero
i telecromomero
d Telechromomer

8058 telluric acid
f acide tellurique
e ácido telúrico
i acido tellurico
d Tellursäure

8059 tellurium
f tellurium
e telurio
i tellurio
d Tellur

8060 tellurium dioxide
f bioxyde de tellure
e dióxido de telurio
i biossido di tellurio
d Tellurdioxyd

8061 tellurium lead
f plomb telluré
e plomo-telurio
i piombo al tellurio
d Tellurblei

8062 telolecithal
f télolécithe
e telolecito
i telolecito
d telolezithal

8063 telomere
f télomère
e telómero
i telomero
d Telomer

8064 telosynapsis
f télosynapse
e telosinapsis
i telosinapsi
d Telosynapsis

8065 telosyndetic
f télosyndétique
e telosindético
i telosindetico
d telosyndetisch

8066 temper v
f traiter par recuit
e recocer; revenir; templar
i temperare
d tempern

8067 temperature
f température
e temperatura
i temperatura
d Temperatur

8068 tempering
f gâchage
e amasado
i ricottura; rinvenimento
d Tonkneten; Mörtelmischen

8069 temporary hardness
f crudité temporaire
e dureza temporal
i durezza temporanea
d vorübergehende Härte

8070 tenacity
f ténacité; résistance
e tenacidad; dureza tensil
i tenacità
d Festigkeit; Zähfestigkeit

8071 teniafuge
f ténifuge
e tenífugo
i tenifugo
d Bandwurmmittel

8072 tennantite
f tennantite
e tenantita
i tennantite
d Graukupfererz; Tennantit

8073 terbium
f terbium
e terbio
i terbio
d Terbium

8074 terebene
f térébène
e terebeno
i terebene
d Terebin

8075 terephthalic acid
f acide téréphtalique
e ácido tereftálico
i acido tereftalico
d Terephthalsäure

8076 terephthaloyl chloride
f chlorure de téréphtaloyle
e cloruro de tereftaloílo
i cloruro di tereftaloile
d Terephthaloylchlorid

8077 terminal chiasma
f chiasma terminal
e quiasma terminal
i chiasma terminale
d Terminalchiasma

*   **termolecular** → 8516

8078 termone
f termone
e termón
i termone
d Termon

8079 terpene
f terpène
e terpeno
i terpene
d Terpen

8080 terphenyl
f terphényle
e terfenilo
i terfenile
d Terphenyl

8081 terpineol
f terpinéol
e terpineol
i terpineolo
d Terpineol

8082 terpin hydrate
f hydrate de terpine
e hidrato de terpina
i idrato di terpina
d Terpinhydrat

8083 terpinolene
f terpinolène
e terpinoleno
i terpinolene
d Terpinolen

8084 terpinyl acetate
f acétate de terpinyle
e acetato de terpinilo
i acetato di terpinile
d Terpinylacetat

8085 terra alba
f terre de pipe
e tierra blanca

i terra alba
d Porzellanerde

* terra verte → 3867

8086 tertiary alcohol
f alcool tertiaire
e alcohol terciario
i alcool terziario
d tertiärer Alkohol

8087 tertiary amide
f amide tertiaire
e amida terciaria
i ammide terziaria
d tertiäres Amid

* tervalent → 8550

8088 testosterone
f testostérone
e testosterona
i testosterone
d Testosteron

8089 testosterone cyclopentylpropionate
f cyclopentylpropionate de testostérone
e ciclopentilpropionato de testosterona
i ciclopentilpropionato di testosterone
d Testosteroncyclopentylpropionat

8090 testosterone propionate
f propionate de testostérone
e propionato de testosterona
i propionato di testosterone
d Testosteronpropionat

8091 test tube
f burette
e probeta; tubo de ensayos
i provetta
d Reagenzglas

8092 test-tube brush
f brosse à tubes
e escobilla para tubos
i spazzola per pulire provette
d Reagenzglasbürste

8093 tetrabromoethylene
f tétrabrométhylène
e tetrabromoetileno
i tetrabromoetilene
d Tetrabromäthylen

8094 tetrabutylthiuram disulphide
f bisulfure de tétrabutylthiuram
e disulfuro de tetrabutiltiouramilo

i bisolfuro de tetrabutiltiuram
d Tetrabutylthiuramdisulfid

8095 tetrabutyltin
f tétrabutylétain
e tetrabutilestaño
i tetrabutilstagno
d Tetrabutylzinn

8096 tetrabutyl titanate
f titanate de tétrabutyle
e titanato de tetrabutilo
i titanato di tetrabutile
d Tetrabutyltitanat

8097 tetrabutyl urea
f tétrabutylurée
e tetrabutilurea
i tetrabutilurea
d Tetrabutylharnstoff

8098 tetrabutyl zirconate
f zirconate de tétrabutyle
e zirconato de tetrabutilo
i zirconato di tetrabutile
d Tetrabutylzirkonat

8099 tetracaine hydrochloride; amethocaine hydrochloride
f chlorhydrate de tétracaïne
e hidrocloruro de tetracaína
i cloridrato di tetracaina
d Tetracainhydrochlorid

8100 tetracarboxybutane
f tétracarboxybutane
e tetracarboxibutano
i tetracarbossibutano
d Tetracarboxybutan

8101 tetracene
f tétracène
e tetraceno
i tetracene
d Tetracen

8102 tetrachlorobenzene
f tétrachlorobenzène
e tetraclorobenceno
i tetraclorobenzene
d Tetrachlorbenzol

8103 tetrachlorodifluoroethane
f tétrachlorodifluoroéthane
e tetraclorodifluoetano
i tetraclorodifluoroetano
d Tetrachlordifluoräthan

8104 tetrachloroethane
f tétrachloréthane
e tetracloroetano
i tetracloroetano
d Tetrachloräthan

8105 tetrachlorophenol
f tétrachlorophénol
e tetraclorofenol
i tetraclorofenolo
d Tetrachlorphenol

8106 tetrachlorophthalic acid
f acide tétrachlorophtalique
e ácido tetracloroftálico
i acido tetracloroftalico
d Tetrachlorphthalsäure

8107 tetrachlorophthalic anhydride
f anhydride tétrachlorophtalique
e anhídrido tetracloroftálico
i anidride tetracloroftalica
d Tetrachlorphthalsäureanhydrid

8108 tetracosane
f tétracosane
e tetracosano
i tetracosano
d Tetrakosan

8109 tetracyanoethylene
f tétracyanoéthylène
c tetracianoetileno
i tetracianoetilene
d Tetracyanäthylen

8110 tetracycline
f tétracycline
e tetraciclina
i tetraciclina
d Tetracyclin

8111 tetrad
f tétrade
e tetrada
i tetrade
d Tetrade

8112 tetradecane
f tétradécane
e tetradecano
i tetradecano
d Tetradecan

8113 tetradecene
f tétradécène
e tetradeceno
i tetradecene
d Tetradecen

8114 tetradecylamine
f tétradécylamine
e tetradecilamina
i tetradecilammina
d Tetradecylamin

8115 tetradecyl mercaptan
f tétradécylmercaptan
e tetradecilmercaptano
i tetradecilmercaptano
d Tetradecylmercaptan

8116 tetrads analysis
f analyse des tétrades
e análisis de las tetradas
i analisi delle tetradi
d Tetradenanalyse

8117 tetradymite
f tétradymite
e tetradimita
i tetradimite
d Tetradymit; Wismuttellur

8118 tetraethanolammonium hydroxide
f hydroxyde de tétraéthanolammonium
e hidróxido de tetraetanolamonio
i idrossido di tetraetanolammonio
d Tetraäthanolammoniumhydroxyd

8119 tetraethylammonium chloride
f chlorure de tétraéthylammonium
e cloruro de tetraetilamonio
i cloruro di tetraetilammonio
d Tetraäthylammoniumchlorid

8120 tetraethyl dithiopyrophosphate
f dithiopyrophosphate de tétraéthyle
e ditiopirofosfato de tetraetilo
i ditiopirofosfato di tetraetile
d Tetraäthyldithiopyrophosphat

8121 tetraethylene glycol
f tétraéthylèneglycol
e tetraetilenglicol
i tetraetileneglicolo
d Tetraäthylenglykol

8122 tetraethylene glycol dicaprylate
f dicaprylate de tetraethylèneglycol
e dicaprilato de tetraetilenglicol
i dicaprilato di glicoltetraetilene
d Tetraäthylenglykoldicaprylat

8123 tetraethylene glycol dimethacrylate
f diméthacrylate de tétraéthylèneglycol
e dimetacrilato de tetraetilenglicol
i dimetacrilato di glicoltetraetilene
d Tetraäthylenglykoldimethacrylat

8124 tetraethylene glycol distearate
 f distéarate de tétraéthylèneglycol
 e diestearato de tetraetilenglicol
 i distearato di glicoltetraetilene
 d Tetraäthylenglykoldistearat

8125 tetraethylene glycol monostearate
 f monostéarate de tétraéthylèneglycol
 e monoestearato de tetraetilenglicol
 i monostearato di glicoltetraetilene
 d Tetraäthylenglykolmonostearat

8126 tetraethylenepentamine
 f tétraéthylènepentamine
 e tetraetilenpentamina
 i tetraetilenepentammina
 d Tetraäthylenpentamin

8127 tetraethylhexyl titanate
 f titanate de tétraéthylhexyle
 e titanato de tetraetilhexilo
 i tetraetilessiltitanato
 d Tetraäthylhexyltitanat

8128 tetraethyl lead
 f plomb tétraéthylique
 e plomo tetraetílico
 i piombo tetraetile
 d Tetraäthylblei

8129 tetraethylpyrophosphate
 f pyrophosphate de tétraéthyle
 e pirofosfato de tetraetilo
 i pirofosfato tetraetile
 d Tetraäthylpyrophosphat

8130 tetraethylthiuram disulphide
 f bisulfure de tétraéthylthiuram
 e disulfuro de tetraetiltiouramilo
 i bisolfuro di tetraetiltiuram
 d Tetraäthylthiuramdisulfid

8131 tetraethylthiuram sulphide
 f sulfure de tétraéthylthiuram
 e sulfuro de tetraetiltiouramilo
 i solfuro di tetraetiltiuram
 d Tetraäthylthiuramsulfid

8132 tetrafluoroethylene
 f tétrafluoroéthylène
 e tetrafluoetileno
 i tetrafluoetileno
 d Tetrafluoräthylen

8133 tetrafluorohydrazine
 f tétrafluorohydrazine
 e tetrafluohidracina
 i tetrafluoroidrazina
 d Tetrafluorhydrazin

8134 tetrafluoromethane
 f tétrafluorométhane
 e tetrafluometano
 i tetrafluorometano
 d Tetrafluormethan

8135 tetraglycol dichloride
 f bichlorure de tétraglycol
 e dicloruro de tetraglicol
 i bicloruro di tetraglicolo
 d Tetraglykoldichlorid

8136 tetrahedrite
 f tétraédrite; tétrahédrite
 e tetrahedrita
 i tetraedrite
 d Tetraedrit; Fahlerz

8137 tetrahydrofuran
 f tétrahydrofuranne
 e tetrahidrofurano
 i tetraidrofurano
 d Tetrahydrofuran

8138 tetrahydrofurfuryl alcohol
 f alcool tétrahydrofurfurylique
 e alcohol tetrahidrofurfurílico
 i alcool tetraidrofurfurilico
 d Tetrahydrofurfurylalkohol

8139 tetrahydrofurfuryl laevulinate
 f lévulinate de tétrahydrofurfuryle
 e levulinato de tetrahidrofurfurilo
 i levulinato di tetraidrofurfurile
 d Tetrahydrofurfuryllävulinat

8140 tetrahydrofurfuryl laurate
 f laurate de tétrahydrofurfuryle
 e laurato de tetrahidrofurfurilo
 i laurato di tetraidrofurfurile
 d Tetrahydrofurfuryllaurat

8141 tetrahydrofurfuryl oleate
 f oléate de tétrahydrofurfuryle
 e oleato de tetrahidrofurfurilo
 i oleato di tetraidrofurfurile
 d Tetrahydrofurfuryloleat

8142 tetrahydrofurfuryl palmitate
 f palmitate de tétrahydrofurfuryle
 e palmitato de tetrahidrofurfurilo
 i tetraidrofurfurilpalmitato
 d Tetrahydrofurfurylpalmitat

8143 tetrahydrofurfuryl phthalate
 f phtalate de tétrahydrofurfuryle
 e ftalato de tetrahidrofurfurilo
 i ftalato di tetraidrofurfurile
 d Tetrahydrofurfurylphthalat

8144 tetrahydronaphthalene; tetralin
 f tétrahydronaphtalène; tétraline
 e tetrahidronaftaleno; tetralina
 i tetraidronaftalene; tetralina
 d Tetrahydronaphthalin; Tetralin

8145 tetrahydrophthalic anhydride
 f anhydride tétrahydrophtalique
 e anhídrido tetrahidroftálico
 i anidride tetraidroftalica
 d Tetrahydrophthalsäureanhydrid

8146 tetrahydropyranmethanol
 f tétrahydropyranméthanol
 e tetrahidropiranmetanol
 i metanol-tetraidropirano
 d Tetrahydropyranmethanol

8147 tetrahydropyridine
 f tétrahydropyridine
 e tetrahidropiridina
 i tetraidropiridina
 d Tetrahydropyridin

8148 tetrahydrothiophene
 f tétrahydrothiophène
 e tetrahidrotiofeno
 i tetraidrotiofene
 d Tetrahydrothiophen

8149 tetrahydroxyethylethylenediamine
 f tétrahydroxyéthyléthylènediamine
 e tetrahidroxietiletilendiamina
 i tetraidrossietilenediammina
 d Tetrahydroxyäthyläthylendiamin

 *   **tetrahydroxyflavonol → 6859**

8150 tetraiodoethylene
 f tétraiodoéthylène
 e tetrayodoetileno
 i tetraiodoetilene
 d Tetrajodäthylen

8151 tetraisopropylthiuram disulphide
 f bisulfure de tétraisopropylthiuram
 e disulfuro de tetraisopropiltiouramilo
 i bisolfuro di tetraisopropiltiuram
 d Tetraisopropylthiuramdisulfid

8152 tetraisopropyl titanate
 f titanate de tétraisopropyle
 e titanato de tetraisopropilo
 i titanato di tetraisopropile
 d Tetraisopropyltitanat

8153 tetraisopropyl zirconate
 f zirconate de tétraisopropyle
 e zirconato de tetraisopropilo

 i zirconato di tetraisopropile
 d Tetraisopropylzirkonat

 *   **tetralin → 8144**

8154 tetramer
 f tétramère
 e tetramero
 i tetramero
 d Tetramer

8155 tetramethylammonium chlorodibromide
 f chlorodibromure de tétraméthyl-
    ammonium
 e clorodibromuro de tetrametilamonio
 i clorodibromuro di tetrametilammonio
 d Tetramethylammoniumchlordibromid

8156 tetramethylbutanediamine
 f tétraméthylbutanediamine
 e tetrametilbutandiamina
 i tetrametilbutanediammina
 d Tetramethylbutandiamin

8157 tetramethyldiaminobenzhydrol
 f tétraméthyldiaminobenzhydrol
 e tetrametildiaminobenzhidrol
 i tetrametildiaminobenzidrolo
 d Tetramethyldiaminobenzhydrol

8158 tetramethyldiaminobenzophenone
 f tétraméthyldiaminobenzophénone
 e tetrametildiaminobenzofenona
 i tetrametildiaminobenzofenone
 d Tetramethyldiaminobenzophenon

8159 tetramethyldiaminodiphenylmethane
 f tétraméthyldiaminodiphénylméthane
 e tetrametildiaminodifenilmetano
 i tetrametildiaminodifenilmetano
 d Tetramethyldiaminodiphenylmethan

8160 tetramethyldiaminodiphenylsulphone
 f tétraméthyldiaminodiphénylsulfone
 e tetrametildiaminodifenilsulfona
 i tetrametildiaminodifenilsolfone
 d Tetramethyldiaminodiphenylsulfon

8161 tetramethylenediamine
 f tétraméthylènediamine
 e tetrametilendiamina
 i tetrametilenediammina
 d Tetramethylendiamin

8162 tetramethylethylenediamine
 f tétraméthyléthylènediamine
 e tetrametiletilendiamina
 i tetrametiletilenediammina
 d Tetramethyläthylendiamin

8163 tetramethyl lead
 f plomb tétraméthylique
 e plomo tetrametílico
 i piombo tetrametilico
 d Tetramethylblei

8164 tetramethylthiuram disulphide
 f bisulfure de tétraméthylthiuram
 e bisulfuro de tetrametiltiouramilo
 i bisolfuro di tetrametiltiuram
 d Tetramethylthiuramdisulfid

8165 tetramethylthiuram monosulphide
 f monosulfure de tétraméthylthiuram
 e monosulfuro de tetrametiltiouramilo
 i monosolfuro di tetrametiltiuram
 d Tetramethylthiurammonosulfid

8166 tetramorphous
 f tétramorphe
 e tetramorfo
 i tetramorfo
 d tetramorph

8167 tetranitroaniline
 f tétranitroaniline
 e tetranitroanilina
 i tetranitroanilina
 d Tetranitroanilin

8168 tetranitromethane
 f tétranitrométhane
 e tetranitrometano
 i tetranitrometano
 d Tetranitromethan

8169 tetraphenylsilane
 f tétraphénylsilane
 e tetrafenilsilano
 i tetrafenilsilano
 d Tetraphenylsilan

8170 tetraphenyltin
 f tétraphénylétain
 e tetrafenilestaño
 i tetrafenilstagno
 d Tetraphenylzinn

8171 tetrapropenylsuccinic anhydride
 f anhydride tétrapropénylsuccinique
 e anhídrido tetrapropenilsuccínico
 i anidride tetrapropenilsuccinica
 d Tetrapropenylbernsteinsäureanhydrid

8172 tetrapropylene
 f tétrapropylène
 e tetrapropileno
 i tetrapropilene
 d Tetrapropylen

8173 tetrapropylthiuram disulphide
 f bisulfure de tétrapropylthiuram
 e bisulfuro de tetrapropiltiouramilo
 i bisolfuro di tetrapropiltiuram
 d Tetrapropylthiuramdisulfid

8174 tetrasomaty
 f tétrasomatie
 e tetrasomatia
 i tetrasomatia
 d Tetrasomatie

8175 tetravalent; quadrivalent
 f tétravalent; quadrivalent
 e tetravalente; cuatrivalente
 i tetravalente
 d vierwertig

8176 tetryl
 f tétryle
 e tetrilo
 i tetrile
 d Tetryl

8177 textile oils
 f huiles textiles
 e aceites textiles
 i oli tessili
 d Textilöle

8178 thalline
 f thalline
 e talina
 i tallina
 d Thallin

8179 thallium
 f thallium
 e talio
 i tallio
 d Thallium

8180 thallium acetate
 f acétate de thallium
 e acetato de talio
 i acetato di tallio
 d Thalliumacetat

8181 thallium carbonate
 f carbonate de thallium
 e carbonato de talio
 i carbonato di tallio
 d Thalliumcarbonat

8182 thallium chloride
 f chlorure de thallium
 e cloruro de talio
 i cloruro di tallio
 d Thalliumchlorid

8183 thallium hydroxide
  f hydroxyde de thallium
  e hidróxido de talio
  i idrossido di tallio
  d Thalliumhydroxyd

8184 thallium monoxide
  f monoxyde de thallium
  e monóxido de talio
  i monossido di tallio
  d Thalliummonoxyd

8185 thallium nitrate
  f nitrate de thallium
  e nitrato de talio
  i nitrato di tallio
  d Thalliumnitrat

8186 thallium sulphate
  f sulfate de thallium
  e sulfato de talio
  i solfato di tallio
  d Thalliumsulfat

8187 thallium sulphide
  f sulfure de thallium
  e sulfuro de talio
  i solfuro di tallio
  d Thalliumsulfid

8188 thaw *v*
  f dégeler
  e descongelar
  i disgelare
  d abtauen; tauen

8189 thebaine
  f thébaïne
  e tebaína
  i tebaina
  d Thebain

8190 Thelan pan
  f pan de Thelan
  e caldero de Thelan
  i vasca di Thelan
  d Thelan-Napf; Thelan-Pfanne

8191 thelycaryon; thelykaryon
  f thélycaryon
  e telicarión
  i telicarion
  d Thelykaryon

8192 thenardite
  f thénardite
  e tenardita
  i tenardite
  d Thenardit

8193 Thenard's blue
  f bleu de Thénard
  e azul de Thénard
  i blu di Thenard
  d Thenardsblau

8194 thenyldiamine hydrochloride
  f chlorhydrate de thényldiamine
  e hidrocloruro de tenildiamina
  i cloridrato di tenildiammina
  d Thenyldiaminhydrochlorid

8195 theobromine
  f théobromine
  e teobromina
  i teobromina
  d Theobromin

8196 theobromine calcium salicylate
  f salicylate de théobromine et de calcium
  e salicilato teobromina-cálcico
  i salicilato di calcio e teobromina
  d Theobromincalciumsalicylat

8197 theobromine sodium salicylate
  f salicylate de théobromine et de sodium
  e salicilato teobromina-sódico
  i salicilato di sodio e teobromina
  d Theobrominnatriumsalicylat

8198 theophylline
  f théophylline
  e teofilina
  i teofillina
  d Theophyllin

8199 theophyllinemethylglucamine
  f théophyllineméthylglucamine
  e teofilinametilglucamina
  i teofillinametilglucamina
  d Theophyllinmethylglucamin

8200 theophylline sodium glycinate
  f théophyllineglycinate sodique
  e teofilina glicinato sódico
  i glicinato di sodio e teofillina
  d Theophyllinnatriumglycinat

8201 theoretical air requirement for combustion
  f air théorique; facteur de combustion
  e aire teórico
  i aria teorica
  d theoretischer Luftbedarf

8202 thermal black
  f noir de charbon
  e negro térmico

i nerofumo termico
d Kohleschwarz

8203 thermal conductivity
f conductivité thermique
e conductividad térmica
i conduttività termica
d Wärmeleitfähigkeit

8204 thermal decomposition
f décomposition calorifique
e descomposición térmica
i decomposizione termica
d Wärmezersetzung

8205 thermal degradation
f dégradation thermique
e degradación termal
i deterioramento termico
d Wärmezersetzung

8206 thermal diffusion plant
f installation de diffusion thermique
e instalación de difusión térmica
i impianto di diffusione termica
d Thermodiffusionsanlage

8207 thermal diffusion separation
f séparation par diffusion thermique
e separación por difusión térmica
i separazione mediante diffusione termica
d Trennung durch Diffusionswärme

8208 thermal diffusity
f diffusivité thermique
e difusibilidad térmica
i costante di diffusione termica
d Wärmeleitvermögen

8209 thermal embrittlement
f élévation de la fragilité sous l'effet de la chaleur
e fragilidad térmica
i infrangilimento termico
d Wärmeversprödung

8210 thermal expansion coefficient
f coefficient d'expansion thermique
e coeficiente de expansión térmica
i coefficiente di espansione termica
d Wärmeausdehnungskoeffizient

8211 thermal glass
f verre résistant à la chaleur
e vidrio térmico
i vetro termico
d hitzebeständiges Glas

8212 thermit
f thermite
e termita
i termite
d Thermit

8213 thermochemistry
f thermochimie
e termoquímica
i termochimica
d Thermochemie; Wärmechemie

8214 thermolabile
f thermolabile
e termolábil
i termolabile
d thermolabil

8215 thermoluminescence
f thermoluminescence
e termoluminiscencia
i termoluminescenza
d Thermolumineszenz

8216 thermolysis
f thermolyse
e termólisis
i termolisi
d Thermolyse

8217 thermometer
f thermomètre
e termómetro
i termometro
d Thermometer

8218 thermophilic organism
f organisme thermophile
e organismo termófilo
i organismo termofilo
d thermophiler Organismus

8219 thermopile
f pile; thermo-électrique; thermopile
e pila; termoeléctrica; termopila
i termopila
d Thermosäule

8220 thermoplastic
f thermoplastique
e termoplástico
i termoplastico
d Thermoplast

8221 thermoplasticity
f thermoplasticité
e termoplasticidad
i termoplasticità
d Thermoplastizität

* **thermoplastic polymers** → **2129**

8222 thermoplastic resins
 f résines thermoplastiques
 e resinas termoplásticas
 i resine termoplastiche
 d thermoplastische Harze

8223 thermosetting
 f thermodurcissable
 e termoestable; termoendurecible
 i termoindurente
 d wärmehärtbar; hitzehärtbar

8224 thermosetting adhesive
 f colle thermodurcissable
 e adhesivo termoendurecible
 i adesivo termoindurente
 d hitzehärtbare Harze

8225 thermostable
 f thermostabile
 e termoestable
 i termostabile
 d hitzebeständig

8226 thermostat
 f thermostat
 e termostato
 i termostato
 d Thermostat; Temperaturregler

* **thiacetazone** → **8253**

* **thiambutosine** → **8237**

* **thiambutoxine** → **8237**

8227 thiamine hydrochloride; vitamin B$_1$
 f chlorhydrate de thiamine
 e hidrocloruro de tiamina
 i cloridrato di tiammina
 d Thiaminhydrochlorid

8228 thiamylal sodium
 f thiamylal sodique
 e tiamilal sodico
 i sodio di tiamilale
 d Thiamylalnatrium

8229 thiazine
 f thiazine
 e tiacina; tiazina
 i tiazina
 d Thiazin

8230 thiazole
 f thiazole
 e tiazol

 i tiazolo
 d Thiazol

8231 thiazosulphone
 f thiazosulfone
 e tiazosulfona
 i tiazosolfone
 d Thiazosulfon

8232 thickener
 f épaississant
 e concentrador
 i concentratore; addensatore
 d Verdickungsmittel

* **thinner** → **2662**

8233 thinning out
 f dilution
 e reducción de viscosidad; dilución
 i fluidificazione; diluizione
 d Verdünnen

8234 thio-
  Prefix denoting "sulphur".
 f thio-
 e tio-
 i tio-
 d Thio-

8235 thioacetic acid
 f acide thioacétique
 e ácido tioacético
 i acido tioacetico
 d Thioessigsäure

8236 thiobenzoic acid
 f acide thiobenzoïque
 e ácido tiobenzoico
 i acido tiobenzoico
 d Thiobenzoesäure

**8237 thiocarbanilide; thiambutosine;
      thiambutoxine**
 f thiocarbanilide
 e tiocarbanilida
 i tiocarbanilide
 d Thiocarbanilid

* **thiocyanates** → **7929**

8238 thiocyanic acid
 f acide thiocyanique
 e ácido tiociánico
 i acido tiocianico
 d Thiocyansäure

8239 thiodiglycol
 f thiodiglycol

    e tiodiglicol
    i tiodiglicolo
    d Thiodiglykol

8240 thiodiglycolic acid
    f acide thiodiglycolique
    e ácido tiodiglicólico
    i acido tiodiglicolico
    d Thiodiglykolsäure

8241 thioethers; alkyl sulphides
    f sulfures d'alcoyles
    e tioéteres
    i tioeteri
    d Thioäther

    * thiofuran → 8246

8242 thioglycerol
    f thioglycérol
    e tioglicerina
    i tioglicerolo
    d Thioglycerin

8243 thioglycolic acid
    f acide thioglycolique
    e ácido tioglicólico
    i acido tioglicolico
    d Thioglykolsäure

8244 thiohydantoin
    f thiohydantoïne
    e tiohidantoína
    i tioidantoina
    d Thiohydantoin

    * thiols → 5225

8245 thionyl chloride
    f chlorure de thionyle
    e cloruro de tionilo
    i cloruro di tionile
    d Thionylchlorid

8246 thiophene; thiofuran
    f thiophène
    e tiofeno
    i tiofene
    d Thiophen

8247 thiophenol; phenyl mercaptan
    f thiophénol
    e tiofenol
    i tiofenolo
    d Thiophenol

8248 thiophosgene
    f thiophosgène
    e tiofosgeno

    i tiofosgene
    d Thiophosgen

8249 thiopropazate hydrochloride
    f chlorhydrate de thiopropazate
    e hidrocloruro de tiopropazato
    i cloridrato di tiopropazato
    d Thiopropazathydrochlorid

8250 thioridazine hydrochloride
    f chlorhydrate de thioridazine
    e hidrocloruro de tioridacina
    i cloridrato di tioridazina
    d Thioridazinhydrochlorid

8251 thiosalicylic acid
    f acide thiosalicylique
    e ácido tiosalicílico
    i acido tiosalicilico
    d Thiosalicylsäure

8252 thiosemicarbazide
    f thiosemicarbazide
    e tiosemicarbacida
    i tiosemicarbazide
    d Thiosemicarbazid

8253 thiosemicarbazone; thiacetazone; amithiozone
    f thiosemicarbazone
    e tiosemicarbazona
    i tiosemicarbazone
    d Thiosemicarbazon

8254 thiouracil
    f thiouracile
    e tiouracilo
    i tiouracile
    d Thiouracil

8255 thiourea
    f thiourée
    e tiourea
    i tiourea
    d Thioharnstoff

8256 thiourea formaldehyde resins
    f résines de thiourée-formaldéhyde
    e resinas de tioureaformaldehído
    i resine di tioureaformaldeide
    d Thioharnstofformaldehydharze

8257 thiourea resins
    f résines de thiourée
    e resinas de tiourea
    i resine tioureiche
    d Thioharnstoffharze

8258 thistle funnel
 f tube de sûreté
 e tubo de seguridad
 i imbuto a rubinetto
 d Sicherheitsröhre

8259 thiuram
 f thiuram
 e tiouramilo; tiocarbamoilo
 i thiuram
 d Thiuram

8260 thiuram sulphides
 f thiurames
 e sulfuros de tiuramilo
 i solfuri di thiuram
 d Thiuramsulfide

8261 thixotrope
 f thixotrope
 e tixotropo
 i tixotropo; tissotropo
 d Thixotrop

8262 thixotropy
 f thixotropie
 e tixotropía
 i tixotropia
 d Thixotropie

8263 thomsonite
 f thomsonite
 e tomsonita
 i tomsonite
 d Thomsonit

**8264 thonzylamine hydrochloride;
     histazylamine hydrochloride**
 f chlorhydrate de thonzylamine
 e hidrocloruro de toncilamina
 i cloridrato di tonsilammina
 d Thonzylaminhydrochlorid

8265 thorite
 f thorite
 e torita
 i torite
 d Thorit

8266 thorium
 f thorium
 e torio
 i torio
 d Thor

8267 thorium content meter
 f teneurmètre en thorium
 e torímetro
 i toriotenorimetro

 d Gerät zur Bestimmung des
     Thoriumgehaltes

8268 thorium cycle
 f cycle du thorium
 e ciclo del torio
 i ciclo del torio
 d Thoriumzyklus

8269 thorium dioxide
 f bioxyde de thorium
 e dióxido de torio
 i biossido di torio
 d Thordioxyd

8270 thorium fluoride
 f fluorure de thorium
 e fluoruro de torio
 i fluoruro di torio
 d Thorfluorid

8271 thorium nitrate
 f nitrate de thorium
 e nitrato de torio
 i nitrato di torio
 d Thornitrat

8272 thorium oxalate
 f oxalate de thorium
 e oxalato de torio
 i ossalato di torio
 d Thoroxalat

8273 thoron
 f thoron
 e torón
 i torone
 d Thoron

8274 three-stage compressor
 f compresseur à trois étages
 e compresor trietápico
 i compressore a tre stadi
 d dreistufiger Kompressor

8275 threonine
 f thréonine
 e treonina
 i treonina
 d Threonin

8276 throwing
 f éjection de particules
 e eyección centrífuga de partículas
 i eiezione di particelle di vernici
 d Streuung

8277 throwing power
 f pouvoir de pénétration

e poder de penetración
i potere ricoprente
d Streuvermögen

8278 thuja oil
f essence de thuya
e aceite de tuya
i essenza di tuia
d Thujaöl

8279 thujone
f thuyone
e tuyona
i tuione
d Thuyon

* **thulite** → **9003**

8280 thulium
f thulium
e tulio
i tulio
d Thulium

8281 thyme oil
f essence de thym
e aceite de tomillo
i essenza di timo
d Thymianöl

8282 thymol; thymic acid
f thymol; acide thymique
e timol; ácido tímico
i timolo; acido timico
d Thymol; Thymiankampfer

8283 thymol iodide
f iodure de thymol
e yoduro de timol
i ioduro di timolo
d Thymoljodid

8284 thymophthalein
f thymophtaléine
e timoftaleína
i timoftaleina
d Thymophthalein

8285 thymoquinone
f thymoquinone
e timoquinona
i timochinone
d Thymochinon

8286 thyroxine
f thyroxine
e tiroxina
i tirossina
d Thyroxin

8287 tiglic acid
f acide tiglique
e ácido tíglico
i acido tiglico
d Tiglinsäure

8288 tin
f étain
e estaño
i stagno
d Zinn

* **tin bichloride** → **7751**

8289 tincal
f tincal
e atíncar
i tincal; borace greggio
d Tinkal

8290 tincture
f teinture
e tintura
i tintura
d Tinktur

8291 tinge
f teinte
e tinte; matiz
i tinta
d Färbung

8292 tin octoate
f octoate d'étain
e octoato de estaño
i ottoato di stagno
d Zinnoctoat

* **tin pyrites** → **7750**

* **tinstone** → **1497**

8293 tinting strength
f pouvoir de coloration
e poder de coloración; resistencia a la coloración
i potere colorante
d Färbekraft

8294 tissue
f papier fin; papier de soie
e tisú papel muy fino; papel de seda
i carta di seta
d Seidenpapier

8295 titanic acid
f acide titanique
e ácido titánico

i acido titanico
d Titansäure

8296 titanite
f titanite
e titanita
i titanite
d Titanit

8297 titanium
f titane
e titanio
i titanio
d Titan

8298 titanium boride
f borure de titane
e boruro de titanio
i boruro di titanio
d Titanborid

8299 titanium carbide
f carbure de titane
e carburo de titanio
i carburo di titanio
d Titancarbid

8300 titanium dioxide
f bioxyde de titane
e dióxido de titanio
i biossido di titanio
d Titandioxyd

8301 titanium peroxide
f peroxyde de titane
e peróxido de titanio
i perossido di titanio
d Titanperoxyd

8302 titanium-potassium oxalate
f oxalate de titane et de potassium
e oxalato de titanio y potasio
i ossalato di potassio e titanio
d Titankaliumoxalat

8303 titanium sulphate
f sulfate de titane
e sulfato de titanio
i solfato di titanio
d Titansulfat

8304 titanium tetrachloride
f tétrachlorure de titane
e tetracloruro de titanio
i tetracloruro di titanio
d Tetrachlortitan

8305 titanium trichloride
f trichlorure de titane

e tricloruro de titano
i tricloruro di titanio
d Titantrichlorid

8306 titanium white
f blanc de titane
e blanco de titanio
i bianco di titanio
d Titanweiß

8307 titanous sulphate
f sulfate de titane
e sulfato titanoso
i solfato titanoso
d Titanosulfat

8308 titanyl acetylacetonate
f acétylacétonate de titanyle
e acetilacetonato de titanilo
i acetilacetonato di titanile
d Titanylacetylacetonat

8309 titration
f titrage
e valoración
i titolazione
d Titrierung

8310 titre
f titre
e valoración; concentración
i titolo
d Titer; Gehalt

8311 titrimetry
f titrimétrie
e titrimetría
i titrimetria
d Titrieranalyse

*   TNB → 8518

*   TNT → 8520

8312 Tobias acid
f acide de Tobias
e ácido de Tobias
i acido di Tobias
d Tobias-Säure

8313 tocopherols
f tocophérols
e tocoferoles
i tocoferoli
d Tocopherole

8314 tolan
f tolane
e tolano

i tolano
d Tolan

8315 tolazoline hydrochloride
f chlorhydrate de tolazoline
e hidrocloruro de tolazolina
i cloridrato di tolazolina
d Tolazolinhydrochlorid

8316 tolerance
f tolérance
e tolerancia
i tolleranza
d Verträglichkeit

8317 tolidine
f tolidine
e tolidina
i tolidina
d Tolidin

8318 tolonium chloride
f chlorure de tolonium
e cloruro de tolonio
i cloruro di tolonio
d Toloniumchlorid

8319 tolu balsam
f baume de tolu
e bálsamo de tolú
i balsamo di toluifera
d Tolubalsam

8320 toluene; toluol
f toluène; toluol
e tolueno; toluol
i toluene; toluolo
d Toluol

8321 toluene diamine
f toluène-diamine
e toluendiamina
i diammina di toluene
d Toluoldiamin

8322 toluene diisocyanate
f toluène-diisocyanate
e toluendiisocianato
i diisocianato di toluene
d Toluoldiisocyanat

8323 toluenesulphanilide
f toluène-sulfanilide
e toluensulfanilida
i toluenesolfanilida
d Toluolsulfanilid

8324 toluene sulphochloride
f sulfochlorure de toluène

e toluensulfocloruro; sulfocloruro de
    tolueno
i solfocloruro di toluene
d Toluolsulfochlorid

8325 toluenesulphonamide
f toluène-sulfonamide
e toluensulfonamida
i toluenesolfonammide
d Toluolsulfonamid

8326 toluenesulphonic acid
f acide toluènesulfonique
e ácido toluensulfónico
i acido toluenesolfonico
d Toluolsulfonsäure

8327 toluhydroquinone
f toluhydroquinone
e toluhidroquinona
i toluidrochinone
d Toluhydrochinon

8328 toluic acid
f acide toluique
e ácido toluico
i acido toluico
d Toluylsäure

8329 toluidine
f toluidine
e toluidina
i toluidina
d Toluidin

8330 toluidine sulphonic acid
f acide toluidinesulfonique
e ácido toluidinasulfónico
i acido toluidinsolfonico
d Toluidinsulfonsäure

* **toluol** → **8320**

8331 toluquinone
f toluquinone
e toluquinona
i toluchinone
d Toluchinon

8332 toluylene red; neutral red
f rouge de toluylène; rouge neutre
e rojo de toluileno; rojo neutro
i rosso di toluilene
d Toluylenrot; Neutralrot

8333 tolylaldehyde
f tolylaldéhyde
e tolilaldehído

i tolilaldeide
d Tolylaldehyd

8334 tolyldiethanolamine
f tolyldiéthanolamine
e tolildietanolamina
i tolildietanolammina
d Tolyldiäthanolamin

8335 tolylethanolamine
f tolyléthanolamine
e toliletanolamina
i toliletanolammina
d Tolyläthanolamin

8336 tolyl naphthylamine
f tolylnaphtylamine
e tolilnaftilamina
i tolilnaftilammina
d Tolylnaphthylamin

8337 tone *v*
f virer
e entonar
i virare
d tönen

8338 toner
f toner; colorant organique
e color orgánico
i pigmento organico
d organischer Pigmentfarbstoff

8339 tonka bean
f tonka
e haba de tonca
i cumarina
d Tonkabohne

8340 topochemical reaction
f réaction topochimique
e reacción topoquímica
i reazione topochimica
d topochemische Reaktion

8341 topped crude
f résidu de première distillation
e crudo de destilación primaria
i residuo di prima distillazione
d Rückstand der ersten Destillation

8342 top roller
f cylindre supérieur
e rodillo superior
i cilindro superiore
d Oberwalze

8343 top yeast
f levure haute

e levadura de fermentación alta
i lievito per alta fermentazione
d Oberhefe

8344 torbanite
f torbanite
e torbanita
i torbanite
d Torbanit

8345 torbernite
f torbernite
e torbenita
i torbenite
d Kupferuranglimmer; Uranglimmer

8346 total ammonia
f ammoniac total
e amoníaco total
i ammoniaca totale
d gesamter Ammoniakgehalt

8347 total cyanide
f cyanure total
e cianuro total
i cianuro totale
d gesamtes Cyanid

8348 tourmaline
f tourmaline
e turmalina
i tormalina
d Turmalin

* **tower → 1990**

8349 town gas
f gaz de ville
e gas de ciudad
i gas di città
d Stadtgas; Versorgungsgas

8350 toxaphene
f toxaphène
e toxafeno
i toxafene
d Toxaphen

8351 toxic
f toxique
e tóxico
i tossico
d giftig

8352 toxicity
f toxicité
e toxicidad
i tossicità
d Giftigkeit

8353 toxin
f toxine
e toxina
i tossina
d Toxin

8354 trace
f trace
e traza
i traccia
d Spur

8355 tracer chemistry
f chimie des indicateurs isotopiques
e química de los trazadores isotópicos
i chimica dei traccianti isotopici
d Indikatorenchemie

8356 tracer compound
f composé traceur
e compuesto trazador
i composto tracciante
d Indikatorverbindung

8357 traction battery
f batterie de traction
e batería de tracción
i batteria di trazione
d Traktionsbatterie

8358 traction fibre
f fibre de traction
e fibra de tracción
i fibra di trazione
d Zugfaser

* **tragacanth gum** → **3903**

8359 trans-
Prefix denoting an isomer in which like atoms lie on opposite sides of the molecule.
f trans-
e trans-
i trans-
d Trans-

8360 transcurium elements
f transcuriens
e elementos transcúricos
i elementi trascurici
d Transcuriumelemente

8361 transduction clone
f clône de transduction
e clon de transducción
i clone di trasduzione
d Transduktionsklon

8362 transferase
f transférase
e transferasa
i enzima trasferitore
d Transferase

8363 transfer chamber
f chambre de transfert
e cámara de transferencia
i camera di riscaldamento
d Füllraum

8364 transfer cull
f culot
e residuo de transferencia
i materozza
d Angusstetzen

* **transference number** → **8375**

8365 transfer RNA
f ARN de transfert
e transducción ARN
i RNA di trasferimento
d Transfer-RNS

8366 transfer unit
f élément de transport
e elemento de transporte
i elemento di trasporto
d Austauscheinheit

8367 transformer compound
f composé pour transformateur
e compuesto para transformador
i miscela per trasformatori
d Transformatorenmischung

8368 transfuse *v*
f transvaser
e transvasar
i transfondere
d umfüllen

8369 transition temperature
f température de transition
e temperatura de transición
i temperatura di transizione
d Übergangstemperatur

8370 translocation
f translocation
e translocación
i traslocazione
d Translokation

8371 translucency
f translucidité
e translucidez

i translucentezza
d Durchsichtigkeit

8372 translucent
  f translucide
  e translúcido
  i translucido
  d durchscheinend

8373 transmutation
  f transmutation
  e transmutación
  i trasmutazione
  d Transmutation

8374 transpecific
  f transspécifique
  e transespecífico
  i transpecifico
  d transspezifisch

8375 transport number; transference number
  f nombre de transport des ions
  e número de transporte de los iones
  i numero di trasporto di ioni
  d Überführungszahl

8376 transuranic elements
  f éléments transuriens
  e elementos transuránicos
  i elementi transuranici
  d Transurane

8377 transvasate
  f liquide de transvasement
  e transvasado
  i travasato
  d Transvasat; dekantierte Flüssigkeit

8378 Trauzl test
  f essai de Trauzl
  e prueba de Trauzl
  i prova di Trauzl
  d Trauzlscher Versuch

  * **treacle** → 5529

8379 trees
  f aborescences
  e arborescencias
  i arborescenze
  d Bäumchen

8380 tremolite
  f trémolite
  e tremolita
  i tremolite
  d Tremolit

8381 triacetin
  f triacétine
  e triacetina
  i triacetina
  d Triacetin

8382 triacetyloleandomycin
  f triacétyloléandomycine
  e triacetiloleandomicina
  i triacetiloleandomicina
  d Triacetyloleandomycin

8383 triacid
  f triacide
  e triácido
  i triacido
  d dreisäurig

8384 triad
  f triade
  e triada
  i triade
  d Triade

8385 triallyl cyanurate
  f cyanurate de triallyle
  e cianurato de trialilo
  i triallilcianurato
  d Triallylcyanurat

8386 triaminotoluene trihydrochloride
  f chlorhydrate de triaminotoluène
  e trihidrocloruro de triaminotolueno
  i triidrocloruro di triamminotoluene
  d Triaminotoluoltrihydrochlorid

8387 triaminotriphenylmethane
  f triaminotriphénylméthane
  e triaminotrifenilmetano
  i triamminotrifenilmetano
  d Triaminotriphenylmethan

8388 triamylamine
  f triamylamine
  e triamilamina
  i triamilammina
  d Triamylamin

8389 triamyl borate
  f borate de triamyle
  e borato de triamilo
  i borato di triamile
  d Triamylborat

8390 triamylphenyl phosphate
  f phosphate de triamylphényle
  e fosfato de triamfenilo
  i fosfato di triamilfenile
  d Triamylphenylphosphat

8391 triatomic
f triatomique
e triatómico
i triatomico
d Dreiatom-

8392 tribasic
f tribasique
e tribásico
i tribasico
d dreibasisch

8393 tribasic calcium phosphate
f phosphate tricalcique
e fosfato cálcico tribásico
i fosfato di calcio tribasico
d Tricalciumphosphat

8394 tribasic copper sulphate
f sulfate tribasique de cuivre
e sulfato tribásico de cobre
i solfato di rame tribasico
d dreibasisches Kupfersulfat

8395 tribasic magnesium phosphate
f phosphate tribasique de magnésium
e fosfato magnésico tribásico
i fosfato di magnesio tribasico
d Trimagnesiumphosphat

8396 tribasic potassium phosphate
f phosphate tribasique de potassium
e fosfato potásico tribásico
i fosfato di potassio tribasico
d Trikaliumphosphat

8397 tribasic sodium phosphate
f phosphate tribasique de sodium
e fosfato sódico tribásico
i fosfato tribasico di sodio
d Trinatriumphosphat

8398 triboluminescence
f triboluminescence
e triboluminiscencia
i triboluminescenza
d Tribolumineszenz

8399 tribromoacetaldehyde
f tribromoacétaldéhyde
e tribromoacetaldehído
i tribromoacetaldeide
d Tribromacetaldehyd

8400 tribromoacetic acid
f acide tribromacétique
e ácido tribromoacético
i acido tribromoacetico
d Tribromessigsäure

8401 tribromoethanol; tribromoethyl alcohol
f tribromoéthanol
e tribromoetanol
i tribromoetanolo
d Tribromäthanol

8402 tributoxyethyl phosphate
f phosphate de tributoxyéthyle
e fosfato de tributoxietilo
i fosfato di tributossietile
d Tributoxyäthylphosphat

8403 tributyl aconitate
f aconitate de tributyle
e aconitato de tributilo
i aconitato di tributile
d Tributylaconitat

8404 tributylamine
f tributylamine
e tributilamina
i tributilammina
d Tributylamin

8405 tributyl borate
f borate de tributyle
e borato de tributilo
i borato di tributile
d Tributylborat

8406 tributyl citrate
f citrate de tributyle
e citrato de tributilo
i citrato di tributile
d Tributylcitrat

8407 tributylphenyl phosphate
f phosphate de tributylphényle
e fosfato de tributilfenilo
i fosfato di tributilfenile
d Tributylphenylphosphat

8408 tributyl phosphate
f phosphate de tributyle
e fosfato de tributilo
i fosfato di tributile
d Tributylphosphat

8409 tributyl phosphine
f tributylphosphine
e tributilfosfina
i tributilfosfina
d Tributylphosphin

8410 tributyl phosphite
f phosphite de tributyle
e fosfito de tributilo
i fosfito di tributile
d Tributylphosphit

8411 tributyltin acetate
  f acétate de tributylétain
  e acetato de tributilestaño
  i acetato di tributilstagno
  d Tributylzinnacetat

8412 tributyltin chloride
  f chlorure de tributylétain
  e cloruro de tributilestaño
  i cloruro di tributilstagno
  d Tributylzinnchlorid

8413 tributyltin oxide
  f oxyde de tributylétain
  e óxido de tributilestaño
  i ossido di tributilstagno
  d Tributylzinnoxyd

8414 tributyl tricarballylate
  f tricarballylate de tributyle
  e tricarbalilato de tributilo
  i tricarballilato di tributile
  d Tributyltricarballylat

8415 trichlorethyl phosphate
  f phosphate de trichloroéthyle
  e fosfato de tricloroetilo
  i fosfato di tricloroetile
  d Trichloräthylphosphat

8416 trichloroacetic acid
  f acide trichloroacétique
  e ácido tricloroacético
  i acido tricloroacetico
  d Trichloressigsäure

8417 trichlorobenzene
  f trichlorobenzène
  e triclorobenceno
  i triclorobenzene
  d Trichlorbenzol

8418 trichloroborazole
  f trichloroborazol
  e tricloroborazol
  i tricloroborazolo
  d Trichlorborazol

8419 trichlorocarbanilide
  f trichlorocarbanilide
  e triclorocarbanilida
  i triclorocarbanilide
  d Trichlorcarbanilid

8420 trichloroethane
  f trichloroéthane
  e tricloroetano
  i tricloroetano
  d Trichloräthan

8421 trichloroethanol
  f trichloroéthanol
  e tricloroetanol
  i tricloroetanolo
  d Trichloräthanol

8422 trichloroethylene
  f trichloroéthylène
  e tricloroetileno
  i tricloroetilene
  d Trichloräthylen

8423 trichlorofluoromethane
  f trichlorofluorométhane
  e triclorofluometano
  i triclorofluorometano
  d Trichlorfluormethan

8424 trichloroisocyanuric acid
  f acide trichloroisocyanurique
  e ácido tricloroisocianúrico
  i acido tricloroisocianurico
  d Trichlorisocyanursäure

8425 trichloromelamine
  f trichloromélamine
  e tricloromelamina
  i tricloromelammina
  d Trichlormelamin

8426 trichloromethyl chloroformate
  f chloroformiate de trichlorométhyle
  e cloroformiato de triclorometilo
  i cloroformiato di triclorometile
  d Trichlormethylchloroformiat

8427 trichloromethyl phenyl carbinyl acetate
  f acétate de trichlorométhyl-phényl-carbinyle
  e acetato de triclorometilfenilcarbinilo
  i acetato di triclorometilfenilcarbinile
  d Trichlormethylphenylcarbinylacetat

8428 trichloromethylphosphonic acid
  f acide trichlorométhylphosphonique
  e ácido triclorometilfosfónico
  i acido triclorometilfosfonico
  d Trichlormethylphosphonsäure

8429 trichloronitrosomethane
  f trichloronitrosométhane
  e tricloronitrosometano
  i tricloronitrosometano
  d Trichlornitrosomethan

8430 trichlorophenol
  f trichlorophénol
  e triclorofenol

i triclorofenolo
d Trichlorphenol

8431 trichlorophenoxyacetic acid
f acide trichlorophénoxyacétique
e ácido triclorofenoxiacético
i acido triclorofenossiacetico
d Trichlorphenoxyessigsäure

8432 trichloropropane
f trichloropropane
e tricloropropano
i tricloropropano
d Trichlorpropan

8433 trichlorosilane
f trichlorosilane
e triclorosilano
i triclorosilano
d Trichlorsilan

8434 trichlorotrifluoroacetone
f trichlorotrifluoroacétone
e triclorotrifluoacetona
i triclorotrifluoroacetone
d Trichlortrifluoraceton

8435 trichlorotrifluoroethane
f trichlorotrifluoroéthane
e triclorotrifluoetano
i triclorotrifluoroetano
d Trichlortrifluoräthan

8436 triclinic
f triclinique
e triclínico
i triclino
d triklin

8437 tricosane
f tricosane
e tricosano
i tricosano
d Trikosan

8438 tricresyl phosphate
f phosphate de tricrésyle
e fosfato de tricresilo
i fosfato di tricresile
d Trikresylphosphat

8439 tricresyl phosphite
f phosphite de tricrésyle
e fosfito de tricresilo
i fosfito di tricresile
d Trikresylphosphit

8440 tricyclamol chloride
f chlorure de tricyclamol

e cloruro de triciclamol
i cloruro di triciclamolo
d Tricyclamolchlorid

8441 tridecane
f tridécane
e tridecano
i tridecano
d Tridecan

8442 tridecyl alcohol
f alcool tridécylique
e alcohol tridecílico
i alcool tridecilico
d Tridecylalkohol

8443 tridecylaluminium
f tridécylaluminium
e tridecilaluminio
i tridecilalluminio
d Tridecylaluminium

8444 tridecyl phosphite
f phosphite de tridécyle
e fosfito de tridecilo
i fosfito di tridecile
d Tridecylphosphit

8445 tridimethylphenylphosphate
f phosphate de tridiméthylphényle
e fosfato de tridimetilfenilo
i fosfato di tridimetilfenile
d Tridimethylphenylphosphat

8446 triethanolamine
f triéthanolamine
e trietanolamina
i trietanolammina
d Triäthanolamin

8447 triethanolamine lauryl sulphate
f triéthanolamine-laurylsulfate
e sulfato de trietanolamina y laurilo
i solfato di laurile e trietanolammina
d Triäthanolaminlaurylsulfat

* **triethanolamine stearate** → **8484**

8448 triethoxyhexane
f triéthoxyhexane
e trietoxihexano
i trietossiesano
d Triäthoxyhexan

8449 triethoxymethoxypropane
f triéthoxyméthoxypropane
e trietoximetoxipropano
i trietossimetossipropano
d Triäthoxymethoxypropan

8450 triethyl aconitate
f aconitate de triéthyle
e aconitato de trietilo
i anonitato di trietile
d Triäthylaconitat

8451 triethyl aluminium
f triéthylaluminium
e trietilaluminio
i alluminio trietilico
d triäthylaluminium

8452 triethylamine
f triéthylamine
e trietilamina
i trietilammina
d Triäthylamin

8453 triethylborane
f triéthylborane
e trietilborano
i trietilborano
d Triäthylboran

8454 triethyl citrate
f citrate de triéthyle
e citrato de trietilo
i citrato di trietile
d Triäthylcitrat

8455 triethylene glycol
f triéthylèneglycol
e trietilenglicol
i glicolo di trietilene
d Triäthylenglykol

8456 triethylene glycol diacetate
f diacétate de triéthylèneglycol
e diacetato de trietilenglicol
i diacetato di glicolo e trietilene
d Triäthylenglykoldiacetat

8457 triethylene glycol dicaprylate
f dicaprylate de triéthylèneglycol
e dicaprilato de trietilenglicol
i dicaprilato di glicolo e trietilene
d Triäthylenglykoldicaprylat

8458 triethylene glycol dicaprylate-caprate
f dicaprylate-caprate de triéthylèneglycol
e dicaprilato-caprato de trietilenglicol
i caprato dicaprilato del glicol trietilenico
d Triäthylenglykoldicaprylat-caprat

8459 triethylene glycol didecanoate
f didécanoate de triéthylèneglycol
e didecanoato de trietilenglicol
i didecanoato di glicolo e trietilene
d Triäthylenglykoldidecanoat

8460 triethylene glycol diethylhexoate
f diéthylhexoate de triéthylèneglycol
e diatilhexoato de trietilenglicol
i dietilesoato di glicolo e trietilene
d Triäthylenglykoldiäthylhexoat

8461 triethylene glycol dihydroabietate
f dihydroabiétate de triéthylèneglycol
e dihidroabietato de trietilenglicol
i diidroabietato di glicolo e trietilene
d Triäthylenglykoldihydroabietat

8462 triethylene glycol dimethyl ether
f diméthyléther de triéthylèneglycol
e dimetiléter del trietilenglicol
i etere dimetilico di glicolo e trietilene
d Triäthylenglykoldimethyläther

8463 triethylene glycol dipelargonate
f dipélargonate de triéthylèneglycol
e dipelargonato de trietilenglicol
i trietileneglicoldipelargonato
d Triäthylenglykoldipelargonat

8464 triethylene glycol dipropionate
f dipropionate de triéthylèneglycol
e dipropionato de trietilenglicol
i dipropionato di glicolo e trietilene
d Triäthylenglykoldipropionat

8465 triethylene glycol propionate
f propionate de triéthylène-glycol
e propionato de trietilenglicol
i propionato del glicol trietilenico
d Triäthylenglykolpropionat

8466 triethylenemelamine
f triéthylènemélamine
e trietilenmelamina
i trietilenemelammina
d Triäthylenmelamin

8467 triethylenetetramine
f triéthylènetétramine
e trietilentetramina
i trietilenetetrammina
d Triäthylentetramin

8468 triethylene thiophosphoramide
f triéthylènethiophosphoramide
e trietilenfosforamida
i tiofosforammide di trietilene
d Triäthylenthiophosphoramid

8469 triethyl orthoformate
f orthoformiate de triéthyle
e ortoformiato de trietilo
i ortoformato di trietile
d Triäthylorthoformiat

8470 triethyl phosphate
f phosphate de triéthyle
e fosfato de trietilo
i fosfato di trietile
d Triäthylphosphat

8471 triethyl phosphite
f phosphite de triéthyle
e fosfito de trietilo
i fosfito di trietile
d Triäthylphosphit

8472 triethyl phosphorothioate
f phosphorothioate de triéthyle
e fosfotioato de trietilo
i fosfotioato di trietile
d Triäthylphosphothioat

8473 triethyl tricarballylate
f tricarballylate de triéthyle
e tricarbalilato de trietilo
i tricarballilato di trietile
d Triäthyltricarballylat

8474 trifluopromazine hydrochloride
f chlorhydrate de trifluopromazine
e hidrocloruro de trifluopromacina
i cloridrato di trifluopromazina
d Trifluopromazinhydrochlorid

8475 trifluoracetic acid
f acide trifluoroacétique
e ácido trifluoacético
i acido trifluoroacetico
d Trifluoressigsäure

8476 trifluoronitrosomethane
f trifluoronitrosométhane
e trifluonitrosometano
i trifluoronitrosometano
d Trifluornitrosomethan

8477 triglycol dichloride
f bichlorure de triglycol
e dicloruro de triglicol
i bicloruro di triglicolo
d Triglykoldichlorid

8478 trigonelline
f trigonelline
e trigonelina
i trigonellina
d Trigonellin

8479 trihaploid
f trihaploïde
e trihaploide
i triaploide
d trihaploid

**8480 trihexylphenidyl hydrochloride;
benzhexol hydrochloride**
f chlorhydrate de trihexylphénidyle
e hidrocloruro de trihexilfenidilo
i idrocloruro di triesilfenidile
d Trihexylphenidylhydrochlorid

8481 trihexyl phosphite
f phosphite de trihexyle
e fosfito de trihexilo
i fosfito di triesile
d Trihexylphosphit

8482 trihydric alcohol
f trialcool
e trialcohol
i alcool triidrico
d dreiwertiger Alkohol

8483 trihydroxyethylamine oleate
f oléate de trihydroxyéthylamine
e oleato de trihidroxietilamina
i oleato di triidrossietilammina
d Trihydroxyäthylaminoleat

**8484 trihydroxyethylamine stearate;
triethanolamine stearate**
f stéarate de trihydroxyéthylamine
e estearato de trihidroxietilamina
i stearato di triidrossietilammina
d Trihydroxyäthylaminstearat

**8485 triiodothyronine sodium; liothyronine
sodium**
f triiodothyronine sodique
e triyodotironina sódica
i sodio di triiodotironina
d Trijodthyroninnatrium

8486 triisobutylaluminium
f triisobutylaluminium
e triisobutilaluminio
i triisobutilalluminio
d Triisobutylaluminium

8487 triisobutylene
f triisobutylène
e triisobutileno
i triisobutilene
d Triisobutylen

8488 triisooctyl phosphite
f phosphite de triisooctyle
e fosfito de triisooctilo
i fosfito di triisoottile
d Triisooctylphosphit

8489 triisopropanolamine
f triisopropanolamine

e triisopropanolamina
i triisopropanolammina
d Triisopropanolamin

8490 triisopropyl phosphite
f phosphite de triisopropyle
e fosfito de triisopropilo
i fosfito di triisopropile
d Triisopropylphosphit

8491 trilauryl amine
f trilaurylamine
e trilaurilamina
i ammina trilaurilica
d Trilaurylamin

8492 trimer
f trimère
e trímero
i trimero
d Trimer

8493 trimerical
f trimérique
e tremérico
i trimero
d trimer

8494 trimery
f trimérie
e trimería
i trimeria
d Trimerie

8495 trimesic acid
f acide trimésique
e ácido trimésico
i acido trimesico
d Trimesinsäure

8496 trimetaphan camphorsulphonate
f camphorsulfonate de triméthaphane
e canfosulfonato de trimetafán
i canforsolfonato di trimetafan
d Trimetaphankampfersulfonat

8497 trimethadione; troxidone
f triméthadione
e trimetadiona
i trimetadione
d Trimethadion

8498 trimethylacetic acid
f acide triméthylacétique
e ácido trimetilacético
i acido trimetilacetico
d Trimethylessigsäure

8499 trimethylaluminium
f triméthylaluminium
e trimetilaluminio
i trimetilalluminio
d Trimethylaluminium

8500 trimethylamine
f triméthylamine
e trimetilamina
i trimetilammina
d Trimethylamin

8501 trimethyl borate
f borate de triméthyle
e borato de trimetilo
i borato di trimetile
d Trimethylborat

8502 trimethylbutane
f triméthylbutane
e trimetilbutano
i trimetilbutano
d Trimethylbutan

8503 trimethylchlorosilane
f triméthylchlorosilane
e trimetilclorosilano
i trimetilclorosilano
d Trimethylchlorsilan

8504 trimethyl dihydroquinoline polymer
f triméthyldihydroquinone polymère
e polímero de trimetildihidroquinoleina
i polimero trimetildiidrochinolina
d Trimethyldihydrochinolinpolymer

8505 trimethylene bromide
f bromure de triméthylène
e bromuro de trimetileno
i bromuro di trimetilene
d Trimethylenbromid

8506 trimethylene glycol
f triméthylèneglycol
e trimetilenglicol
i glicolo di trimetilene
d Trimethylenglykol

8507 trimethylhexane
f triméthylhexane
e trimetilhexano
i trimetilesano
d Trimethylhexan

8508 trimethylnonanone
f triméthylnonanone
e trimetilnonanona
i trimetilnonanone
d Trimethylnonanon

8509 trimethylolethane
f triméthyloléthane
e trimetiloletano
i trimetiloletano
d Trimethyloläthan

8510 trimethylolpropane
f triméthylolpropane
e trimetilolpropano
i trimetilolpropano
d Trimethylolpropan

8511 trimethylolpropane monooleate
f monooléate de triméthylolpropane
e monoleato de trimetilolpropano
i monooleato di trimetilolpropano
d Trimethylolpropanmonooleat

8512 trimethylpentane
f triméthylpentane
e trimetilpentano
i trimetilpentano
d Trimethylpentan

8513 trimethylpentene
f triméthylpentène
e trimetilpenteno
i trimetilpentene
d Trimethylpenten

8514 trimethylpentylaluminium
f triméthylpentylaluminium
e trimetilpentilaluminio
i trimetilpentilalluminio
d Trimethylpentylaluminium

8515 trimethyl phosphite
f phosphite de triméthyle
e fosfito de trimetilo
i fosfito di trimetile
d Trimethylphosphit

8516 trimolecular; termolecular
f trimoléculaire
e trimolecular
i trimolecolare
d trimolekular

8517 trimonoecious
f trimonoïque
e trimonoico
i trimonoico
d trimonoezisch

8518 trinitrobenzene; TNB
f trinitrobenzène
e trinitrobenceno
i trinitrobenzene
d Trinitrobenzol

8519 trinitrobenzoic acid
f acide trinitrobenzoïque
e ácido trinitrobenzoico
i acido trinitrobenzoico
d Trinitrobenzoesäure

* trinitrophenol → 6387

8520 trinitrotoluene; TNT
f trinitrotoluène
e trinitrotolueno
i trinitrotoluene
d Trinitrotoluol

8521 trioctyl phosphate
f phosphate de trioctyle
e fosfato de trioctilo
i fosfato di triottile
d Trioctylphosphat

8522 trioctyltrimellitate
f trioctyltrimellitate
e trimelitato de trioctilo
i triottiltrimellitato
d Trioctyltrimellitat

8523 trioses
f trioses
e triosas
i triosi
d Triosen

8524 trioxane; metaformaldehyde
f trioxane
e trioxano
i triossano
d Trioxan

8525 tripalmitin
f tripalmitine
e tripalmitina
i tripalmitina
d Tripalmitin

8526 triparanol
f triparanol
e triparanol
i triparanolo
d Triparanol

8527 triphenylguanidine
f triphénylguanidine
e trifenilguanidina
i trifenilguanidina
d Triphenylguanidin

8528 triphenyl phosphate
f phosphate de triphényle
e fosfato de trifenilo

i fosfato di trifenile
d Triphenylphosphat

8529 triphenyl phosphite
f phosphite de triphényle
e fosfito de trifenilo
i fosfito di trifenile
d Triphenylphosphit

8530 triphenylstibine
f triphénylstibine
e trifenilestibina
i trifenilstibina
d Triphenylstibin

8531 triphenyltin chloride
f chlorure de triphénylétain
e cloruro de trifenilestaño
i cloruro di trifenilstagno
d Triphenylzinnchlorid

8532 triple-roller mill
f laminoir triple
e molino de triple rodillo
i molino a tre rulli
d Dreiwalzenmahlwerk

8533 triplet
f triplet
e triplete
i tripletto
d Triplett

8534 triploidy
f triploïdie
e triploidia
i triploidismo
d Triploidie

8535 tripod stand
f trépied
e trípode
i treppiede
d Dreifuß

8536 tripoli
f tripoli
e trípoli; tierra de Trípoli
i tripoli
d Tripelerde

8537 triprolidine hydrochloride
f chlorhydrate de triprolidine
e hidrocloruro de triprolidina
i cloridrato di triprolidina
d Triprolidinhydrochlorid

8538 tripropylene
f tripropylène

e tripropileno
i tripropilene
d Tripropylen

8539 tripropylene glycol
f tripropylèneglycol
e tripropilenglicol
i glicolo di tripropilene
d Tripropylenglykol

8540 tripsometer
f tripsomètre
e tripsómetro
i tripsometro; misuratore di resilienza
d Tripsometer

8541 tris(diethylene glycol monoethyl ether) citrate
f citrate de tris(diéthylèneglycol-monoéthyl-éther)
e citrato de tris(dietilenglicolmonoetiléter)
i triscitrato di etere dietileneglicol-monoetilico
d Tris(diäthylenglykolmonoäthyläther)citrat

8542 trisethylhexyl phosphite
f phosphite de triséthylhexyle
e fosfito de trisetilhexilo
i trisetilsilfosfito
d Trisäthylhexylphosphit

8543 tris(hydroxymethyl) aminomethane
f tris(hydroxyméthyl)aminométhane
e tris(hidroximetil)aminometano
i tris(idrossimetil)amminometano
d Tris(hydroxymethyl)aminomethan

8544 tris(hydroxymethyl) nitromethane
f tris(hydroxyméthyl)nitrométhane
e tris(hidroximetil)nitrometano
i tris(idrossimetil)nitrometano
d Tris(hydroxymethyl)nitromethan

8545 trisomic
f trisomique
e trisómico
i trisomico
d trisom

8546 trisomy
f trisomie
e trisomia
i trisomia
d Trisomie

8547 tritium
f tritium
e tritio

   i tritio
   d Tritium

8548 trituration
   f trituration
   e trituración
   i triturazione
   d Zerreibung

8549 triuranium octoxide
   f octoxyde de triuranium
   e octaóxido de triuranio
   i ottossido di triuranio
   d Triuranoctoxyd

8550 trivalent; tervalent
   f trivalent
   e trivalente
   i trivalente
   d dreiwertig

8551 trommel
   f cible rotatif
   e cilindro clasificador; criba giratoria
   i classifica a tamburo
   d Siebtrommel

8552 trona
   f trona
   e trona
   i trona
   d Trona; Tronasalz

8553 tropacocaine hydrochloride
   f chlorhydrate de tropacocaïne
   e hidrocloruro de tropacocaína
   i cloridrato di tropacocaina
   d Tropacocainhydrochlorid

8554 trophicity
   f trophicité
   e troficidad
   i processo nutritivo
   d Trophizität

8555 trophic nucleus
   f noyau trophique
   e núcleo trófico
   i nucleo trofico
   d Makronukleus

8556 trophoplasm
   f trophoplasme
   e trofoplasma
   i trofoplasma
   d Trophoplasma

8557 trophotropism
   f trophotropisme

   e trofotropismo
   i trofotropismo
   d Trophotropismus

8558 tropic acid
   f acide tropique
   e ácido trópico
   i acido tropico
   d Tropasäure

8559 tropine diphenylmethyl ether
   f éther diphénylméthylique de tropine
   e difenilmetiléter de tropina
   i difenilmetiletere di tropina
   d Tropindiphenylmethyläther

* troxidone → 8497

8560 tryparsamide
   f tryparsamide
   e triparsamida
   i triparsamide
   d Tryparsamid

8561 trypsin
   f trypsine
   e tripsina
   i tripsina
   d Trypsin

8562 tryptophane
   f tryptophane
   e triptófano
   i triptofan
   d Tryptophan

8563 tube mill
   f broyeur à tube
   e molino de tambor
   i molino a tubo
   d Rohrmühle

8564 tube-type plate
   f plaque à tubes
   e placa de tubos
   i piastra a tubi
   d Röhrchenplatte

8565 tubocurarine chloride
   f chlorure de tubocurarine
   e cloruro de tubocurarina
   i cloruro di tubocurarina
   d Tubocurarinchlorid

8566 tub sizing
   f imperméabilisation
   e satinado impermeable
   i imbozzimatura a umido
   d Leimung

* **tumbling** → 7155

8567 tuna oil
 f huile de thone
 e aceite de atún
 i olio di fegato di tonno
 d Thunfischöl

8568 tung oil; abrasin oil; China wood oil; mu oil
 f huile d'abrasin; huile de tung
 e aceite de tung; aceite de palo
 i olio di legno della Cina; olio essicante cinese
 d Tungöl; Holzöl

8569 tungsten
 f tungstène
 e tungsteno; wolframio
 i tungsteno
 d Wolfram

8570 tungsten carbide
 f carbure de tungstène
 e carburo de tungsteno
 i carburo di wolframio
 d Wolframcarbid; Wildiametall

8571 tungstic acid
 f acide tungstique
 e ácido túngstico
 i acido tungstico
 d Wolframsäure

8572 tungstic oxide
 f oxyde tungstique
 e óxido túngstico
 i ossido tungstico
 d Wolframoxyd

8573 tunnel dryer
 f tunnel de séchage
 e túnel secador
 i galleria di essicamento
 d Trockentunnel

8574 tunnel tray dryer
 f tunnel de séchage à plateaux
 e tubo secador de bandejas
 i galleria di essicamento a vassoi
 d Trockentunnel

8575 turbid
 f trouble
 e turbio
 i torbido
 d trübe

8576 turbidimeter
 f opacimètre
 e turbidímetro
 i torbidimetro
 d Trübungsmesser

8577 turbulent
 f turbulent
 e turbulento
 i torbolento
 d wirbelnd

8578 turbulent heating
 f chauffage turbulent
 e calentamiento turbulento
 i riscaldamento turbolento
 d Wirbelheizung

8579 turgescence
 f turgescence
 e turgescencia
 i turgescenza
 d Quellung

8580 Turkey red
 f rouge turc
 e rojo de Turquía
 i rosso turco
 d Türkischrot

8581 Turkey red oil; sulphonated castor oil
 f huile de ricin sulfonée
 e aceite rojo de Turquía
 i olio di ricino solfonato
 d Türkischrotöl

8582 Turkey umber
 f ombre turque
 e tierra de sombra turca
 i terra d'ombra turca
 d türkische Umbra

8583 turnover
 f cycle métabolique
 e ciclo metabólico
 i ciclo metabolico
 d Umwandlung

* **turnsol paper** → 4977

8584 turpentine
 f essence de térébenthine
 e trementina
 i trementina
 d Terpentin

8585 turpentine substitute
 f essence de térébenthine artificielle
 e substituto de la trementina

i succedaneo della trementina
d Terpentinölersatz

8586 turquoise
f turquoise
e turquesa
i turchese
d Türkis

8587 tutocaine hydrochloride
f chlorhydrate de tutocaïne
e hidrocloruro de tutocaína
i cloridrato di tutocaina
d Tutocainhydrochlorid

8588 two-fluid cell
f pile à deux liquides
e pila de dos líquidos
i pila a due elettroliti
d Volta-Element

8589 two-roller mill
f laminoir à deux cylindres
e molino de doble rodillo
i molino a due cilindri
d Zweiwalzenmühle

8590 two-stage compressor
f compresseur à deux étages
e compresor trietápico
i compressore a due stadi
d zweistufiger Kompressor

8591 typolysis
f typolyse
e tipólisis
i tipolisi
d Typolyse

8592 tyramine; hydroxyphenylamine
f tyramine
e tiramina
i tirammina
d Tyramin

8593 tyrocidine
f tyrocidine
e tirocidina
i tirocidina
d Tyrocidin

8594 tyrosine
f tyrosine
e tirosina
i tirosina
d Tyrosin

8595 tyrothricin
f tyrothricine

e tirotricina
i tirotricina
d Tyrothricin

U

8596 ulexite
 f ulexite
 e ulexita
 i ulexite
 d Ulexit

8597 ullage
 f vidange
 e merma
 i perdità
 d Flüssigkeitsmanko

8598 ultra-accelerator
 f ultraaccélerateur
 e ultracelerador
 i ultraccelerante
 d Ultrabeschleuniger

8599 ultrafiltration
 f ultrafiltration
 e ultrafiltración
 i ultrafiltrazione
 d Ultrafiltration

8600 ultramarine
 f outremer
 e ultramar
 i oltremare
 d Ultramarin

8601 ultramarine blue; French blue
 f bleu d'outremer
 e azul ultramar
 i blu oltremare; azzurro di Parigi
 d Ultramarinblau

8602 ultramicrosome
 f ultramicrosome
 e ultramicrosoma
 i ultramicrosoma
 d Ultramikrosom

8603 ultraviolet
 f ultraviolet
 e ultravioleta
 i ultravioletto
 d ultraviolett

8604 ultraviolet absorber
 f absorbant d'ultraviolet
 e absorbente de rayos ultravioletos
 i assorbente di raggi ultravioletti
 d ultraviolettes Absorptionsmittel

8605 ultraviolet inhibitor
 f inhibiteur d'ultraviolet
 e inhibidor ultravioleta
 i protezione contro i raggi ultravioletti
 d Ultraviolettinhibitor

8606 umber
 f terre d'ombre
 e tierra de sombra
 i terra d'ombra
 d Umbra

8607 unbalanced polyploidy
 f polyploïdie déséquilibrée
 e poliploidia desequilibrada
 i poliploidismo squilibrato
 d unbalancierte Polyploidie

8608 unctuousness; greasiness
 f onctuosité
 e untuosidad
 i untuosità
 d Öligkeit; Fettigkeit

8609 uncured
 f non vulcanisé
 e no vulcanizado
 i non vulcanizzato
 d unvulkanisiert

8610 undecalactone
 f undécalactone
 e undecalactona
 i undecalattone
 d Undecalacton

8611 undecane
 f undécane
 e undecano
 i undecano
 d Undecan

8612 undecanoic acid
 f acide undécanoïque
 e ácido undecanoico
 i acido undecanoico
 d Undecansäure

8613 undecanol
 f undécanol
 e undecanol
 i undecanolo
 d Undecanol

8614 undecylenic acid
 f acide undécylénique
 e ácido undecilénico
 i acido undecilenico
 d Undecylensäure

8615 **undecylenic alcohol**
  f alcool undécylénique
  e alcohol undecilénico
  i alcool undecilenico
  d Undecylenalkohol

8616 **undecylenic aldehyde**
  f aldéhyde undécylénique
  e aldehído undecilénico
  i aldeide undecilenica
  d Undecylensäurealdehyd

8617 **undecylenyl acetate**
  f acétate d'undécylényle
  e acetato de undecilenilo
  i acetato di undecilenile
  d Undecylenylacetat

8618 **undercoat**
  f première couche
  e aparejo
  i prima mano
  d Grundierungsmittel

8619 **undersize**
  f trop petit
  e menudos de criba; partículas subtamaño
  i sottomisura
  d Unterkorn; Untermaß

8620 **unimolecular**
  f monomoléculaire
  e monomolecular
  i monomolecolare
  d monomolekular

8621 **uninflammable**
  f ininflammable
  e ininflamable
  i ininfiammabile
  d unentzündlich

8622 **unit**
  f unité
  e unidad
  i unità
  d Einheit

8623 **unit process**
  f procédé unitaire
  e procedimiento unitario
  i processo unitario
  d Einheitsverfahren

  * **univalent** → 5583

8624 **univalent alcohol**
  f alcool monovalent
  e alcohol monovalente

  i alcool monovalente
  d einwertiger Alkohol

8625 **unloaded vulcanizate**
  f vulcanisat sans charge
  e vulcanizado sin carga
  i vulcanizzato senza carico
  d Reinkautschukvulkanisat

  * **uns-** → 8630

8626 **unsaturated**
  f non saturé
  e no saturado
  i non saturato
  d ungesättigt

8627 **unsaturated alcohol**
  f alcool non saturé
  e alcohol no saturado
  i alcool non saturo
  d ungesättigter Alkohol

8628 **unsaturated standard cell**
  f pile étalon non saturée
  e pila patrón no saturada
  i pila campione non satura
  d ungesättigtes Normalelement

8629 **unselected markers**
  f marqueurs non sélectionnés
  e marcadores no seleccionados
  i marcatori non selezionati
  d unselektierte Markierungsgene

8630 **unsym-; uns-**
  Prefix denoting unsymmetrical structure
  in an organic compound.
  f asym-
  e asim-
  i a-
  d Unsym-

8631 **unwrapped construction**
  f montage sans habillage
  e montaje sin recubrimiento
  i montaggio senza rivestimento
  d wickellose Bauweise

8632 **uracil**
  f uracile
  e uracilo
  i uracile
  d Uracil

8633 **uranine**
  f uranine
  e uranina

i uranina
d Uranin

* **uraninite → 6421**

8634 uranium
  f uranium
  e uranio
  i uranio
  d Uran

8635 uranium barium oxide
  f oxyde d'uranium et de barium
  e óxido de uranio y bario
  i ossido di uranio e bario
  d Uranbariumoxyd

8636 uranium carbide
  f carbure d'uranium
  e carburo de uranio
  i carburo di uranio
  d Urancarbid

8637 uranium chalcogenide
  f chalcogénure d'uranium
  e calcogenuro de uranio
  i calcogenuro d'uranio
  d Uranchalkogenid

8638 uranium concentrate
  f concentré uranifére
  e concentrado uranifero
  i concentrato uranifero
  d Urankonzentrat

8639 uranium content
  f teneur en uranium
  e contenido de uranio
  i contenuto d'uranio
  d Urangehalt

8640 uranium dioxide
  f bioxyde d'uranium
  e dióxido de uranio
  i biossido di uranio
  d Urandioxyd

8641 uranium halide
  f halogénure d'uranium
  e halogenuro de uranio
  i alogenuro d'uranio
  d Uranhalogenid

8642 uranium hydride
  f hydrure d'uranium
  e hidruro de uranio
  i idruro di uranio
  d Uranhydrid

8643 uranium monocarbide
  f monocarbure d'uranium
  e monocarburo de uranio
  i monocarburo di uranio
  d Uranmonocarbid

8644 uranium oxide
  f oxyde d'uranium
  e óxido de uranio
  i ossido di uranio
  d Uranoxyd

8645 uranium oxosalts
  f sels des oxo-acides d'uranium
  e sales de los oxoácidos de uranio
  i sali degli ossoacidi d'uranio
  d Oxosalze des Urans

8646 uranium series
  f famille de l'uranium
  e familia del uranio
  i famiglia dell'uranio
  d Uranreihe

8647 uranium tetrafluoride
  f tétrafluorure d'uranium
  e tetrafluoruro de uranio
  i tetrafluoruro di uranio
  d Urantetrafluorid

8648 uranium trioxide
  f trioxyde d'uranium
  e trióxido de uranio
  i triossido di uranio
  d Urantrioxyd

8649 uranocircite
  f uranocircite
  e uranocircita
  i uranocircite
  d Uranozirzit

8650 uranyl acetate
  f acétate d'uranyle
  e acetato de uranilo
  i acetato d'uranile
  d Uranylacetat

8651 uranyl ammonium carbonate
  f carbonate d'uranyle et d'ammonium
  e carbonato de uranilo y amonio
  i carbonato di uranile e ammonio
  d Uranylammoniumcarbonat

8652 uranyl nitrate
  f nitrate d'uranyle
  e nitrato de uranilo
  i nitrato di uranile
  d Uranylnitrat

8653 **uranyl sulphate**
  f sulfate d'uranyle
  e sulfato de uranilo
  i solfato di uranile
  d Uranylsulfat

  * **urea** → **1428**

8654 **urea adducts**
  f produits d'insertion avec l'urée
  e aductos de urea
  i prodotti aggiuntivi di urea
  d Harnstoffaddukte

8655 **urea enzyme**
  f uréase
  e enzima urea
  i ureasi
  d Urease

8656 **urea-formaldehyde resins**
  f résines d'uréeformaldéhyde
  e resinas de ureaformaldehído
  i resine di ureaformaldeide
  d Harnstoff-Formaldehydharze

8657 **urea peroxide**
  f peroxyde d'urée
  e peróxido de urea
  i perossido di urea
  d Harnstoffperoxyd

8658 **urea resin**
  f résine urée
  e resina urea
  i resine ureiche
  d Harnstoffharz

8659 **urease**
  f uréase
  e ureasa
  i urease
  d Urease

8660 **urethane**
  f uréthane
  e uretano
  i uretano
  d Urethan

8661 **uric acid**
  f acide urique
  e ácido úrico
  i acido urico
  d Harnsäure

8662 **uridine**
  f uridine
  e uridina
  i uridina
  d Uridin

8663 **uridylic acid**
  f acide uridylique
  e ácido uridílico
  i acido uridilico
  d Uridylsäure

  * **urotropine** → **4066**

8664 **U-tube manometer**
  f manomètre à tube en U
  e manómetro de tubo en U
  i manometro a tubo a U
  d U-Rohr-Manometer

V

8665 vaccine
 f vaccin
 e vacuna
 i vaccino
 d Kuhpockenlymphe

8666 vacuole
 f vacuole
 e vacúolo
 i vacuolo
 d Vakuole

8667 vacuome
 f vacuome
 e vacuoma
 i vacuoma
 d Vakuom

8668 vacuum distillation
 f distillation dans le vide
 e vacuodestilación
 i distillazione nel vuoto
 d Vakuumdestillation

8669 vacuum drying oven
 f four à sécher dans le vide
 e horno de secar en el vació
 i fornello essiccatore a vuoto
 d Vakuumtrockenofen

8670 vacuum filter
 f entonnoir filtrant
 e filtro de vacío
 i filtro aperto a vuoto
 d Saugfilter

8671 vacuum filtration
 f filtration par le vide
 e vacuofiltración
 i filtrazione a vuoto
 d Vakuumfiltrierung

8672 vacuum gauge
 f indicateur de vide
 e vacuómetro
 i vuotometro
 d Manometer; Unterdruckmesser

8673 vacuum metallizing
 f métallisation par le vide
 e vacuometalización
 i metallizzazione sotto vuoto
 d Vakuummetallisierung

8674 vacuum pan
 f évaporateur à vide
 e tacho de vació
 i evaporatore a compressione
 d Vakuumpfanne

8675 vacuum shelf dryer
 f armoire à sécher par le vide
 e armario vacuosecador
 i essicatore nel vuoto
 d Vakuumtrockenschrank

8676 vacuum still
 f alambic à basse pression
 e alambique de vacío
 i alambicco a bassa pressione
 d Vakuumdestillierapparat

8677 valence; valency
 f valence
 e valencia
 i valenza
 d Valenz; Wertigkeit

8678 valence bond
 f liaison de valence
 e enlace de valencia
 i legame di valenza
 d Valenzbindung

8679 valence number
 f nombre de valence
 e número de valencia
 i numero di valenza
 d Valenzzahl

 * valency → 8677

8680 valentinite
 f valentinite
 e valentinita
 i valentinite
 d Antimonblüte; Valentinit

8681 valerian
 f valériane
 e valeriana
 i valeriana
 d Baldrian

8682 valerian oil
 f essence de valériane
 e aceite de valeriana
 i essenza di valeriana
 d Baldrianöl; Valerianöl

8683 valeric acid
 f acide valérique
 e ácido valérico
 i acido valerico
 d Baldriansäure; Valeriansäure

8684 **valerolactone**
  f valérolactone
  e valerolactona
  i valerolattone
  d Valerolacton

8685 **valine**
  f valine
  e valina
  i valina
  d Valin

  * **vanadic acid** → 8693

8686 **vanadinite**
  f vanadinite
  e vanadinita
  i vanadinite
  d Vanadinbleierz

8687 **vanadium**
  f vanadium
  e vanadio
  i vanadio
  d Vanadium; Vanad

8688 **vanadium carbide**
  f carbure de vanadium
  e carburo de vanadio
  i carburo di vanadio
  d Vanadincarbid

8689 **vanadium dichloride**
  f bichlorure de vanadium
  e dicloruro de vanadio
  i bicloruro di vanadio
  d Vanadindichlorid

8690 **vanadium driers**
  f siccatifs au vanadium
  e secadores de vanadio
  i essicatori al vanadio
  d Vanadintrockenmittel

8691 **vanadium ethylate**
  f éthylate de vanadium
  e etilato de vanadio
  i etilato di vanadio
  d Vanadinäthylat

8692 **vanadium oxytrichloride**
  f oxytrichlorure de vanadium
  e oxitricloruro de vanadio
  i ossitricloruro di vanadio
  d Vanadinoxytrichlorid

8693 **vanadium pentoxide; vanadic acid**
  f pentoxyde de vanadium
  e pentóxido de vanadio

  i pentossido di vanadio
  d Vanadinpentoxyd

8694 **vanadium sulphide**
  f sulfure de vanadium
  e sulfuro de vanadio
  i solfuro di vanadio
  d Vanadinsulfid

8695 **vanadium tetrachloride**
  f tétrachlorure de vanadium
  e tetracloruro de vanadio
  i tetracloruro di vanadio
  d Vanadintetrachlorid

8696 **vanadium tetraoxide**
  f tétraoxyde de vanadium
  e tetróxido de vanadio
  i tetrossido di vanadio
  d Vanadintetraoxyd

8697 **vanadium trichloride**
  f trichlorure de vanadium
  e tricloruro de vanadio
  i tricloruro di vanadio
  d Vanadintrichlorid

8698 **vanadium trioxide**
  f trioxyde de vanadium
  e trióxido de vanadio
  i triossido di vanadio
  d Vanadiumtrioxyd

8699 **vanadyl chlorid**
  f chlorure de vanadyle
  e cloruro de vanadilo
  i cloruro di vanadile
  d Vanadylchlorid

8700 **vanadyl sulphate**
  f sulfate de vanadyle
  e sulfato de vanadilo
  i solfato di vanadile
  d Vanadylsulfat

8701 **vancomycin hydrochloride**
  f chlorhydrate de vancomycine
  e hidrocloruro de vancomicina
  i cloridrato di vancomicina
  d Vancomycinhydrochlorid

8702 **Vandyke brown**
  f brun foncé; brun Van Dyke
  e pardo de Van-Dyke
  i marrone di Van Dyke
  d Van-Dyke-Braun

8703 **vane-type draught gauge**
  f déprimomètre à volet

e deprimómetro de batiente
i deprimometro a battente
d Klappenzugmesser

8704 vanillin; vanillic aldehyde
f vanilline
e vainillina
i vaniglina
d Vanillin

8705 vaporimeter
f vaporimètre
e vaporímetro
i vaporimetro
d Vaporimeter; Verdampfungsmesser

8706 vaporization
f vaporisation
e vaporización
i evaporazione; vaporizzazione
d Verdampfung

8707 vaporization heat
f chaleur de vaporisation
e calor de vaporización
i calore di vaporizzazione
d Verdampfungswärme

8708 vaporize *v*
f vaporiser
e vaporizar; nebulizar
i vaporizzare
d verdampfen

8709 vapour
f vapeur
e vapor
i vapore
d Dampf

8710 vapour bonnet
f capot de vapeur
e cámara de vapor
i camera di vapore
d Brüdenraum

8711 vapour degreasing
f dégraissage à la vapeur
e desengrase por vapor
i sgrassamento a vapore
d Dampfentfetten

8712 vapour degreasing chamber
f chambre de dégraissage à la vapeur
e cámara de desengrase por vapor
i camera di sgrassamento a vapore
d Dampfentfettungsraum

8713 vapour density
f densité de vapeur
e densidad de vapor
i densità di vapore
d Dampfdichte

8714 vapour lock
f poche de vapeur
e bolsa de vapor
i bolla di vapore
d Dampfsack

8715 vapour pressure; steam pressure
f tension de vapeur; pression de vapeur
e tensión del vapor; presión del vapor
i tensione di vapore; pressione di vapore
d Dampfdruck

8716 varnish
f vernis
e laca; barniz
i vernice
d Lack

8717 vasopressin
f vasopressine
e vasopresina
i vasopressina
d Vasopressin

8718 vatted
f mûr
e hecho
i maturato
d abgelagert

8719 vegetable black
f noir végétal
e negro vegetal
i nero vegetale
d Pflanzenkohle

8720 vegetable charcoal
f charbon végétal
e carbón vegetal
i carbone vegetale
d Pflanzenkohle; Holzkohle

8721 vehicle
f véhicule; milieu de suspension
e medio; excipiente líquido
i veicolo
d Bindemittel; Medium

8722 Venetian red
f rouge vénitien
e rojo veneciano
i rosso di Venezia
d Zementrot

8723 Venice turpentine
f térébenthine de Venise; térébenthine du mélèze
e trementina de Venecia
i trementina di Venezia
d Lärchenterpentin; Venezianeröl

8724 veratrine
f vératrine
e veratrina
i veratrina
d Veratrin

8725 veratrol
f vératrol
e veratrol
i veratrolo
d Veratrol

8726 veratrum
f vératre
e veratro
i veratro
d Veratrum

8727 verbena oil
f essence de verveine
e esencia de verbena
i essenza di verbena
d Verbenaöl

8728 verdigris
f vert-de-gris
e cardenillo verde
i verderame
d Grünspan

8729 verdunization
f verdunisation
e verdunización
i verdunizzazione
d Chlorsterilisierung von Wasser

8730 vermicide; vermifuge; helminthicide
f vermicide; vermifuge
e vermicida
i vermicida; vermifugo
d Wurmmittel

8731 vermiculite
f vermiculite
e vermiculita
i vermiculite
d Vermiculit

* **vermifuge** → 8730

8732 vermilion
f vermillon

e bermellón
i vermiglione cinabro
d Vermillon; Zinnoberfarbe

* **veronal** → 783

8733 vetiverol
f vétivérol
e vetiverol
i vetiverolo
d Vetiverol

8734 vetivert acetate
f acétate de vétiver
e acetato de vetiver
i acetato di andropogon muricatus
d Vetiveracetat

8735 vetivert oil
f essence de vétiver
e aceite de vetiver
i olio di andropogon muricatus
d Vetiveröl

8736 viability
f viabilité
e viabilidad
i viabilità
d Lebensfähigkeit

8737 viable
f viable
e viable
i vitale
d lebensfähig

8738 vibrating screen; vibrating sieve
f crible oscillant
e criba oscilante
i staccio oscillante
d Schüttelsieb

8739 viburnum
f viorne
e viburno
i viburno
d Viburnumrinde

8740 vinal
f vinal
e vinal
i vinale
d Vinal

8741 vinbarbital
f vinbarbital
e vinbarbital
i vinbarbital
d Vinbarbital

8742 vinegar
  f vinaigre
  e vinagre
  i aceto
  d Essig

8743 vinegar generator
  f générateur de vinaigre
  e generador de vinagre
  i generatore di aceto
  d Essigerzeuger

  * **vinometer** → 5932

8744 vinyl acetate
  f acétate de vinyle
  e acetato de vinilo
  i acetato di vinile
  d Vinylacetat

8745 vinylacetylene
  f vinylacétylène
  e vinilacetileno
  i vinilacetilene
  d Vinylacetylen

8746 vinylation
  f vinylation
  e vinilación
  i vinilazione
  d Vinylation

8747 vinyl butyl ether
  f vinylbutyléther
  e vinilbutiléter
  i etere vinilbutilico
  d Vinylbutyläther

8748 vinyl butyrate
  f butyrate de vinyle
  e butirato de vinilo
  i vinilbutirato
  d Vinylbutyrat

8749 vinylcarbazole
  f vinylcarbazole
  e vinilcarbazol
  i vinilcarbazolo
  d Vinylcarbazol

8750 vinyl chloride; chloroethylene
  f chlorure de vinyle
  e cloruro de vinilo
  i cloruro di vinile
  d Vinylchlorid

8751 vinyl compounds
  f composés vinyliques
  e compuestos de vinilo

  i composti vinilici
  d Vinylverbindungen

8752 vinyl cyclohexene
  f vinylcyclohexène
  e vinilciclohexeno
  i cicloesene di vinile
  d Vinylcyclohexen

8753 vinyl cyclohexene monoxide
  f monoxyde de vinylcyclohexène
  e monóxido de vinilciclohexeno
  i monossido di vinilcicloesene
  d Vinylcyclohexenmonoxyd

8754 vinyl ether
  f vinyléther; éther vinylique
  e viniléter
  i etere vinilico
  d Vinyläther

8755 vinyl ethoxyethyl sulphide
  f sulfure de vinyléthoxyéthyle
  e sulfuro de viniletoxietilo
  i solfuro di viniletossietile
  d Vinyläthoxyäthylsulfid

8756 vinyl ethyl ether
  f vinyléthyléther; éther vinyléthylique
  e viniletiléter
  i etere viniletilico
  d Vinyläthyläther

8757 vinyl ethyl hexoate
  f hexoate de vinyléthyle
  e hexoato de viniletilo
  i esoato di viniletile
  d Vinyläthylhexoat

8758 vinyl ethylhexyl ether
  f vinyléthylhexyléther
  e viniletilhexiléter
  i etere viniletilesilico
  d Vinyläthylhexyläther

8759 vinyl ethylpyridine
  f vinyléthylpyridine
  e viniletilpiridina
  i viniletilpiridina
  d Vinyläthylpyridin

8760 vinyl fluoride
  f fluorure de vinyle
  e fluoruro de vinilo
  i fluoruro di vinile
  d Vinylfluorid

8761 vinylidene chloride
  f chlorure de vinylidène

e cloruro de vinilideno
i cloruro di vinilidene
d Vinylidenchlorid

8762 vinylidene fluoride
f fluorure de vinylidène
e fluoruro de vinilideno
i fluoruro di vinilidene
d Vinilidenfluorid

8763 vinylidene resins
f résines de vinylidène
e resinas de vinilideno
i resine di vinilidene
d Vinylidenkunstharze

8764 vinyl isobutyl ether
f éther vinylisobutylique
e vinilisobutiléter
i etere vinilisobutilico
d Vinylisobutyläther

8765 vinyl methyl ether
f vinylméthyléther; éther vinylméthylique
e vinilmetiléter
i etere vinilmetilico
d Vinylmethyläther

8766 vinyl plastics
f plastiques vinyliques
e plásticos de vinilo
i plastiche viniliche
d Vinylkunststoffe

8767 vinyl propionate
f propionate de vinyle
e propionato de vinilo
i propionato di vinile
d Vinylpropionat

8768 vinyl pyridine
f pyridine de vinyle
e vinilpiridina
i vinilpiridina
d Vinylpyridin

8769 vinyl resins
f résines vinyliques
e resinas vinílicas
i resine viniliche
d Vinylharze

8770 vinyl stabilizers
f stabilisants de vinyle
e estabilizadores de vinilo
i stabilizzatori di vinile
d Vinylstabilisatoren

8771 vinyl stearate
f stéarate de vinyle
e estearato de vinilo
i stearato di vinile
d Vinylstearat

8772 vinyl toluene
f toluène de vinyle
e viniltolueno
i toluene di vinile
d Vinyltoluol

8773 vinyl trichlorosilane
f trichlorosilane de vinyle
e viniltriclorosilano
i triclorosilano di vinile
d Vinyltrichlorsilan

8774 vinyon
f vinyon
e vinión
i vinion
d Vinyon

8775 viomycin
f viomycine
e viomicina
i viomicina
d Viomycin

* **viridian green** → 3895

8776 virogenetic segment
f segment virogénétique
e segmento virogenético
i segmento virogenetico
d virogenetisches Segment

* **visbreaking** → 8782

8777 viscometer; viscosimeter
f viscosimètre
e viscosímetro
i viscosimetro
d Viskosimeter; Viskositätsmesser

8778 viscose
f viscose
e viscosa
i viscosa
d Viscose

8779 viscose process
f procédé à la viscose
e método de la viscosa
i procedimento alla viscosa
d Viscoseverfahren

8780 viscose rayon
f rayonne de viscose
e rayón viscosa
i rayon viscosa
d Viscoseseide

* **viscosimeter** → 8777

8781 viscosity
f viscosité
e viscosidad
i viscosità
d Viskosität; Zähigkeit

8782 viscosity-breaking; visbreaking
f réduction de viscosité
e reducción de viscosidad por descomposición
i riduzione di viscosità
d Viskositätsbrechen

8783 viscosity depressant
f réducteur de viscosité
e depresor de viscosidad
i riduttore di viscosità
d Viskositätsverminderer

8784 viscous flow
f écoulement visqueux
e flujo viscoso
i flusso viscoso
d plastisches Fließen

8785 vitamin
f vitamine
e vitamina
i vitamina
d Vitamin

* **vitamin B$_1$** → 8227

* **vitamin B$_2$** → 4677, 7090

* **vitamin B$_6$** → 6820

* **vitamin C** → 685

* **vitamin H** → 970

8786 vitrification
f vitrification
e vitrificación
i vitrificazione
d Glasierung

8787 vivianite
f vivianite
e vivianita

i vivianite
d Blaustein; Eisenblau; Vivianit

8788 volatile constituents
f constituents volatils
e constituyentes volátiles
i costituenti volatili
d flüchtige Bestandteile

8789 volatility
f volatilité
e volatilidad
i volatilità
d Flüchtigkeit

8790 volume percentage
f pourcentage en volume
e porcentaje en volumen
i percentuale di volume
d Volumenprozentzahl

8791 volumetric analysis
f analyse volumétrique
e análisis volumétrico
i analisi volumetrica
d Maßanalyse; volumetrische Analyse

8792 vortex
f vortex
e vórtice
i vortice
d Wirbel

8793 vortex separator
f séparateur turbulent
e separador turbulento
i separatore turbolento
d Wirbeltrenner

8794 vulcanite
f vulcanite
e vulcanita
i vulcanite
d Vulkanit

8795 vulcanization
f vulcanisation
e vulcanización
i vulcanizzazione
d Vulkanisierung

8796 vulcanization rate
f vitesse de vulcanisation
e relación de vulcanización; velocidad de vulcanización
i velocità di vulcanizzazione
d Vulkanisiergeschwindigkeit

8797 vulcanized fibre
  f fibre vulcanisée
  e fibra vulcanizada
  i fibra vulcanizzata
  d Vulkanfaser

W

8798 walnut oil
 f huile de noix
 e aceite de nuez
 i olio di noce
 d Walnußöl

8799 warfarin
 f warfarine
 e warfarina; varfarina
 i varfarina
 d Warfarin

8800 warfarin sodium
 f warfarine sodique
 e varfarina sódica
 i varfarina sodica
 d Warfarinnatrium

8801 wash-bottle
 f pissette
 e matraz de lavado
 i spruzzetta
 d Spritzflasche

8802 washing room
 f chambre de lavage
 e cámara de lavado
 i camera di lavaggio
 d Spülraum

8803 waste-heat boiler
 f récupérateur
 e caldera de recuperación
 i caldaia di ricupero
 d Abgaskessel

8804 watch glass
 f verre de montre
 e vidrio de reloj
 i vetro da orologio
 d Uhrglas

8805 water
 f eau
 e agua
 i acqua
 d Wasser

8806 water absorption
 f absorption d'eau
 e absorción de agua
 i assorbimento d'acqua
 d Wasseraufnahme

8807 water bath
 f bain-marie

 e baño María
 i bagnomaria; bagno d'acqua
 d Wasserbad

8808 water colours
 f couleurs à l'eau; couleurs à l'aquarelle
 e acuarelas
 i acquarelli
 d Wasserfarben

8809 water equivalent
 f équivalent en eau
 e equivalente en agua
 i equivalente in acqua
 d Wasserwert

8810 water gas
 f gaz à l'eau
 e gas de agua
 i gas d'acqua
 d Wassergas

8811 water-gauge pressure
 f pression de colonne d'eau
 e presión de columna de agua
 i pressione a colonna d'acqua
 d Wasserstandsdruck

* water glass → 6645, 7602

8812 watermark
 f filigrane
 e corondel; filigrana
 i filigrana
 d Wasserzeichen

8813 water of crystallization
 f eau de cristallisation
 e agua de cristalización
 i acqua di cristallizzazione
 d Kristallwasser

8814 water-repellent
 f hydrofuge
 e impenetrable al agua; repelente de agua
 i idrorepellente
 d wasserabstoßend

8815 water resistance
 f résistance à l'eau
 e resistencia líquida
 i resistenza all'acqua
 d Wasserbeständigkeit

8816 water-tube boiler
 f chaudière aquatubulaire; chaudière à
    tubes d'eau
 e caldera acuotubular

i caldaia a tubi d'acqua
d Siederohrkessel; Wasserrohrkessel

8817 water-vapour absorption
f absorption de vapeur d'eau
e absorción del vapor de agua
i assorbimento di vapor d'acqua
d Wasserdampfaufnahme

8818 wavellite
f wavellite
e wavelita
i wavellite
d Wavellit

8819 wax
f cire
e cera; parafina
i cera
d Wachs

8820 waxing
f cirage
e encerado
i inceratura
d Wachsen

8821 weathering test
f essai de vieillissement aux intempéries
e ensayo de envejecimiento para intemperie
i assaggio d'invecchiamento per intemperie
d Bewetterungsprüfung

8822 webbing
f craquelure
e aspecto agrietado
i ragnatura
d Rißbildung

8823 wedge-wire screen
f surface criblante à fils métalliques
e superficie cribante de hilos perfilados
i setaccio a fili sagomati
d Profildrahtsieb

8824 weighing bottle
f verre de pesée
e frasquito para pesadas
i bicchiere per pesare
d Wiegeflasche

8825 weight percentage
f pourcentage en poids
e porcentaje en peso
i percentuale in peso
d Gewichtsprozentzahl

8826 Weldon's process
f procédé Weldon
e método de Weldon
i processo di Weldon
d Weldonsches Verfahren

8827 Weston normal cell
f pile étalon Weston
e pila patrón Weston
i pila campione Weston
d Weston-Normalelement

8828 wet assay
f analyse par voie humide
e análisis por vía húmeda
i saggio per via umida
d Naßprobe

8829 wet binder
f liant humide
e aglomerante acuoso
i agglutinante acquoso
d flüssiges Bindemittel

8830 wet cell
f pile liquide
e pila líquida
i pila a liquido
d Naßelement

8831 wet process
f procédé humide; voie humide
e método húmedo
i processo per via umida
d Naßverfahren

8832 wet spinning
f filage au mouillé
e hilatura en húmedo
i filatura a umido
d Naßspinnen

8833 wet steam
f vapeur humide
e vapor húmido
i vapore umido
d nasser Dampf

8834 wet strength
f résistance en condition mouillée
e resistencia en condiciones húmedas
i resistenza in condizioni umide
d Naßfestigkeit

8835 wetting agent
f agent mouillant
e agente humector
i agente bagnante
d Netzmittel

8836 wetting power
f pouvoir mouillant; pouvoir humectant
e poder humectante
i potere umettante
d Benetzungsfähigkeit

8837 whale oil
f huile de baleine
e aceite de ballena
i olio di balena
d Walfischtran

* **wheel ore → 1148**

8838 white arsenic
f arsenic blanc
e arsénico blanco
i triossido di arsenico
d Arsenblüte; Giftstein; weißer Arsenik

8839 white French polish
f vernis au tampon blanc
e barniz blanco de muñeca
i lacca limpida a tampone
d Klarlack

8840 white lac
f laque blanchie
e laca blanca
i vernice bianca
d gebleichter Schellack

8841 white lead
f blanc de céruse; plomb blanc
e plomo blanco
i biacca di piombo
d Bleiweiß; Deckweiß

8842 white liquor
f liqueur blanche
e licor blanco
i liscivia bianca
d Weißlauge

8843 white oils
f huiles paraffinées
e aceites blancos
i oli bianchi
d weiße Öle

8844 white spirit
f essence de térébenthine artificielle;
white-spirit
e espíritu blanco
i acqua ragia minerale
d Terpentinölersatz

* **white vitriol → 8970**

8845 white water
f eau blanche
e agua blanca
i acqua di ricupero
d Siebwasser; Kreidewasser; Abwasser der
Papierfabrikation

8846 whiting
f blanc d'Espagne; blanc de Meudon
e blanco de España
i gesso in polvere; bianco di Spagna
d Kalktünche; Schlämmkreide

8847 willemite
f willémite
e wilemita
i willemite
d Kieselzinkerz

8848 wine vinegar
f vinaigre de vin
e vinagre de vino
i aceto di vino
d Weinessig

8849 wintergreen oil
f essence de gaulthéria; essence de
wintergreen
e aceite de pirola
i olio di sempreverdi
d Wintergrünöl

8850 wire
f tamis métallique; toile métallique
e malla metálica
i tela
d Drahtgaze

8851 wire gauze
f tissu métallique
e tela metálica
i rete metallica
d Drahtgewebe

8852 wire screen
f tamis métallique
e tamiz de tela metálica
i rete metallica
d Drahtsieb

8853 witherite
f withérite
e witerita
i viterite
d Witherit

8854 wolframite
f wolframite
e wolframita

i volframite
d Wolframit

8855 wollastonite
f wollastonite
e wolastonita; volastonita
i vollastonite
d Kieselkalkspat; Wollastonit

* **wood alcohol** → **8860**

8856 wood flour
f farine de bois
e aserrín
i farina di legno
d Holzmehl

8857 wood gas
f gaz de bois
e gas de madera
i gas di legno
d Holzgas

8858 wood pulp
f pâte de bois
e pulpa de madera; pasta de madera
i pasta di legno
d Holzzellstoff

8859 wood pulp black
f noir de pâte de bois
e negro de pulpa de madera
i nero di pasta di legno
d Holzstoffschwarz

8860 wood spirit; wood alcohol
f esprit de bois; alcool de bois
e espíritu de madera; alcohol de madera
i spirito di legno; alcool di legno
d Holzgeist

* **wool fat** → **4713**

8861 Woolff's bottle
f flacon de Woolff
e frasco de Woolff
i bottiglia di Woolff
d Woolffsche Flasche

8862 worm knotter
f compresseur rotatif
e compresor rotatorio de pulpa
i separanodi elicoidale
d Knotenfänger

8863 wormwood oil; absinthe oil
f essence d'absinthe
e aceite de ajenjo

i olio d'assenzio
d Wermutöl

8864 wort
f moût
e mosto
i mosto
d Würze

8865 wulfenite
f wulfénite; plomb jaune
e wulfenita
i wulfenite
d Gelbbleierz; Wulfenit

8866 Wulff process
f méthode Wulff
e método de Wulff
i processo di Wulff
d Wulff-Verfahren

8867 wurtzite
f wurtzite
e wurtzita
i wurtzite
d Wurtzit

8868 Wurtz synthesis
f synthèse de Wurtz
e síntesis de Wurtz
i sintesi di Wurtz
d Wurtzsche Synthese

X

8869 xanthene
f xanthène
e xanteno
i xantene
d Xanthen

8870 xanthene carboxylic acid
f acide carboxylique de xanthène
e ácido xantenocarboxílico
i acido xantenocarbossilico
d Xanthencarboxylsäure

8871 xanthene dyestuffs
f colorants de xanthène
e colorantes de xanteno
i coloranti di xantene
d Xanthenfarbstoffe

* xanthenol → 8878

8872 xanthic acid
f acide xanthique
e ácido xántico
i acido xantico
d Xanthosäure

8873 xanthine; dioxopurine
f xanthine
e xantina
i xantina
d Xanthin

8874 xanthocilin
f xanthocilline
e xantocilina
i xantocillina
d Xanthocillin

8875 xanthophyll
f xanthophylle
e xantofila
i xantofilla
d Xanthophyll

8876 xanthopterin
f xanthoptèrine
e xantopterina
i xantopterina
d Xanthopterin

8877 xanthotoxin; methoxsalen
f xanthotoxine
e xantotoxina
i xantossina
d Xanthotoxin

8878 xanthydrol; xanthenol
f xanthydrol
e xantidrol
i xantidrolo
d Xanthydrol

8879 xenon
f xénon
e xenón
i xeno
d Xenon

8880 xenoplastic
f xénoplastique
e xenoplástico
i xenoplastico
d xenoplastisch

8881 xenotime
f xénotime
e xenotima
i xenotimo
d Xenotim; Ytterspat

8882 xylene; dimethylbenzene
f xylène
e xileno
i xilene
d Xylol

8883 xylenol; hydroxydimethol benzene
f xylénol
e xilenol
i xilenolo
d Xylenol

8884 xylenol resins
f résines de xylénol
e resinas de xilenol
i resine di xilenolo
d Xylenolharze

8885 xylidine; aminodimethyl benzene
f xylidine
e xilidina
i xilidina
d Xylidin

8886 xylitol; pentanepentol
f xylithol
e xilitol
i xilitolo
d Xylithol

8887 xylol
f xylol
e xilol
i xilolo
d Xylol

8888 xylometaloline hydrochloride
 f chlorhydrate de xylométaloline
 e hidrocloruro de xilometalolina
 i cloridrato di xilometalolina
 d Xylometalolinhydrochlorid

8889 xylose
 f xylose; sucre de bois
 e xilosa
 i xilosio
 d Xylose; Holzzucker

8890 xylyl bromide
 f bromure de xylyle
 e bromuro de xililo
 i bromuro di xilile
 d Xylylbromid

Y

8891 yacht varnish
f vernis pour yachts
e barniz para yates
i vernice per scafi
d Yachtfarbe; Bootslack

8892 yeast
f levure
e levadura; fermento
i lievito
d Hefe

8893 yeast food
f aliment de la levure
e nutrición de la levadura
i alimento per lievito
d Hefenahrung

8894 yeast head
f couvercle
e cubierta
i coperta
d Gärdecke

8895 yeast pressings
f bière de levure
e cerveza por prensa
i resti ricuperati dal lievito
d Abpreßbier

8896 yeast strain
f race de levure
e raza de levadura
i razza di lievito
d Heferasse

* **yellow chrome** → 1800

8897 yellowing
f jaunissement
e cambio de un color al amarillo
i ingiallimento
d Vergelbung

8898 yellowing resistance
f résistance au jaunissement
e resistencia a amarillearse
i resistenza all'ingiallimento
d Beständigkeit gegen Vergilbung

8899 yellow mercuric oxide
f oxyde de mercure jaune
e óxido mercúrico amarillo
i ossido mercurico giallo
d gelbes Quecksilberoxyd

8900 yield point
f limite élastique
e límite aparente de elasticidad
i limite di carico di snervamento
d Streckgrenze

8901 ylang ylang oil
f essence d'ylang-ylang
e aceite de ilang ilang
i essenza di ylang ylang
d Ylang-Ylangöl

8902 yohimbine; quebrachine
f yohimbine
e yohimbina
i yoimbina
d Yohimbin

8903 young fustic
f fustet
e fustete joven
i colorante di cotino
d Fustikholz

8904 Young's modulus
f module d'Young
e módulo de Young
i modulo di Young
d Youngscher Elastizitätsmodul

8905 ytterbium
f ytterbium
e iterbio
i itterbio
d Ytterbium

8906 yttrium
f yttrium
e itrio
i ittrio
d Yttrium

8907 yttrium acetate
f acétate d'yttrium
e acetato de itrio
i acetato di ittrio
d Yttriumacetat

8908 yttrium chloride
f chlorure d'yttrium
e cloruro de itrio
i cloruro di ittrio
d Yttriumchlorid

8909 yttrium oxide
f oxyde d'yttrium
e óxido de itrio
i ossido di ittrio
d Yttriumerde; Yttriumoxyd

8910 yttrium sulphate
   f sulfate d'yttrium
   e sulfato de itrio
   i solfato di ittrio
   d Yttriumsulfat

Z

8911 z; zygotic chromosome number
 f z; nombre chromosomique du zygote
 e z; número cromosómico zigótico
 i z; numero cromosomico zigotico
 d z; zygotische Chromosomenzahl

8912 zein
 f zéine
 e zeína
 i zeina
 d Zein

8913 zeolites; molecular sieves
 f zéolithes
 e zeolitas; cribas moleculares
 i zeoliti
 d Zeolithe

8914 Ziegler catalyst
 f catalyseur de Ziegler
 e catalizador de Ziegler
 i catalizzatore di Ziegler
 d Ziegler-Katalysator

8915 Ziegler process
 f procédé Ziegler
 e método Ziegler
 i processo di Ziegler
 d Ziegler-Verfahren

8916 zinc
 f zinc
 e zinc; cinc
 i zinco
 d Zink

8917 zinc acetate
 f acétate de zinc
 e acetato de zinc
 i acetato di zinco
 d Zinkacetat

8918 zinc ammonium chloride
 f chlorure de zinc et d'ammonium
 e cloruro de zinc y amonio
 i cloruro di zinco e ammonio
 d Zinkammoniumchlorid

8919 zinc arsenate
 f arséniate de zinc
 e arseniato de zinc
 i arseniato di zinco
 d Zinkarsenat

8920 zinc arsenite
 f arsénite de zinc

 e arsenito de zinc
 i arsenito di zinco
 d Zinkarsenit

* zinc blende → 1041

8921 zinc borate
 f borate de zinc
 e borato de zinc
 i borato di zinco
 d Zinkborat

8922 zinc bromide
 f bromure de zinc
 c bromuro de zinc
 i bromuro di zinco
 d Bromzink

8923 zinc caprylate
 f caprylate de zinc
 e caprilato de zinc
 i caprilato di zinco
 d Zinkcaprylat

8924 zinc carbonate
 f carbonate de zinc
 e carbonato de zinc
 i carbonato di zinco
 d kohlensaures Zink; Zinkspat

8925 zinc chloride
 f chlorure de zinc
 e cloruro de zinc
 i cloruro di zinco
 d Chlorzink

8926 zinc chloroiodide
 f chloroiodure de zinc
 e cloroyoduro de zinc
 i cloroioduro di zinco
 d Zinkchlorjodid

8927 zinc chromate
 f chromate de zinc
 e cromato de zinc
 i cromato di zinco
 d Zinkchromat

8928 zinc chrome; zinc yellow
 f chromate jaune de zinc; jaune de zinc
 e cromo-zinc; amarillo de zinc
 i giallo di zinco
 d Zinkchromgelb; Zinkgelb

8929 zinc cyanide
 f cyanure de zinc
 e cianuro de zinc
 i cianuro di zinco
 d Zinkcyanid; Cyanzink

8930 zinc dibutyldithiocarbamate
 f dibutyldithiocarbamate de zinc
 e dibutilditiocarbamato de zinc
 i dibutilditiocarbammato di zinco
 d Zinkdibutyldithiocarbamat

8931 zinc dichromate
 f bichromate de zinc
 e dicromato de zinc
 i bicromato di zinco
 d Zinkdichromat

8932 zinc diethyl; diethylzinc
 f zinc diéthylique
 e zinc dietilo
 i zinco dietilico
 d Zinkdiäthyl

8933 zinc diethyldithiocarbamate
 f diéthyldithiocarbamate de zinc
 e dietilditiocarbamato de zinc
 i dietilditiocarbammato di zinco
 d Zinkdiäthyldithiocarbamat

8934 zinc dimethyldithiocarbamate
 f diméthyldithiocarbamate de zinc
 e dimetilditiocarbamato de zinc
 i dimetilditiocarbammato di zinco
 d Zinkdimethyldithiocarbamat

8935 zinc driers
 f siccatifs de zinc
 e desecantes de zinc
 i essicatori di zinco
 d Zinktrockenstoffe

8936 zinc dust
 f poudre de zinc
 e polvo de zinc
 i zinco in polvere
 d Zinkmehl; Zinkstaub

8937 zinc ethylenebisdithiocarbamide
 f éthylènebisdithiocarbamide de zinc
 e etilenbisditiocarbamida de zinc
 i etilenebisditiocarbammide di zinco
 d Zinkäthylenbisdithiocarbamid

8938 zinc ethylsulphate
 f éthylsulfate de zinc
 e etilsulfato de zinc
 i etilsolfato di zinco
 d Zinkäthylsulfat

8939 zinc fluoride
 f fluorure de zinc
 e fluoruro de zinc
 i fluoruro di zinco
 d Fluorzink; Zinkfluorid

8940 zinc formaldehyde sulphoxylate
 f sulfoxylate de formaldéhyde et de zinc
 e sulfoxilato de formaldehído y zinc
 i solfoxilato di formaldeide e zinco
 d Zinkformaldehydsulfoxylat

8941 zinc formate
 f formiate de zinc
 e formiato de zinc
 i formiato di zinco
 d Zinkformiat

* zinc green → 1901

8942 zinc hydrosulphite
 f hydrosulfite de zinc
 e hidrosulfito de zinc
 i idrosolfito di zinco
 d Zinkhydrosulfit

8943 zinc iodide
 f iodure de zinc
 e yoduro de zinc
 i ioduro di zinco
 d Jodzink; Zinkjodid

8944 zincite
 f zincite
 e zincita
 i zincite
 d Zinkit; rotes Zinkerz

8945 zinc laurate
 f laurate de zinc
 e laurato de zinc
 i laurato di zinco
 d Zinklaurat

8946 zinc linoleate
 f linoléate de zinc
 e linoleato de zinc
 i linoleato di zinco
 d Zinklinoleat

8947 zinc naphthenate
 f naphténate de zinc
 e naftenato de zinc
 i naftenato di zinco
 d Zinknaphthenat

8948 zinc nitrate
 f nitrate de zinc
 e nitrato de zinc
 i nitrato di zinco
 d Zinknitrat

8949 zinc octoate
 f octoate de zinc
 e octoato de zinc

i ottoato di zinco
d Zinkoctoat

8950 zinc oleate
f oléate de zinc
e oleato de zinc
i oleato di zinco
d Zinkoleat

8951 zinc oxalate
f oxalate de zinc
e oxalato de zinc
i ossalato di zinco
d Zinkoxalat

8952 zinc oxide; zinc white
f oxyde de zinc; blanc de zinc
e óxido de zinc; blanco de zinc
i ossido di zinco; bianco di zinco
d Zinkoxyd; Zinkweiß

8953 zinc palmitate
f palmitate de zinc
e palmitato de zinc
i palmitato di zinco
d Zinkpalmitat

8954 zinc perborate
f perborate de zinc
e perborato de zinc
i perborato di zinco
d Zinkperborat

8955 zinc permanganate
f permanganate de zinc
e permanganato de zinc
i permanganato di zinco
d Zinkpermanganat

8956 zinc peroxide
f peroxyde de zinc
e peróxido de zinc
i perossido di zinco
d Zinkperoxyd

8957 zinc phenate
f phénate de zinc
e fenato de zinc
i fenato di zinco
d Zinkphenat

8958 zinc phenolsulphonate
f phénolsulfonate de zinc
e fenolsulfonato de zinc
i fenolsolfonato di zinco
d Zinkphenolsulfonat

8959 zinc phosphate
f phosphate de zinc

e fosfato de zinc
i fosfato di zinco
d Zinkphosphat

8960 zinc phosphide
f phosphure de zinc
e fosfuro de zinc
i fosfuro di zinco
d Zinkphosphid

8961 zinc phosphite
f phosphite de zinc
e fosfito de zinc
i fosfito di zinco
d Zinkphosphit

8962 zinc propionate
f propionate de zinc
e propionato de zinc
i propionato di zinco
d Zinkpropionat

8963 zinc pyrophosphate
f pyrophosphate de zinc
e pirofosfato de zinc
i pirofosfato di zinco
d Zinkpyrophosphat

8964 zinc resinate
f résinate de zinc
e resinato de zinc
i resinato di zinco
d Zinkresinat

8965 zinc ricinoleate
f ricinoléate de zinc
e ricinoleato de zinc
i ricinoleato di zinco
d Zinkricinoleat

8966 zinc salicylate
f salicylate de zinc
e salicilato de zinc
i salicilato di zinco
d Zinksalicylat

8967 zinc silicofluoride
f silicofluorure de zinc
e silicofluoruro de zinc
i silicofluoruro di zinco
d Zinksilicofluorid

8968 zinc stearate
f stéarate de zinc
e estearato de zinc
i stearato di zinco
d Zinkstearat

8969 zinc sulphanilate
f sulfanilate de zinc
e sulfanilato de zinc
i solfanilato di zinco
d Zinksulfanilat

8970 zinc sulphate; white vitriol
f sulfate de zinc
e sulfato de zinc
i solfato di zinco
d Bergbutter; weißer Vitriol; Zinksulfat

8971 zinc sulphide
f sulfure de zinc
e sulfuro de zinc
i solfuro di zinco
d Zinksulfid

8972 zinc sulphite
f sulfite de zinc
e sulfito de zinc
i solfito di zinco
d Zinksulfit

8973 zinc sulphoxylate
f sulfoxylate de zinc
e sulfoxilato de zinc
i solfoxilato di zinco
d Zinksulfoxylat

8974 zinc tallate
f tallate de zinc
e talato de zinc
i tallato di zinco
d Zinktallat

8975 zinc thiocyanate
f thiocyanate de zinc
e tiocianato de zinc
i tiocianato di zinco
d Zinkthiocyanat

8976 zinc undecylenate; zinc undecanoate
f undécylénate de zinc
e undecilenato de zinc
i undecilenato di zinco
d Zinkundecylenat

8977 zinc valerate
f valérianate de zinc
e valerianato de zinc
i valerianato di zinco
d Zinkvalerianat

* **zinc white** → 8952

* **zinc yellow** → 8928

8978 zinc zirconium silicate
f silicate de zinc et de zirconium
e silicato de zinc y zirconio
i silicato di zirconio e zinco
d Zinkzirkoniumsilicat

8979 zinkenite
f zinkénite
e zinkenita
i zinchenite
d Bleiantimonerz

8980 zippeite
f zippéite
e zipeita
i zippeite
d Zippeit

8981 zircon
f zircon
e zircón
i zircone
d Zirkon

8982 zirconium
f zirconium
e zirconio; circonio
i zirconio
d Zirkonium; Zirkon

8983 zirconium acetate
f acétate de zirconium
e acetato de zirconio
i acetato di zirconio
d Zirkonacetat

8984 zirconium acetylacetonate
f acétylacétonate de zirconium
e acetilacetonato de zirconio
i acetilacetonato di zirconio
d Zirkonacetylacetonat

8985 zirconium boride
f borure de zirconium
e boruro de zirconio
i boruro di zirconio
d Zirkonborid

8986 zirconium carbide
f carbure de zirconium
e carburo de zirconio
i carburo di zirconio
d Zirkoncarbid

8987 zirconium carbonate
f carbonate de zirconium
e carbonato de zirconio
i carbonato di zirconio
d Zirkoncarbonat

8988 zirconium dioxide
 f bioxyde de zirconium
 e dióxido de zirconio
 i biossido di zirconio
 d Zirkondioxyd

8989 zirconium glycolate
 f glycolate de zirconium
 e glicolato de zirconio
 i glicolato di zirconio
 d Zirkonglycolat

8990 zirconium hydride
 f hydrure de zirconium
 e hidruro de zirconio
 i idruro di zirconio
 d Zirkonhydrid

8991 zirconium hydroxide
 f hydroxyde de zirconium
 e hidróxido de zirconio
 i idrossido di zirconio
 d Zirkonhydroxyd

8992 zirconium lactate
 f lactate de zirconium
 e lactato de zirconio
 i lattato di zirconio
 d Zirkonlactat

8993 zirconium naphthenate
 f naphténate de zirconium
 e naftenato de zirconio
 i naftenato di zirconio
 d Zirkonnaphthenat

8994 zirconium nitrate
 f nitrate de zirconium
 e nitrato de zirconio
 i nitrato di zirconio
 d Zirkonnitrat

8995 zirconium nitride
 f nitrure de zirconium
 e nitruro de zirconio
 i nitruro di zirconio
 d Zirkonnitrid

8996 zirconium oxide
 t oxyde de zirconium
 e óxido de zirconio
 i ossido di zirconio
 d Zirkonoxyd

8997 zirconium oxychloride
 f oxychlorure de zirconium
 e oxicloruro de zirconio
 i ossicloruro di zirconio
 d Zirkonoxychlorid

8998 zirconium pyrophosphate
 f pyrophosphate de zirconium
 e pirofosfato de zirconio
 i pirofosfato di zirconio
 d Zirkonpyrophosphat

8999 zirconium sulphate
 f sulfate de zirconium
 e sulfato de zirconio
 i solfato di zirconio
 d Zirkonsulfat

9000 zirconium tetraacetylacetonate
 f tétraacétylacétonate de zirconium
 e tetraacetilacetonato de zirconio
 i tetraacetilacetonato di zirconio
 d Zirkontetraacetylacetonat

9001 zirconium tetrachloride
 f tétrachlorure de zirconium
 e tetracloruro de zirconio
 i tetracloruro di zirconio
 d Zirkontetrachlorid

9002 zirconyl hydroxychloride
 f hydroxychlorure de zirconium
 e clorhidrato de zirconio
 i idrossicloruro di zirconio
 d Zirkonylhydroxychlorid

9003 zoisite; thulite
 f zoïsite
 e zoisita
 i zoisite
 d Zoisit

9004 zygosome
 f zygosome
 e zigosoma
 i zigosoma
 d Zygosom

9005 zygote
 f zygote
 e zigote
 i zigote
 d Zygote

9006 zygotene
 f zygotène
 e cigotena
 i zigotene
 d Zygotän

9007 zygotenic
 f zygoténique
 e zigoteno
 i zigotenico
 d zygotänisch

9008 zygotic
  f zygotique
  e zigótico
  i zigotico
  d zygotisch

  * **zygotic chromosome number** → 8911

9009 zymase
  f zymase
  e zimasa
  i zimasi alcoolica
  d Zymase

9010 zymogen
  f zymogène
  e zimógeno
  i zimogeno
  d Zymogen; Gärungsstoff

9011 zymology
  f zymologie
  e zimología
  i zimologia
  d Gärungslehre; Zymotechnik

9012 zymometer
  f zymosimètre
  e zimómetro
  i zimometro
  d Gärungsmesser

9013 zymosis
  f fermentation
  e zímosis
  i zimosi
  d Gärung

FRANÇAIS

fragmentation nucléolaire 5873
francium 3589
frangé 3473
franklinite 3591
freibergite 3601
freinant la fermentation 3385
friabilité 3604
frittage 7424
fritte 3607
fritte bleue 1067
fritter 3608, 7422
fructose 3611
fuchsine 3612
fuel-oil 2567
fugitomètre 3615
fulminate de mercure 5250
fulminates 3619
fumarate de dibutyle 2516
fumarate de dioctyle 2725
fumarate ferreux 3431
fumée 7463
fumées 3531
fungicides 3628
furanne 3631
furfural 3633
furoate de méthyle 5384
fusain 3639, 5498
fuseau achromatique 101
fuseau central 1572
fusiforme 3645
fusion 3646
fusions en tandem 8030
fustet 3649, 8903
fût 1494

gâchage 8068
gâcher 1072
gadolinite 3650
gadolinium 3651
galactose 3653
galène 3654
galipot 3655
gallium 3658
gallocyanine 3659
galvanisation 3660
galvanoplastie 2998
galvanoplastie au rhodium 7082
galvanoplastie au tambour 814
galvanoplastie automatique 3616
galvanoplastie mécanique 5198
galvanoplastie semi-automatique 7306
gamète 3662
gamétique 3663
gamétocyte 3665
gamétoïde 3666
gamodème 3670
ganister 3672
garance 5044
garniérite 3678
garniture 6068
garniture chromosomique 3728
garniture en cuir 4814
gas des marais 5284
gas étouffant 1770
gasoil 3695
gasoline naturelle 1493
gaylussite 3710

gaz 3679
gaz à l'eau 1068, 8810
gaz à l'eau carburé 1463
gaz, à l'épreuve des ~ 3699
gaz ammoniac 450
gaz bleu 1068
gaz de bois 8857
gaz d'éclairage 1890, 4856
gaz de coupage 6858
gaz de gasogène 6724
gaz de haut-fourneau 1026
gaz de houille 1890
gaz de Mond 5559
gaz de raffinerie 6999
gaz de ville 8349
gaz double 1893
gazéification de la houille 1891
gaz hilarant 5802
gaz inertes 4336, 5806
gaz moutarde 2536
gaz naturel 5662
gaz naturel acide 7673
gaz nobles 5806
gaz occlus 5898
gazogène à aspiration 7900
gazole 3695
gaz parfait 6203
gaz pauvre 6724
gaz riche 7096
gaz sans goudrons 1868
gaz synthèse 7996
gel 3712
gélatine 3713
gélatine bichromatée 938
gélatine explosive 1028
gélatine explosive sous-marine 7885
gélatinisation 7639
gel de silice 7379
gel irréversible 4454
gélose inclinée 228
gel réversible 7067
gelsémine 3714
gène 3716
gène inhibiteur 4349
gène majeur 5095
gène-mutateur 5622
général 8007
générateur d'acétylène 91
générateur de vapeur électrique 2952
générateur de vinaigre 8743
générateur d'oxygène "de chantier" 5970
générateur pour galvanoplastie 6467
génératrice pour galvanoplastie 2999
gène restrictif 7058
gènes d'incompatibilité 4305
génie chimique 1650
génoblaste 3726
génoïd 3727
génome 3728
génosome 3730
gentiane 3731
gentisate de sodium 7540
géraniol 3733
germanium 3738
germe 3737
germen 3742
germer 7731

germicide 3741
gigantocyte 3745
gilsonite 3746
gingembre 3747
gitoxine 3749
glace carbonique 2882
glaçure 3763
glande 3751
glandes endocrines 3039
glasérite 3752
glaubérite 3759
glaucodot 3761
glauconite 3762
globule 3765
globule polaire 6499
globule sanguin 3923
globules sanguins 1051
globulines 3766
glover 3773
gluant 8015
glucagon 3774
glucoheptinate de sodium 7541
gluconate de calcium 1355
gluconate de manganèse 5131
gluconate de potassium 6620
gluconate de sodium 7542
gluconate ferreux 3432
glucosécrétoire 3815
glucosides 3816
glucuronolactone 3776
glutamate de sodium 7543
glutamine 3780
glutaronitrile 3784
gluten 3785
glycéraldéhyde 3786
glycéride 3787
glycérine 3789
glycérol 3789
glycérophosphate de calcium 1356
glycérophosphate de magnésium 5063
glycérophosphate de manganèse 5132
glycérophosphate de sodium 7544
glycérophosphate ferrique 3401
glycérylacétal de méthylcyclohexanone 5361
glycide 3807
glycidol 3807
glycinate de cuivre 2093
glycine 3808
glycogène 3810
glycolate de butyle et phtalylbutyle 1279
glycolate de sodium et de zirconium 7628
glycolate de zirconium 8989
glycolonitrile 3813
glycols de polyoxypropylène 6549
glycyrrhizine 3817
glyodine 3818
glyoxal 3819
glyoxaline 4280
godet à fusion 1995
goethite 3820
golgiolyse 3837
gomme 7144
gomme adragante 3903
gomme arabique 3898

gomme de genévrier 7206
gomme de karaya 4605
gomme-gutte 3661
gomme-laque 7358
gomme mastic 5182
gommes 3901
gommes esters 3114
gommeux 3899
gonade 3838
gonflement 4394
gonie 3839
gonocyte 3841
gonomère 3842
gonosome 3843
gossypol 3846
goudron 8043
goudron de bois de sapin 6403
goudron de houille 1892
goudron de pin 6403
goulot d'embouteillage 1134
gouttelette 2869
gouttiforme 3908
gradient de potentiel 6660
grahamite 3849
graine de lin 4916
grains de cacao 1294
grains flottants 7977
grains plongeurs 7421
graissage par trempage 4283
graisse 3355
graisse de laine 4713
graisses de pétrole 6231
graisses lubrifiantes 4999
graisseux 3360
gramicidine 3854
gramme-atome 3852
gramme-équivalent 3853
Gram-négatif 3857
Gram-positif 3858
grande dureté 3962, 7808
grandeur 5094
grand teint 3352
granule 3861
granule basophile 838
granulés 3860
graphite 1015, 3862
graphite à creuset 2186
graphite en flocons 3494
gras 3360, 4920, 6404
gravimètre 3863
greenockite 3869
grenaillement 7364
grenat 3676
grès 7209, 7822
grès résistant aux acides et aux
    bases 1678
grillage à mort 2324
grille 3870
grille à barreaux 3876
grimpement des sels 2157
grindélia 3874
grisou 3482, 5284
grossissement 5091, 5093
groupe 3345
groupe acétoxyle 80
groupe acétyle 93
groupe acyle 176
groupe amido 442

groupe amino 445
groupe amyle 504
groupe azoïque 747
groupe carbonyle 1457
groupe de séparation 7321
groupe dibenzylique 2504
grue de coulée 4683
grumeler, se ~ 1322
guaïacol 3883
guanidine 3884
guanine 3887
guano 3888
guanosine 3889
guanylnitrosoamino-
    guanylidènehydrazine 3891
guanylnitrosoaminoguanyltétracène
    3892
gutta-percha 3907
gynocardate de sodium 7545
gynogénèse 3909
gypse 3911, 5503
gypsite 3910

hafnium 3931
halazone 3932
halite 3941
halloysite 3942
halogénation 3945
halogènes 3946
halogénure 3940
halogénure d'uranium 8641
halogénures d'argent 7403
haploïde 3951
haploïde dominant 3952
haploïdie 3953
haplomitose 3954
haplontique 3955
haplopolyploïde 3956
haschisch 5162
haute fluidité 7634
haut polymère 4094
hazard, par ~ 6926
héliorouge 3992
héliotropine 3993
hélium 3994
helvite 3999
hémagglutinine 3918
hématimètre 3915
hématine 3920
hématite 3921
hématoblaste 3922
hématopoïèse 3924
hématoporphyrine 3925
hématoxyline 3926
hémicellulose 4000
hémichromatidique 4001
hémihaploïde 4002
hémimorphite 4003
hémixie 4004
hémizygotique 4005
hémo-agglutination 3917
hémoculture 3927
hémoglobine 3928
hémolitique 3930
hémolysine 3929
hénéicosanoate 5387
héparine 4010
heptabarbital 4011

heptachlore 4012
heptadécanoate de méthyle 5388
heptadécanol 4013
heptadécylglyoxalidine 4014
heptanal 4016
heptane 4017
heptanol 4018
heptavalent 4019
heptène 4020
heptoses 4021
herbe 4023
herbicides 4024
héréditaire 4025
hérédité 4348
héroïne 4026
hespéridine 4028
hétéro- 4030
hétérochromatine 4032
hétérochromatinosome 4033
hétérochromatique 4031
hétérochromatisme 4034
hétérochromosome 4035
hétéroécique 4039
hétérogame 4038
hétérogamète 4037
hétérogamétique 2606
hétérolécithe 4040
hétéromolybdates 4041
hétéropolaire 4042
hétéropolymère 4044
hétéropycnose reverse 7063
hétérosomal 4045
hétérotopique 4046
hétérozygote résiduel 7037
heulandite 4047
hexabromoéthane 4048
hexachloréthane 1449
hexachlorobenzène 4049
hexachlorobutadiène 4050
hexachlorocyclohexane 4051
hexachlorocyclopentadiène 4052
hexachloropropylène 4054
hexachlorure de benzène 882
hexadécane 4055
hexadécène 4056
hexadécyltrichlorosilane 4058
hexafluorure de soufre 7944
hexaldéhyde 4062
hexaméthylènediamine 4063
hexaméthylènetétramine 4064, 4066
hexane 4067
hexanetriol 4068
hexanitrate de mannitol 5151
hexanitrodiphénylamine 4069
hexanol 4070
hexavalent 4071, 7348
hexène 4072
hexénol 4073
hexoate de vinyléthyle 8757
hexobarbital 4074
hexokinase 4075
hexoses 4076
hexylèneglycol 4079
hexylèneglycolmonoborate de
    sodium 7546
hexylphénol 4083
hexylrésorcine 4084
hibernation 4086

hydroxyéthylhydrazine 4227
hydroxyéthylpipérazine 4228
hydroxyisobutyrate d'éthyle 3231
hydroxylamine 4230
hydroxymercurichlorophénol 4234
hydroxymercuricrésol 4235
hydroxymercurinitrophénol 4236
hydroxyméthylbutanone 4237
hydroxynaphtoquinone 4240
hydroxyphénylglycine 4241
hydroxyproline 4243
hydroxypropylglycérine 4244
hydroxypropyltoluidine 4245
hydroxyquinoléine 4247
hydroxystéarate d'aluminium 413
hydrozincite 4252
hydrure de bore 1125
hydrure de calcium 1357
hydrure de lithium 4957
hydrure de lithium et d'aluminium
    4945
hydrure de sodium 7547
hydrure de sodium et d'aluminium
    7480
hydrure d'étain 7744
hydrure de zirconium 8990
hydrure d'uranium 8642
hydrures 4170
hygrographe 4253
hygromètre 4254
hygromètre d'absorption 34
hygroscopique 4255
hyperchromasie 4256
hyperpolyploïdie 4257
hypertélie 4258
hypochlorite de calcium 1359
hypochlorite de lithium 4959
hypochlorite de sodium 7550
hypochromaticité 4261
hypohaploïde 4262
hypohaploïdie 4263
hypophosphite de calcium 1360
hypophosphite de manganèse 5134
hypophosphite de sodium 7551
hypophosphite ferrique 3403
hypopituitarisme 4265
hypostase 4266
hypostatique 4267
hyposurrénalisme 4259
hypothèse d'Avogadro 736

iconogène 2940
idiomutation 4271
idiopathique 4272
idioplasma 4273
idiosome 4274
îles d'isomérie 4462
ilménite 4278
image latente 4723
imidazole 4280
imides 4281
iminobispropylamine 4282
immunisation 4288
immunité 4287
immunologie 4289
imperméabilisation 8566
imperméable 4292
imprégnation 4294, 6219

imprégné de résine 7044
imprégner 4293
impsonite 4295
inactinique 5812
inassimilable 4298
incidence 4299
incinérateur 2436
inclusion 4300
inclusions cellulaires 4301
incompatibilité 4303
incrustation 7238
indamines 4308
indanthrène 4309
indène 4310
indicateur 4314
indicateur à ferroxyle 3442
indicateur chimique 1658
indicateur d'adsorption 207
indicateur de vide 8672
indicateur externe 3321
indicateur intérieur 4386
indicateurs d'oxydoréduction 6045
indice d'acétyle 98
indice d'acide 136
indice de combustion 1996
indice de coordination 2073
indice de démulsification 2393
indice de fluage 5217
indice de fluidité 7635
indice de mousse 4731
indice d'éther 3129
indice d'iode 4410
indice d'octane 5909
indices des facettes 4315
indigo bleu 4316
indium 4322
indol 4323
indophénols 4325
indulines 4331
inerte 4335
inertie chimique 1659
inexplosible 5823
infection 4338
infection réductive 6996
infestation 4339
infiltration 1039
inflammabilité 3497, 4340
inflammable 4276
inflammatoire 6316
influencé par le milieu 3061
infraprotéines 4342
infrarouge 4343
infuser 4346
infusion 4347
inhibiteur catalytique 1512
inhibiteur de décapage 6383
inhibiteur de la préprophase 6692
inhibiteur d'ultraviolet 8605
inhibiteurs 4351
inhibiteurs anodiques 576
inhibiteurs cathodiques 1523
inhibition 4350
ininflammable 5826, 8621
initiateur 4352
injecteur à vapeur 7772
inoculable 4353
inoculation 4355
inoculer 4354, 6980

inodore 5931
inorganique 4356
inosine 4358
inositol 4359
insecticide 4360
insectifuge 4361
insoluble 5026
instabilité de plasma 6441
installation à lixiviation 4983
installation d'échantillonnage actif
    171
installation de concentration 2021
installation de diffusion thermique
    8206
installation de séparation d'isotopes
    en cascade 1484
installation d'essai 6398
installation de triage de coke 1946
installation pour l'embouteillage 1138
instrument à lecture directe 2783
instrumentation 4366
instrumentation rhéologique 7075
instrument de lecture à distance
    7025
instrument de précision 6678
insuline 4369
intensité lumineuse 5007
interaction des gènes 3719
interbande 4372
interchange 4373
interchromomère 4375
interchromosomique 4376
interface 4378
interface liquide/liquide 4937
interférence chiasmatique 1696
intergénique 4381
intermédiaire 5205
intermédine 4384
interspécifique 4388
intragénique 4389
intrahaploïde 4390
intransparence 5977
inuline 4396
invariant 4397
inversion 4398
inversion du type d'émulsion 6242
invertase 4399
inviabilité 4401
in vitro 4402
in vivo 4403
iodargyrite 4404
iodate de potassium 6622
iodate de sodium 7552
iode 4407
iodéosine 4405
iodoforme 4414
iodosuccinimide 4415
iodure cuivreux 2236
iodure d'acétyle 94
iodure d'argent 7404
iodure de cadmium 1305
iodure de calcium 1361
iodure de lithium 4960
iodure de mercure et de barium 5231
iodure de mercure et de potassium
    5239
iodure de méthyle 5396

iodure de méthyle et de magnésium 5408
iodure de méthylène 5377
iodure de phosphonium 6329
iodure de plomb 4782
iodure de potassium 6623
iodure de sodium 7553
iodure d'éthyle 3234
iodure de thymol 8283
iodure de zinc 8943
iodure d'isopropyle 4553
iodure ferreux 3433
iodure mercurique 5235
iodure octyle 5923
iodure palladeux 6074
ion 4416
ion amphotérique 494
ion carbénium 1451
ion complexe 2010
ionimètre 4428
ionisation 4427
ionisation minimale 5505
ionium 4426
ion négatif 558
ionogène 4435
ionone 4436
ipécacuanha 4437
iridium 4440
iridosmine 6016
irritant 4457
isatine 4459
iséthionate de stilbamidine 7810
isinglass 3490
iso- 4463
isoallèle 4464
isoalloxazine 4465
isoamylbenzyléther 4469
isoamylènes 4472
isoautopolyploïdie 4475
isobornéol 4476
isobutane 4480
isobutène 4481
isobutylamine 4484
isobutylundécylénamide 4490
isobutyraldéhyde 4491
isobutyrate de phényléthyle 6295
isobutyrate d'éthyle 3235
isobutyronitrile 4494
isochromatide 4498
isochromosome 4499
isocyanate de phényle 6298
isocyanate d'éthyle 3236
isocyanates 4500
isodécaldéhyde 4503
isodécanol 4505
isodurène 4508
isoeugénol 4510
isoeugénol de benzyle 916
isogénie 4512
isoheptane 4513
isohexane 4514
isolation 7825
isolation cellulaire 1549
isolécithe 4515
isolement biologique 955
isoler 7320
isoleucine 4516
isologues 4517

isomère 4518
isomères optiques 5988
isomérie géométrique 3732
isomérie nucléaire 5864
isomérisation 4519
isomérisme dynamique 8054
isomorphe 4520
isomorphie 4521
isomorphisme 4521
isooctane 4523
isooctène 4524
isopentaldéhyde 4530
isopentane 4531
isophène 4533
isophorone 4534
isophtalate de diallyle 2468
isophtalate de diméthyle 2695
isophtalate de dioctyle 2726
isoploïdie 4537
isopolyploïde 4538
isopral 4539
isoprène 4540
isopropanolamine 4541
isopropénylacétylène 4542
isopropylamine 4545
isopropylaminodiphénylamine 4546
isopropylaminoéthanol 4547
isopropylate d'aluminium 414
isopropylmercaptan 4554
isopropylméthylpyrazolyldiméthyl-
    carbamate 4555
isopropylphénol 4561
isopulégol 4563
isoquinoléine 4564
isosafrol 4565
isostérie 4567
isotactique 4568
isotherme 4569, 4570
isothiocyanate d'allyle 381
isotonique 4571
isotope 4572
isotypie 4575
isovaléraldéhyde 4576
isovalérate d'éthyle 3237
isovalérianate de bornyle 1122
itaconate de diméthyle 2697

jade 4583
jamésonite 4585
jauge 3705
jauge de pression d'huile 5945
jaugeur 5197
jaune brillant 4590
jaune citron 4821
jaune de benzidine 891
jaune de cadmium 1316
jaune de Cassel 1495
jaune de chaux 4890
jaune de chrome 1800, 4820
jaune de cobalt 4313
jaune de fer 4445
jaune de Leipzig 4820
jaune de Naples 5652
jaune de Paris 6126
jaune de quinoléine 6876
jaune des Indes 4313
jaune de strontium 7853
jaune de zinc 8928

jaune d'or 4591
jaune Hansa 3949
jaune royal 4642
jaunes de ferrite 3416
jaune serin de litharge 1412
jaunissement 8897
jet 4594
joint de vapeur 7776

kaïnite 4600
kaolin 4601
kaolinite 4602
kapok 4603
kauri 4612
kératine 4615
kermès 4616
kermésite 4617
kernite 4618
kérogène 4619
kérosène 4620
kétobenzotriazine 4622
kieselgur 2489
kiesérite 4632
kinétochore 4641
krypton 4654
kyanisation 4655

labile 4660
labilité 4661
laboratoire 4662
labyrinthe d'entrée 259
lacmoïde 4668
lacrymogène 4667
lactalbumines 4670
lactame 4671
lactase 4672
lactate 4673
lactate d'amyle 505
lactate de butyle 1269
lactate de calcium 1362
lactate de cuivre 2095
lactate de méthyle 5402
lactate de sodium 7556
lactate de sodium et de zirconium
    7629
lactate d'éthyle 3238
lactate d'eucaïne 3274
lactate de zirconium 8992
lactate ferreux 3434
lactobutyromètre 4676
lactoflavine 4677
lactonisation 4679
lactonitrile 4678
lactophénine 4680
lactose 4681
laine de scorie 5504
lait 7456
lait de chaux 4888
laiteux 4674
laiton 1155
laiton gamma 3667
lamellaire 4695, 4697
lamelle 4694, 4696
laminer 4699
laminoir à deux cylindres 8589
laminoir triple 8532
lampe 4705
lampe à vapeur de mercure 5253

lanatoside C 4709
lancéolé 4710
langbéinite 4712
lanoléine 4713
lanoline 4713
lanostérol 4714
lanthane 4716
lanthanides 4715
lanthionine 4719
lapis lazuli 4720, 4754
laque 4666, 4669, 4691
laque blanchie 8840
laque carminée 2169
laque de garance 5045
laque écarlate 7242
laque en baton 7805
laque en écailles orangé 5994
laque florentine 2169
laque grenat 3677
latex 4725
latex ammonié 452
laudanine 4733
laudanosine 4734
laudanum 4735
laurate butoxyéthylique 1244
laurate d'ammonium 468
laurate d'amyle 506
laurate de butyle 1270
laurate de diglycol 2614
laurate de méthyle 5403
laurate de monoéthyléther
    d'éthylèneglycol 3204
laurate de tétrahydrofurfuryle 8140
laurate de zinc 8945
laurate d'isocétyle 4495
laurier 4737
lauroléate de méthyle 5404
laurylmercaptan 4746
laurylsulfate de sodium 7557
lavage 7266
lavage par inversion de courant 753
lavande 4749
laveur de gaz 3700
laveuse de bouteilles 1135
lawrencium 4752
laxatif 4753
lazulite 4754
"leached rubber" 4755
leadhillite 4780
lécithine 4817
lenticulaire 4825
lentille 4824
lentille convergente 2032
léonite 4826
lépidine 4827
lépidolite 4828
leptocyte 4829
leptomètre 4830
leptonème 4831
leptotène 4832
lessivage 4982
lessive 5021
lessive liquide 4933
lessiver 4981
lessive résiduaire 7697
lessive sulfitée 7925
leucine 4834
leucite 4835

leucobase 4836
leucocyte 4838
leucodérivés 4837
levain 808, 6423
lévigation 4841
lévogyre 4684
lévulinate de calcium 1363
lévulinate de tétrahydrofurfuryle
    8139
lévulinate d'éthyle 3239
lévulose 4686
levure 8892
levure basse 1141
levure de bière 808
levure haute 8343
levurer 6420
léwisite 4842
liaison 1096, 1097, 4906
liaison chimique 1642
liaison covalente 2139
liaison de valence 8678
liaison d'hydrogène 4191
liaison hétéropolaire 4043
liaison ionique polarisée 6492
liaison moléculaire 5533
liaison polaire 6488
liaison semi-polaire 2316
liaison transversale 2178
liant 944, 1100
liant à sec 2880
liant humide 8829
libérer 4844
lichen 4846
lichénine 4847
lié 1143
liège 2112
lié par la résine 7042
lignée de substitution compensatoire
    pour un chromosome 7418
ligneux 4868
lignification 4869
lignine 4870
lignite 4871
lignocelluloses 4874
lignocérate de méthyle 5405
lignosulfate de sodium 7558
ligroïne 4876
limite de fluage 2158
limite élastique 8900
limnimètre 4935
limonite 4892
limpidité 4893
linalol 4896
liniment 4905
linnéite 4907
linoléate cobalteux 1913
linoléate d'ammonium 469
linoléate de calcium 1364
linoléate de cobalt 1913
linoléate de manganèse 5135
linoléate de méthyle 5406
linoléate de plomb 4783
linoléate de zinc 8946
linoléine 4910
linoxyne 4915
linters 4918
lipases 4921

lipide 193
lipidose 4923
lipochromes 4924
lipocyte 4925
lipoïde 4922
lipolyse 4926
lipoxydase 4929
liquation 4931
liquéfaction 4932
liqueur blanche 8842
liqueur de Fehling 3375
liqueur de Fowler 3583
liqueur de lessive régénérée 7011
liqueur des Hollandais 3187
liqueur d'extraction 3328
liqueur d'Igewesky 4275
liqueur-mère 5595
liquide de refroidissement 2069
liquide de transvasement 8377
liquide polaire 6494
litharge 4940
lithine 4941
lithine caustique 4958
lithium 4942
lithiumamide 4946
lithopone 4975
litre 4978
lixiviation 4982
lobélie 4987
lobéline 4988
local de soutirage 6887
loi de Blagden 1021
loi de Boyle 1149
loi de Dalton 2311
loi de Dulong et Petit 2895
loi de Faraday de l'électrolyse 3348
loi de Gay-Lussac 3708
loi d'Einstein sur l'équivalence
    photochimique 2942
loi de Joule 4597
loi de Kick 4630
loi de Raoult 6928
loi d'Henry 4009
loi d'Hess 4029
loi d'Hooke 4128
lot 840
loupe 5092
lubrifiant 4998
lumen 5003
luminescence 5004
lupanine 5010
lupuline 5011
lustrage 5012
lustre 3767
lutécium 5015
lutéine 5016
lutidine 5017
lymphe 5022
lymphoblaste 5023
lyolyse 5024
lyophile 5025
lyophobe 5026
lyse 5028
lysine 5027
lysogénique 5029
lysogénisation 5030
lysozime 5031
lytique 5032

macérer 7787
mâchefer 1871
machine à rincer les bouteilles 1135
machine à rouleaux inversés 7065
machine d'expansion à compresseur 1113
machine pour briquettes 1169
maclure 5952
macro-axe 5036
macromolécule 5038
macromolécule linéaire 4901
macronucléus 5040
macroscopique 5041
macrospore 5042
macrostructure 5043
magnésie 5068
magnésie anglaise 5492
magnésie blanche 5492
magnésie calcinée 1330
magnésite 5047
magnésium 5048
magnétisme 5085
magnétite 5086
magnétomètre 5089
magnétostriction 5090
maille 5256
maillechort 5723
maische 5166
malachite 5098
malathion 5101
malaxeur 6794, 7115
malaxeur à cuve mobile 1628
maléate de chlorphéniramine 1766
maléate de diallyle 2469
maléate de dibutyle 2517
maléate de dibutylétain 2527
maléate de diéthyle 2591
maléate de méthylergonovine 5378
maléate de pyrilamine 6821
malléable 5108
malonate de diéthyle 2592
malonate d'éthyle 3242
malt 5110
maltase 5112
malter 5111
malterie 5114
malthènes 5113
maltose 5115
malt torréfié 1016
malt touraillé 4638
malt vert 3868, 4990
mamelon 5736
maneb 5120
manganate de baryum 797
manganate de potassium 6625
manganèse 5121
manganite 5141
manganite de lithium 4961
manne 5149
mannitol 5150
mannose 5152
manomètre 5153
manomètre à décharge lumineuse 3689
manomètre à membrane 2483
manomètre anéroïde 537
manomètre à poids mort 2325
manomètre à tube en U 8664

manomètre enregistreur 6971
manque de lustre 7447
manteau chondriosomal 1779
maquette 5517
marbrage 3850
marbrure 5597, 5598
marcasite 5158
marche, en ~ 5971
margarine 5159
margarite 5160
marmite 4627
marmite à pression 6695
marmite de Papin 6099
marquer 4658, 8017
marqueurs non sélectionnés 8629
marron d'alizarin 319
marron de manganèse 5125
martensite 5164
masse 5171
masse active 168
masse moléculaire 5541
masse volumique apparente 1221
massicot 5175
mastic 3461, 5180, 6808
mastic à forte résilience 1142
masticage 5019
masticateur 5181
mastication à mort 2323
mastic bouche-pores 3673
mastice 7824
mat 2894
matage 3610
mat de surface 7963
mater 3507
matériau réfractaire basique 830
matériaux réfractaires 7005
matière active 169
matière carbonisée 1630
matière cellulosique 7137
matière d'enrobage 3016, 3033
matière de protection 7048
matière plastique 6456
matière plastique de caséine 1490
matière première 823
matière primaire 6941
matières plastiques à base de cellulose 1559
matrice 5185
matte 5188
maturation 5189
mauvaise odeur 3448
mauvéine 5191
maximum 6148
mazout 2567, 3613
médicament 2875
mégasporocyte 5206
méiocyte 5207
méiosome 5208
mélamine 5210
mélange 1043, 5515
mélange additionnel 196
mélange à froid 1955
mélange anticorrosif de trempage 4139
mélange de bière et levure 7435
mélange de chlorures d'amyle 499
mélange dépolarisant 2418
mélange de trempage 2773

mélange dystectique 2911
mélange eutectique 3286
mélange maître 5178
mélange pauvre 4810
mélanger 1040
mélanges à base d'oxyde de butylène 1261
mélange trop mastiqué 6035
mélangeur 1632, 2191, 5513
mélangeur à cylindres 7115
mélangeur à dispersion 2804
mélangeur à tonneau 813
mélangeur centrifuge 1574
mélangeur-clarificateur à pompes 6798
mélangeur de boisson d'épreuve du baryum 798
mélangeur en discontinu 843
mélangeuse 1042
mélanine 5213
mélasse 5529
meldomètre 5214
mélitose 5216
membrane 2480
membrane semi-perméable 7314
mendelévium 5220
menthanediamine 5221
menthol 5222
menthone 5223
mercaptan amylique 507
mercaptan d'hexadécyle 4057
mercaptan d'hexyle 4081
mercaptans 5225
mercaptobenzothiazol 5226
mercaptoéthanol 5227
mercaptothiazoline 5228
mercure 5248
mésitylène 5257
mésomérie 5259
méson 5260
mésothorium-I 5261
mesure 3705
métabisulfite de potassium 6626
métabolisme 5263
métabolisme de base 817
métabolite 5264
métaborate de cuivre 2096
métaborate de sodium 7559
métaldéhyde 5265
métal delta 2390
métal léger 4857
métallisation 5269
métallisation par le vide 8673
métalloïde 5271
métallurgie 5275
métallurgie des poudres 6666
métal non ferreux 5825
métamorphisme 5276
métanilate de sodium 7560
métaphase 5278
métaphosphate d'aluminium 415
métaphosphate de béryllium 926
métaphosphate de potasse 6627
métaphosphate de sodium 7561
métasilicate de lithium 4962
métasilicate de sodium 7562
métastable 5279
métaux alcalins 323

métaux nobles 5807
métaux vils 824
métavanadate d'ammonium 470
métavanadate de sodium 7563
méthacroléine 5280
méthacrylate décyle-octylique 2350
méthacrylate de lauryle 4747
méthacrylate de méthyle 5410
méthacrylate de stéaryle 7785
méthacrylate d'éthyle 3246
méthacrylate d'hexyle 4082
méthacrylate d'octyle 5925
méthane 5284
méthanediamine 5285
méthanol 5287
méthionine 5288
méthode colorimétrique 1980
méthode cryoscopique 2194
méthode de fabrication de mousse
  3557
méthode de Kjeldahl 4647
méthode de sels de plomb-éther 4795
méthode Wulff 8866
méthoxyacétophénone 5290
méthoxybutanol 5291
méthoxychlore 5292
méthoxyméthylpentanol 5298
méthoxyméthylpentanone 5299
méthoxypropylamine 5300
méthylacétanilide 5303
méthylacétone 5306
méthylacétophénone 5307
méthylacétylène 5308
méthylacétylsalicylate 5310
méthylal 5312
méthylamine 5315
méthylaminophénol 5316
méthylamylcarbinol 5319
méthylamylcétone 5320
méthylaniline 5321
méthylanthracène 5322
méthylanthraquinone 5324
méthylate de magnésium 5065
méthylate de sodium 7564
méthylbenzylamine 5330
méthylbenzyldiéthanolamine 5331
méthylbenzyldiméthylamine 5332
méthylbutanol 5336
méthylbutène 5337
méthylbuténol 5338
méthylbutylbenzène 5339
méthylbutylcétone 5340
méthylbutynol 5341
méthylcellulose 5346
méthylclothiazide 5354
méthylcoumarine 5355
méthylcyclohexane 5358
méthylcyclohexanol 5359
méthylcyclohexanone 5360
méthylcyclohexènecarboxaldéhyde
  5362
méthylcyclohexylamine 5363
méthylcyclopentane 5366
méthyldiéthanolamine 5369
méthyldioxolane 5370
méthylènebisacrylamide 5371
méthylènedianiline 5375
méthylèneditannine 5376

méthyléthylcétone 5379
méthyléthylpyridine 5380
méthylformanilide 5381
méthylfuranne 5383
méthylglucoside 5386
méthylheptane 5389
méthylhepténone 5390
méthylhexane 5391
méthylhexaneamine 5392
méthylhexylcétone 5393
méthylhydrazine 5394
méthylhydroxybutanone 5395
méthylionone 5397
méthylisoamylcétone 5398
méthylisobutylcétone 5399
méthylisoeugénol 5400
méthylisopropénylcétone 5401
méthylmercaptan 5409
méthylmorpholine 5411
méthylnaphtalène 5414
méthylnaphtylcétone 5415
méthylnonylacétaldéhyde 5419
méthylnonylcétone 5420
méthyloldiméthylhydantoïne 5421
méthyloléoyltaurate de sodium 7565
méthylolurée 5423
méthylorthoanisidine 5424
méthylpentadiène 5429
méthylpentaldéhyde 5430
méthylpentane 5431
méthylpentanediol 5432
méthylpentène 5433
méthylpentynol 5434
méthylphényldichlorosilane 5436
méthylphloroglucine 5437
méthyl-phtalyl-éthyl-glycolate 5439
méthylpipérazine 5440
méthylpropylcétone 5443
méthylpyrrole 5444
méthylpyrrolidone 5445
méthylstyrolène 5450
méthylsulfate de diphémanile 2740
méthyltaurine 5451
méthyltétrahydrofuranne 5452
méthylthiouracile 5453
méthyltrichlorosilane 5454
méthylvinyldichlorosilane 5457
méthylvinylpyridine 5459
meule verticale 2928
miasme 5461
mica 5462
mica aggloméré 5463
micelle 5465
microanalyse 5466
microbalance 5468
microballons 5469
microbe 5470
microbiologie 5471
microchimie 5473
microchromosome 5474
microgliacyte 5478
microglie 5477
microhétérochromatique 5479
microlite 5480
micromanomètre 5481
micromètre 5482
micron 5483
micronucléus 5484

microporeux 5485
migration 5488
migration atomique 710
migration de couleur 1985
migration électrique 2950
migration électrochimique 2956
milieu 3060
milieu aqueux 651
milieu de suspension 8721
milieu génétique 3724
milieu ionisant 4433
mimétène 5496
mimétèse 5496
mimétite 5496
mine de plomb 1015
mine d'étain 1497
minerai 5996
minerai en rognons 4631
minium 5506
mirabilite 5507
mispickel 671
mitose réductionnelle 6994
mitosome 5508
mitotique 5509
mixochromosome 5514
mobilité 5596
mobilité électrophorétique 2996
mobilité ionique 4424
modèle 5517
modification d'adaptation 180
modifié chimiquement 1661
module d'élasticité 5520
module d'Young 8904
moelle osseuse 1110
mofette 1012
moisi, à odeur de ~ 5620
molasses noires 1018
mole 5530
molécule 5548
molécule activée 162
molécule-gramme 3856
molécule homonucléaire 4123
molécule isostérique 4566
molécule neutre 5706
molybdate d'ammonium 471
molybdate de lithium 4963
molybdate de plomb 4784
molybdate de potassium 6628
molybdate de sodium 7566
molybdène 5551
molybdénite 5550
molybdine 5556
molybdite 5556
moment de liaison 1102
monatomique 5557
monazite 5558
mongolisme 5561
monoacétate de glycérine 3793
monoacétate de résorcine 7054
monoacétate d'éthylèneglycol 3196
monobasique 5563
monobenzoate de résorcine 7055
monoblaste 5570
monobromure d'iode 4408
monobutyléther-acétate d'éthylèneglycol 3199
monobutyléther d'éthylèneglycol
  3198

nigrosine 5731
nikéthamide 5732
niobium 5733
nitrate 5738
nitrate cérique d'ammonium 1598
nitrate cobalteux 1915
nitrate d'aluminium 417
nitrate d'amidon 5794
nitrate d'ammonium 472
nitrate d'argent 7405
nitrate de baryum 799
nitrate de bismuth 985
nitrate de cadmium 1307
nitrate de calcium 1367
nitrate de calcium et d'ammonium 1338
nitrate de chaux 1367
nitrate de cuivre 2098
nitrate de fer 3405
nitrate de guanidine 3886
nitrate de lanthane 4717
nitrate de lithium 4964
nitrate de magnésium 5066
nitrate de méthyle 5416
nitrate de nickel 5721
nitrate de palladium 6077
nitrate de plomb 4788
nitrate de potassium 6630
nitrate de sodium 7571
nitrate de strontium 7855
nitrate de thallium 8185
nitrate de thorium 8271
nitrate d'éthyle 3250
nitrate de zinc 8948
nitrate de zirconium 8994
nitrate d'uranyle 8652
nitrate manganeux 5143
nitrate mercureux 5246
nitrate mercurique 5236
nitration 5739
nitre 6630
nitrer 5737
nitreur 5740
nitrification 5746
nitrifier 5747
nitrile 5748
nitrile acrylique 149
nitrile benzoïque 898
nitriles aliphatiques 3366
nitrite 5749
nitrite d'amyle 509
nitrite d'argent 7406
nitrite de cobalt et de potassium 1923
nitrite de potassium 6631
nitrite de sodium 7572
nitrite d'éthyle 3251
nitroacétanilide 5750
nitroamidon 5794
nitroaminophénol 5751
nitroaniline 5752
nitroanisole 5753
nitrobenzaldéhyde 5754
nitrobenzène 5755
nitrobenzèneazorésorcinol 5756
nitrobiphényle 5760
nitrocellulose 1557, 5761
nitrodiphénylamine 5762

nitroéthane 5763
nitroferricyanure de sodium 7573
nitrofurantoïne 5764
nitrofurazone 5765
nitrogénase 5767
nitroglycérine 5773
nitroguanidine 5774
nitromersol 5775
nitrométhane 5776
nitromètre de Lunge 5009
nitron 5777
nitronaphtaline 5778
nitroparacrésol 5779
nitroparaffines 5780
nitrophénétol 5781
nitrophénide 5782
nitrophénol 5783
nitropropane 5785
nitroprussiate de sodium 7574
nitrosite de styrène 7873
nitrosodiméthylaniline 5788
nitrosodiphénylamin 5789
nitrosoguanidine 5790
nitrosonaphtol 5791
nitrosophénol 5792
nitrostyrène 5795
nitrosulfathiazol 5796
nitrotoluène 5798
nitrotoluidine 5799
nitrotrifluorométhylbenzonitrile 5800
nitrourée 5801
nitroxylène 5803
nitruration 5745
nitrure de bore 1126
nitrure de silicium 7389
nitrure de zirconium 8995
nitrures 5744
niveau d'énergie d'impurité 4297
nivénite 5804
nobélium 5805
nocif 5853
nocivité 5808
nodal 5809
nodule 5810
nodules 5811
nœud chromatique 1819
noir acide 5636
noir animal 556, 1106
noir au tunnel 1629
noir d'acétylène 90
noir d'alizarine 318
noir d'Allemagne 3590
noir de benzol 896
noir de carbone 1443
noir de charbon 8202
noir de fumée 4706
noir de gaz 3682
noir de lampe 4706
noir de magnétite 5087
noir de pâte de bois 8859
noir de platine 6472
noir de Suède 7972
noir d'ilménite 4279
noir d'ivoire 4579
noir d'os 1106
noir fourneau 3636
noir minéral 5497
noir pour cuir 4812

noir végétal 8719
noix vomique 5884
nombre cétane 1610
nombre chromosomique du zygote 8911
nombre d'Avogadro 735
nombre de chromosomes du gamète 3664
nombre de Reynolds 7072
nombre de saponification 7216
nombre de transport des ions 8375
nombre de valence 8679
nonadécane 5813
nonadécanoate de méthyle 5417
nonanal 5815
nonane 5816
nonanoate de méthyle 5418
non conducteur 5818
non congelable 5819
non ferreux 5824
non miscible 4285, 5829
nonoses 5830
non polaire 5831
non poreux 5832
non saturé 8626
non volatil 5834
non vulcanisé 8609
nonylamine 5837
nonylbenzène 5838
nonylène 5839
nonyllactone 5840
nonylphénol 5841
noradrénaline 5843
norleucine 5844
normal 5846
normalisation 5847
norme 5845
normocyte 5850
novobiocine 5852
noyage 3525
noyau 5879
noyau à chromocentre 1586
noyau définitif 2356
noyau de fusion 3647
noyau en terre 4986
noyau génératif 3721
noyau métabolique 5262
noyau quiescent 3048
noyau trophique 8555
nucléaire 5855
nucléine 5868
nucléisation 5869
nucléoalbumine 5870
nucléole 5874
nucléole satélite 7227
nucléoprotéine 5877
nucléotide 5878
nullisomique 5882
numération globulaire 3916
nylon 5885
nylon élastique 2944
nylon sous forme de résin soluble 7644
nytril 5891

objectif 5893, 5894
obsidiane 5895
obsidienne 5895

oxyde de zinc plombifère 4778
oxyde de zirconium 8996
oxyde de zirconium pur 769
oxyde d'hexachlorodiphényle 4053
oxyde d'hydroxypyridine 4246
oxyde d'octylène 5922
oxyde d'or 3827
oxyde d'uranium 8644
oxyde d'uranium et de barium 8635
oxyde d'yttrium 8909
oxyde éthylarsénieux 3145
oxyde ferreux 3435
oxyde ferrique 3407
oxyde nitreux 5802
oxyde noir d'uranium 1020
oxyde rouge de fer 6982
oxyde rouge d'Espagne 7677
oxydes 6046
oxydes de fer jaunes 3416, 4450
oxydes de fer rouge 4449
oxyde stanneux 7756
oxyde stannique 7749
oxyde tungstique 8572
oxydichlorophosphorobenzène 885
oxygène 6054
oxygène dissous 2809
oxyhémoglobine 6055
oxyméthandrolone 6057
oxyproline 6060
oxysulfure de diméthyle 2709
oxytétracycline 6061
oxytocine 6062
oxytrichlorure de vanadium 8692
ozokérite 6063
ozone 6064
ozonides 6065

palladium 6075
palmitate d'aluminium 420
palmitate de calcium 1370
palmitate de magnésium 5069
palmitate de méthyle 5425
palmitate de tétrahydrofurfuryle
    8142
palmitate de zinc 8953
palmitate d'isopropyle 4558
palmitate isooctylique 4528
palmitoléate de méthyle 5426
panallèle 6084
panchromatique 6086
pancréatine 6087
pancytolyse 6088
pan de Thelan 8190
pangène 6089
pangénosome 6090
panier 834
panmictique 6091
pantothénate de calcium 1371
papaïne 6094
papavérine 6096
papier de soie 8294
papier de tournesol 4977
papier-filtre sans cendres 688
papier fin 8294
papyracé 6100
para- 6101
parachromatine 6102
paraffine brute 7440

paraffine fluorinée 3540
paraffines 6105
paraformaldéhyde 6107
parahélium 6108
parahydrogène 6109
paraldéhyde 6110
parallaxe 6111
paramagnétique 6113
paraméiose 6114
paraméthadione 6116
paramètre 6115
paramictique 6117
paramitose 6118
paranucléine 6119
paranucléole 5875
parasiticide 6120
paratrophe 6121
parchemin 6123
particule 6130
particule élémentaire 3626
particules colloïdales 1969
passer 6201
passif 6134
passivation 6136
passivation anodique 577
passivation chimique 1663
passivation électrochimique 2958
passivité 6135
pasteurisation 6142
pastillage 6160
pastille 2002, 6158
pâte 2854, 6137, 6138, 7450, 7819
pâte à papier 6098
pâte de bois 8858
pâte mi-chimique 7308
"path-coefficient" 6144
pattinsonage 6146
pavot 6095
pechblende 6421
pectines 6153
peinture 6070
peinture à base d'émulsion 3027
peinture à base de plomb 4790
peinture à l'huile 5943
peinture anticondensation 600
peinture antidérapante 635
peinture à séchage rapide 7354
peinture au pinceau 1204
peinture d'impression 6707
peinture d'insonorisation 628
peinture-émulsion 3027
peinture isolante 7270
peinture luminescente 5008
peinture radioactive 6900
peintures au silicate 7381
peinture sensible à la chaleur 3984
peintures fungicides 3627
pélargonate d'éthyle 3255
pellicule 3462
pellicule à autosupport 6161
pellicule cellulosique 1547
pellicule d'acétocellulose 57
pellicule rétrécissable 7366
pénétration 6219
pénétration de colle 3778
pénétromètre 6163
pénicillinase 6165
pénicilline 6164

pénicilline-résistant 6166
pentaborane 6168
pentaborate de sodium 7577
pentabromure de phosphore 6339
pentachloréthane 6169
pentachloronitrobenzène 6170
pentachlorophénate de sodium 7578
pentachlorophénol 6171
pentachlorostéarate de méthyle 5427
pentachlorure d'antimoine 619
pentachlorure de molybdène 5553
pentachlorure de phosphore 6340
pentadécane 6172
pentadécanoate de méthyle 5428
pentaérythritol 6174
pentafluorure de brome 1178
pentandiol 6177
pentane 6176
pentanol 6178
pentasomique 6179
pentasulfure d'antimoine 620
pentasulfure d'arsenic 673
pentasulfure de phosphore 6341
pentatriacontane 6180
pentavalent 6181
pentlandite 6182
pentobarbital 6183
pentolite 6184
pentose 6185
pentoxyde de phosphore 6342
pentoxyde de vanadium 8693
pentylènetétrazol 6186
pépite 5881
pepsine 6188
peptisation 6189
peptiser 6190
peptone 6192
peracide 6194
perbenzoate de butyle 1275
perborate de magnésium 5070
perborate de sodium 7579
perborate de zinc 8954
perbromure de bromure de
    pyridinium 6818
percarbonate de potassium 6635
percarbonate d'isopropyle 4559
perchlorate d'ammonium 474
perchlorate de magnésium 5071
perchlorate de potassium 6636
perchlorate de sodium 7580
perchlorates 6195
perchloréther 6196
perchloréthylène 6198
perchlorométhylmercaptan 6199
percolation 6202
percoler 6201
péricarpe 6206
périclase 6207
périkinétique 6208
periodate de potassium 6637
periodate de sodium 7581
période de demi-vie 3937
période de séchage à vitesse
    décroissante 3341
période d'induction 4329
périplasme 6212
perlite 6151, 6214
permanganate de calcium 1372

plomb telluré 8061
plomb tétraéthylique 8128
plomb tétraméthylique 8163
plumbo-solubilisation 6480
plutonium 6482
poche de coulée 4682
poche de vapeur 8714
poids atomique 714
poids de formule 3577
poids équivalent 3093
poids moléculaire 5547
poids net 5697
poids sec absolu 6029
poids spécifique 7687
poils de nylon 5886
point critique 2170
point d'achèvement 3046
point d'aniline 555
point d'ébullition 1090
point de congélation 3600
point de cristallisation 3851
point de feu 3483
point de fusion 5218
point de granulation 3851
point de nuage d'une huile 1877
point de ramollissement 7632
point de rosée 2453
point de rupture 1157
point de stagnation 7736
point d'ignition 4277
point d'inflammabilité 3504
point d'inflammation 3483
point eutectique 3288
point isoélectrique 4509
points brillants 4093
poison 6487
poison mitotique 5510
poix 6419
poix d'acides gras 3362
poix de Bourgogne 1232, 3655
poix de colophane 7126
polarimètre 6489
polarisant 6493
polarisation 6490
polarisation cathodique 1525
polarisation électrolytique 2983
poli 7451
polimérisation en masse 1223
polissage électrolytique 3000
polissage par l'attaque à l'acide 127
polonium 6500
poly- 6501
polyacrylamide 6502
polyacrylate 6503
polyalcane 6505
polyallomère 6506
polyamide 6507
polyaminotriazoles 6510
polybasique 6511
polybasite 6512
polybutadiène 6513
polybutènes 6515
polybutylènes 6515
polychloroprène 6516
polychlorotrifluoroéthylène 6517
polychondrique 6518
polycondensat 6519
polycondensation 6520

polycyclique 6521
polydispersé 6523
polyester 6524
polyester-amide 6525
polyéther éthylvinylique 6567
polyéthylène 6529
polyéthylèneglycol 6530
polyformaldéhydes 6532
polygala de Virginie 7317
polyglycols 6534
polyhexaméthylène-adipamide 6535
polyisoprène 6536
polymère 6537
polymère en blocs isotactiques 7792
polymère en masse 1048
polymère linéaire 4903
polymères stéréospécifiques 7795,
  7797
polymérique 6538
polymérisation 6540
polymérisation à froid 1957
polymérisation de condensation 2029
polymérisation en continu 2063
polymérisation par rayonnement
  6894
polyméthacrylates 6543
polymorphisme 6544
polymyxines 6545
polyoléfines 6546
polyoses 6547
polyploïdie 6550
polyploïdie déséquilibrée 8607
polyploïdie secondaire 7279
polypropylène 6551
polypropylène chloré 1723
polystyrène 6552
polysulfure de potassium 6641
polysulfure de sodium 7591
polytétrafluoroéthylène 6555
polyvinylcarbazol 6564
polyvinylméthyléther 6573
polyvinylpyrrolidone 6574
pompe à air 262
pompe à boue 7457
pompe à diaphragme 2482
pompe à filtrer 3469
pompe à membrane 2482
pompe centrifuge 1578
pompe de circulation 1853
pompe de dosage 6752
pompe élévatoire 258
pompe Geissler 3711
pontil 6799
pontis 6799
porcelaine diélectrique 2561
porcelaine dure 3965
porcelaine isolante 2949
porcelaine phosphatée 1108
poreux 6582
poromètre 6579
porosimètre 6580
porosité 6581
porphyrines 6584
porte-objet 5486
porteur de germes 3740
position 7767
posologie 2845
postchromer 224

post-refroidisseur 225
potable 6589
potasse 6607
potasse alcoolique 280
potasse caustique 1534
potassium 6591
potentiel biotique 969
potentiel de Donnan 2843
potentiel de limite 1145
potentiel de polarisation 6491
potentiel de sédimentation 7284
potentiel de séparation 7324
potentiel d'ionisation 4431
potentiel électroosmotique 2994
potentiel électrophorétique 2997
potentiomètre 6661
potion 6662
poudre 3905
poudre à blanchir 1037
poudre d'ardoise 7446
poudre de fer carbonylé 1458
poudre de lycopode 5020
poudre de zinc 8936
poudres à mouler de nylon 5889
poudre sans fumée 7464
poudres de mine 1029
poulie électromagnétique 2988
poulie magnétique 5082
pourcentage en poids 8825
pourcentage en volume 8790
pourpre de bromocrésol 1185
pourpre de Cassius 3834
poussière d'anthracite 2217
poutre à treillis 4732
pouvoir absorbant 39
pouvoir calorifique 1401
pouvoir courant 7730
pouvoir couvrant 4088
pouvoir d'amertume 1002
pouvoir de coloration 8293
pouvoir de pénétration 8277
pouvoir de saigner 1038
pouvoir de séchage 2878
pouvoir diastatique 2486
pouvoir fécondant 3444
pouvoir humectant 8836
pouvoir mouillant 8836
pozzolane 6668
praséodyme 6671
préchauffage 6687
préchauffeur 4138, 6686
précipitable 6672
précipitation 6676
précipitation électrostatique 3003
précipitation par un sel 7194
précipité 3524, 6673
précipité préformé 6683
précipiter 6674
prednisolone 6680
prednisone 6681
prégnandiol 6684
prégnenolone 6685
préhnitène 6688
préimmunisation 6689
prélèvement d'échantillons 7202
prémélange 6690
première couche 8618
préparation de cyanamide 2254

ptomaïnes 6793
ptyaline 7186
puiser 2863
puissance 7832
puissance nominale 6937
pulpe de chiffons 6920
pulpe de paille 7830
pulvérisateur de plomb 4763
pulvérisation 6796
pulvérisation d'acide 134
pulvérulent 6667
pumite 6797
pureté chimique 1672
pureté du plasma 6444
pureté radiochimique 6902
purgatif violent 2861
purge de chaudière 1057
purification par centrifugation 1575
purifier 1864
purine 6801
puromycine 6802
purpurine 6803
putréfaction 6804
putréfier, se ~ 6805
putrescence 6806
putride 6807
PVC rigide 7102
pycnose 6810
pycnotique 6811
pyrargyrite 6812
pyrazinamide 6814
pyrazolone 6815
pyréthrine 6816
pyrétogène 3369
pyridine 6817
pyridine de vinyle 8768
pyridoxine 6820
pyridyléthylacrylate d'éthyle 3262
pyriméthamine 6822
pyrimidines 6823
pyrite 6824
pyrite de cuivre 1617
pyrite jaune 4453
pyro- 6826
pyrocatéchine 1517, 6828
pyrochimie 6829
pyrogénation 2435
pyrolignite de fer 4444
pyrolusite 6832
pyrolyse 6833
pyromètre 6835
pyromètre optique 5990
pyrones 6836
pyrophorique 6837
pyrophosphate de calcium 1376
pyrophosphate de potasse 6642
pyrophosphate de sodium 7594
pyrophosphate de tétraéthyle 8129
pyrophosphate de zinc 8963
pyrophosphate de zirconium 8998
pyrophosphate sodique de fer 7554
pyrophyllite 6839
pyroxyle 6841
pyroxyline 6841
pyrrhotine 6842
pyrrole 6843
pyrrolidone 6844

quadrivalent 8175
quadroxalate de potassium 6643
quantité d'élément non-actif 1657
quantités molaires 5527
quantum 6849
quartz 6851
quassie 6852
québrachitol 6854
québracho 6855
quercétine 6859
quercite 6860
quercitol 6860
quercitron 6861
quinaldine 6863
quinhydrone 6864
quinidine 6868
quinine 6870
quinoléine 6875
quinone 6877
quinoxalines 6878
quintessence 6879

race de levure 8896
racémisation 6882
râcle 2832
racloir 2832
radiateur 6895
radiateur intégral 1007
radiation actinique 151
radiation infrarouge 4344
radiation thermique 3981
radical 6896
radical acide 128
radical libre 3597
radio- 6897
radioactif 6899
radioactinium 6898
radioactivité 6901
radiochimie 6903
radioélément 6904
radiothorium 6909
radium 6910
radon 6915
rafaélite 6916
raffinat 6917
raffinerie 6998
raffinose 6919
ralentissement 3340
ramie 6923
ramifié 1153
rance 6924
rancidité 6925
randomisation 6927
rapport de courant primaire 6703
rapport de distribution du métal 5266
rapport de soupape électrolytique 2986
rapport nucléoplasmatique 5876
raréfaction 6931
raréfaction de l'air 6932
raréfier 6933
rauwolfia 6939
rayonnage 1928, 2184
rayonne 6945
rayonne cupro-ammoniacale 2229
rayonne de viscose 8780
rayonnement 6889
rayonnement alpha 392

rayonnement de corps noir 1008
rayonnement ionisant 4434
rayons de Becquerel 860
réacteur 6952
réacteur catalytique 1513
réactif 6950, 6953
réactif de Bettendorf 929
réactif de Carnot 1473
réactif de Fischer 3488
réactif de Griess 3871
réactif de Karl Fischer 4606
réactif de Nessler 5693
réactif d'Huber 4146
réactif microanalytique 5467
réactifs cationiques 1530
réactifs de Grignard 3872
réaction 6946
réaction à échange chimique 1654
réaction anodique 578
réaction cathodique 1527
réaction chimique 1673
réaction d'Abderhalden 1
réaction de Baudouin 850
réaction de Diels-Alder 2562
réaction d'électrode 2967
réaction de sédimentation 3339
réaction de substitution 3298
réaction d'Hofmann 4103
réaction en chaîne 1614
réaction Friedel-Crafts 3606
réaction induite 4326
réaction irréversible 4456
réaction réversible 7068, 7069
réaction secondaire 7280
réaction sérologique faussement
    positive 956
réaction topochimique 8340
réactivation 6949
réactivité 6951
réalgar 6955
rebobineuse 7031
récessif 6958
récessivité 6959
récipient 6957
récipient à chlorure de calcium 774
récipient Cellarius 1541
récipient de solution pour plaques
    enrichies 7438
récipient poreux 6583
recombinaison 6966
recombiner 6967
recristallisation 6973
recristalliser 6974
rectification 6975
rectifier 6977
recuit 568
récupérateur 7014, 8803
récupération 6972
récupération de l'huile 6922
récupération du solvant 7658
récupération thermique 3982
récupérer 6962, 7030
recyclage 6961
redissolution 7050
redresseur 6976
redresseur à vapeur de mercure 5254
redresseur électrolytique 2984
rédruthite 6987

spectroscope 7692
spermaceti 7698
spermatide 7699
spermatocèle 3840
spermatocyte 7700
spermatozoïde 7701
sperme 7303
spermine 7702
sphalérite 1041
sphérome 7705
sphéroplaste 7706
sphérosome 7707
spinelle 7711
spirale contactrice de Podbielniak
    6485
spirale moléculaire 5544
spirème 7713
splénocyte 7718
spodumène 7719
spore 7724
sporocyte 7725
sporophyte 7726
squalène 7732
stabilisant 7733
stabilisants de vinyle 8770
stabilité chimique 1677
stabilité de couleur 1987
stabilité marginale 5161
stabilité sous l'effet de la chaleur
    3974
stade de dispersion 2805
stade dictyotique 2548
stannate de calcium 1381
stannate de plomb 4798
stannate de potassium 6647
stannate de sodium 7604
stannates 7745
stannine 7750
stannite 7750
stéaramide de butyle 1282
stéarate 7780
stéarate butoxyéthylique 1246
stéarate d'ammonium 477
stéarate d'amyle 520
stéarate de baryum 803
stéarate de butyle 1283
stéarate de calcium 1382
stéarate de cyclohexyle 2286
stéarate de diglycol 2618
stéarate de fer 3413
stéarate de lithium 4968
stéarate de magnésie 5077
stéarate de mercure 5240
stéarate de méthoxyéthyle 5297
stéarate de méthyle 5449
stéarate de plomb 4799
stéarate de potassium 6648
stéarate de sodium 7605
stéarate de trihydroxyéthylamine
    8484
stéarate de vinyle 8771
stéarate de zinc 8968
stéarate d'hydroxytitane 4251
stéarate d'isocétyle 4497
stéarine 7782
stéarylmercaptan 7784
stéatite 7786
stéphanite 7791

stéréochimie 7793
stéréoïsomères 7794
stérile 7799
stéroïdes 7801
stérols 7802
stibine 614, 7803
stibnite 614
stibophène 7804
stigmastérine 7809
stigmastérol 7809
stilbène 7811
stilbite 7812
stoechiométrie 7820
stramonine 7829
strate 4703
stratification 4701
stratifié-tissu 4700
strepsinème 7833
streptodornase 7834
streptoduocine 7835
streptokinase 7836
streptoline 7837
streptomycine 7838
streptothricine 7839
stroboscope 7847
strontianite 7848
strontium 7849
strophantine 7863
structure de molécule 5545
structure extranucléaire 3332
structure polymère 6542
structure submicroscopique 7888
strychnine 7865
studtite 7866
styrax 7870
styrolène 7871
subatomique 7877
sublimation 7883
sublimé 7882
sublimé corrosif 2119
submicron 7886
substance adsorbée 203
substance bactéricide 758
substance de base 833
substance dissoute 7645
substance isotropique 4574
substance ostéophile 1112
substance pure 6800
substitution 7889
subtiline 7890
succinate de diéthyle 2597
succinate de sodium 7606
succinimide 7893
succinylsulfathiazole 7895
suc nucléaire 7890
sucre centrifuge 1579
sucre colonial 6432
sucre de betteraves 865
sucre de bois 8889
sucre de canne 1414
sucre de lait 4681
sucre de malt 5115
sucre de manne 5150
sucre de saturne 4758
sucre inverti 4400
sucre non cristallisé 5817
suie 7666
suif 8025

suintement 5976, 6202, 7971
suinter 7286
sulfacétamide 7905
sulfadiazine 7906
sulfadiméthoxine 7907
sulfaguanidine 7908
sulfamate de calcium 1383
sulfamérazine 7909
sulfaméthazine 7910
sulfaméthiazole 7911
sulfaméthoxypyridazine 7912
sulfamide de butylbenzène 1251
sulfanilamide 7914
sulfanilate de sodium 7607
sulfanilate de zinc 8969
sulfapyridine 7916
sulfaquinoxaline 7917
sulfarsphénamine 7918
sulfatage 7922
sulfate acide d'hydroxylamine 4231
sulfate basique de plomb 828
sulfate cérique 1601
sulfate chromique 1811
sulfate cobalteux 1921
sulfate d'aluminium 424
sulfate d'ammonium 478
sulfate d'ammonium cobalteux 1907
sulfate d'argent 7415
sulfate de baryum 804
sulfate de butacaïne 1238
sulfate de cadmium 1312
sulfate de calcium 1384
sulfate de cuivre 2107
sulfate de dibutoline 2511
sulfate de diéthyle 2598
sulfate de diméthyle 2707
sulfate de guanylurée 3893
sulfate de magnésium 5078
sulfate de nickel 5725
sulfate de nickel ammoniacal 5715
sulfate de pipérazine-oestrone 6409
sulfate de plomb 4800
sulfate de potassium 6649
sulfate de potassium et magnésium
    6624
sulfate de quinidine 6869
sulfate de quinine 6874
sulfate de radium 6914
sulfate de sodium 7608
sulfate de sodium brut 7190
sulfate de spartéine 7681
sulfate de strontium 7858
sulfate de thallium 8186
sulfate de titane 8303, 8307
sulfate de vanadyle 8700
sulfate de zinc 8970
sulfate de zirconium 8999
sulfate d'hydrazine 4166
sulfate d'hydroxylamine 4233
sulfate d'hydroxyquinoléine 4249
sulfate double d'aluminium et
    d'ammonium 406
sulfate double d'aluminium et de
    potassium 421
sulfate double d'aluminium et de
    sodium 423
sulfate double d'ammonium et de
    chrome 1815

sulfate d'uranyle 8653
sulfate d'yttrium 8910
sulfate ferreux 3439
sulfate ferreux ammoniacal 3427
sulfate ferrique 3414
sulfate manganeux 5145
sulfate mercureux 5247
sulfate mercurique 5241
sulfates 7920
sulfate stanneux 7757
sulfate tribasique de cuivre 8394
sulfathiazole 7921
sulfénamide d'oxydiéthylène-
benzothiazol 6053
sulfinylpyrazone 7924
sulfite de calcium 1386
sulfite de magnésium 5079
sulfite de potassium 6651
sulfite de sodium 7610
sulfite de zinc 8972
sulfobromophtaléine-sodium 7928
sulfocarbonate de potassium 6652
sulfochlorure de toluène 8324
sulfocyanures 7929
sulfonate d'éthyle chloré 5352
sulfonates d'alkylaryl 342
sulfonateur 7932
sulfonation 7931
sulfonéthylméthane 7933
sulfonméthane 7934
sulfonyldianiline 7935
sulforicinoléate de sodium 7611
sulfoxylate de formaldéhyde et de
zinc 8940
sulfoxylate de zinc 8973
sulfure cuivreux 2237
sulfure d'ammonium 479
sulfure de baryum 805
sulfure de cadmium 1313
sulfure de calcium 1385
sulfure de diamyle 2477
sulfure de dibromodiéthyle 2506
sulfure de dibutylétain 2529
sulfure de dilauryle 2657
sulfure de diméthyle 2708
sulfure de mercure rouge 6984
sulfure de plomb 4801
sulfure de potassium 6650
sulfure de sélénium 7297
sulfure de sodium 7609
sulfure de strontium 7859
sulfure de tétraéthylthiuram 8131
sulfure de thallium 8187
sulfure d'éthyle 3266
sulfure d'éthyle dichloré 2536
sulfure de vanadium 8694
sulfure de vinyléthoxyéthyle 8755
sulfure de zinc 8971
sulfure dipyridyléthylique 2782
sulfure ferreux 3440
sulfure mercurique 5242
sulfures 7923
sulfures d'alcoyles 8241
sulfure stanneux 7758
sumbul 7949
supercalandrage 7952
superoxyde de magnésium 5073
superphosphate 7958

superpolymère linéaire 4904
support d'accrochage 6468
support de cornue 7062
support de filtre 3630
suractivité 7951
suramine de sodium 7612
surchauffeur 7955
surcuisson 6031
surépaisseur en bordure 3356
surface active d'une électrode 2963
surface criblante 2342
surface criblante à fils métalliques
8823
surface de contact 4378
surface ondulée 7965
surfusion 7953
sursaturation 7959
sursoufflage 223
surtension 6037
surtension de concentration 2020
surtension ohmique 5933
survolteur 5491
susceptibilité 7967
suspension 7968
suspension colloïdale 1975
suspensoïde 7969
sylvanite 7979
sylvinite 7980
sylvite 7981
sym- 7982
symbiose 7983
symétrie 7984
sympatholytique 7985
synapsis 7986
synaptique 7987
syncyte 7989
syndiotactique 7990
synérèse 7991
synergiste 7992
synkaryon 7988
synthèse 7995
synthèse de Diels-Alder 2564
synthèse de Fittig 3491
synthèse de l'ammoniac 454
synthèse de Wurtz 8868
synthétique 7997
syntrophe 8003
syrosingopine 8005
système asymétrique 700
système chromaffine 1785
système multiple 5615
système série 7328
système triclinique 700

table de laboratoire 4663
table périodique 6211
table physique de poids atomiques
6371
taches de cristaux 2212
tachetures 7727
tachystérine 8011
tactique 8016
talbutal 8022
talc 3602, 8023
tallate de zinc 8974
tamis 7263, 7413
tamiser 7371
tamis métallique 8850, 8852

tanin 8035
tank de soutirage 1139
tannage 2249, 8036
tannate de bismuth 992
tannate de pelletiérine 6159
tanner 8027
tannoforme 8037
tantale 8039
tantalite 8038
taraxacum 8045
tare 8046
tartrate acide de potassium 6604
tartrate antimonico-potassique 621
tartrate d'amyle 521
tartrate de dibutyle 2523
tartrate de diéthyle 2599
tartrate de phénindamine 6251
tartrate de potassium 6653
tartrate de potassium et sodium 6646
tartrate de sodium 7613
tartrate double de sodium et de
bismuth 988
tartrate ferrico-potassique 4452
tartrate ferrique de potassium 3410
tartrate stanneux 7759
tartre 1085, 8050
tartre émétique 8051
tas de malt 5116
taurine 8053
tautomérie 8054
taux de cendres 687
taux de compression 2014
taux de reflux 7002
technécium 8056
technologie nucléaire 5866
teinte 8291
teinture 8290
teinture d'iode 4411
teintures 2904
teintures azoïques 4268
teintures de kétonimine 4624
teintures métallisées 5270
téléchromomère 8057
tellure noir 1019
tellurium 8059
tellurure de bismuth 993
tellurures d'or 3833
télolécithe 8062
télomère 8063
télosynapse 8064
télosyndétique 8065
température 8067
température absolue 20
température ambiante 434, 7118
température critique 2171
température d'auto-inflammation
7302
température d'écoulement 3530
température de couleur 1988
température de four 4639
température de prise 7338
température de séchage accéléré
3564
température de transition 8369
température du corps noir 1009
temps de duplication 2853
temps de gélification 3715
temps de réaction 6947

vératrine 8724
vératrol 8725
verdunisation 8729
vermicide 8730
vermiculite 8731
vermifuge 8730
vermillon 8732
vermillon d'antimoine 625
vernis 4669, 8716
vernis à l'asphalte 1200
vernis à masquer les nœuds 4649
vernis antitrace 638
vernis au tampon blanc 8839
vernis au trempe 2774
vernis carmoisi 2169
vernis cristallisé 2213
vernis d'asphalte 1014
vernis de polissage 6498
vernis doré à l'étuvage 3832
vernis du Japon 4586
vernis dur brillant 3963
vernis imperméable 7682
vernis isolant 4368, 7823
vernis léger 7363
vernis lithographique 4974
vernis marine 1074
vernis mat 3511
vernis mat d'apprêt 3510
vernis pour yachts 8891
vernissage par salage 7192
vernis séchant à l'air 254
véronal 783
verre 3753
verre à boudines 2183
verre à pellicule d'or 3824
verre de chaux sodée 7471
verre de montre 8804
verre de pesée 8824
verre gradué 5195
verre ne produisant pas d'éclat 633
verre résistant à la chaleur 8211
verre soluble 6645, 7602
vert de Brême 1159
vert de cadmium 1303
vert de chrome 1796, 3895
vert de cobalt 1901
vert-de-gris 8728
vert de Guinée 3897
vert de manganèse 5133
vert de Rinmann 1901
vert de Schweinfurt 7250
vert de zinc 1901
vert émeraude 3018
vert Guignet 3895
vert malachite 5099
vétivérol 8733
viabilité 8736
viable 8737
vibrateur de trémie 4130
vibreur de filtre 7374
vibreur magnétique 5084
vidange 8597
vide préalabe 7138
vide préalable à basse pression 3568
vieillissement 230
vieillissement naturel 5658
vinaigre 8742
vinaigre de vin 8848

vinaigre glacial 3750
vinal 8740
vinbarbital 8741
vin doux 5618
vinomètre 5932
vinylacétylène 8745
vinylation 8746
vinylbutyléther 8747
vinylcarbazole 8749
vinylcyclohexène 8752
vinyléther 8754
vinyléthyléther 8756
vinyléthylhexyléther 8758
vinyléthylpyridine 8759
vinylméthyléther 8765
vinyon 8774
violet de cobalt 1927
violet de méthyle 5460
violet minéral 5502
viomycine 8775
viorne 8739
virage au titrage 3046
virer 8337
virus de la poliomyélite 6495
viscose 8778
viscosimètre 5516, 8010, 8777
viscosimètre à chute de bille 779
viscosimètre capillaire 1419
viscosité 8781
viscosité cinématique 4640
viscosité de la matrice 5186
viscosité élevée, à ~ 4095
viscosité intrinsèque 4393
viscosité relative 7021
viscosité spécifique 7690
vitamine 8785
vitesse critique 2172
vitesse d'absorption d'eau 7694
vitesse d'accroissement de la
    viscosité 1078
vitesse de combustion 1233
vitesse de réaction 6948
vitesse de vulcanisation 8796
vitesse spatiale 7676
vitrification 8786
vitriol bleu 1071
vivianite 8787
voie humide 8831
voile 3503
volatilité 8789
volatilité relative 7022
volume atomique 713
volume d'un gramme-molécule 3855
volume incompressible 4307
volume moléculaire 5546
volume spécifique 7691
vortex 8792
voyant au travers, se ~ 3875
vulcanisation 8795
vulcanisation à froid 1954
vulcanisation à haute fréquence 4092
vulcanisation à la vapeur 7779
vulcanisat sans charge 8625
vulcanite 8794

warfarine 8799
warfarine sodique 8800
wavellite 8818

white-spirit 8844
willémite 8847
withérite 8853
wolframite 8854
wollastonite 8855
wulfénite 8865
wurtzite 8867

xanthate de cellulose 1563
xanthate isopropylique de sodium
    7555
xanthène 8869
xanthine 8873
xanthocilline 8874
xanthogène de biséthyle 976
xanthophore 4927
xanthophylle 8875
xanthoptérine 8876
xanthotoxine 8877
xanthydrol 8878
xénon 8879
xénoplastique 8880
xénotime 8881
xylène 8882
xylène-sulfonate de sodium 7627
xylénol 8883
xylidine 8885
xylithol 8886
xylol 8887
xylose 8889

yohimbine 8902
ypérite 2536
ytterbium 8905
yttrium 8906

z 8911
zéine 8912
zéolithe feuilletée 4047
zéolithes 8913
zéro absolu 21
zinc 7695, 8916
zinc diéthylique 8932
zincite 8944
zinc sulfuré 1041
zinkénite 8979
zippéite 8980
zircon 8981
zirconate de lithium 4971
zirconate de tétrabutyle 8098
zirconate de tétraisopropyle 8153
zirconium 8982
zoïsite 9003
zone d'enrichissement 3055
zone de plastification 5219
zone pauvre en résine 7047
zone riche en résine 7046
zygosome 9004
zygote 9005
zygotène 9006
zygoténique 9007
zygotique 9008
zymase 9009
zymogène 9010
zymologie 9011
zymosimètre 9012

ESPAÑOL

E

abelito 3
abelmoscho 4
abertura 6005
abietato de glicerilo 3799
abietato de metilo 5302
abietato potásico 6592
abietato sódico 7474
abieteno 7
abietinato 6
abiureto 9
ablandamiento 7630
ablastina 10
abono 5156
abortivo 12
abrasivos 13
abrazadera de compresión 6399
abrotina 14
absintina 15
absorbedor 23, 24
absorbente 23
absorbente de rayos ultravioletos 8604
absorber 5897
absorbibilidad 22
absorciómetro 26
absorción 27, 5899
absorción aparente 645
absorción de aceite 5934
absorción de agua 8806
absorción defectuosa 5097
absorción de haz ancho 1171
absorción del vapor de agua 8817
absorptancia 25
abundancia molecular 5531
abundancia relativa isotópica 40
acabado cristalino 2208
acabado de aspecto aterciopelado 7902
acabado ondulado 7106
acacatechina 41
acacetina 42
acaciina 43
acaricida 44
acariocito 265
acción anticoagulante 599
acción de masa 5173
aceite animal 1111
aceite bruto 2188
aceite cocido 1082
aceite contraido 7743
aceite crudo 6942
aceite de ajenjo 8863
aceite de ajo 3675
aceite de alcanfor 1408
aceite de anís 559
aceite de atún 8567
aceite de ballena 8837
aceite de cacahuete 654
aceite de cada 1297
aceite de canela 1851
aceite de cáñamo 4008
aceite de cardamomo 1466
aceite de cártamo 7171
aceite de casia 1496
aceite de castor 1502
aceite de cayeput 1321
aceite de coco 1934
aceite de colofonia 1053, 7125

aceite de comino 2221
aceite de copaiba 2074
aceite de Copaifera officinalis 2074
aceite de Dippel 1111
aceite de enebro 4599
aceite de esperma 7703
aceite de espliego 4750, 7710
aceite de esquistosa 7352
aceite de estilingia 7815
aceite de eucalipto 3276
aceite de fusel 3643
aceite de girasol 7950
aceite de hígado de bacalao 1936
aceite de hígado de halibut 3939
aceite de hígado de tiburón 7353
aceite de huesos 1111
aceite de ilang ilang 8901
aceite de jazmín 4589
aceite de ládano 4657
aceite de laurel 4738
aceite de lavanda 4750
aceite de lignito 4873
aceite de lima 4883
aceite de linaza 4917
aceite de linaza cocido 1081
aceite de linaza crudo 6940
aceite de linaza soplado 1060
aceite de lino refinado por proceso alcalínico 332
aceite de macis 5033
aceite de madera de cedro 1538
aceite de madera de sándalo 7205
aceite de manteca de cerdo 4721
aceite de mejorana 5163
aceite de menta piperita 6187
aceite de mirra 5631
aceite de mostaza 5619
aceite de neroli 5691
aceite de nuez 8798
aceite de nuez de palma 6082
aceite de oliva 5965
aceite de palmarosa 6079
aceite de palo 8568
aceite de pie de buey 5665
aceite de pino 6402
aceite de pirola 8849
aceite de poleo 6167
aceite de quenopodio 1692
aceite de resina 8024
aceite de ricino 1502
aceite de romero 7122
aceite de ruda 7154
aceite de sasafrás 7224
aceite de sebo vegetal 8026
aceite de semilla de colza 1992
aceite de semilla de soja 7675
aceite de semillas de algodón 2125
aceite de semillas de caucho 7148
aceite de sésamo 7335
aceite de tomillo 8281
aceite de trementina 3902
aceite de tung 8568
aceite de tuya 8278
aceite de valeriana 8682
aceite de vetiver 8735
aceite de vitriolo 7945
aceite diesel 2567
aceite hervido 1082

aceite no secante 5822
aceite para algodón 2124
aceite para barnizar 3769
aceite pesado 3613
aceite rojo de Turquía 8581
aceites 5000
aceites absorbentes 35
aceites blancos 8843
aceites de acetona 77
aceites de cloronaftaleno 1748
aceite secante 2883, 7743
aceites esenciales 3112
aceites hidrogenados 4187
aceites ligeros 4859
aceites litográficos 4973
aceites para cortar 2252
aceites para regenerar la goma 6965
aceites polimerizados 6541
aceites polimerizados por calor 3971
aceites segregados 7290
aceites soplados 1061
aceites textiles 8177
aceite sulfonado 7930
aceites volátiles 3112
aceite yodado 4413
aceitoso 5953, 6404
acelerador de temperatura baja 4994
aceleradores del caucho 7145
acelerante 46
acenafteno 48
acenaftenoquinona 49
aceptador de iones 4417
acero al níquel 5724
acetal 50
acetaldehidasa 51
acetaldehído 52
acetales de polivinilo 6560
acetamida 53
acetamidina 54
acetanilida 55
acetato 56
acetato alumínico 405
acetato amónico 456
acetato bárico 788
acetato butirato de celulosa 1553
acetato cálcico 1333
acetato cobaltoso 1906
acetato crómico 1801
acetato de amilo 496
acetato de bencilo 904
acetato de bornilo 1121
acetato de cadmio 1299
acetato de celulosa 1552
acetato de ciclohexanol 2280
acetato de cinamilo 1852
acetato de cobre 2082
acetato de dodecilo 2837
acetato de etilhexilo 3224
acetato de etilo 3134
acetato de eugenol 3281
acetato de feniletilo 6290
acetato de fenilmetilo 5435
acetato de fenilo 6275
acetato de fenilpropilo 6312
acetato de furfurilo 3634
acetato de geranilo 3736
acetato de hexilo 4077
acetato de isoamilo 4466

ácido heptafluobutírico 4015
ácido hexahidrobenzoico 4060
ácido hexahidroftálico 4061
ácido hialurónico 4152
ácido hidnocárpico 4157
ácido hidrocinámico 4177
ácido hidroxiacético 4218
ácido hidroxibutírico 4221
ácido hidroxinaftoico 4238
ácido hipocloroso 4260
ácido hipofosforoso 4264
ácido hipúrico 4097
ácido indolbutírico 4324
ácido isetiónico 4460
ácido isobutírico 4492
ácido isocianúrico 4501
ácido isodecanoico 4504
ácido isoftálico 4535
ácido isonicotínico 4522
ácido isopentanoico 4532
ácido isovaleriánico 4577
ácido itacónico 4578
ácido láctico 4675
ácido láurico 4739
ácido levúlico 4685
ácido lignocerínico 4875
ácido linoleico 4909
ácido linolénico 4911
acidólisis 124
ácido litocólico 4972
ácido maleico 5103
ácido málico 5107
ácido malónico 5109
ácido mandélico 5118
ácido mecónico 5204
ácido melísico 5215
ácido metacrílico 5283
ácido metanílico 5277
ácido metilclorofenoxipropiónico
    5351
ácido metilfosfórico 5438
ácido metoxiacético 5289
ácido mirístico 5628
ácido monobásico 5564
ácido múcico 5603
ácido muriático 4176
ácido naftalendisulfónico 5637
ácido naftalensulfónico 5638
ácido naftilaminodisulfónico 5644
ácido naftilaminosulfónico 5645
ácido naftilaminotrisulfónico 5646
ácido naftiónico 5639
ácido naftoldisulfónico 5641
ácido nicotínico 5729
ácido nítrico 5741
ácido nitrobenzoico 5757
ácido nitrofenilacético 5784
ácido nitrosalicílico 5786
ácido nonaldecanoico 5814
ácido nonoico 6156
ácido oleico 5958
ácido ortofosforoso 6012
ácido ósmico 6015
ácido oxálico 6040
ácido palmítico 6080
ácido palmitoleico 6081
ácido pantoténico 6093
ácido pelargónico 6156

ácido pentadecanoico 6173
ácido peracético 6193
ácido perclórico 6197
ácido perfórmico 6205
ácido peryódico 6209
ácido picrámico 6386
ácido pícrico 6387
ácido picrolónico 6388
ácido pirofosfórico 6838
ácido pirogálico 6830
ácido piroleñoso 6831
ácido piromelítico 6834
ácido pirotartárico 6840
ácido poliacrílico 6504
ácido propiónico 6749
ácido quínico 6867
ácido racémico 6880
ácido resorcílico 7057
ácido ribonucleico 7092
ácido ricinoleico 7098
ácidorresistente 111
ácido rubeánico 7151
ácidos aldehídicos 286
ácido salicílico 7182
ácidos del alquitrán 8044
ácidos dibásicos 2495
ácido sebácico 7271
ácido selenioso 7298
ácidos estánnicos 7746
ácidos haloideos 3947
ácido silícico 7382
ácido silicotúngstico 7392
ácidos nucleicos 5867
ácido sórbico 7667
ácido subérico 7880
ácido succínico 7891
ácido sulfámico 7913
ácido sulfanílico 7915
ácido sulfobenzoico 7927
ácido sulfoftálico 7936
ácido sulfosalicílico 7937
ácido sulfúrico 7945
ácido sulfúrico fumante 3624
ácido sulfuroso 7946
ácidos verdes 3866
ácido tánico 8034
ácido tartárico 8052
ácido telúrico 8058
ácido tereftálico 8075
ácido tetracloroftálico 8106
ácido tíglico 8287
ácido tímico 8282
ácido tioacético 8235
ácido tiobenzoico 8236
ácido tiociánico 8238
ácido tiodiglicólico 8240
ácido tioglicólico 8243
ácido tiosalicílico 8251
ácido titánico 8295
ácido toluensulfónico 8326
ácido toluico 8328
ácido toluidinasulfónico 8330
ácido tribromoacético 8400
ácido tricloroacético 8416
ácido triclorofenoxiacético 8431
ácido tricloroisocianúrico 8424
ácido triclorometilfosfónico 8428
ácido trifluoacético 8475

ácido trimésico 8495
ácido trimetilacético 8498
ácido trinitrobenzoico 8519
ácido trópico 8558
ácido túngstico 8571
ácido undecanoico 8612
ácido undecilénico 8614
ácido úrico 8661
ácido uridílico 8663
ácido valérico 8683
ácido xantenocarboxílico 8870
ácido xántico 8872
ácido yodhídrico 4171
ácido yódico 4406
acidular 135
acilación 175
aciloina 177
aciloxi- 178
aclarar 3474
aclimatización 47
aconitato de tributilo 8403
aconitato de trietilo 8450
aconitina 138
acoplamiento de valor intermedio
    4383
acridina 139
acridona 141
acriflavina 142
acrilamida 144
acrilato cálcico 1335
acrilato de butilo 1250
acrilato de cianoetilo 2260
acrilato de etilhexilo 3225
acrilato de etilo 3137
acrilato de metilo 5311
acrilato de polidihidroperfluobutilo
    6522
acrilonitrilo 149
acrodextrina 99
acroleína 143
acromático 100
acromia 102
actínidos 152
actinio 153
actinolita 156
actinología 157
actinómetro 158
actinón 159
actinoquímica 155
activador 165, 6737
activar la fermentación 7140
actividad 172
actividad iónica 4418
actividad óptica 5985
acuarelas 8808
acumulador 7826
acumulador alcalino 330
acumulador de cadmio-níquel 1306
acumulador de ferro-níquel 4446
acumulador de plata 7414
acumulador de plomo 4759
achicoria 1698
achiote 567
adaptación 179
adaptación ecoclimática 2918
adenina 182
adenocromo 183
adenosina 184

albúmina de sangre 1049
albuminados 273
albuminoides 274
albumosas 275
álcali 320
álcalicelulosa 321
alcalimetría 325
alcalímetro 324
alcalinidad 331
alcalinizar 334
alcaloides 335
alcalosis 336
alcanfor 1406
alcano 337
alcohilación 344
alcohilato 343
alcohilo 340
alcohol 276
alcohol absoluto 16
alcohol alílico 373
alcohol amílico 497
alcoholatos de litio 4943
alcoholatura 277
alcohol bencílico 905
alcohol cetílico 1612
alcohol cinámico 1849
alcohol colonial 5287
alcohol de Colonia 1978
alcohol de madera 8860
alcohol de metilo 5287
alcohol de quemar 2396
alcohol desnaturalizado 2396, 4332
alcohol dihidroabietílico 2621
alcohol erucílico 3100
alcoholes grasos 3363
alcoholes secundarios 7276
alcohol esteárilico 7783
alcohol estirálico 7869
alcohol etilbutílico 3154
alcohol etilhexílico 3226
alcohol etílico 3138
alcohol feniletílico 6292
alcohol fenquílico 3377
alcohol furfurílico 3635
alcohol hidroxiestearílico 4250
alcoholimetría 283
alcoholímetro 282
alcohol isoamílico 4467
alcohol isobutílico 4483
alcohol isooctílico 4526
alcohol isopropílico 4544
alcoholización 281
alcohol láurico 4740
alcohol laurílico 4743
alcohol linoleílico 4913
alcohol linolenílico 4912
alcohol metilado 5326
alcohol metilado industrial 4334
alcohol metilamílico 5318
alcohol miristílico 5630
alcohol monovalente 8624
alcohol nonílico 5836
alcohol no saturado 8627
alcohol oleílico 5963
alcohol polivinílico 6562
alcohol propargílico 6743
alcohol propílico 6754
alcohol ricinoleílico 7099

alcohol salicílico 7177
alcohol terciario 8086
alcohol tetrahidrofurfurílico 8138
alcohol tridecílico 8442
alcohol undecilénico 8615
alcosol 284
aldehidasa 285
aldehidato amónico 287
aldehído acético 52
aldehído butírico 1286
aldehído cinámico 1850
aldehído fenilpropílico 6313
aldehído fórmico 3569
aldehído glutárico 3781
aldehído isobutírico 4491
aldehído isovaleriánico 4576
aldehído laurílico 4744
aldehído octadecenílico 5903
aldehído pirúvico 6845
aldehídos 289
aldehído undecilénico 8616
aldiminas 290
aldocetenas 292
aldohexosas 291
aldol 293
aldolasa 294
aldosas 295
aldosterona 296
aldoximas 297
aldrina 298
aleación 371
aleaciones de cobre-berilio 2085
aleaciones fusibles 3644
aleación ligera 4850
aleato alumínico 418
alelismo 350
alelo 349
alelomorfismo gradual 7790
alelos de posición 6585
aleno 351
aletrina 352
aleurona 300
alexina 301
alexina de la sangre 3919
alfa 387
alfacelulosa 388
alfarería 1591, 2913
alfatrón 394
alfilos 396
alga 303
algesiógeno 304
alginato de amonio 457
alginato de calcio 1336
alginatos 305
alginato sódico 7478
algodón cianoetilado 2261
algodonita 307
algodón pólvora 3904
aliciclico 309
aliesterasa 312
alifático 313
alilamina 374
aliltiourea 383
alimentador a correa 871
alisado 2251
alita 316
alizarina 317
alkilato 343

alkilización 344
alkilo 340
almacenamiento de calor interno
    3972
almáciga 5182
almandina 384
almendra amarga 1001
almidón 7760
alobara 353
alocimeno 363
aloestérico 367
alogamia 354
alogenético 356
alógeno 355
aloína 386
aloiogénesis 357
aloisomerismo 358
alomérico 359
alomorfismo 361
alomorfo 360
alomórfosis 362
alopátrico 364
alopoliploide segmental 7288
alosa 365
alosoma 366
alotropía 369
alotrópico 368
aloxacina 370
alpaca 5723
alquileno 345
alquitrán 8043
alquitrán de hulla 1892
alquitrán de pino 6403
alta fluidez 7634
alta. viscosidad, de ~ 4095
altrosa 397
alumbrado por los bordes 4860
alumbre 399
alumbre crómico 1795
alumbre potásico 6595
alúmina 400
alúmina activada 160
aluminato de calcio 1337
aluminato de litio 4944
aluminato potásico 6596
aluminato sódico 7479
aluminio 404
aluminita 403
alunita 428
alunógeno 429
alveolación mecánica 5199
alveolación química 1655
alveolar 1548
alveolización por ionización 4430
allanar 3507
amagnético 5828
amalgama 430
amalgamación 431
amalgama de sodio 7482
amarillo brillante 4590
amarillo de bencidina 891
amarillo de cadmio 1316
amarillo de cal 4890
amarillo de Cassel 1495
amarillo de cromo 1800, 4820
amarillo de estroncio 7853
amarillo de hierro 4445
amarillo de la India 4313

amarillo de Nápoles 5652
amarillo de oro 4591
amarillo de París 6126
amarillo de quinolina 6876
amarillo de zinc 8928
amarillo hansa 3949
amarillo real 4642
amarillos de ferrita 3416
amarillos de óxido de hierro 4450
amasado 8068
amasador-dispersador 2804
amasar 1072, 5167
amasijo 5165
ámbar 432
ámbar gris 433
ambarina 4
ambiental 3061
ambiente genético 3724
ameiosis 436
americio 437
amfibolita 486
amianto 682
amianto paladinizado 6073
amianto platinado 6469
amiba 435
amicrón 438
amíctico 439
amida 440
amida poliéster 6525
amida primaria 6701
amidas de ácidos 104
amida secundaria 7277
amida sódica 7483
amida terciaria 8087
amidinas 441
amilasa 498
amileno 501
amilmercaptano 507
amilnaftaleno 508
amilo 495
amiloide 511
amiloidosis 512
amilólisis 514
amilopsina 515
amilosa 516
amina 443
aminas alifáticas 3364
aminoacetato dihidroxialumínico 2627
aminoácidos 444
aminobenzoato de etilo 3142
aminobenzoato de isobutilo 4485
para-aminobenzoato sódico 7484
para-aminohipurato sódico 7485
aminorar 4833
para-aminosalicilato sódico 7486
amitosis 447
amixia 448
amixis 448
amminas 449
amoníaco 450
amoníaco anhidro 549
amoniaco combinado 3493
amoníaco libre 3594
amoníaco total 8346
amonio 455
amonólisis 482
amorfo 483

ana- 522
anabasina 523
anabergita 566
anacromasia 524
anaeróbico 525
anaerobio 525
anafase 530
anafóresis 532
analcima 526
analcita 526
analgésico 527
análisis 528, 695
análisis conductométrico 2033
análisis cualitativo 6847
análisis cuantitativo 6848
análisis de las tetradas 8116
análisis genómico 3729
análisis granulométrico 7372
análisis gravimétrico 3864
análisis por absorción 28
análisis por activación 164
análisis por dilución isotópica 4573
análisis por vía húmeda 8828
análisis por vía seca 2879
análisis radiométrico 6907
análisis volumétrico 8791
analizador de gas 3680
anaranjado de antimonio 616
anaranjado de cromo 1797
anaranjado de molibdato 5549
anaranjado mineral 5499
anato 567
andalucita 534
androsterona 535
anemómetro 536
anemómetro térmico 4144
anetol 538
aneuploide 539
aneuploidia 540
anfiaster 484
anfíboles 485
anfidiploidia 487
anfigonia 488
anfimixis 489
anfogenia 490
anfólito 491
anfolitoide 492
anfótero 493
angioblasto 541
anglesita 542
anhídrido acético 59
anhídrido carbónico 1012
anhídrido cloromaleico 1744
anhídrido dodecenilsuccínico 2836
anhídrido ftálico 6365
anhídrido glutárico 3783
anhídrido isobutírico 4493
anhídrido maleico 5104
anhídrido propiónico 6750
anhídridos 545
anhídrido succínico 7892
anhídrido tetraclorofatálico 8107
anhídrido tetrahidroftálico 8145
anhídrido tetrapropenilsuccínico 8171
anhidrita 546
anhidro 548
anidación 569

anilida hidroxinaftoica 4239
anilina 554
anillo de benceno 886
anillos Raschig 6935
anillo tándem 8031
anión 558
anisidinas 560
anisol 561
anisomérico 562
anisotónico 563
anisotrópico 564
anodización 579
anodización al ácido crómico 1803
anodizado 579
ánodo 570
ánodo preformado 2037
anólito 580
anortosita 582
anquerita 565
antiadhesivo 2439
antialbumosas 590
antialdoximas 591
antibariona 592
antibiosis 593
antibiótico de amplio espectro 1173
antibióticos 594
anticatalizador 596
anticloro 597
anticoagulante 598
anticuerpo 595
antidetonante 601
antídoto 603
antienvejecedor contra temperaturas altas 3970
antienzima 604
antihistamina 608
antihormona 609
antimagnético 5828
antimicina 627
antimonial 611
antimoniato de plomo 4760
antimoniatos 613
antimoniato sódico 7488
antimonio 615
antimonita 614
antimonita de isopropilo 4548
antioxidante 629
antiozonizante 630
antipirina 631
antisépticos 632
antofilita 583
antraceno 584
antraflavina 585
antranilato de dimetilo 2680
antranilato de feniletilo 6293
antranilato de metilo 5323
antranol 587
antraquinona 588
antrarrobina 589
anzulete 4736
apagar (la cal) 7444
aparato bimetálico 943
aparato de contacto por etapa 7735
aparato de Dean y Stark 2330
aparato de destilación 7813
aparato de evaporación 3295
aparato de extracción 3326
aparato de Golgi 3836

aparato de Jolly 4596
aparato de Kipp 4645
aparato de Landsberger 4711
aparato para emulsionar 3024
aparato para retirar catalizador de
  cobalto 2343
aparejo 3460, 8618
apatita 639
apocrino 641
apocromático 640
apoenzima 642
apogamia 643
apomorfina 644
aporoso 5832
apresto 7430
aprestos para correas 869
aptitud por la combinación 1993
arabinosa 652
aragonito 655
araquidato de metilo 5325
arborescencias 8379
arborescente 656
arcilla 1867, 4601
arcilla calcinada 3877
arcilla grasa 778
arcilla Gross-Almerode 3878
arcilla plástica 4985
arcilla refractaria 3480
arcilloso 661
arena 7204
areómetro 4208
areopicnómetro 659
areopicnómetro de Eichhorn 2938
argamasa de bario 802
argentita 660
arginina 662
argol 663
argón 604
armario vacuosecador 8675
arrastre 3058
arrollamiento recíproco 7019
arrugamiento 2149, 2156
arsanilato sódico 7489
arsenamina 676
arseniato cálcico 1339
arseniato cobaltoso 1908
arseniato de cobre 2084
arseniato de níquel 5716
arseniato de plomo 4761
arseniato de zinc 8919
arseniato férrico 3395
arseniato ferroso 3428
arseniato magnésico 5051
arseniato mercúrico 5230
arseniato potásico 6598
arseniato sódico 7490
arsénico 669
arsénico blanco 8838
arsénico sulfurado rojo 6955
arsenito cálcico 1340
arsenito de plomo 4762
arsenito de zinc 8920
arsenito potásico 6599
arsenito sódico 7491
arsenopirita 671
arsenorresistente 668
arsfenamina 677
arsfenamina de plata 7397

arsina 676
asbesto 682
asbolana 683
asbolita 683
ascaridol 684
ascorbato sódico 7492
aséptico 686
aserrín 8856
asfaltita 690
asfalto 689, 1004
asfalto oxidado 1058
asfalto soplado 1058
asim- 8630
asimilable 308
asimilación incompleta 5100
asociación 696
asociación molecular 5532
asociación nuclear 5856
asociación por aglutinación 7806
asociación tándem 8028
aspecto agrietado 8822
aspecto deslustrado 7447
aspecto granuloso 7285
aspecto lechoso 7653
aspecto nuboso superficial 7963
aspecto sedoso 7393
aspirador 691
aspirador accionado por vapor 7770
aspirar 7896
astatino 697
áster 698
astrocito 699
atacamita 701
atáctico 702
atapulguita 717
atenuación de haz ancho 1172
aterciopelado 3522
aterciopelar 3520
atíncar 8289
atmólisis 703
atomicidad 709
átomo-gramo 3852
átomo hidrogenoide 4200
átomo marcado 8018
átomo pesado 3987
atropina 716
aumentar a escala 7240
aumentar la viscosidad 1075
aumento 5091, 5093
aumento de viscosidad 3374
aurinas 721
auroso 722
austenita 723
autocatálisis 724
autoclave 725, 6695
autofértil 7300
autofertilidad 7301
autólisis 726
autooxidación 730
autoploidia 727
autorradiólisis 728
autosoma 3279
autotetraploidia 729
autunita 731
auxiliares de blanqueo 1036
auxocito 733
auxocromos 732
avidina 734

azabache 4594
azafràn 7172
azar, al ~ 6926
azaserina 737
azelato de diisobutilo 2634
azelato de diisooctilo 2642
azeotropo 739
azidas 742
azidas de ácidos 105
azida sódica 7493
azinas 744
azobenceno 745
azofenina 749
azoimida 748
azotómetro 750
azúcar blanco 6432
azúcar centrífugo 1579
azúcar colonial 6432
azúcar de caña 1414
azúcar de plomo 4758
azúcar de remolacha 865
azúcar invertido 4400
azúcar no centrífugo 5817
azufre 7938
azufre combinado 1147
azufre hepático 4980
azufre libre 3598
azufre moldeado 7117
azul a la cal 4878
azul Berlín 922
azul brema 1158
azul Brunswick 1201
azul chino 1703
azul de Brema 1158
azul de bromofenol 1189
azul de bromotimol 1192
azul de cobalto 1897
azul de lavandería 4736
azul de metileno 5372
azul de París 6782
azul de Prusia 6782
azul de Thénard 8193
azul pompeyano 6575
azul real 1897
azul ultramar 8601
azurita 1694

bacilo 751
bacitracina 752
bacteria 767
bacteria donante competente 2006
bacteria recipiente 6960
bacterias acéticas 60
bactericida 758
bacteriocina 759
bacteriofago 763
bacteriógeno 760
bacteriolisina 761
bacteriólisis 762
bacteriostasis 764
bacteriostático 766
bacteriostato 765
baculiforme 768
badeleyita 769
bagazo 771
baja fluidez 7808
balance de materia 5174
balanza de ensayo 692

blanco chino 1705
blanco de antimonio 626
blanco de España 987, 8846
blanco de estroncio 7862
blanco de París 6125
blanco de titanio 8306
blanco de zinc 8952
blanco fijo 1023
blanco fijo de alúmina 401
blanco interrumpido (por otros
. pigmentos) 1175
blanco mineral 5503
blanco quebrado 1175
blanco satín 7228
blandura 1634
blanqueador 1034
blanqueadores ópticos 5987
blanquear 1033
blanqueo 1035
blastema 1025
blastocito 1031
blastomero 1032
blefaroplasto 1044
blenda 1041
blenda oscura 1041
bloque 1879
bloque de carbón 1444
bloque polímero 1048
boca del caldero 4919
bodega de fermentación 3387
bolita 6158
bolómetro 1093
bolsa de vapor 8714
bomba catalítica 1510
bomba centrífuga 1578
bomba de aire 262
bomba de circulación 1853
bomba de diafragma 2482
bomba de dosificación 6752
bomba de filtración 3469
bomba de Geissler 3711
bomba de lechada 7457
bomba de membrana 2482
bomba elevadora de líquidos por aire
258
bomba para trasvase de ácidos 109
bombear por medio de una vasija de
aire comprimido 2935
boquilla 5854
borato de etilo 3152
borato de manganeso 5124
borato de plomo 4767
borato de triamilo 8389
borato de tributilo 8405
borato de trimetilo 8501
borato de zinc 8921
borato fenilmercúrico 6300
borato magnésico 5052
borato sódico 7501
borato sódico perhidratado 7502
bórax 1114
bórax anhidro 550
bórax pentahidratado 1115
borde craso 3356
boriborato de glicerina 3790
borneol 1119
bornita 1120
boro 1123

borohidruro alumínico 407
borohidruro de litio 4947
borohidruro sódico 7503
borosilicato de plomo 4768
borra de algodón 4918
bort 1130
boruro de titanio 8298
boruro de zirconio 8985
botella 1132
bournonita 1148
braceaje 5169
braquimeiosis 1150
brazo 665
brazo de balanza 855
brazos de chapoteo 7717
brea 6419
brillancímetro 3768
brillante 5014
brillantez 1165
brillantez comparada con óxido de
magnesio puro 1163
brillo 3767, 5012
brillo, sin ~ 5013
briqueteadora 1169
brocantita 1174
bromato magnésico 5054
bromato sódico 7504
bromirita 1194
bromo 1177
n-bromoacetamida 69
bromoacetato de metilo 5335
bromoacetona 1181
bromobenceno 1182
bromoclorodimetilhidantoína 1183
bromocloroetano 1184
bromodietilacetilurea 1186
bromoestirol 1191
bromoformo 1187
bromofosgeno 1190
bromometiletilcetona 1188
bromotrifluorometano 1193
bromuro alumínico 408
bromuro cálcico 1342
bromuro de acetilo 85
bromuro de alilo 375
bromuro de bencilo 908
bromuro de cadmio 1300
bromuro de cianógeno 2264
bromuro de ciclopentilo 2293
bromuro de domifeno 2842
bromuro de estiroleno 7876
bromuro de estroncio 7850
bromuro de etilhexilo 3228
bromuro de etilo 3153
bromuro de hexilo 4078
bromuro de hidrógeno 4192
bromuro de isopropilo 4549
bromuro de linoleiltrimetilamonio
4914
bromuro de litio 4948
bromuro de magnesio y etilo 3240
bromuro de metileno 5373
bromuro de metilmagnesio 5407
bromuro de metilo 5334
bromuro de octilo 5918
bromuro de oxifenonio 6059
bromuro de piridostigmina 6819
bromuro de plata 7398

bromuro de propantelina 6742
bromuro de radio 6911
bromuro de succinilcolina 7894
bromuro de trimetileno 8505
bromuro de xililo 8890
bromuro de zinc 8922
bromuro potásico 6605, 6606
bromuros 1176
bromuro sódico 7505
bronce 1195
bronce dorado 3822
bronce plomoso 4777
bronces fosforosos 6331
brote 7362
brucina 1198
brucita 1199
bruñido con bola 777
buque cisterna 8032
bulbocapnina 1219
burbuja 1206
burbujeo 1213
bureta 1231
bureta de explosión 3312
busolfan 1236
butabarbital sódico 1237
butadieno 1239
butano 1240
buteno 1241
butilbencenosulfonamida 1251
butilcloral 1255
butilenglicol 1259
butilestearamida 1282
butilftalilglicolato de butilo 1279
butil-litio 1271
butinodiol 1284
butinol 1285
butiral de polivinilo 6563
butirato de bencilo 909
butirato de etilo 3160
butirato de geraniol 3734
butirato de isoamilo 4470
butirato de isopropilo 4550
butirato de manganeso 5126
butirato de metilo 5342
butirato de propilo 6756
butirato de vinilo 8748
butirolactona 1288
butironitrilo 1289
butonato 1242
butopironoxilo 1243

cabinete 1485
cable de calefacción 3975
cacodilato férrico 3396
cacodilato sódico 7506
cadena 1613
cadena lateral 7368
cadena molecular lineal 4902
cadmio 1298
caduco 2341
cafeato de etilo 3162
cafeína 1320
caída 2340
caída de tensión óhmica 4438
caja refractaria de loza 7175
cal 4877
calamina 1323
calandra 1390

cloruro de cobre 2088
cloruro de cromilo 1833
cloruro de dietilaluminio 2571
cloruro de diisobutilaluminio 2632
cloruro de dimetiltubocurarina 2711
cloruro de dodeciltrimetilamonio
  2840
cloruro de endrofonio 3047
cloruro de etilbencilo 3151
cloruro de etilhexilo 3229
cloruro de etilo 3168
cloruro de fenarsacina 6248
cloruro de fenilcarbilamina 6282
cloruro de fenilo 6283
cloruro de fumarilo 3621
cloruro de hidrógeno 4193
cloruro de iridio y potasio 4441
cloruro de isoamilo 4471
cloruro de isodecilo 4506
cloruro de isoftaloílo 4536
cloruro de isopropilo 4551
cloruro de itrio 8908
cloruro de laurilo 4745
cloruro de laurilo y piridinio 4748
cloruro de lauroílo 4741
cloruro de litio 4951
cloruro de magnesio y etilo 3241
cloruro de metilalilo 5313
cloruro de metileno 5374
cloruro de metilo 5348
cloruro de níquel 5719
cloruro de níquel y amonio 5714
cloruro de nitrobenzoílo 5758
cloruro de nitrosilo 5797
cloruro de octanoílo 5911
cloruro de oleoílo 5962
cloruro de oro 3823
cloruro de oro y potasio 3828
cloruro de oro y sodio 3830
cloruro de paladio 6076
cloruro de pelargonilo 6157
cloruro de pirvinio 6846
cloruro de plata 7399
cloruro de platino 6473
cloruro de plomo 4771
cloruro de polivinilideno 6570
cloruro de polivinilo 6565
cloruro de potasio 6609
cloruro de radio 6913
cloruro de rubidio 7153
cloruro de rutenio 7160
cloruro de sulfurilo 7948
cloruro de talio 8182
cloruro de tántalo 8041
cloruro de tereftaloílo 8076
cloruro de tetraetilamonio 8119
cloruro de tionilo 8245
cloruro de tolonio 8318
cloruro de tributilestaño 8412
cloruro de triciclamol 8440
cloruro de trifenilestaño 8531
cloruro de tubocurarina 8565
cloruro de vanadilo 8699
cloruro de vinilideno 8761
cloruro de vinilo 8750
cloruro de zinc 8925
cloruro de zinc y amonio 8918
cloruro estánnico 7747

cloruro estannoso 7751
cloruro etilmercúrico 3244
cloruro fenilmercúrico 6301
cloruro férrico 3397
cloruro ferroso 3429
cloruro hidroxifenilmercúrico 4242
cloruro irídico 4439
cloruro magnésico 5058
cloruro manganoso 5142
cloruro mercúrico 5232
cloruro mercúrico-amónico 5229
cloruro mercurioso 1398, 5244
cloruro metacrilatocrómico 5282
cloruros 1722
cloruros de ácidos 106
cloruro sódico 7512
coagulación 1875, 1885
coagulación bacteriana 756
coagulación enzímica 3066
coagulación por alta frecuencia 4091
coagulación por sistema doble 2890
coagulante 1883
coagulasa 1884
coágulo de espuma de látex 7269
coalescencia 1889
cobaltita 1904
cobaltita de litio 4954
cobalto 1895
cobalto brillante 1900
cobalto oxidado negro 683
cobalto tetracarbonilo 1924
cobre 2080
cobre arsenical 670
cobre de cadmio 1302
cobre electrolítico 2978
coca 1929
coca de Levante 1931
cocaína 1930
cocción de vidriado 3771
cocido 3485
cocoinato de etilo 3173
cochinilla 1932
codeína 1935
coeficiente de absorción atómica 704
coeficiente de acidez 122
coeficiente de actividad 173
coeficiente de aire 257
coeficiente de cobre 2099
coeficiente de difusión 2604
coeficiente de distribución 2820
coeficiente de expansión térmica
  8210
coeficiente de resistencia a la
  desintegración 7020
coeficiente de trayectoria 6144
coenzima 1940
cofinita 1941
cohesión 1942
cok 1943
cola 3777
cola a base de caseína 1488
colada 1499
cola de espuma 3555
cola de látex 4726
cola de pescado 3476, 4461
cola de pez 3490
coladura 1950
cola en hojas 3464

cola en solución 7661
colágeno 1962
colagogo 1771
cola para brochas 1203
cola para dorado 3829
cola para unir los bordes 2926
colapez 4461
colcótar 1953
colemanita 1960
colepoyesis 1772
colerético 1773
colesterasa 1774
colesterol 1775
colicina 1961
colicuativo 1964
colina 1777
colinesterasa 1778
colirio 1977
colodión 1966
colofonia 7124
colofonia endurecida 3959, 4880
colofonia líquida 4938
coloide 1967
coloidoquímica 1965
colonia 1979
coloración alternada de los
  segmentos cromosómicos 1828
coloración general 3720
colorante acrídico 140
colorante adjetivo 195
colorante de desarrollo 2447
colorante insoluble 4365
colorantes 2904
colorantes ácidos 108
colorantes azoicos 746, 4268
colorantes de cetonimina 4624
colorantes de xanteno 8871
colorantes luminiscentes 5005
colorantes metalizados 5270
colorante soluble 7643
color del barniz anterior visto a
  través de la nueva capa 3875
coloreado químico 1643
colores al hielo 4268
colores consistentes 1077
colores de dispersión 2806
colorimetría 1981
colorímetro de Guild 3896
color orgánico 8338
colquicina 1952
columbita 1989
columna 1990, 1991
columna a testado de contracorriente
  2134
columna de burbujeo 1210, 1212
columna de Clusius 1881
columna de condensación 2028
columna de destilación 2819
columna de pulsaciones y de
  contracorriente 2135
columna de relleno 6066
columna de rotación 7128
columna de rotación y de
  contracorriente 2136
columna de separación 7323
columna rectificadora 6978
colutorio 1976
combustible 1994, 4276

criptomero 2198
criptómetro 2200
criptón 4654
crisarrobina 1834
criseno 1835
crisocola 1836
crisol 6588
crisol de filtración 3466
crisol de Gooch 3844
crisol de porcelana 6577
crisol de Rose 7121
crisol para filtrar de vidrio fritado 3609
crisol para fusión 2185
crisótilo 1837
cristal 2201
cristales de cámaras 1624
cristalinidad 2205
cristalito 2206
cristalización 2207
cristalización fraccionada 3584
cristalograma 2210
cristaloluminiscencia 2211
crocidolita 1065
crocoita 2175
cromado 1817
cromaticidad 1784
cromatidio 1787
cromatidio anular 7103
cromatidio hijo 2318
cromatidio plumoso 4707
cromatidios hermanos 7428
cromatina 1789
cromatina génica 3717
cromatina nucleolar asociada 5871
cromatina sexual 7345
cromato amónico 463
cromato bárico 792
cromato cálcico 1349
cromato cobaltoso 1911
cromato de bismuto 982
cromato de cobre 2089
cromato de estroncio 7853
cromato de litio 4952
cromato de plata 7401
cromato de plomo 4772, 4773
cromato de zinc 8927
cromato estánnico 7748
cromato estannoso 7752
cromato férrico 3398
cromatoforo 1793
cromatografía 1790
cromatografía a gas 3686
cromatoide 1791
cromato mercurioso 5245
cromatómetro 1792
cromato potásico 6612
cromatoptómetro 1794
cromatos 1786
cromato sódico 7516
cromato sódico tetrahidratado 7517
cromidio 1812
cromita 1813
cromo 1814
cromocentro 1819
cromocito 1820
cromo de amarillo 1800
cromo escarlato 1799

cromofílico 1824
cromofobio 1825
cromóforos 1826
cromogénico 1821
cromoisomerismo 1822
cromo limón 4821
cromomero 1823
cromosoma 1827
cromosoma anular 1829, 7104
cromosoma gigante 3743
cromosoma hijo 2319
cromosoma lineal 4899
cromosoma plumoso 4708
cromosoma polímero 6539
cromosoma ramificado 1154
cromosoma residual 7035
cromosoma SAT 7226
cromosomas cubiertos de genes 3723
cromosoma sexual 7346
cromosomina 1830
cromos primavera 6708
cromotropía 1822
cromo-zinc 8928
croquesita 2176
crotamitón 2180
crotonaldehido 2181
crotonato de etilo 3174
crudo de destilación primaria 8341
cruzamiento doble 3582
cuajar 2242
cuanto 6849
cuarteado 1633
cuartear 6850
cuarzo 6851
cuasia 6852
cuatrivalente 8175
cuba comercial 1999
cuba de absorción 30
cuba de clarificación 7341
cuba de depósito total 2420
cuba de envasado 6888
cuba de fermentación 3386
cuba de inmersión 2781
cuba de liberación 4845
cuba de medida 5197
cuba de mosto 5170
cuba de reducción 6989
cuba de regulación 7015
cuba electrolítica 2976
cuba mezcladora 5168
cuba para líquidos fermentados 863
cuba separadora 7101
cubierta 8894
cubilete 854
cubreobjetos 2142
cubrir con algodón 3520
cuchara 7254
cucharilla de deflagración 2357
cuchillo de balanza 4648
cuchillo rascador 2832
cuerno de ciervo 3969
cuerpo 1076
cuerpo celular 1542
cuerpo ecuatorial 3084
cuerpo negro 1007
cuerpos acétonicos 76
cuerpos de inclusión 4301
cuerpos semilunares 2392

culicida 2215
cumarina 2127
cumarona 2128
cumeno 2219
cumilfenol 2222
cúprico 2230
cuprita 2232
cuproamoníaco 2227
cuprogalvanoplastia 2103
cuproso 2233
curación 2245
curación excesiva 6031
curare 2238
curarina 2239
curarización 2240
curcumina 2241
curie 2244
curio 2247
curtido 8036
curtimiento 8036
curtir 8027
curva de absorción 31
curva de solubilidad 7642

chalona 609
chamosita 1627
chamuscar 7258
chesilita 1694
chicana 770
chicle 1697
choque molecular 5534
chorreo con granalla 7364

damajuana 1460
damara 2312
dañino 5853
dar cuerpo 1075
DDT 2320
decaborano 2332
decahidronaftaleno 2333
decaimiento 2340
decalcificación 2334
decalescencia 2335
decantación 2337, 7340
decantador 2338
decantar 2336
decapado catódico 1524
decapado con ácido 6382
decapado electrolítico 2982
decapado químico 1664
decapante 6380
decapar al ácido 6381
deciduo 2341
decilmercaptano 2349
decocción 2344
decomposición por radiación 6892
decoración imitando madera 3850
defecación 2353
defibrilación 2355
deflagración 2358
defloculación 2359
deformación permanente 6215
degeneración 2365
degeneración vacuolar 1963
degradación 2366
degradación termal 8205
dehidroacetato de calcio 1353
dehidroacetato sódico 7522

diastema 2487
diatermano 2488
diatilhexoato de trietilenglicol 8460
diatomita 2489
diatrizoato de metilglucamina 5385
diatrizoato sódico 7524
diazometano 2492
diazotización 2494
dibenzantrona 2502
diborano 1125
dibromodifluometano 2507
dibromopropanol 2509
dibromuro de etileno 3186
dibucaína 2510
dibutilditiocarbamato de níquel 5720
dibutilditiocarbamato de zinc 8930
di-ter-butilmetacresol 2518
dibutiltiourea 2524
dibutirato de etilenglicol 3191
dibutoxitetraglicol 2514
dicaprilato-caprato de trietilenglicol 8458
dicaprilato de tetraetilenglicol 8122
dicaprilato de trietilenglicol 8457
dicarboximida de octilbiciclohepteno 5917
dicarión 2653
dicetena 2654
dicetiléter 2533
dicetonas 2655
diciandiamida 2550
diciclohexilamina 2551
diciclopentadieno 2554
diclona 2534
diclorfenamida 2546
dicloroacetato de metilo 5367
diclorobenceno 2535
diclorodifluometano 2537
dicloroestearato de metilo 5368
dicloroetilformal 2539
diclorofeniltriclorosilano 2545
diclorofeno 2544
diclorofenoxiacetato sódico 7526
diclorofluometano 2540
dicloroisocianurato sódico 7525
diclorometano 2542
dicloropentano 2543
dicloruro clorometilfosfónico 1747
dicloruro de azufre 7942
dicloruro de benceno fosforoso 884
dicloruro de bencilo 913
dicloruro de estaño 7751
dicloruro de etileno 3187
dicloruro de etilo y aluminio 3139
dicloruro de polivinilo 6566
dicloruro de propileno 6762
dicloruro de tetraglicol 8135
dicloruro de triglicol 8477
dicloruro de vanadio 8689
dicromato amónico 464
dicromato de zinc 8931
dicromato potásico 6615
dicromato sódico 7527
didecanoato de trietilenglicol 8459
dideciléter 2557
didimio 2559
dieldrina 2560
dienestrol 2566

dieno 2563
diestearato de poliglicol 6533
diestearato de tetraetilenglicol 8124
dietanolamina 2569
dietética 2568
dietilamina 2572
dietilaminoetanol 2573
dietilbenceno 2574
dietildifenilurea 2577
dietilditiocarbamato de selenio 7294
dietilditiocarbamato de zinc 8933
dietilenglicol 2579
dietilentriamina 2589
dietilestilbestrol 2595
dietiletilfosfonato 2590
difenilacetonitrilo 2744
difenilamina 2745
difenilbencidina 2746
difenilcarbacida 2747
difenildiclorosilano 2751
difenilguanidina 2752
difenilhidantoína 2754
difenilmetano 2755
difenilmetiléter de tropina 8559
difenilnaftilendiamina 2757
difenilo 2743
difenilsilanodiol 2763
difenilurea 1432
diferencia de potencial 6659
difilina 2907
difluente 2600
diformiato de etilenglicol 3193
difusado 2602
difusibilidad térmica 8208
difusión 2603
difusión continua 2060
difusión electroquímica 2954
difusión molecular 5536
difusión química 1646
digametia 2607
digamético 2606
digestor de Papín 6099
digestor de regeneración 6964
digital 2609
digitalina 2608
digitoxina 2610
dihidroabietato de trietilenglicol 8461
dihidrocolesterol 2622
dihidroestreptomicina 2625
dihidroxiacetona 2626
dihidroxidifenilsulfona 2628
diisobutilamina 2633
diisobutilcetona 2636
diisobutileno 2635
diisocianato de difenilmetano 2756
diisocianatos 2638
diisopropanolamina 2645
diisopropilbenceno 2646
diisopropilcarbinol 2647
diisopropilcresol 2648
diisopropiltiourea 2652
dilatación lineal 4900
dilaurato de dibutilestaño 2526
dilución 2665, 8233
diluente 2661
diluir 2663
diluir un ingrediente 3317
diluyente 2662, 3318

dimaleato de proclorperacina 6722
dimenhidrinato 2666
dimercaptoacetato de glicol 3811
dimeria 2670
dimérico 2669
dímero 2667
dímero de metilciclopentadieno 5365
dimetacrilato de tetraetilenglicol 8123
dimeticona 2671
dimetilacetal 2674
dimetilacetamida 2675
dimetilamina 2676
dimetilaminoazobenceno 2677
dimetilaminopropilamina 2678
dimetilanilina 2679
dimetilbencilcarbinol 2681
dimetilciclohexano 2684
dimetildiclorosilano 2685
dimetildifenilurea 2686
dimetilditiocarbamato de zinc 8934
dimetilditiocarbamato sódico 7528
dimetiléter de la hidroquinona 4213
dimetiléter del dietilenglicol 2582
dimetiléter del trietilenglicol 8462
dimetilglioxima 2689
dimetilhexanodiol 2690
dimetilhexinol 2691
dimetilhidantoína 2692
dimetilisopropanolamina 2696
dimetiloctanodiol 2698
dimetiloctanol 2699
dimetiloctinodiol 2700
dimetiloletilenurea 2701
dimetilolurea 2702
dimetilpiperacina 2705
dimetoxitetraglicol 2673
dimorfismo 2713
dimorfismo nuclear 5859
dimorfismo sexual 7347
dimorfo 2712
dinamita 2906
dínamo para galvanostegia 6467
dinérico 2716
dinitrato de dietilenglicol 2583
dinitrobenceno 2717
dinitrofenol 2719
dinitronaftaleno 2718
dinitroortocresilato sódico 7529
dinitrotolueno 2723
dinonilfenol 2721
diorita 2730
dioxano 2731
dióxido de azufre 7943
dióxido de carbono 1447
dióxido de cloro 1727
dióxido de diciclopentadieno 2555
dióxido de manganeso 5123, 5129
dióxido de plomo 4774
dióxido de selenio 7295
dióxido de telurio 8060
dióxido de titanio 8300
dióxido de torio 8269
dióxido de uranio 8640
dióxido de zirconio 8988
dioxolano 2732
dipelargonato de trietilenglicol 8463
dipentaeritritol 2737

espesamiento 4979
espinela 7711
espiral contactora de Podbielniak 6485
espiral molecular 5544
espirema 7713
espíritu 7714
espíritu blanco 8844
espíritu de colofonia 7127
espíritu de cuerno de ciervo 7715
espíritu de madera 8860
espíritu de sal 7716
espíritu de vino 3138
espíritu dulce de nitro 7975
espitón de trasvase 6885
esplenócito 7718
espliego 4749
espodúmeno 7719
esponja 7720
esponja celulósica 1561
esponja de caucho 7150
esponja de platino 6474
espora 7724
esporócito 7725
esporófito 7726
espuma 4727, 7268
espuma de látex microporosa 5476
espuma de poliuretano 6556
espumadera 6104
espumas de poliéter 6528
esquelita 7246
esquema de fabricación 3528
esquisto aceitoso 5948
estabilidad de los colores 1987
estabilidad marginal 5161
estabilidad química 1677
estabilizador 7733
estabilizadores de vinilo 8770
estable 3352
estado coloidal 1972
estado de disperción 2805
estado dictiótico 2548
estado duro y seco 3957
estado enérgico 3049
estado fundamental 3880
estado sólido 7641
estado viscoso 1863
estameña 1094
estannato cálcico 1381
estannato de plomo 4798
estannato potásico 6647
estannatos 7745
estannato sódico 7604
estannita 7750
estanque de evaporación 3294
estanqueidad por vapor 7776
estante 7816
estante para pipetas 6418
estaño 8288
estearato 7780
estearato amónico 477
estearato bárico 803
estearato de amilo 520
estearato de butilo 1283
estearato de butilo epóxido 1263
estearato de butoxietilo 1246
estearato de calcio 1382
estearato de ciclohexilo 2286

estearato de diglicol 2618
estearato de hidroxititanio 4251
estearato de isocetilo 4497
estearato de litio 4968
estearato de metilo 5449
estearato de metoxietilo 5297
estearato de plomo 4799
estearato de trihidroxietilamina 8484
estearato de vinilo 8771
estearato de zinc 8968
estearato férrico 3413
estearato magnésico 5077
estearato mercúrico 5240
estearato potásico 6648
estearato sódico 7605
estearilmercaptano 7784
estearina 7782
esteatita 7786
estefanita 7791
estequiometría 7820
éster alílico 379
esterasa 3113
éster de celulosa 1555
estereoisómeros 7794
estereoquímica 7793
ésteres 3116
ésteres ácidos 110
ésteres de colofonia 3114
ésteres de metacrilato 5281
ésteres grasos 3365
esterificación 3115
estéril 7799
esterilizado 7799
esterilizador de vapor 7777
éster nítrico 5742
esteroides 7801
esteroles 7802
estibina 7803
estibnita 614
estibofeno 7804
estigmasterol 7809
estilbeno 7811
estilbita 7812
estirado 2864
estireno 7871
estoraque 7870
estradiol 3117
estragol 3118
estramonio 7829
estratificado a base de tejido 4700
estrepsinema 7833
estreptodornasa 7834
estreptoduocina 7835
estreptolina 7837
estreptomicina 7838
estreptoquinasa 7836
estreptotricina 7839
estribos 7818
estricnina 7865
estriol 3120
estroboscopio 7847
estrofantina 7863
estrógeno 3121, 3122
estrón 3123
estroncianita 7848
estroncio 7849
estructura de polímeros 6542
estructura extranuclear 3332

estructura molecular 5545
estudtita 7866
estufa 2884
estufa de aire caliente 260
estufa de envejecimiento 231
etano 3124
etanol 3138
etanolamina 3125
etapa de adsorción 208
etenoide 3126
éter 3128
éter acético 61
éter acetoacético 66
éter amílico 502
éter butílico 1264
éter de celulosa 1556
éter dibencílico 2503
éter dibutílico del dietilenglicol 2580
éter dibutílico del etilenglicol 3190
éter dicetílico 2533
éter dicloroetílico 2538
éter dicloroisopropílico 2541
éter didecílico 2557
éter diestearílico 2811
éter dietílico del dietilenglicol 2581
éter dietílico del etilenglicol 3192
éter diglicidílico de resorcina 7053
éter dilaurílico 2656
éter dimetílico 2687
éter dimetílico del etilenglicol 3194
éter dimiristílico 2714
éter dioctílico 2724
éteres 3130
éter etilbutílico 3156
éter etilglicolmonohexílico 3222
éter etílico del isoeugenol 4511
éter fenílico del propilenglicol 6765
éter hexílico 4080
éter isopropílico 4552
éter metilbencílico 5333
éter metilpropílico 5442
éter monobencílico de etilenglicol 3197
éter monobutílico de dietilenglicol 2584
éter monoetílico del dietilenglicol 2586
éter monofenílico del etilenglicol 3213
éter monohexílico del etilenglicol 3206
éter monometílico de dipropilenglicol 2777
éter monometílico del dietilenglicol 2588
éter monooctílico del etilenglicol 3212
éter naftiletílico 5648
éter naftilmetílico 5650
etilacetanilida 3133
etilacetileno 3136
etilamina 3141
etilaminocetona 3143
etilantranilato 3144
etilato de vanadio 8691
etilato sódico 7532
etilbenceno 3146
etilbencilanilina 3150

fosfato de tridimetilfenilo 8445
fosfato de trietilo 8470
fosfato de trifenilo 8528
fosfato de trioctilo 8521
fosfato de zinc 8959
fosfato dibásico de plomo 2498
fosfato dibásico magnésico 2499
fosfato etilmercúrico 3245
fosfato férrico 3408
fosfato ferroso 3436
fosfato magnésico-amónico 5050
fosfato magnésico tribásico 8395
fosfato monobásico de magnesio 5567
fosfato potásico dibásico 2500
fosfato potásico monobásico 5568
fosfato potásico tribásico 8396
fosfatos 6322
fosfato sódico-amónico 7487
fosfato sódico dibásico 2501
fosfato sódico monobásico 5569
fosfato sódico tribásico 8397
fosfina 6324
fosfito de dietilo 2593
fosfito de difenildecilo 2750
fosfito de dimetilo 2703
fosfito de dioctilo 2727
fosfito de etilhexiloctilfenílo 3230
fosfito de plomo 4792
fosfito de tributilo 8410
fosfito de tricresilo 8439
fosfito de tridecilo 8444
fosfito de trietilo 8471
fosfito de trifenilo 8529
fosfito de trihexilo 8481
fosfito de triisooctilo 8488
fosfito de triisopropilo 8490
fosfito de trimetilo 8515
fosfito de trisetilhexilo 8542
fosfito de zinc 8961
fosfitos 6325
fosfolípido 6326
fosfomolibdato sódico 7588
fosfonato de dibutilbutilo 2515
fosfopenia 6330
fosforescencia 6332
fósforo 6337
fósforos 6336
fosfotioato de trietilo 8472
fosfotungstato sódico 7589
fosfuranilita 6347
fosfuro cálcico 1374
fosfuro de zinc 8960
fosfuro ferroso 3437
fosgeno 6320
fotocatálisis 6348
fotodesdoblamiento 6350
fotodisociación 6350
fotofóresis 6355
fotólisis 6353
fotómetro 6354
fotonegativo 4858
fotopolímero 6356
fotoquímica 6349
fotoreactivación 6357
fotorresistente 4858
fotosensible 4862
fotosíntesis 6359
fraccionamiento 3586

fracción ligera 4855
fragilidad 1170, 3587
fragilidad por el hidrógeno 4197
fragilidad térmica 8209
fragmentación 3588, 7355
fragmentación mediocromatidio 3936
fragmentación nuclear 5862
fragmentación nucleolar 5873
fragmoplasto 6360
fragmosoma 6361
francio 3589
franklinita 3591
frasco 3505
frasco cuentagotas 2870
frasco de agitación 7351
frasco de Claisen 1862
frasco de destilación 2817
frasco de Engler 3052
frasco de filtrar 3467
frasco de Kjeldahl 4646
frasco Dewar 2451
frasco de Woolff 8861
frasco para determinación de
  densidades 2400
frasco para reactivos 6954
frasquito para pesadas 8824
freibergita 3601
friabilidad 3604
frita 3607
frita azul 1067
fritar 3608, 7422
fructosa 3611
frutos del cacao 1294
ftalamida 6362
ftalato de butilbencilo 1253
ftalato de butilciclohexilo 1256
ftalato de butildecilo 1257
ftalato de butiletilhexilo 1265
ftalato de butilisodecilo 1267
ftalato de butilisohexilo 1268
ftalato de butiloctilo 1274
ftalato de dialilo 2470
ftalato de diamilo 2476
ftalato de dibutilo 2521
ftalato de dibutoxietilo 2513
ftalato de dicaprilo 2531
ftalato de diciclohexilo 2552
ftalato de didecilo 2558
ftalato de dietilo 2594
ftalato de difenilguanidina 2753
ftalato de difenilo 2761
ftalato de dihexilo 2619
ftalato de dihidrodietilo 2623
ftalato de diisobutilo 2637
ftalato de diisodecilo 2640
ftalato de diisooctilo 2643
ftalato de dimetilglicol 2688
ftalato de dimetilisobutilcarbinol
  2694
ftalato de dimetilo 2704
ftalato de dinonilo 2722
ftalato de dioctilo 2728
ftalato de dipropilo 2780
ftalato de ditridecilo 2827
ftalato de glicerilo 3800
ftalato de isooctiloisodecilo 4527
ftalato de metilciclohexilisobutilo
  5364

ftalato de octildecilo 5920
ftalato de tetrahidrofurfurilo 8143
ftaleínas 6363
ftalilsulfacetamida 6369
ftalilsulfatiazol 6370
ftalimida 6367
ftalonitrilo 6368
fucsina 3612
fuente luminosa 4864
fuerza 7832
fuerza de defenza 2354
fuerza de enlace 1104
fuerza intramolecular 4391
fuerza iónica 4425
fugitómetro 3615
fulminante 2443
fulminato de mercurio 5250
fulminatos 3619
fumarato de dibutilo 2516
fumarato de dioctilo 2725
fumarato ferroso 3431
fundente 3553
fundición en hueco 7458
fungicidas 3628
fungistatos 3629
furano 3631
furfural 3633
furoato de metilo 5384
fuseno 3639
fusiforme 3645
fusión 3646
fusiones tándem 8030
fustete 3649
fustete joven 8903
fustete viejo 5952

gadolinio 3651
gadolinita 3650
galactosa 3653
galena 3654
galio 3658
galipodio 3655
galocianina 3659
galones de aceite por 100 libras de
  pigmento 5941
galvanización 3660
galvanización química 1666
galvanoplastia 2998
galvanoplastia automática 3616
galvanoplastiado con rodio 7082
galvanoplastia en tambor 814
galvanoplastia mecánica 5198
galvanoplastia semiautomática 7306
gamético 3663
gameto 3662
gametocito 3665
gametoide 3666
gamodemo 3670
gangliocito 3671
ganíster 3672
garnierita 3678
garrafa 1460
gas 3679
gas, a prueba de ~ 3699
gas azul 1068
gas de agua 1068, 8810
gas de agua carburado 1463
gas de agua y hulla 1893

gas de alto horno 1026
gas de alumbrado 4856
gas de ciudad 8349
gas de extinción 6858
gas de gasógeno 6724
gas de hulla 1890
gas de madera 8857
gas de pantano 5284
gas de refinería 6999
gas de síntesis 7996
gas encerrado 5898
gases inertes 4336
gases nobles 5806
gas hilarante 5802
gas ideal 6203
gasificación de la hulla 1891
gas Mond 5559
gas natural 5662
gas natural ácido 7673
gas ocluso 5898
gasógeno 7900
gasoil 3695
gasolina 3696
gasolina de gas natural 1493
gasolina desmercaptanizada 7973
gasolina mezclada con mercaptanos
    7674
gas pobre 6724
gas rico 7096
gas sin aditivos 1868
gaylusita 3710
gel 3712
gelatina 3713
gelatina bicromatada 938
gelatina explosiva 1028
gelatina para explosiones
    submarinas 7885
gelatinización 7639
gel de sílice 7379
gel irreversible 4454
gel reversible 7067
gelsemina 3714
gen 3716
genciana 3731
generador de acetileno 91
generador de vapor eléctrico 2952
generador de vinagre 8743
generador para galvanoplastia 2999
general 8007
genes de incompatibilidad 4305
gen inhibidor 4349
gen mayor 5095
gen mutador 5622
genoblasto 3726
genoide 3727
genomia 3728
genomio 3728
genosoma 3730
gen restrictivo 7058
gentisato sódico 7540
geraniol 3733
germanio 3738
germen 3737
germicida 3741
germinar 7731
gigantocito 3745
gilsonita 3746
ginebra 3747

ginocardato sódico 7545
ginogénesis 3909
gipsita 3910
girado 3912
giste 808
gitoxina 3749
glándula 3751
glándulas de secreción interna 3039
glaserita 3752
glauberita 3759
glaucodota 3761
glauconita 3762
gliceraldehído 3786
glicerido 3787
glicerilacetal de la
    metilciclohexanona 5361
glicerina 3789
glicerofosfato cálcico 1356
glicerofosfato de manganeso 5132
glicerofosfato férrico 3401
glicerofosfato magnésico 5063
glicerofosfato sódico 7544
glicerol 3789
glicidol 3807
glicinato de cobre 2093
glicirrina 3817
glicocola 3808
glicolato de sodio y zirconio 7628
glicolato de zirconio 8989
glicoles de polioxipropileno 6549
glicolonitrilo 3813
glicosecretorio 3815
glicósidos 3816
gliodina 3818
glioxal 3819
globulinas 3766
glóbulo 3765
glóbulo polar 6499
glóbulos sanguíneos 1051
glucagón 3774
glucógeno 3810
glucoheptinato sódico 7541
gluconato cálcico 1355
gluconato de manganeso 5131
gluconato ferroso 3432
gluconato potásico 6620
gluconato sódico 7542
glucósidos 3816
glucuronolactona 3776
glutamato sódico 7543
glutamina 3780
glutaronitrilo 3784
gluten 3785
goetita 3820
golgiólisis 3837
goma 7144
goma arábica 3898
goma benjuí 895
goma conductora de electricidad 2035
goma copal 2075
goma crepe 2162
goma de diente de león 2314
goma de estireno-butadieno 7872
goma de guayaco 3882
goma de karaya 4605
goma de kauri 4612
goma de silicona 1142
goma etilenpropilénica 3218

goma fría 1958
gomaguta 3661
goma laca 7358
goma laca natural 7805
goma microcelular 5472
goma plombífera 4794
goma regenerada 6963
gomas 3901
goma sandáraca 7206
gomas éster 3114
goma silicónica 7387
goma sintética 1224
goma tragacanto 3903
gomoso 3899
gónada 3838
gonio 3839
gonocele 3840
gonócito 3841
gonómero 3842
gonosoma 3843
gordo 3360
gosipol 3846
goteo, a prueba de ~ 2867
gotita 2869
gradiente potencial 6660
grado al que una substancia puede
    secarse 2878
grado de desdoblamiento 2369
grado de disociación 2369
grado de fluidez 2370
grado de ionización 2372
grado de molienda 2371
grado de polimerización 2373
grado de sobrecalentamiento 2374
grados de libertad 2375
gráfico exotérmico 3309
grafito 3862
grafito cristalino 3494
grafito lamelar 3494
grafito para crisoles 2186
grahamita 3849
gramicidina 3854
Gram-negativo 3857
gramo-equivalente 3853
Gram-positivo 3858
granate 3676
granos finos 3475
granos flotantes de cebada 7977
granos pesados 7421
granulados 3860
gránulo 3861
gránulo basófilo 838
grasa 3355
grasa de ballena 1064
grasas del petróleo 6231
grasas endurecidas 3958
grasas lubricantes 4999
grasiento 6404
graso 3360
gravímetro 3863
greenoquita 3869
gres 7822
gres químico 1678
gres sin vidriar negro y de grano
    fino 818
grifo de boca curva 933
grifo de descarga 2788
grifo de ensayo 3706

grifo de muestreo 7203
grifo de unión 2048
grindelia 3874
grisú 3482
grúa de cucharón 4683
grupo 3345
grupo acetilo 93
grupo acetoxilo 80
grupo acilo 176
grupo amido 442
grupo amilo 504
grupo amino 445
grupo azo 747
grupo carbonilo 1457
grupo de separación 7321
grupo dibencílico 2504
guanidina 3884
guanilnitrosaminoguaniliden-
     hidracina 3891
guanilnitrosaminoguaniltetraceno
     3892
guanina 3887
guano 3888
guanosina 3889
guarnición de cuero 4814
guayacol 3883
gutapercha 3907

haba de tonca 8339
hafnio 3931
halazona 3932
halita 3941
halogenación 3945
halógenos 3946
halogenuro 3940
halogenuro de uranio 8641
haloideo 3940
haloysita 3942
haluro 3940
haluros de plata 7403
haploide 3951
haploide-suficiente 3952
haploidia 3953
haplomitosis 3954
haplóntico 3955
haplopoliploide 3956
haz de tubos 5694
haz tubular 5694
hecho 8718
hedor 3448
helecho 3390
helio 3994
heliotropina 3993
helvita 3999
hemacito 3923
hemaglutinina 3918
hematimetría 3916
hematímetro 3915
hematina 3920
hematita 3921
hematoblasto 3922
hematoporfirina 3925
hematoxilina 3926
hemicelulosa 4000
hemicigótico 4005
hemicromatídico 4001
hemihaploide 4002
hemimorfita 4003

hemixis 4004
hemoaglutinación 3917
hemocultura 3927
hemoglobina 3928
hemolisina 3929
hemolítico 3930
hemopoyesis 3924
heneicosanoato de metilo 5387
heparina 4010
heptabarbital 4011
heptacloro 4012
heptadecanoato de metilo 5388
heptadecanol 4013
heptadecilglioxalidina 4014
heptanal 4016
heptano 4017
heptanol 4018
heptavalente 4019
hepteno 4020
heptosas 4021
herbicidas 4024
hereditario 4025
herencia 4348
hermético 4292
heroína 4026
herradura, en forma de ~ 4134
hervir 7236
hervir parcialmente 6122
hesperidina 4028
hetero- 4030
heterocigoto residual 7037
heterocromático 4031
heterocromatina 4032
heterocromatinosoma 4033
heterocromatismo 4034
heterocromosoma 4035
heteroecio 4039
heterogamético 2606
heterogameto 4037
heterógamo 4038
heterolecito 4040
heteromolibdatos 4041
heteropicnosis de inversión 7063
heteropolar 4042
heteropolímero 4044
heterosomal 4045
heterotópico 4046
heulandita 4047
hexabromoetano 4048
hexaclorobenceno 4049
hexaclorobutadieno 4050
hexaclorociclohexano 4051
hexaclorociclopentadieno 4052
hexacloroetano 1449
hexacloropropileno 4054
hexacloruro de benceno 882
hexacloruro de carbono 1449
hexadecano 4055
hexadeceno 4056
hexadeciltriclorosilano 4058
hexafluoruro de azufre 7944
hexaldehído 4062
hexametilendiamina 4063
hexametilenotetramina 4066
hexametilentetramina 4064
hexanitrato de manitol 5151
hexanitrodifenilamina 4069
hexano 4067

hexanol 4070
hexanotriol 4068
hexavalente 4071, 7348
hexeno 4072
hexenol 4073
hexilenglicol 4079
hexilenglicolmonoborato sódico 7546
hexilfenol 4083
hexilresorcina 4084
hexoato de viniletilo 8757
hexobarbital 4074
hexoquinasa 4075
hexosas 4076
hialina 4151
hialuronidasa 4153
hibernación 4086
hibridación 4155
híbrido 4154
hidantoína 4156
hidracida del ácido maleico 5105
hidracidas ácidas 114
hidracina 4164
hidratación 4160
hidratado 4217
hidrato de cloral 1711
hidrato de hidracina 4165
hidrato de piperacina 6408
hidrato de terpina 8082
hidratos 4159
hidratos de carbono 1439
hidratos de gas 3692
hidrazidas 4163
hidrazobenceno 4167
hidrazonas 4169
hidrocarburo 4173
hidrocarburo polímero 4174
hidrocarburos saturados 7231
hidrocelulosas 4175
hidrocincita 4252
hidrocloruración anhídrido de
     combustibles zirconíferos 551
hidrocloruro de caucho 7146
hidrocloruro de ciclicina 2273
hidrocloruro de ciclopentolato 2292
hidrocloruro de cicrimina 2296
hidrocloruro de cloroguanidina 1741
hidrocloruro de cloroprocaína 1757
hidrocloruro de diaminofenol 2473
hidrocloruro de diciclomina 2553
hidrocloruro de diclonina 2903
hidrocloruro de difenhidramina 2741
hidrocloruro de difenilpiralina 2762
hidrocloruro de dihidromorfinona
     2624
hidrocloruro de etilmorfina 3248
hidrocloruro de fenilpropanolamina
     6311
hidrocloruro de fentolamina 6272
hidrocloruro de hidrastina 4158
hidrocloruro de hidroquinona 4214
hidrocloruro de hidroxilamina 4232
hidrocloruro del etiléster de la
     glicina 3809
hidrocloruro de orfenadrina 6007
hidrocloruro de oximorfona 6058
hidrocloruro de piperocaína 6412
hidrocloruro de piperoxano 6414
hidrocloruro de piratiacina 6813

hidrocloruro de pramoxina 6670
hidrocloruro de procaína 6715
hidrocloruro de procainamida 6714
hidrocloruro de promacina 6733
hidrocloruro de prometacina 6735
hidrocloruro de quinina 6873
hidrocloruro de racefedrina 6883
hidrocloruro de semicarbacida 7307
hidrocloruro de tenildiamina 8194
hidrocloruro de tetracaína 8099
hidrocloruro de tiamina 8227
hidrocloruro de tiopropazato 8249
hidrocloruro de tioridacina 8250
hidrocloruro de tolazolina 8315
hidrocloruro de toncilamina 8264
hidrocloruro de trifluopromacina 8474
hidrocloruro de trihexilfenidilo 8480
hidrocloruro de triprolidina 8537
hidrocloruro de tropacocaína 8553
hidrocloruro de tutocaína 8587
hidrocloruro de vancomicina 8701
hidrocloruro de xilometalolina 8888
hidrocortisona 4178
hidrófilo 4209
hidrofisuración 4179
hidrófobo 4210
hidroformación 4182
hidrofuramida 4184
hidrogela 4185
hidrogenación 4189
hidrogenación alicíclica 311
hidrogenado 4201
hidrógeno 4186
hidrógeno activo 167
hidrógeno sulfurado 4203
hidrolasa 4205
hidrólisis 4206
hidrólisis alcalina 327
hidrometalurgia 4207
hidrómetro 4208
hidrómetro de Hicks 4087
hidrómetro de succión 7901
hidroperóxido de cumeno 2220
hidroperóxido de metano 5286
hidroquinona 4211
hidroso 4217
hidrosulfito de zinc 8942
hidrosulfito sódico 7549
hidrosulfuro sódico 7548
hidrótropos 4216
hidroxiadipaldehído 4219
hidroxibenzaldehído 4220
hidroxibenzoato de propilo 6768
hidroxicitronelal 4222
hidroxidibenzofurano 4223
hidroxidifenilamina 4224
hidróxido amónico 467
hidróxido bárico 796
hidróxido cálcico 1358
hidróxido cérico 1599
hidróxido ceroso 1606
hidróxido cobáltico 1902
hidróxido cobaltoso 1912
hidróxido crómico 1807
hidróxido de aluminio 412
hidróxido de cadmio 1304
hidróxido de cobre 2094

hidróxido de estroncio 7854
hidróxido de litio 4958
hidróxido de oro 3825
hidróxido de plomo 4781
hidróxido de talio 8183
hidróxido de tetraetanolamonio 8118
hidróxido de zirconio 8991
hidróxido fenilmercúrico 6302
hidróxido férrico 3402
hidróxido magnésico 5064
hidróxido mangánico 5140
hidróxido potásico 6621
hidróxido sódico 1535
hidroxiestearato alumínico 413
hidroxietilcelulosa 4225
hidroxietilendiamina 4226
hidroxietilhidracina 4227
hidroxietilpiperazina 4228
hidroxifenilglicina 4241
hidroxiisobutirato de etilo 3231
hidroxilamina 4230
hidroximercuriclorofenol 4234
hidroximercuricresol 4235
hidroximercurinitrofenol 4236
hidroximetilbutanona 4237
hidroxinaftoquinona 4240
hidroxiprolina 4243
hidroxipropilglicerina 4244
hidroxipropiltoluidina 4245
hidroxiquinoleína 4247
hidruro cálcico 1357
hidruro de aluminio y litio 4945
hidruro de boro 1125
hidruro de carbono 4173
hidruro de estaño 7744
hidruro de litio 4957
hidruro de uranio 8642
hidruro de zirconio 8990
hidruros 4170
hidruro sódico 7547
hidruro sódico-alumínico 7480
hielo seco 2882
hierba 4023
hierro 4443
hierro alfa 389
hierro arsenical 671
hierro espático 7369
hierro especular 7709
hierro gamma 3669
hierro pentacarbonilo 4451
higrógrafo 4253
higrómetro 4254
higrómetro de absorción 34
higrómetro registrador 4253
higroscópico 4255
hilado en seco 2888
hilatura en húmedo 8832
hilatura en seco 2888
hilo de nylon 5890
hinojo 3378
hipercromasia 4256
hiperpoliploidia 4257
hipertelia 4258
hipoadrenia 4259
hipoclorito cálcico 1359
hipoclorito de litio 4959
hipoclorito sódico 7550
hipocromaticidad 4261

hipofosfito cálcico 1360
hipofosfito de manganeso 5134
hipofosfito férrico 3403
hipofosfito sódico 7551
hipohaploide 4262
hipohaploidia 4263
hipopituitarismo 4265
hipostasia 4266
hipostático 4267
hipótesis de Avogadro 736
histamina 4098
histidina 4100
histiocito 4101
histoquímica 4102
hoja de oro 3826
holmio 4104
holocrino 4107
hollín 7666
homatropina 4108
homeosis 4114
homeostasis 4115
homeótico 4116
homoalelo 4111
homocarión 4112
homocíclico 4113
homogamético 4117
homogenización 4118
homolecito 4119
homólogos 4121
homomorfos 4122
homoploidia 4124
homopolar 4125
homopolímero 4126
hordenina 4132
hormigón 2023
hormigón armado 7016
hormigón de bario 794
hormigón ligero 4867
hormona adrenocorticotrópica 200
hormona cromatoforotrópica 4384
hormona folicular 3119
hormonas 4133
horno 6028
horno de arco indirecto 4318
horno de atmósfera controlada 678
horno de baño 3986
horno de baño de plomo 4766
horno de baño de sales 7189
horno de baño salino de electrodos 2968
horno de Birkeland-Eyde 974
horno de cal 4882
horno de calentamiento directo por resistencia 2784
horno de calentamiento indirecto por resistencia 4320
horno de cemento 1569
horno de cochura 7827
horno de copela 5609
horno de copelar 2224
horno de coque 1948
horno de ensayo 694
horno de esmaltar 5609
horno de fundición 7461
horno de inducción con canal 4327
horno de mufla 5608
horno de piritas 6825
horno de recocer continuo 4819

horno de recuperación 7013
horno de secar en el vació 8669
horno de sinterizar 7425
horno de tubo de combustión 1997
horno de vidriar 3772
horno Heroult 4027
horno incinerador de basuras 2436
horno Keller 4613
horno rotatorio 7133
huésped 4135
huevo 6038
hulla 1887
hulla bituminosa 1005
humectante 4147
humedad 4148
humo 7463
humor 4149
humos 3531
humus 4150
huso acromático 101
huso central 1572

identificar 8017
idiomutación 4271
idiopático 4272
idioplasma 4273
idiosoma 4274
igualar 4839
ilmenita 4278
iluminación de fondo oscuro 2315
imagen latente 4723
imantación 5088
imidas 4281
imidazol 4280
iminobispropilamina 4282
impedimento estérico 7798
impenetrable 4292
impenetrable al agua 8814
impermeable 4292
imporoso 5832
impregnación 4294, 6219, 7788
impregnado por resina 7044
impregnar 4293
impsonita 4295
inactínico 5812
inactivo 6134
inasimilable 4298
incidencia 4299
inclusión 4300
incompatibilidad 4303
incongelable 5819
incrementado de cuerpo 1079
incrustación 7238
incrustación de calderas 1085
indaminas 4308
indantreno 4309
indeleble 3352
indeno 4310
indetonante 5823
indicador 4314
indicador al ferroxilo 3442
indicador de adsorción 207
indicador de nivel 4935
indicador de vacío 2862
indicadores de oxidación-reducción 6045
indicador externo 3321
indicador interior 4386

indicador químico 1658
índice de acetilo 98
índice de acidez 136
índice de combustión 1996
índice de coordinación 2073
índice de desmulsificación 2393
índice de éter 3129
índice de fusión 5217
índice de plasticidad 7635
índice de yodo 4410
índice espumante 4731
índices de las caras 4315
índigo 4316
indio 4322
indofenoles 4325
indol 4323
indulinas 4331
induración 3961
inercia química 1659
inerte 4335, 6134
inestabilidad de plasma 6441
inestable 4660
inexplosible 5823
infección 4338
infección reductiva 6996
infestación 4339
infiltración 6202
infiltrarse 7286
inflamabilidad 3497, 4340
inflamable 4276
inflamarse 8021
influenciado por el medio 3061
infraproteínas 4342
infrarroja 4343
infusión 4347
ingeniería química 1650
inhibición 4350
inhibidor catalítico 1512
inhibidor de decapado 6383
inhibidor de la preprofase 6692
inhibidores 4351
inhibidores anódicos 576
inhibidores catódicos 1523
inhibidor ultravioleta 8605
iniciación de baño primario 7842
iniciador de reacción 4352
ininflamable 5826, 8621
inmergir 2733
inmersión en ácido 107
inmiscible 4285, 5829
inmunidad 4287
inmunización 4288
inmunología 4289
inoculable 4353
inoculación 4355
inocular 4354, 6980
inocular con levadura 6420
inodoro 5931
inorgánico 4356
inosina 4358
inositol 4359
insecticida 4360
insectífugo 4361
insolubilizar 4364
instalación de clasificación de coque 1946
instalación de concentración 2021
instalación de difusión térmica 8206

instalación de embotellado 1138
instalación de ensayo 6398
instalación de lixiviación 4983
instalación de oxígeno "en el sitio" 5970
instalación piloto 6398
instrumentación 4366
instrumentación reológica 7075
instrumento de cero al centro 1573
instrumento de lectura directa 2783
instrumento de precisión 6678
instrumento medidor de amplificación 2147
instrumento medidor teleaccionado 7025
insuficiencia en ácido nucleico 7763
insulina 4369
intensidad luminosa 5007
interacción de los genes 3719
interacción nuclear 5863
intercambio 4373
intercambio de iones 4420
intercambio químico 1653
intercesor 5205
intercromómero 4375
intercromosómico 4376
interespecífico 4388
interfase 4378
interfase de un líquido con otro 4937
interferencia quiasmática 1696
intergénico 4381
intoxicación por respirar gases 3698
intragénico 4389
intrahaploide 4390
intumescencia 4394
inulina 4396
inundación 3525
invariante 4397
inversión 4398
inversión de fase del tipo de emulsión 6242
invertasa 4399
inviabilidad 4401
in vitro 4402
in vivo 4403
involátil 5834
ión 4416
ión anfotérico 494
ión carbonio 1451
ión complejo 2010
ionio 4426
ionización 4427
ionización mínima 5505
ión negativo 558
ionógeno 4435
ionómetro 4428
ionona 4436
ipecacuana 4437
iridio 4440
iridosmina 6016
irritante 4457
isatina 4459
isetionato de estilbamidina 7810
islas de isomería 4462
iso- 4463
isoalelo 4464
isoaloxacina 4465
isoamilbenciléter 4469

leucita 4835
leucobase 4836
leucocito 4838
leucocito basófilo 835
leucocompuestos 4837
leucolina 6875
levadura 8892
levadura de fermentación alta 8343
levadura de fermentación baja 1141
levadura de siembra 6423
levigación 4841
levógiro 4684
levorrotatorio 4684
levulinato cálcico 1363
levulinato de etilo 3239
levulinato de tetrahidrofurfurilo 8139
levulosa 4686
lewisita 4842
ley de Blagden 1021
ley de Boyle 1149
ley de Dalton 2311
ley de Dulong y Petit 2895
ley de Einstein de la equivalencia
    fotoquímica 2942
ley de Gay-Lussac 3708
ley de Henry 4009
ley de Hess 4029
ley de Hooke 4128
ley de Joule 4597
ley de Kick 4630
ley de Raoult 6928
leye de Faraday sobre la electrólisis
    3348
liberación 7024
liberar 4844
licopodio 5020
licor blanco 8842
licor de acetato de hierro 4444
licor de sulfitos 7925
licuación 4931
licuar 4930
licuefacción 4932
ligadura doble 2850
ligazón 1096, 1097, 1099
lignificación 4869
lignina 4870
lignito 4871
lignocelulosas 4874
lignocerato de metilo 5405
lignosulfato sódico 7558
ligroína 4876
límite aparente de elasticidad 8900
límite de termodeformación 2158
limo 7448
limonita 4892
limpiar con agua hirviente 7237
limpidez 4893
linalol 4896
linaza 4916
línea de substitución compensatoria
    a cromosoma simple 7418
línea isoterma 4570
lineíta 4907
linfa 5022
linfoblasto 5023
linimento 4905
linoleato amónico 469
linoleato cálcico 1364

linoleato cobaltoso 1913
linoleato de manganeso 5135
linoleato de metilo 5406
linoleato de plomo 4783
linoleato de zinc 8946
linoleína 4910
linoxina 4915
linters 4918
liofílico 5025
liófobo 5026
liólisis 5024
liparoideo 4920
lipasas 4921
lípido 193, 4922
lipídosis 4923
lipocito 4925
lipocromos 4924
lipoforo 4927
lipólisis 4926
lipoxidasa 4929
liquen 4846
liquenina 4847
líquido condensado 2025
líquido de extracción 3328
líquido de Holanda 3187
líquido de solución regenerada 7011
líquido filtrado 3471
líquido polar 6494
lisina 5027
lisis 5028
lisogénico 5029
lisogenisación 5030
lisozima 5031
litargirio 4940
litargirio amarillo 1412
lítico 5032
litio 4942
litioamida 4946
litopón 4975
litro 4978
lixiviación 4982
lixiviar 4981
lobelia 4987
lobelina 4988
lodo 5606
lote 840
loza de barro 2913
lubricación por baño de aceite 4283
lubricante 4998
lubricante de petróleo residual 1164
lubrificación con neblina de aceite
    5942
lucidificación 5001
lumen 5003
luminiscencia 5004
lupa 5092
lupanina 5010
lupulina 5011
lúpulo 4131
lustre 3767, 5012
lustroso 5014
lutecio 5015
luteína 5016
lutidina 5017

llama desoxidante 6991
llama oxidante 6049
llama reductora 6991

llenadora de botellas 1137

maceración 7788
macerar 5034
macroestructura 5043
macromolécula 5038
macromolécula lineal 4901
macronúcleo 5040
macroscópico 5041
macrospora 5042
maculado en gota 3908
machacadora 7821
machacadora de azúcar 7903
machacadora de mordazas 4593
macho de arcilla 4986
madera contrachapada encolada con
    resina sintética 7043
madescente 5046
madreperla 5632
maduración 5189
magnesia 5068
magnesia calcinada 1330
magnesio 5048
magnesita 2841, 5047
magnetismo 5085
magnetita 5086
magnetoestricción 5090
magnetómetro 5089
magnitud 5094
maillechort 5723
mala fluidez 3962
malaquita 5098
malatión 5101
maleabilidad 2891
maleable 5108
maleato de clorofeniramina 1766
maleato de dialilo 2469
maleato de dibutilestaño 2527
maleato de dibutilo 2517
maleato de dietilo 2591
maleato de metilergonovina 5378
maleato de pirilamina 6821
malonato de dietilo 2592
malonato de etilo 3242
malta 5110
malta colorante 1016
maltasa 5112
malta tostada 4638
malta verde 3868, 4990
maltear 5111
maltenos 5113
maltería 5114
maltosa 5115
malla 5256
malla metálica 8850
maná 5149
manchas cristalizadas 2212
maneb 5120
manganato bárico 797
manganato potásico 6625
manganeso 5121
manganita 5141
manganita de litio 4961
manitol 5150
mano de mortero 6226
manómetro 5153
manómetro aneroide 537

molino triturador 1218
momento de enlace 1102
monacita 5558
mongolismo 5561
monoacetato de etilenglicol 3196
monoacetato de glicerina 3793
monoacetato de resorcina 7054
monoatómico 5557
monobásico 5563
monobenzoato de resorcina 7055
monoblasto 5570
monobromuro de yodo 4408
monobutiléter de etilenglicol 3198
monobutiléter del acetato de
  etilenglicol 3199
monobutiléter del laurato de
  etilenglycol 3200
monobutiléter del oleato de
  etilenglicol 3201
monocarburo de uranio 8643
monocelular 5571
monoclínico 5572
monocloruro de yodo 4409
monodisperso 5573
monoecia 5574
monoestearato de diglicol 2615
monoestearato de glicerina 3796
monoestearato de sacarosa 7897
monoestearato de tetraetilenglicol
  8125
monoetiléter de etilenglicol 3202
monoetiléter del acetato de
  etilenglicol 3203
monoetiléter del laurato de
  etilenglicol 3204
monoetiléter del ricinoleato de
  etilenglicol 3205
monofilamento 5575
monofilamentos de nylon 5888
monolaurato de glicerina 3794
monoleato de trimetilolpropano 8511
monomería 5577
monómero 5576
monometiléter de etilenglicol 3207
monometiléter del acetato de
  etilenglicol 3208
monometiléter del acetilricinoleato
  de etilenglicol 3209
monometiléter de la hidroquinona
  4215
monometiléter del estearato de
  etilenglicol 3211
monometiléter del ricinoleato de
  etilenglicol 3210
monomolecular 8620
monomorfo 5579
mononitrorresorcinato de plomo 4785
monorricinoleato de etilenglicol 3214
monorricinoleato de glicerina 3795
monorricinoleato de propilenglicol
  6764
monosacarosas 5580
monosalicilato de dipropilenglicol
  2778
monosas 5580
monosoma 5581
monostearato alumínico 416

monosulfuro de tetrametiltiouramilo
  8165
monotrópico 5582
monovalente 5583
monóxido de carbón 1453
monóxido de dipenteno 2739
monóxido de litio 4941
monóxido de plomo 4940
monóxido de talio 8184
monóxido de vinilciclohexeno 8753
monóxido sódico 7567
monta-ácidos 109
montaje en paralelo 6112
montaje recubierto de papel 6097
montaje sin recubrimiento 5827, 8631
montaje tipo saco 772
montajugos 1054
montmorillonita 5585
mordentes 3492
mordiente 5586
morfina 5588
morfismo 5589
morfolina 5590
morfología 5591
morfoplasma 5592
morina 5587
mortero 5593
mortero de ágata 229
mortero de bario 802
mosto 5166, 5618, 8864
mosto acidulado 7672
mosto de cerveza 861
mosto dulce 7976
motilidad 5596
movilidad iónica 4424
movimiento anafásico 531
movimiento browniano 1197
mucilago 5604
muela vertical 2928
muesca 5711
muestra 7199
mufla 5607
multivalente 5616
mullita 5610
muscovita 5617
musgo de Irlanda 1475
mutación 5621
mutación somática 7663
mutante reparable 7027
mutante supresible 7960

nácar 5632
nacarado 5633
nacarino 5633
nafta 5634
nafta disolvente 7657
naftaleno 5635
naftalenosulfonato de plomo 4786
naftalenosulfonato sódico 7568
nafta pesada 3990
naftenato cálcico 1366
naftenato cobaltoso 1914
naftenato de cerio 1604
naftenato de cobre 2097
naftenato de cromo 1816
naftenato de manganeso 5136
naftenato de mercurio 5251
naftenato de plomo 4787

naftenato de zinc 8947
naftenato de zirconio 8993
naftenato fenilmercúrico 6303
naftenato férrico 3404
naftenato potásico 6629
naftenato sódico 7569
naftilamina 5643
naftilendiamina 5647
naftilmetilcarbamato 5649
naftiltiourea 5651
naftionato sódico 7570
naftol 5640
naftoquinona 5642
nagiagita 1019
narceína 5653
narcótico 5654
narcotina 5655
nativo 5656
nauseabundo 5664
navecilla de fusión 1995
neblina de ácido crómico 1804
nebulizar 8708
necrobiosis 5666
necrobiótico 5667
nefelita 5686
nefelómetro 5687
negro ácido 5636
negro animal 556
negro Brunswick 1200
negro de acetileno 90
negro de alizarina 318
negro de benceno 896
negro de canal 1629
negro de carbón 1443
negro de Franckfort 3590
negro de gas 3682
negro de horno 3636
negro de humo 4706
negro de ilmenita 4279
negro de lámpara 4706
negro de magnetita 5087
negro de marfil 4579
negro de París 4706
negro de platino 6472, 7722
negro de pulpa de madera 8859
negro de túnel 1629
negro mineral 5497
negro por cuero 4812
negro sueco 7972
negro térmico 8202
negro vegetal 8719
nematicida 5676
neoarsfenamina 5677
neocentrómero 5678
neocincófeno 5679
neodimio 5680
neohexano 5681
neomicina 5682
neón 5683
neopreno 5684
neostigmina 5685
neptunio 5688
nerol 5689
nerolidol 5690
neumoelutriación 255
neurina 5698
neurógeno 5699
neurólisis 5702

neuroqueratina 5701
neutralización 5704
neutralizar 5705
neutro 5703
neutrones C 1882
neutrones epitérmicos 3078
nicol 5726
nicoles cruzados 2177
nicolita 5710
nicotina 5728
nicotinamida 5727
nidación 5730
niebla luminosa 4854
niebla metálica 5267
nigrosina 5731
nilón 5885
niobio 5733
niple 5736
níquel 5712
niquelado electrolítico 5722
níquel carbonilo 5718
niquetamida 5732
nitración 5739
nitradora 5740
nitrar 5737
nitrato 5738
nitrato alumínico 417
nitrato amónico 472
nitrato bárico 799
nitrato cálcico 1367
nitrato cálcico amónico 1338
nitrato cérico amónico 1598
nitrato cobaltoso 1915
nitrato de bismuto 985
nitrato de cadmio 1307
nitrato de celulosa 1557
nitrato de cobre 2098
nitrato de estroncio 7855
nitrato de etilo 3250
nitrato de guanidina 3886
nitrato de lantano 4717
nitrato de litio 4964
nitrato de metilo 5416
nitrato de níquel 5721
nitrato de paladio 6077
nitrato de plata 7405
nitrato de plomo 4788
nitrato de talio 8185
nitrato de torio 8271
nitrato de uranilo 8652
nitrato de zinc 8948
nitrato de zirconio 8994
nitrato férrico 3405
nitrato magnésico 5066
nitrato manganoso 5143
nitrato mercúrico 5236
nitrato mercurioso 5246
nitrato potásico 6630
nitrato sódico 7571
nitrificación 5746
nitrificar 5747
nitrilo 5748, 5891
nitrilos alifáticos 3366
nitrito 5749
nitrito de amilo 509
nitrito de cobalto y potasio 1923
nitrito de etilo 3251
nitrito de plata 7406

nitrito potásico 6631
nitrito sódico 7572
nitro 6630
nitroacetanilida 5750
nitroalgodón 3904
nitroalmidón 5794
nitroaminofenol 5751
nitroanilina 5752
nitroanisol 5753
nitrobenceno 5755
nitrobencenoazorresorcina 5756
nitrobenzaldehído 5754
nitrobifenilo 5760
nitrocelulosa 5761
nitrodifenilamina 5762
nitroestireno 5795
nitroetano 5763
nitrofenetol 5781
nitrofenida 5782
nitrofenol 5783
nitroferricianuro sódico 7573
nitrofurantoína 5764
nitrofurazona 5765
nitrogenado 5769
nitrogenasa 5767
nitrógeno 5766
nitrógeno activo 170
nitroglicerina 5773
nitroguanidina 5774
nitromersol 5775
nitrometano 5776
nitrómetro de Lunge 5009
nitrón 5777
nitronaftaleno 5778
nitroparacresol 5779
nitroparafinas 5780
nitropropano 5785
nitroprusiato sódico 7574
nitrosita de estireno 7873
nitrosodifenilamina 5789
nitrosodimetilanilina 5788
nitrosofenol 5792
nitrosoguanidina 5790
nitrosonaftol 5791
nitrosulfatiazol 5796
nitrotolueno 5798
nitrotoluidina 5799
nitrotrifluometilbenzonitrilo 5800
nitrourea 5801
nitroxileno 5803
nitruración 5745
nitruro de boro 1126
nitruro de silicio 7389
nitruro de zirconio 8995
nitruros 5744
nivel de energía de impureza 4297
nivenita 5804
nobelio 5805
nocividad 5808
nocivo 5853
no conductor 5818
nodal 5809
nódulo 5810
nódulos 5811
no ferroso 5824
no irritante 1024
nombre de cromosomas gaméticos 3664

nombre genérico de polímeros y
   copolímeros de cloruro de
   vinilideno 7220
nonadecano 5813
nonadecanoato de metilo 5417
nonanal 5815
nonano 5816
nonanoato de metilo 5418
nonilamina 5837
nonilbenceno 5838
nonileno 5839
nonilfenol 5841
nonil-lactona 5840
nonosas 5830
no pegajoso 8013
no polar 5831
noradrenalina 5843
norleucina 5844
norma 5845
normal 5846
normalización 5847
normocito 5850
no saturado 8626
no sujeto al arrugamiento 5820
novobiocina 5852
no vulcanizado 8609
nuclear 5855
nucleína 5868
nucleinación 5869
núcleo 5879
nucleoalbúmina 5870
núcleo de cristalización 5880
núcleo definitivo 2356
núcleo de fusión 3647
núcleo enérgico 3048
núcleo generativo 3721
nucléolo 5874
nucléolo satélite 7227
núcleo metabólico 5262
nucleoproteína 5877
nucleótida 5878
núcleo trófico 8555
nuez vómica 5884
nulisómico 5882
número cromosómico zigótico 8911
número de Avogadro 735
número de cetano 1610
número de elementos 1545
número de octano 5909
número de Reynolds 7072
número de saponificación 7216
número de transporte de los iones
   8375
número de valencia 8679
nutrición de la levadura 8893
nylon 5885
nylon elástico 2944

objetivo 5893, 5894
obsidiana 5895
obturación 5896
ocluir 5897
oclusión 5899
ocre 3603, 5900
ocre de bismuto 986
octaacetato de sacarosa 7898
octadecano 5901
octadeceno 5902

octaédrico 5904
octaedro 5905
octametilpirofosforamida 5906
octanal 5907
octano 5908
octanol 5910
octaóxido de triuranio 8549
octeno 5913
octilamina 5916
octilfenol 5926
octilfenoxipolietoxietanol 5927
octilmercaptano 5924
octoato de estaño 8292
octoato de zinc 8949
octosas 5914
octovalente 5912
ocular 5928
odorante 5929
odorífero 5929
odorimetría 5930
oleaginoso 5953
oleamida 5954
oleato cobaltoso 1916
oleato de amilo 513
oleato de amonio 473
oleato de butilo 1262
oleato de butoxietilo 1245
oleato de cobre 2100
oleato de diglicol 2616
oleato de etilo 3253
oleato de isopropilo 4557
oleato de manganeso 5137
oleato de metilo 5422
oleato de metoxietilo 5295
oleato de plomo 4789
oleato de tetrahidrofurfurilo 8141
oleato de trihidroxietilamina 8483
oleato de zinc 8950
oleato estannoso 7754
oleato magnésico 5067
oleato mercúrico 5237
oleato potásico 6632
oleato sódico 7575
olefinas 5957
oleína 5959
oleoandomicina 5955
oleómetro 5960
oleorresina 5961
oleorresistente 5947
olivenita 5964
olivino 5966
olla de barro 2174
oncógeno 5967
ondulaciones debidas a los gases 3691
ondulación superficial 7965
ontogénesis 5972
ontogenia 5972
oocito 5973
oosoma 5974
oótida 5975
opacidad 5977
opaco 5981
opalescencia 5978
opalescente 5979
ópalo de tinte negro 1017
operación por lotes 846
opiatos 5982

opio 5983
órbita de enlace 1103
órbita electrónica común 1101
ordenada 5995
organismo aerobio 212
organismo termófilo 8218
organosol 6002
orientación 6004
orientar 6003
orificio 6005
ornitina 6006
oro 3821
oropimente 6008
ortocromático 6009
ortoformiato de trietilo 8469
ortohelio 6010
ortohidrógeno 6011
ortoploidia 6013
ortorrómbico 6014
osificación 6021
osmiato potásico 6633
osmio 6017
osmiridio 6016
osmómetro 6018
ósmosis 6019
osteoblasto 6022
oxalacetato de sodio y etilo 3265
oxalato cálcico 1368
oxalato cobaltoso 1917
oxalato de amilo 517
oxalato de dibutilo 2519
oxalato de etilo 3254
oxalato de titanio y potasio 8302
oxalato de torio 8272
oxalato de zinc 8951
oxalato estannoso 7755
oxalato férrico 3406
oxalato férrico-amónico 3393
oxalato férrico-sódico 3412
oxalato potásico 6634
oxalatos 6039
oxalato sódico 7576
oxamida 6041
oxicelulosas 6052
oxicianuro mercúrico 5238
oxicloruro de antimonio 618
oxicloruro de bismuto 987
oxicloruro de carbono 1454
oxicloruro de cobre 2102
oxicloruro de fósforo 6338
oxicloruro de plomo 1495
oxicloruro de zirconio 8997
oxicrinina 6056
oxidación 6044
oxidación a baja temperatura 4997
oxidación electroquímica 2957
oxidante 6042, 6048
oxidasa 6043
oxidicloruro de benceno fosforoso 885
oxidietilenbenzotiazolsulfenamida 6053
óxido alumínico 419
óxido áurico 720
óxido básico 829
óxido bórico 1118
óxido cálcico 1369
óxido cérico 1600

óxido cobáltico 1903
óxido cobaltoso 1918
óxido cobaltoso-cobáltico 1905
óxido crómico 1808
óxido de antimonio 617
óxido de bario 800
óxido de berilio 927
óxido de butileno 1260
óxido de cadmio 1308
óxido de cobre 2101, 2231
óxido de dibutilestaño 2528
óxido de difenilo 2760
óxido de estireno 7874
óxido de estroncio 7856
óxido de etileno 3217
óxido de etileno de polioxipropilen-glicol 6548
óxido de gadolinio 3652
óxido de germanio 3739
óxido de hexaclorodifenilo 4053
óxido de hidroxipiridina 4246
óxido de hierro micáceo 5464
óxido de hierro usado 7696
óxido de holmio 4105
óxido de itrio 8909
óxido de lantano 4718
óxido de mesitilo 5258
óxido de octileno 5922
óxido de oro 3827
óxido de paladio 6078
óxido de plata 7407
óxido de propileno 6766
óxido de tántalo 8042
óxido de tributilestaño 8413
óxido de uranio 8644
óxido de uranio y bario 8635
óxido de zinc 8952
óxido de zirconio 8996
óxido estánnico 7749
óxido estannoso 7756
óxido etilarsenioso 3145
óxido férrico 1953, 3407
óxido ferroso 3435
óxido magnésico 5068
óxido manganoso 5144
óxido mercúrico amarillo 8899
óxido mercúrico rojo 6983
óxido negro de cobre 1010
óxido negro de uranio 1020
óxido nítrico 5743
óxido nitroso 5802
óxido rojo de cobre 6981
óxido rojo de hierro 6982
óxido rojo español 7677
óxidos 6046
óxido túngstico 8572
oxígeno 6054
oxígeno disuelto 2809
oxihemoglobina 6055
oximas 6050
oximetandrolona 6057
oxiprolina 6060
oxitetraciclina 6061
oxitocina 6062
oxitricloruro de vanadio 8692
oxtrifilina 6051
ozonidas 6065
ozono 6064

ozoquerita 6063

paladio 6075
palmitato alumínico 420
palmitato cálcico 1370
palmitato de isooctilo 4528
palmitato de isopropilo 4558
palmitato de metilo 5425
palmitato de tetrahidrofurfurilo 8142
palmitato de zinc 8953
palmitato magnésico 5069
palmitoleato de metilo 5426
palo campeche 4989
panalelo 6084
pancitólisis 6088
pancreatina 6087
pancromático 6086
pángene 6089
pangenosoma 6090
panmíctico 6091
pantalla 7361
pantotenato cálcico 1371
papaína 6094
papaverácea 6095
papaverina 6096
papel de filtro sin cenizas 688
papel de seda 8294
papel tornasol 4977
papiráceo 6100
para- 6101
paracromatina 6102
paracromia 6103
parafina 8819
parafina cruda 7440
parafina fluorada 3540
parafinas 6105
paraformaldehído 6107
parahello 6108
parahidrógeno 6109
paralaje 6111
paraldehído 6110
paramagnético 6113
parameiosis 6114
parametadiona 6116
parámetro 6115
paramíctico 6117
paramitosis 6118
paranucleína 6119
paranucléolo 5875
parasiticida 6120
parátrofo 6121
par de cilindros de calandria 5735
pardo de Bismarck 979
pardo de Van-Dyke 8702
pared de calefacción 3533
pares alelomórficos 6071
partícula 6130
partícula fundamental 3626
partículas coloidales 1969
partículas subtamaño 8619
partida 840
pasivación 6136
pasivación anódica 577
pasivación electroquímica 2958
pasivación química 1663
pasividad 6135
pasivo 6134
pasta 2854, 5165, 6137, 6138

pasta adhesiva 3777
pasta de aceite 5944
pasta de acrilla acuosa 7450
pasta de madera 8858
pasta de madera desintegrada
    mecánicamente 5203
pasta de madera química 1681
pasta de papel húmeda 7819
pasta de pigmento 6394
pasta de trapos 6920
pasta papelera 6098
pasta semiquímica 7308
pasteurización 6142
patas de gallo 2184
pavonado en azul 1069
pecas 7727
pectinas 6153
pechblenda 6421
pedernal 3516
pegajosidad 8012
pegajosidad remanente 7038
pegajoso 8015
pegamiento 1099
pelargonato de etilo 3255
película 3462
película colada 1498
película de aceite 5937
película de acetato celulósico 57
pelúcido 6162
pelusilla 3521
penetrabilidad 6218
penetración 6219
penetración de la cola 3778
penetración del adhesivo 1039
penetrómetro 6163
penicilina 6164
penicilinasa 6165
pentaborano 6168
pentaborato sódico 7577
pentabromuro de fósforo 6339
pentacloroestearato de metilo 5427
pentacloroetano 6169
pentaclorofenato sódico 7578
pentaclorofenol 6171
pentacloronitrobenceno 6170
pentacloruro de antimonio 619
pentacloruro de fósforo 6340
pentacloruro de molibdeno 5553
pentadecano 6172
pentadecanoato de metilo 5428
pentaeritrita 6174
pentaeritritol 6174
pentafluoruro de bromo 1178
pentano 6176
pentanodiol 6177
pentanol 6178
pentasómico 6179
pentasulfuro de antimonio 620
pentasulfuro de arsénico 673
pentasulfuro de fósforo 6341
pentatriacontano 6180
pentavalente 6181
pentilentetrazol 6186
pentlandita 6182
pentobarbital 6183
pentolita 6184
pentosa 6185
pentóxido de fósforo 6342

pentóxido de vanadio 8693
pepita 5881
pepsina 6188
peptización 6189
peptizar 6190
peptona 6192
perácido 6194
perbenzoato de butilo 1275
perborato de zinc 8954
perborato magnésico 5070
perborato sódico 7579
perbromuro bromuro de piridinio
    6818
percarbonato de isopropilo 4559
percarbonato potásico 6635
perclorato amónico 474
perclorato magnésico 5071
perclorato potásico 6636
percloratos 6195
perclorato sódico 7580
percloréter 6196
percloroetileno 6198
perclorometilmercaptano 6199
percolación 6202
percolar 6201
pérdida de carga por rozamiento
    3605
pérdida de humedad 5524
pérdida por radiación 6893
perdigón 6158
perfenacina 6222
pergamino 6123
pericarpio 6206
pericinético 6208
periclasa 6207
periódico 2270
período biólogico 953
período de disminución del régimen
    de secado 3341
período de inducción 4329
período de vida media 3937
periplasma 6212
perlita 6151, 6214
permanganato cálcico 1372
permanganato de zinc 8955
permanganato magnésico 5072
permanganato potásico 6638
permanganatos 6217
permanganato sódico 7582
permeabilidad 6218
permeabilidad al gas 3697
permeabilidad para al vapor de agua
    5525
permeación 6219
peroxidasa 6220
peroxidicarbonato de isopropilo 4560
peróxido cálcico 1373
peróxido de acetilbenzoílo 84
peróxido de bario 801
peróxido de benzoílo 903
peróxido de carbamida 1429
peróxido de dicumilo 2549
peróxido de di-ter-butilo 2520
peróxido de estroncio 7857
peróxido de hidrógeno 4202
peróxido de lauroílo 4742
peróxido del carbonato sódico 7509
peróxido de litio 4965

plástico rígido de cloruro de
    polivinilo 7102
plásticos 6456
plásticos alveolares sintácticos 7994
plásticos de celulosa 1559
plásticos de vinilo 8766
plásticos fenólicos 6258
plásticos ópticos 5989
plásticos reforzados 7017
plastidio 6457
plastidomio 6458
plastificante 6454
plastisol 6460
plastoma 6461
plastómero 6462
plastómetro 6459, 6463
plastosoma 6464
plata 7396
plata coloidal 1970
plata córnea 1593
plata de níquel 5723
plateado preliminar con mercurio
    6862
platicultivo 6465
platillo de balanza 6083
platina de ajuste 5201
platino 6470
plato de burbujeo 1209, 1211
plato de tostadero 4637
plegabilidad 6476
pleocroísmo 6475
plombagina 3862
plombato cálcico 1375
plombo en bruto 821
plomo 4757
plomo anaranjado 5992
plomo antimoniado 612
plomo blanco 8841
plomo blanco sublimado 7884
plomo comercial purísimo 2115
plomo negro 1015
plomo químico 1660
plomo rojo 2175
plomo-telurio 8061
plomo tetraetílico 8128
plomo tetrametílico 8163
plumas, en ~ 6481
plumbosolvencia 6480
plutonio 6482
pobre 4993
poca ley, de ~ 4993
poción 6662
poder absorbente 39
poder calorífico 1401
poder cubriente 2143, 4088
poder de amargor 1002
poder de coloración 8293
poder de cubrición 7730
poder de penetración 8277
poder diastático 2486
poder fecundante 3444
poder humectante 8836
podrido 6807
polarímetro 6489
polarización 6490
polarización catódica 1525
polarización electrolítica 2983
polarizador 6493

polea electromagnética 2988
polea magnética 5082
poli- 6501
poliacrilamida 6502
poliacrilato 6503
polialcano 6505
polialómero 6506
poliamida 6507
poliaminotriazoles 6510
polibásico 6511
polibasita 6512
polibutadieno 6513
polibutilenos 6515
policíclico 6521
policloropreno 6516
policlorotrifluoroetileno 6517
policondensación 6520
policondensado 6519
policóndrico 6518
polidisperso 6523
poliéster 6524
poliestireno 6552
polietilenglicol 6530
polietileno 6529
poliformaldehídos 6532
poliglicoles 6534
polihexametilenadipamida 6535
polimérico 6538
polimerización 6540
polimerización continua 2063
polimerización de condensación 2029
polimerización en frío 1957
polimerización en masa 1223
polimerización por radiación 6894
polímero 6537
polímero cuya estructura estereo-
    específica está separada por
    secciones de diferente estructura
    7792
polímero de
    trimetildihidroquinoleina 8504
polímero elevado 4094
polímero lineal 4903
polímeros estereoespecíficos 7797
polímeros estereorregulares 7795
polimetacrilatos 6543
polimixinas 6545
polimorfismo 6544
poliolefinas 6546
poliosas 6547
poliovirus 6495
poliploidia 6550
poliploidia desequilibrada 8607
poliploidia secundaria 7279
polipropileno 6551
polipropileno clorado 1723
poliisopreno 6536
polisulfuro potásico 6641
polisulfuro sódico 7591
politetrafluoetileno 6555
polivinilcarbazol 6564
poliviniletiléter 6567
polivinilformals 6569
polivinilisobutiléter 6572
polivinilmetiléter 6573
polivinilpirrolidona 6574
polocito 6499
polonio 6500

polvo de antracita 2217
polvo de hierro carbonilo 1458
polvo de pizarra 7446
polvo de púrpura 2214
polvo de zinc 8936
pólvora 3905
pólvoras explosivas 1029
pólvora sin humo 7464
polvos de moldear de nylon 5889
poner en infusión 4346
pontil 6799
porcelana aislante 2949
porcelana de Holanda 2385
porcelana dieléctrica 2561
porcelana dura 3965
porcelana fosfatada 1108
porcentaje en peso 8825
porcentaje en volumen 8790
porfirinas 6584
porímetro 6579
porosidad 6581
porosímetro 6580
poroso 6582
portador de gérmenes 3740
portaobjeto 5486
portarretorta 7062
posición 7767
posología 2845
post-cocer 222
postchromar 224
postrefrigerador 225
postrefrigerante 225
potable 6589
potasa alcohólica 280
potasa cáustica 1534
potasio 6591
potencia 7832
potencia de régimen 6937
potencial biótico 969
potencial de Donnan 2843
potencial de ionización 4431
potencial de límite 1145
potencial de polarización 6491
potencial de sedimentación 7284
potencial de separación 7324
potencial electroforético 2997
potencial electroosmótico 2994
potencia normal 6937
potenciómetro 6661
pozolana 6668
praseodimio 6671
precalentador 6686
precalentador de aire 4138
precalentamiento 6687
precipitable 6672
precipitación 6676
precipitación de un electrodo 7194
precipitación electrostática 3003
precipitado 3524, 6673
precipitado preformado 6683
precipitador de Cottrell 2126
precipitar 6674
prednisolona 6680
prednisona 6681
preenfriado 6679
preenfriador 3566
pregnanodiol 6684
pregnenolona 6685

prehniteno 6688
preinmunización 6689
premezclador 5168
preniteno 6688
prensa de cocimiento 773
prensado 6694
prensa mecánica 5200
prensa para filtrar 3468
preprofago 6691
prerrefrigerado 6679
presecador 6682
presecadora 6682
preservativo 6693
presión absoluta 18
presión de columna de agua 8811
presión de expulsión 2789
presión de ionización 4432
presión del plasma 6442
presión del vapor 8715
presión de vapor saturado 7234
presión osmótica 6020
prevulcanización 7342
priceíta 6698
primeros productos de la destilación
   fraccionada 3487
primidona 6706
primulina 6709
principio de Le Chatelier 4816
prisma de Nicol 5726
prisómetro 6711
probarbital sódico 6712
probeta 8091
probeta de gas 3683
procedimiento 6716
procedimiento al amoníaco de Haber
   3914
procedimiento Biazzi 931
procedimiento Bicheroux 935
procedimiento continuo 2064
procedimiento Deacon 2321
procedimiento de Carter 1476
procedimiento de producción de
   agua pesada de Clusius y Starke
   1880
procedimiento de regeneración de
   Bemelmans 872
procedimiento químico 1667
procedimientos redox 6986
procedimiento unitario 8623
procesar 6717
proceso 6716
proceso aluminotérmico 426
proceso anhidrítico 547
proceso BASF 825
proceso Betterton-Kroll 930
proceso Bischof 975
proceso Brewster 1162
proceso Cabot 1293
proceso Casale 1482
proceso de Angus Smith 544
proceso de Claude 1866
proceso de contacto 2056
proceso de contracorriente 2132
proceso de cracking 2146
proceso de Edeleanu 2925
proceso de espumación 3557
proceso de fabricación 5155
proceso de Fauser al amoníaco 3367

proceso DeFlorez 2360
proceso de fusión 3648
proceso de hidroformación 4183
proceso de Hoopes 4129
proceso de la cianamida 2254
proceso de la soda al amoníaco 7651
proceso del disco giratorio 7135
proceso de Leblanc 4815
proceso de Parke 6127
proceso de Reppe 7028
proceso de rodillos 7116
proceso de Simons 7417
proceso de Suida 7904
proceso de termofraccionamiento
   catalítico 2146
proceso Downs 2855
proceso en grupos 845
proceso en lotes 845
proceso Euston 3285
proceso extranuclear 3331
proceso Frasch 3592
proceso Gossage 3845
proceso Guggenheim 3894
proceso Harris 3968
proceso Houdry 4145
proceso irreversible 4455
proceso Keyes 4628
proceso Linde 4898
proceso Mond 5560
proceso Orford 5998
proceso Othmer 6024
proceso para la desodorización de
   petróleos 2833
proceso para la producción de sal
   pura 269
proceso Pattinson 6146
proceso por vía seca 2886
proceso Raschig 6934
procesos Fauser 3368
proceso Solvay 7651
proceso Steffens 7789
procromosoma 6723
producción 6027
producir espuma 4728
producir un color uniforme 4839
producir vapor 6921
producto 6725
producto cementador 1464
producto de cabeza 3567
producto de combustión 6726
producto hidrohalogenado 4204
producto intermedio 4382
producto primario 6700
producto secundario 1292
productos fumigantes 3623
producto sinergético 7992
productos para opacificar 3509
productos químicos fluorados 3547
productos refractarios 7004
productos sintéticos 8002
profago 6744
profase 6745
profibrinolisina 6727
profilaxis 6746
proflavina 6728
progesterona 6729
prolactina 6730
prolamina 6731

prolina 6732
promecio 6736
prometafase 6734
prometio 6736
pronúcleo 6739
pronucléolo 6738
propano 6740
propanol 6741
propiedad química 1669
propilacroleína de etilo 3261
propilamina 6755
propilendiamina 6761
propilenglicol 6763
propileno 6758
propilhexedrina 6767
propilmercaptano 6769
propiltiouracilo 6771
propiofenona 6751
propiolactona 6747
propionaldehído 6748
propionato de amilo 518
propionato de celulosa 1560
propionato de etilo 3260
propionato de feniletilo 6296
propionato de isobutilo 4488
propionato de metilo 5441
propionato de propilo 6770
propionato de testosterona 8090
propionato de trietilenglicol 8465
propionato de vinilo 8767
propionato de zinc 8962
propionato fenilmercúrico 6304
propionato sódico 7593
propulsantes de aerosol 219
propulsión por plasma 6443
protactinio 6772
proteasa 6773
protección catódica 1526
proteína argéntica 7412
proteínas 6774
proteolítico 6775
protoblasto 6776
protón negativo 5674
protoplasma 6777
protoplásmico 6778
protótrofo 6779
protoveratrina 6780
proustita 6781
prueba 7199
prueba Abel 5
prueba acidotérmica 113
prueba a la llama 3496
prueba al cadmio 1314
prueba al estiramiento 2159
prueba Almen-Nylander 385
prueba cis-trans 1855
prueba de flexión por choque 4291
prueba de laboratorio 4664
prueba del mandril 5119
prueba de Mackey 5035
prueba de plasticidad 2226
prueba de resorcina 7056
prueba de Trauzl 8378
prueba estadística 7768
pseudoácido 6784
pseudobase 6785
pseudocumeno 6786
pseudomorfo 6787, 6788

pseudoracémico 6789
psilomelano 6790
ptomaínas 6793
pulido electrolítico 3000
pulimento al ácido 127
pulir 7262
pulpa de madera 8858
pulpa de paja 7830
pulverización 6796
pulverización de ácido 134
pulverizador de plomo 4763
pulverizador por aire 250
pulverulento 6667
pulvimetalogía 6666
pulvimetalurgia 6666
puntel 6799
puntero 6799
punto crítico 2170
punto de anilina 555
punto de congelación 3600
punto de cristalización 3851
punto de ebullición 1090
punto de fusión 5218
punto de ignición 4277
punto de inflamabilidad 3504
punto de inflamación 3483
punto de rocío 2453
punto de ruptura 1157
punto de turbiedad 1877
punto de valoración 3046
punto eutéctico 3288
punto final 3046
punto isoeléctrico 4509
punto neutro 7736
puntos brillantes 4093
punzón 5102
pureza del plasma 6444
pureza química 1072
pureza radioquímica 6902
purga de calderas 1057
purgante enérgico 2861
purificador de aceite 5936
purificar 1864
purina 6801
puromicina 6802
púrpura de bromocresol 1185
púrpura de oro y estaño 3834
purpurina 6803
putrefacción 6804
putrescencia 6806
pútrido 6807
putrificar 6805

quadroxalato potásico 6643
quebrachitol 6854
quebracho 6855
quebrantadora de mandíbulas 4593
quebrantadora giratoria 3913
quelación 1637
quelato 1635
quelator 1636
quelpo 4614
quemador de amonio 451
quemador de piritas 6825
quemar 3478
querargirita 1593
queratina 4615
quercetina 6859

quercitol 6860
quercitrón 6861
quermes 4616
quermesita 4617
querógeno 4619
quiasma 1695
quiasma complementario 2007
quiasma de compensación 2005
quiasma lateral 4724
quiasma múltiplo 5612
quiasma parcial 6128
quiasmas de fusión 2179
quiasma terminal 8077
quilate 1426
quilificación 1840
quilo 1839
quimiadsorción 1683
química de los átomos muy excitados 4136
química de los elementos livianos 4852
química de los elementos pesados 3989
química de los trazadores isotópicos 8355
química de radiación 6891
química industrial 4333
química inorgánica 4357
química nuclear 5858
química orgánica 5999
quimiólisis 1685
quimioluminiscencia 1682
quimionuclear 1686
quimiorreceptor 1684
quimiorresistencia 1687
quimiósmosis 1688
quimiotaxis 1690
quimioterapia 1689
quimiotipo 1691
quimosina 1841
quimotripsinas 1842
quinaldina 6863
quinhidrona 6864
quinidina 6868
quinina 6870
quinolina 6875
quinona 6877
quinoxalinas 6878
quintaesencia 6879
quistamina 2298
quitapinturas 7844
quitar la espuma 7434
quitina 1706
quitinoso 1707

racemización 6882
radiación 6889
radiación actínica 151
radiación alfa 392
radiación del cuerpo negro 1008
radiación infrarroja 4344
radiación ionizadora 4434
radiación térmica 3981
radiactinio 6898
radiador 6895
radical 6896
radical ácido 128
radical libre 3597

radio 6910, 6897
radioactividad 6901
radioactivo 6899
radioelemento 6904
radioopaco 6908
radioquímica 6903
radiotorio 6909
radón 6915
rafaelita 6916
rafinosa 6919
ramificado 1153
ramio 6923
rancidez 6925
ráncido 6924
randomización 6927
rarefacción 6931
rarificar 6933
rauwolfia 6939
rayón 6945
rayón cuproamoniacal 2229
rayón viscosa 8780
rayos de Becquerel 860
raza de levadura 8896
reacción 6946
reacción anódica 578
reacción catódica 1527
reacción de Abderhalden 1
reacción de Baudouin 850
reacción de cambio químico 1654
reacción de Diels-Alder 2562
reacción de electrodo 2967
reacción de Friedel-Crafts 3606
reacción de Hofmann 4103
reacción de intercambio 3298
reacción en cadena 1614
reacción inducida 4326
reacción irreversible 4456
reacción química 1673
reacción reversible 7068, 7069
reacción secundaria 7280
reacción serológica falsamente positiva 956
reacción topoquímica 8340
reactivación 6949, 7070
reactividad 6951
reactivo 6950, 6953
reactivo de Bettendorf 929
reactivo de Carnot 1473
reactivo de Fischer 3488
reactivo de Griess 3871
reactivo de Huber 4146
reactivo de Karl Fischer 4606
reactivo de Nessler 5693
reactivo microanalítico 5467
reactivos de Grignard 3872
reactor 6952
reactor catalítico 1513
rebajar 4833
rebanada de remolacha en forma de V 2123
rebose 6032
recalentador 7955
receptor 6957
recesividad 6959
recesivo 6958
recipiente 1410, 4588, 6957
recipiente Cellarius 1541
recipiente de agitación 244

recipiente de cristalización 6384
recipiente de solución para placas
   enriquecidas 7438
recipiente para carbonación 1442
recipiente para solución potásica
   6590
recirculación 6961
recobrado de disolvente 7658
recocer 8066
recocido 568
recombinación 6966
recombinar 6967
recristalización 6973
recristalizar 6974
rectificación 6975
rectificador 6976
rectificador de vapor de mercurio
   5254
rectificador electrolítico 2984
rectificar 6977
recubrimiento 1894
recubrimiento con una máquina de
   extrusión 4704
recuperación 6972
recuperación de aceite de aguas de
   la gamuza 6922
recuperar 6962
red cristalina 2204
red cúbica de caras centradas 3333
red de difracción 2601
redisolución 7050
redrutita 6987
reducción 6993
reducción de la sección transversal
   6995
reducción de viscosidad 8233
reducción de viscosidad por
   descomposición 8782
reducción electroquímica 2959
reducción somática 7664
reducir 4833
reducir a escala 7239
reductasa 6992
reductor 6990
reenfriador 6968
reextracción 755
refinación 221
refinería 6998
reflujo 7000
"reforming" 7003
refracción atómica 711
refractario ácido 129
refractario alto en óxidos básicos 830
refrigeración por aire 252
refrigerante 2069, 7006
regaliz 4848
regeneración 7012
regeneración macronuclear 5039
regenerador 7014
regenerar 7008, 7030
régimen 6936
registrador 6969
regla de Abegg 2
regla de Dühring 2892
regla de las fases 6241
regla graduada 5196
rejalgar 6955
rejilla 3870

rejilla de centrifugación 834
relación de compresión 2014
relación de corriente primaria 6703
relación de distribución del metal
   5266
relación de la densidad de una pieza
   moldeada a la densidad del polvo
   constituyente 1222
relación de reflujo 7002
relación de válvula electrolítica 2986
relación de vulcanización 8796
relación nucleoplasmática 5876
relé fotoeléctrico 4861
relleno 6068
rendimiento anódico 571
rendimiento de corriente 2248
rendimiento de la caldera 1084
rendimiento en cantidad 2931
renina 1841
renio 7074
reología 7076
reopéctico 7078
reopexia 7079
reparar 6033
repelente de agua 8814
represor 7029
reprocesamiento químico 1674
reptación 2148
rescinamina 7032
reserpina 7033
reserva alcalina 333
residuo anódico 573
residuo de tamiz 7373
residuo de transferencia 8364
residuos 8019
residuos de decortización 2444
residuos de destilación 1140
residuos de refinado 6917
resiliencia 7039
resina 7040
resina de anacardo 1492
resina de colada 1500
resina de cresol 2164
resina de jalapa 4584
resina de poliamina-metileno 6509
resina endurecible sin presión 5842
resina modificada 6988
resina para laminados 4702
resina para pastas 6141
resina que reacciona con el aceite
   secante 5946
resina quininacarbacrílica 6871
resinas acrílicas 147
resinas aldehídicas 288
resinas alílicas 382
resinas alquídicas 339
resinas amínicas 446
resinas de acetalpolivinilo 6559
resinas de acrilonitrilo-butadieno-
   estireno 150
resinas de condensación 2030
resinas de contacto 2057
resinas de cumarona 2130
resinas de cumarona e indeno 2129
resinas de dimetilhidantoína 2693
resinas de estireno 7875
resinas de fenolformaldehído 6256
resinas de fluocarbono 3545

resinas de furano 3632
resinas de melamina 5212
resinas de melamina y formaldehído
   5211
resinas de politerpeno 6554
resinas de poliuretano 6557
resinas de resol 7049
resinas de tiourea 8257
resinas de tioureaformaldehído 8256
resinas de ureaformaldehído 8656
resinas de vinilideno 8763
resinas de xilenol 8884
resinas epoxídicas 3080
resinas epóxido 3080
resinas etenoides 3127
resinas fenólicas 6259, 6263
resinas ftálicas 6366
resinas maleicas 5106
resinas naturales 3900, 5663
resina soluble de nylon 7644
resina soluble en aceite 5949
resinas para cambiar iones 4421
resinas poliamídicas 6508
resinas poliésteres 6527
resinas silicónicas 7386
resinas sintéticas 8000
resinas termoplásticas 8222
resinas vinílicas 8769
resinato cálcico 1377
resinato cobaltoso 1920
resinato de cobre 2104
resinato de manganeso 5138
resinato de plomo 4793
resinato de zinc 8964
resinato férrico 3411
resinatos 7041
resina urea 8658
resinoles 7045
resistencia 3354
resistencia a amarillearse 8898
resistencia a la coloración 8293
resistencia a la migración de color
   5489
resistencia al deslucimiento 4853
resistencia al envejecimiento 232
resistencia a los productos químicos
   1675
resistencia a un solvente 7659
resistencia de deslizamiento 7452
resistencia en condiciones húmedas
   8834
resistencia interna 4387
resistencia líquida 8815
resistente a la congelación y deshielo
   cíclicos 3599
resistente a la penicilina 6166
resistente al calor 3980
resistente al fuego 3484
resol 5969
resonancia 7051
resorcina 7052
resplandor 1166
restricción 6400
retardador 7059
reticulación 2178
reticulado 5598
retículo cristalino 2204
retinol 7060

retorta 7061
revelador 2448
reveladores fotográficos 6352
revenir 8066
revestidora de rodillos de rotación
    inversa 7065
revestimiento aislador térmico 4689
revestimiento con la calandra 1391
revestimiento electrolítico de cobre
    2103
revestimiento intumescente 4395
revestimiento por conversión 2066
revestimiento por inmersión 2736
revestimiento protector 7048
revestir 4687
revisar 6033
revistir por inmersión 2735
revulsivo 7071
rezumamiento 5976
rezumar 6201, 7286
rezume 5976
riboflavina 7090
ribonucleasa 7091
ribosa 7093
ribosoma 7095
ribosomal 7094
ricinoleato cádmico 1310
ricinoleato cálcico 1378
ricinoleato de butilo 1280
ricinoleato de cobre 2105
ricinoleato de diglicol 2617
ricinoleato de litio 4966
ricinoleato de metilo 5447
ricinoleato de metoxietilo 5296
ricinoleato de zinc 8965
ricinoleato magnésico 5074
ricinoleato sódico 7595
ristocetina 7108
roca de fosfato 6321
rociadura de pelusilla 3523
rodaminas 7080
rodanina 7081
rodiado 7082
rodillo flotante 2313
rodillo refrigerador 1702
rodillos soporte 8008
rodillo superior 8342
rodinol 7083
rodio 7085
rodocrosita 7086
rodonita 7087
rodopsina 7088
rojo Berlín 923
rojo Congo 2040
rojo chino 1704, 2423
rojo de cadmio 1309
rojo de cal 4884
rojo de clorofenol 1752
rojo de correos 3992
rojo de cromo 1798
rojo de estafeta 3992
rojo de fenol 6262
rojo de la India 4312
rojo de metilo 5446
rojo de Persia 6223
rojo de primulina 6710
rojo de selenio 7296
rojo de toluileno 8332

rojo de Turquía 8580
rojo inglés 3053
rojo neutro 8332
rojo pompeyano 6576
rojos de óxido de hierro 4449
rojo veneciano 8722
rompimiento de la espuma 4730
roscoelita 7120
rotación molecular 5542
rotenona 7136
rotular 4658
rubia 5044
rubidio 7152
ruibarbo 7089
ruptura cromatídica 1788
ruptura de fago 6235
ruptura de inserción 4363
ruptura mediocromatidio 3935
rust 7158
rutenio 7159
rutilo 7161
rutina 7162

sabadilla 7163
sacamuestras 7200
sacar 2863
sacarato potásico ácido 6594
sacarimetría 7165
sacarímetro 7164
sacarina 7166
sacar la espuma 7267
sacarómetro 7167
sacarosa 7168
safraninas 7173
safrol 7174
sal 7188
sala de las balanzas 775
sala de llenado 6087
sala de trasvase 6887
sal amoníaco 462
sal anhidra 552
sal connata 2046
sal de acederas 6643
sal de Glauber 3760
sal de Rochelle 7111
sal de Schäffer 7245
sales ácidas 131
sales básicas 831
sales conductoras 2034
sales crómicas 1810
sales cromosas 1832
sales de diazonio 2493
sales de Epsom 3082
sales de los oxoácidos de uranio 8645
sales de Roussin 7143
sales normales 5848
sales tampones 1216
salicilaldehído 7178
salicilamida 7179
salicilanilida 7180
salicilato cálcico 1379
salicilato de amilo 519
salicilato de bencilo 917
salicilato de diisopropilenglicol 2650
salicilato de etilo 3263
salicilato de glicol 3814
salicilato de isoamilo 4473
salicilato de isobornilo 4478

salicilato de isobutilo 4489
salicilato de litio 4967
salicilato de metilo 5448
salicilato de zinc 8966
salicilato magnésico 5075
salicilato sódico 7596
salicilato teobromina-cálcico 8196
salicilato teobromina-sódico 8197
salicilazosulfapiridina 7181
salicina 7176
salina 7191
salinidad 7184
salino 7183
salinómetro 7185
salinómetro eléctrico 2951
salitre 6630
salivina 7186
sal marina 7638
salmastro 1151
sal microcósmica 5475
salmuera 1167
sal negra de Roussin 7141
salol 7187
sal roja de Roussin 7142
salsedumbre 7184
sal singenética 2046
salvado 1152
samario 7197
samarskita 7198
samarsquita 7198
sangre de drago 2859
santalol 7210
santonina 7212
saponáceo 7214
saponificación 7215
saponificar 7217
saponina 7219
sarcosina 7221
sarcosinato sódico 7597
sardónice 7222
sasolita 7225
satinado 5014
satinado impermeable 8566
saturación 7235
saturante 7229
saturnismo 6479
savia 7213
scoparius 7255
scheelita 7246
schoepita 7248
schroeckingerita 7249
sebacato de butilbencilo 1254
sebacato de dibencilo 2505
sebacato de dibutilo 2522
sebacato de dicaprilo 2532
sebacato de dihexilo 2620
sebacato de diisooctilo 2644
sebacato de dimetilo 2706
sebacato de dioctilo 2729
sebaconitrilo 7272
sebo 8025
sec- 7273
secadero 4634
secado instantáneo 3500
secador 2866
secadora giratoria 7131
secadora neumática 6483
secadora por correa 870

tetrametildiaminobenzhidrol 8157
tetrametildiaminobenzofenona 8158
tetrametildiaminodifenilmetano 8159
tetrametildiaminodifenilsulfona 8160
tetrametilendiamina 8161
tetrametiletilendiamina 8162
tetramorfo 8166
tetranitrato de pentaeritrita 6175
tetranitroanilina 8167
tetranitrometano 8168
tetrapropileno 8172
tetrasomatia 8174
tetrasulfuro sódico 7616
tetravalente 8175
tetrayodoetileno 8150
tetrilo 8176
tetróxido de vanadio 8696
tiacina 8229
tiamilal sodico 8228
tiazina 8229
tiazol 8230
tiazosulfona 8231
tibio 5002
tiempo de detención 2440
tiempo de duplicación 2853
tiempo de gelificación 3715
tiempo de reacción 6947
tiempo de reposo 2440
tiempo de secado con velocidad
    constante 2052
tiempo necesario para la vitrificación
    7466
tierra blanca 8085
tierra cocida pulverizada 3877
tierra decolorante 3618
tierra de infusorios 2489
tierra de Siena calcinada 1234
tierra de sombra 8606
tierra de sombra turca 8582
tierra de Trípoli 8536
tierra verde 3867
timoftaleína 8284
timol 8282
timoquinona 8285
tina de mezcla 5170
tinte 7737, 8291
tintómetro de Lovibond 4991
tintura 8290
tintura de yodo 4411
tio- 8234
tiocarbamoilo 8259
tiocarbanilida 8237
tiocianato amónico 480
tiocianato cálcico 1387
tiocianato de bencilo 918
tiocianato de plomo 4804
tiocianato de zinc 8975
tiocianato mercúrico 5243
tiocianato sódico 7617
tiocianoacetato de isobornilo 4479
tiodiglicol 8239
tiodipropionato de diestearilo 2812
tiodipropionato de dilaurilo 2658
tiodipropionato de ditridecilo 2828
tioéter dimiristílico 2715
tioéteres 8241
tiofeno 8246
tiofenol 8247

tiofosgeno 8248
tioglicerina 8242
tioglicolato de isooctilo 4529
tioglicolato sódico 7618
tiohidantoína 8244
tiomalonato de oro y sodio 3831
tiosemicarbacida 8252
tiosemicarbazona 8253
tiosulfato amónico 481
tiosulfato bárico 806
tiosulfato sódico 7619
tiouracilo 8254
tiouramilo 8259
tiourea 8255
tipólisis 8591
tiramina 8592
tirar de la batidora (en pequeñas
    cantidades) 844
tirocidina 8593
tiro forzado 3563
tirosina 8594
tirotricina 8595
tiroxina 8286
tisú papel muy fino 8294
titanato de estroncio 7860
titanato de litio 4970
titanato de octilenglicol 5921
titanato de plomo 4805
titanato de tetrabutilo 8096
titanato de tetraetilhexilo 8127
titanato de tetraisopropilo 8152
titanato potásico 6654
titanio 8297
titanita 8296
titrimetría 8311
titular 4658
tixotropía 8262
tixotropo 8261
tiza 1620
tizón 7465
tobera 5854
tocoferoles 8313
tolano 8314
tolerancia 8316
tolerancia de solución 7660
tolidina 8317
tolilaldehído 8333
tolildietanolamina 8334
toliletanolamina 8335
tolilnaftilamina 8336
toluendiamina 8321
toluendiisocianato 8322
tolueno 8320
toluensulfanilida 8323
toluensulfocloruro 8324
toluensulfonamida 8325
toluensulfonato de etilo 3270
toluensulfonato sódico 7620
toluhidroquinona 8327
toluidina 8329
toluol 8320
toluquinona 8331
tomsonita 8263
torbanita 8344
torbenita 8345
torímetro 8267
torio 8266
torita 8265

tornasol 4976
torón 8273
torre 1990
torre de absorción 38
torre de Gay-Lussac 3709
torre de Glover 3773
torre de pulverización 7729
torre depuradora 7265
torre de rectificación de relleno 6067
torre de refrigeración 2071
torre de relleno concentradora 6147
torre lavadora 7265
torta de sal 7190
tostación total 2324
tostadero 4634
tostar 4636
toxafeno 8350
toxicidad 8352
tóxico 8351
tóxico mitótico 5510
toxina 8353
toxina de fatiga 3358
trabajo 6719
trans- 8359
transducción ARN 8365
transespecífico 8374
transferasa 8362
transición de fase 6243
translocación 8370
translocación de inserción 4362
translucidez 8371
translúcido 8372
transmisión de la luz 4866
transmutación 8373
transmutación artificial de
    elementos 680
transparente 6162
transporte por cisterna 2058
transposición de Beckmann 858
transvasado 8377
transvasar 6884, 8368
tratamiento acuoso 649
tratamiento con substancias grasas
    2249
tratamiento en tambor giratorio 7155
tratamiento en un baño de cal 4891
tratamiento por arco voltaico 657
tratamiento preliminar 6697
tratamiento térmico 3985
tratar con formaldehído 3572
traza 8354
trementina 8584
trementina de Burdeos 3655
trementina de Venecia 8723
trementina grasa 3359
trementina oxidada 3359
trementina soplada 3359
tremérico 8493
tremolita 8380
tren laminador de redondos 7113
treonina 8275
triacetato de celulosa 1562
triacetato de glicerina 3797
triacetiloleandomicina 8382
triacetilricinoleato de glicerilo 3802
triacetina 8381
triacetoxiestearato de glicerilo 3801
triácido 8383

triada 8384
trialcohol 8482
triamida hexametilfosfórica 4065
triamilamina 8388
triaminotrifenilmetano 8387
triángulo de alambre 6406
triatómico 8391
tribásico 8392
triboluminiscencia 8398
tribromoacetaldehído 8399
tribromoetanol 8401
tribromuro de boro 1127
tribromuro de fósforo 6344
tribromuro de oro 3835
tributilamina 8404
tributilfosfina 8409
tributirato de glicerilo 3803
tricarbalilato de tributilo 8414
tricarbalilato de trietilo 8473
triclínico 8436
tricloroacetato sódico 7621
triclorobenceno 8417
tricloroborazol 8418
triclorocarbanilida 8419
tricloroetano 8420
tricloroetanol 8421
tricloroetileno 8422
triclorofenol 8430
triclorofluometano 8423
tricloromelamina 8425
tricloronitrosometano 8429
tricloropropano 8432
triclorosilano 8433
triclorotrifluoacetona 8434
triclorotrifluoetano 8435
tricloruro de antimonio 622
tricloruro de arsénico 674
tricloruro de boro 1128
tricloruro de fósforo 6345
tricloruro de nitrógeno 5771
tricloruro de titano 8305
tricloruro de vanadio 8697
tricloruro de yodo 4412
tricosano 8437
tridecano 8441
tridecanoato de metilo 5455
tridecilaluminio 8443
trietanolamina 8446
trietilaluminio 8451
trietilamina 8452
trietilborano 8453
trietilenfosforamida 8468
trietilenglicol 8455
trietilenmelamina 8466
trietilentetramina 8467
trietoxihexano 8448
trietoximetoxipropano 8449
trifenilestibina 8530
trifenilguanidina 8527
trifluonitrosometano 8476
trifluoruro de boro 1129
trifluoruro de cloro 1728
trifluoruro de cobalto 1925
trifluoruro de nitrógeno 5772
triformiato alumínico 425
trifosfato de adenosina 186
trigonelina 8478
trihaploide 8479

trihidrato de alúmina 402
trihidrocloruro de triaminotolueno
  8386
trihidroxiestearato de glicerilo 3804
triisobutilaluminio 8486
triisobutileno 8487
triisopropanolamina 8489
trilaurilamina 8491
trimelitato de trioctilo 8522
trimería 8494
trímero 8492
trimetadiona 8497
trimetilaluminio 8499
trimetilamina 8500
trimetilbutano 8502
trimetilclorosilano 8503
trimetilenglicol 8506
trimetilhexano 8507
trimetilnonanona 8508
trimetiloletano 8509
trimetilolpropano 8510
trimetilpentano 8512
trimetilpenteno 8513
trimetilpentilaluminio 8514
trimolecular 8516
trimonoico 8517
trinitrobenceno 8518
trinitrorresorcinato de plomo 4806
trinitrotolueno 8520
triosas 8523
trioxano 8524
trióxido de antimonio 623
trióxido de arsénico 675
trióxido de azufre 7947
trióxido de molibdeno 5555
trióxido de uranio 8648
trióxido de vanadio 8698
tripalmitina 8525
triparanol 8526
triparsamida 8560
triplete 8533
triploidia 8534
trípode 8535
trípoli 8536
tripolifosfato potásico 6655
tripolifosfato sódico 7622
tripropilenglicol 8539
tripropileno 8538
tripropionato de glicerilo 3805
tripsina 8561
tripsómetro 8540
triptófano 8562
trirricinoleato de glicerilo 3806
tris(hidroximetil)aminometano 8543
tris(hidroximetil)nitrometano 8544
trisilicato magnésico 5080
trisomia 8546
trisómico 8545
trisómico de intercambio 4374
trisulfuro de antimonio 624
triterpeno con seis enlaces dobles
  7732
tritio 8547
trituración 8548
triturador 2000
trituradora 4629, 7365
trituradora de coque 1944
trituradora de chorro 4595

trituradora de martillos 3948
trituradora de varillas 7113
triturador para huesos 1109
trivalente 8550
triyodotironina sódica 8485
troficidad 8554
trofoplasma 8556
trofotropismo 8557
trona 8552
tubería de entrega 2388
tubería espiral para la refrigeración
  2070
tubería helicoidal del recalentador
  7956
tubo barométrico 810
tubo de bolas 1220
tubo de ensayos 8091
tubo de fusión 1998
tubo de Pitot 6424
tubo de seguridad 7169, 8258
tubo de vidrio 3758
tubo en U de Ostwald 6023
tubo en U para cloruro de calcio
  1348
tubo graduado 5193
tubo oscilante 7978
tubo para cloruro de calcio 1347
tubo para transvasar 6886
tubo rociador 7679
tubos de viscosidad Gardner-Holt
  3674
tubo secador 2885
tubo secador de bandejas 8574
tulio 8280
túnel secador 8573
tungstato de cadmio 1315
tungstato de calcio 1388
tungstato de cobalto 1922
tungstato de plomo 4807
tungstato magnésico 5081
tungstato sódico 7623
tungsteno 8569
turbidímetro 8576
turbieza 1052, 1073
turbina 2933
turbio 8575
turbulento 8577
turgescencia 8579
turmalina 8348
turquesa 8586
tuyona 8279

uabaína 6025
ulexita 8596
ultracelerador 8598
ultrafiltración 8599
ultramar 8600
ultramar de cobalto 1926
ultramicrosoma 8602
ultravioleta 8603
undecalactona 8610
undecano 8611
undecanoato de metilo 5456
undecanol 8613
undecilenato de zinc 8976
undecilenato potásico 6656
undecilenato sódico 7624
ungüento 5950

ITALIANO

acido caprilico 1423
acido caproico 1421
acido capronico 1421
acido carbamico 1427
acido carbolico 6254
acido carbonico 1450
acido chinico 6867
acido cianidrico 4180, 4195
acido cianoacetico 2259
acido cianurico 2268
acido citrico 1858
acido clorico 1721
acido cloridrico 4176, 4193
acido cloroacetico 1732
acido clorodifluoroacetico 1738
acido clorometilfosfonico 1746
acido clorosolfonico 1759
acido cogico 4651
acido colico 1776
acido cresilico 2167
acido cresotinico 2165
acido cromico 1802
acido crotonico 2182
acido deidroacetico 2380
acido deossiribonucleico 2408
acido diallilbarbiturico 2467
acido dibromomalonico 2508
acido di cocco 1933
acido dietilbarbiturico 782
acido dietilditiocarbamico 2578
acido dietilesobarbiturico 4074
acido difenico 2742
acido di Gay-Lussac 3707
acido diglicolico 2613
acido dilinoleico 2659
acido dimerico 2668
acido di Neville e Winther 5708
acido ditiocarbammico 2823
acido di Tobias 8312
acido dodecenilsuccinico 2835
acido eptafluorobutirrico 4015
acido eritorbico 3101
acido esaidrobenzoico 4060
acido esaidroftalico 4061
acido etilbutirrico 3161
acido etilesoico 3223
acido etilfosforico 3258
acido etilsolforico 3267
acido etilsolforoso 3268
acido fenilacetico 6276
acido fenilamminonaftolsolfonico 6277
acido fenilarsonico 6278
acido feniletilacetico 6291
acido fenoldisolfonico 6255
acido fenolsolfonico 6264
acido fenossiacetico 6269
acidofilo 120, 126
acido fitico 6375
acido fluofosforico 3534
acido fluoridrico 4181
acido fluoroborico 3544
acido fluosilicico 3551
acido fluosolfonico 3552
acido folico 3559
acido formico 3574
acido fosfomolibdico 6327
acido fosforico 6334

acido fosforoso 6335
acido fosfotungstico 6346
acido ftalico 6364
acido fumarico 3620
acido furilacrilico 3638
acido gallico 3657
acido gibberellico 3744
acido glicerofosforico 3798
acido glicolico 3812
acido gluconico 3775
acido glutammico 3779
acido glutarico 3782
acido grasso 3361
acido guanilico 3890
acido ialuronico 4152
acido idnocarpico 4157
acido idrocinnamico 4177
acido idrossiacetico 4218
acido idrossibutirrico 4221
acido idrossinaftoico 4238
acido indolobutirrico 4324
acido iodico 4406
acido iodidrico 4171
acido ipocloroso 4260
acido ipofosforoso 4264
acido ippurico 4097
acido isetionico 4460
acido isobutirrico 4492
acido isocianurico 4501
acido isodecanoico 4504
acido isoftalico 4535
acido isonicotinico 4522
acido isopentanoico 4532
acido isovalerico 4577
acido itaconico 4578
acido lattico 4675
acido laurico 4739
acido levulico 4685
acido lignocerico 4875
acido linoleico 4909
acido linolenico 4911
acido litocolico 4972
acido maleico 5103
acido malico 5107
acido malonico 5109
acido mandelico 5118
acido meconico 5204
acido melissico 5215
acido metacrilico 5283
acido metanilico 5277
acido metilclorofenossipropionico 5351
acido metilfosforico 5438
acido metossiacetico 5289
acidometro 125
acido miristico 5628
acido monobasico 5564
acido mucico 5603
acido muriatico 4176
acido naftalendisolfonico 5637
acido naftalensolfonico 5638
acido naftilammindisolfonico 5644
acido naftilamminsolfonico 5645
acido naftilammintrisolfonico 5646
acido naftionico 5639
acido naftoldisolfonico 5641
acido nicotinico 5729
acido nitrico 5741

acido nitrobenzoico 5757
acido nitrofenilacetico 5784
acido nitrosalicilico 5786
acido nonaldecanoico 5814
acido nonoico 6156
acido oleico 5958
acido ortofosforoso 6012
acido osmico 6015
acido ossalico 6040
acido palmitico 6080
acido palmitoleico 6081
acido pantotenico 6093
acido pelargonico 6156
acido pentadecanoico 6173
acido peracetico 6193
acido perclorico 6197
acido performico 6205
acido periodico 6209
acido picramico 6386
acido picrico 6387
acido picrolonico 6388
acido pirofosforico 6838
acido pirogallico 6830
acido piroligneo 6831
acido piromellitico 6834
acido pirotartarico 6840
acido poliacrilico 6504
acido propionico 6749
acido prussico 4180
acido racemico 6880
acido resorcilico 7057
acido ribonucleico 7092
acido ricinoleico 7098
acido rubianico 7151
acido salicilico 7182
acido sebacico 7271
acido selenioso 7298
acido silicico 7382
acido silicotungstico 7392
acido solfamico 7913
acido solfanilico 7915
acido solfasalicilico 7937
acido solfobenzoico 7927
acido solfoftalico 7936
acido solforico 849, 7945
acido solforico (delle camere di piombo) 1623
acido solforico fumante 3624
acido solforoso 7946
acido sorbico 7667
acido stearico 7781
acido stifnico 7867
acido suberico 7880
acido succinico 7891
acido tannico 8034
acido tartarico 8052
acido tellurico 8058
acido tereftalico 8075
acido tetracloroftalico 8106
acido tiglico 8287
acido timico 8282
acido tioacetico 8235
acido tiobenzoico 8236
acido tiocianico 8238
acido tiodiglicolico 8240
acido tioglicolico 8243
acido tiosalicilico 8251
acido titanico 8295

amilopsina 515
amilosio 516
para-aminobenzoato di sodio 7484
para-aminoippurato di sodio 7485
para-aminosalicilato di sodio 7486
amissia 448
amitosi 447
amittico 439
amixi 448
ammide 440
ammide di litio 4946
ammide poliesterica 6525
ammide primaria 6701
ammide secondaria 7277
ammide sodica 7483
ammide terziaria 8087
ammidi acidi 104
ammidine 441
ammina 443
ammina ottilica 5916
ammina trilaurilica 8491
ammine 449
ammine alifatiche 3364
amminoacetato di diidrossialluminio 2627
amminoacidi 444
amminobenzoato d'etile 3142
amminobenzoato di isobutile 4485
amminofenolo metilico 5316
ammoniaca 450, 453
ammoniaca anidra 549
ammoniaca combinata 3493
ammoniaca in soluzione acquosa 7715
ammoniaca libera 3594
ammoniaca totale 8346
ammonio 455
ammonolisi 482
amorfo 483
ana- 522
anabasina 523
anacromasi 524
anaerobio 525
anafase 530
anaforesi 532
analcime 526
analcite 526
analgesico 527
analisi 528, 695
analisi al cannello ferruminatorio 1063
analisi alla fiamma 3496
analisi al setaccio 7372
analisi conduttometrica 2033
analisi delle tetradi 8116
analisi elettrolitica 2953
analisi genomica 3729
analisi gravimetrica 3864
analisi per assorbimento 28
analisi per attivazione 164
analisi per diluizione isotopica 4573
analisi qualitativa 6847
analisi quantitativa 6848
analisi radiometrica 6907
analisi volumetrica 8791
analizzatore di gas 3680
ancherite 565
andalusite 534

androsterone 535
anelli Raschig 6935
anello cromosomico 1829
anello tandem 8031
anemometro 536
anemometro a filo caldo 4144
anetolo 538
aneuploide 539
aneuploidismo 540
anfiboli 485
anfibolite 486
anfidiploidismo 487
anfigenia 488
anfigonia 488
anfimixi 489
anfogenia 490
anfolito 491
anfolitoide 492
anfoterico 493
angioblasto 541
anglesite 542
anidride acetica 59
anidride borica 1118
anidride cloromaleica 1744
anidride dodecenilsuccinica 2836
anidride ftalica 6365
anidride glutarica 3783
anidride isobutirrica 4493
anidride maleica 5104
anidride propionica 6750
anidride solforica 7947
anidride succinica 7892
anidride tetracloroftalica 8107
anidride tetraidroftalica 8145
anidride tetrapropenilsuccinica 8171
anidridi 545
anidrite 546
anidro 548
anilide idrossinaftoica 4239
anilina 554
anilina alla formaldeide 3570
anione 558
anisidine 560
anisolo 561
anisomerico 562
anisotonico 563
anisotropico 564
annabergite 566
annidamento 5730
annidazione 569
anodo 570
anodo preformato 2037
anodo residuo 573
anolito 580
anonitato di trietile 8450
anortosite 582
antiadesivo 2439
antialbumose 590
antialdossime 591
antibariona 592
antibiosi 593
antibiotici 594
antibiotico a largo spettro 1173
anticatalizzatore 596
anticloro 597
anticoagulante 598
anticorpo 595
antidetonante 601

antidetonanti 610
antidoto 603
antienzima 604
antiistaminico 608
antimagnetico 5828
antimicina 627
antimoniale 611
antimoniati 613
antimoniato di piombo 4760
antimoniato sodico 7488
antimonio 615
antimonio rosso 4617
antimonite 614
antimonite di isopropile 4548
antinvecchiante 235
antinvecchiante contro alte temperature 3970
antiormone 609
antiossidante 629
antipiombo 7243
antipirina 631
antischiuma 2361
antisettici 632
antofillite 583
antracene 584
antrachinone 588
antracite in polvere 2217
antraflavina 585
antragallolo 319
antranilato di feniletile 6293
antranilato di metile 5323
antranilato dimetilico 2680
antranilico 586
antranolo 587
antrarobino 589
apatite 639
aploide 3951
aploidismo 3953
aplomitosi 3954
aplontico 3955
aplopoliploide 3956
aplosufficiente 3952
apocrino 641
apocromatico 640
apofermento 642
apogamia 643
apomorfina 644
appallottolamento 6160
appannamento 8049
apparato d'estrazione 3326
apparato di Golgi 3836
apparato per evaporare 3295
apparato per potassa 6590
apparecchio di contatto per stadio 7735
apparecchio di Dean e Stark 2330
apparecchio di emulsione 3024
apparecchio di Jolly 4596
apparecchio di Kipp 4645
apparecchio di Landsberger 4711
apparecchio essicatore 2430
apparecchio per inclinare le damigiane 1462
apparecchio per la misurazione dello sbiadimento 3615
apparecchio per muovere liquidi mediante pressione d'aria 2936
appiccicoso 8015

barite 816
barodiffusione 809
barrel 811
bartrina 815
basalto 818
base 819, 820
basicità 331, 826
basi di fosfonio 6328
basidiomiceto della ruggine delle
    piante 7158
basi di Schiff 7247
basofilo 836, 837
basso tenore, a ~ 4993
bastnaesite 839
batteri acetici 60
batteria di avviamento 7762
batteria di trazione 8357
batteria donatrice competente 2006
batteria tampone 1214
battericida 758
batterio 767
batteriocina 759
batteriofago 763
batteriogeno 760
batteriolisi 762
batteriolisina 761
batterio ricettore 6960
batteriostasi 764
batteriostatico 766
batteriostato 765
bauxite 852
bebeerina 857
becco Bunsen 1228
becco della siviera 4919
becco Méker 5209
beenato di metile 5327
belladonna 867
bentonite 874
benzalconio 876
benzaldeide 875
benzamide 877
benzamina 878
benzanilide 879
benzedrina 880
benzene 881
benzenefosforodicloruro 884
benzenefosforoossidicloruro 885
benzene idrazoico 4167
benzenesolfonbutilammide 889
benzidina 890
benzilacetato 904
benzilanilina 906
benzilbenzoato 907
benzil formiato 914
benzina 892, 3696
benzina acida 7674
benzina desolforata 7973
benzina pesante 3990
benzoacetato d'etile 3147
benzoato di ammonio 458
benzoato di etile 3148
benzoato di idrossichinolina 4248
benzoato di isoamile 4468
benzoato di isobutile 4486
benzoato di metile 5328
benzoato sodico 7494
benzofenone 899
benzoguanimina 893

benzoilbenzoato di metile 5329
benzolo 881
benzonitrile 898
benzoporporina 900
benzotrifluoro 902
berberamina 920
berchelio 921
berillio 925
berillo 924
betafite 928
bi- 2457
biacca all'ossido di zinco 1705
biacca di piombo 8841
bianca 1035
biancastro 1175
bianco dell'uovo 271
bianco di antimonio 626
bianco di barite 1023
bianco di calcite 7228
bianco di Cina 1022
bianco di Parigi 6125
bianco di Spagna 8846
bianco di stronzio 7862
bianco di titanio 8306
bianco di zinco 8952
bianco fiocco 987
bianco fisso 1023
bianco fisso di allumina 401
bianco minerale 5503
bianco sublimato di piombo 7884
bibasico 932
bibromuro d'etilene 3186
bi-ter-butilemetacresolo 2518
bibutirrato glicoletilenico 3191
bicarbonati 934
bicarbonato di idrossietiltrimetil-
    ammonio 4229
bicarbonato di potassio 6600
bicarbonato di sodio 7495
bicchiere 854, 1410
bicchiere graduato 5193
bicchiere per pesare 8824
bicloruro 936
bicloruro clorometilfosfonico 1747
bicloruro di benzile 913
bicloruro di etilene 3187
bicloruro di polivinile 6566
bicloruro di propilene 6762
bicloruro di stagno 7751
bicloruro di tetraglicolo 8135
bicloruro di triglicolo 8477
bicloruro di vanadio 8689
bicloruro di zolfo 7942
bicromato di ammonio 464
bicromato di sodio 7527
bicromato di zinco 8931
bicromato potassico 6615
bifluoruro di ammonio 459
bifluoruro di sodio 7496
bifosfato di sodio 7497
bigenerico 940
bilancia chimica 1640
bilancia d'assaggio 692
bilancia di precisione 529, 6677
bilancia manometrica 2325
bilancia per gas 3681
bilanciere 855
bilancio di materia 5174

bile 3656
bioblasto 946
biochimica 948
biocora 949
biofisica 961
biofotometro 960
biogenesi 951
biogenetico 952
biogenia 951
biogeno 950
bioluminescenza 957
biometria 958
bionica 959
bioplasma 962
biopolimero 963
biopsia 964
bioritmo 965
biosintesi 967
biossalato potassico 6601
biossido di carbonio 1012, 1447, 1770
biossido di cloro 1727
biossido di manganese 5123, 5129
biossido di piombo 4774
biossido di rame 1010
biossido di selenio 7295
biossido di tellurio 8060
biossido di titanio 8300
biossido di torio 8269
biossido di uranio 8640
biossido di zirconio 8988
biossido di zolfo 7943
biossolano di metile 5370
biostratigrafia 966
biotico 968
biotina 970
biotipo 973
biotite 971
biotopo 972
birra normale 4688
birra pronta per il consumo 3477
birrificio 1160
bisidrossicumarina 977
bisilicato di piombo 4775
bismanolo 978
bismite 986
bismutinite 984
bismutite 994
bismuto 980
bisolfati 998
bisolfato di sodio 7498
bisolfato potassico 6602
bisolfiti 999
bisolfito di calcio 1341
bisolfito di sodio 7499
bisolfito potassico 6603
bisolfuro di tetrabutiltiuram 8094
bisolfuro di arsenico 672
bisolfuro di carbonio 1448
bisolfuro di molibdeno 5552
bisolfuro di tetraetiltiuram 8130
bisolfuro di tetraisopropiltiuram 8151
bisolfuro di tetrametiltiuram 8164
bisolfuro di tetrapropiltiuram 8173
bistriclorosililetano 997
bistro 996
bitartrato di sodio 7500
bitartrato potassico 6604
bitionolo 1000

diuranato d'ammonio 465
diuranato di sodio 7530
dividere in quattro parti 6850
divinilbenzene 2830
divisione cellulare 1546
divisione in due parti uguali 6072
divisione maturativa 5190
divisione nucleare 5861
dixantogenato isopropilico 2649
dodecene 2834
dodecilbenzene 2838
dodecilfenolo 2839
doppia parete, a ~ 4582
doppietto 2852
doppi legami coniugati 2044
doppio incrocio 3582
doppio legame 2850
doratura a fuoco 3832
dosaggio 2845
dosatore 2849
dose 2847
dose emergente 3304
dose in aria 3593
dosimetro 2848
dosimetro a ionizzazione 4428
dosimetro chimico 1647
drenaggio 2452
drocarbile 2868
droga 2875
dulcina 2893
dumortierite 2896
duplicazione in tandem 8029
duraina 2897
durame 719
durene 2898
durezza 3964
durezza Brinell 1168
durezza permanente 6216
durezza temporanea 8069
duttibilità 2891

ebollizione 1086, 2916, 3701
ebonite 2914
ebullioscopia 2915
ecade 2917
eccipiente per polvere 6665
eccipiente per unguenti 5951
eccitazione molecolare 5538
eccitoanabolico 3299
eccitocatabolico 3300
ecocline 2919
economizzatore 2920
ecotipo 2921
ectoplasma 2922
ectosfera 2924
ectosoma 2923
effetto centrifugo 1576
effetto d'agglutinazione 7807
effetto di combinazione in blocco
    1047
effetto di contrazione 6400
effetto di legame chimico 1641
effetto di schermo 7349
effetto Joule-Thomson 4598
effetto piezoelettrico 6391
effimero 3072
efflorescenza 2932
efflorescenza di zolfo 7940

effusione 2934
effusione molecolare 5537
eiconogeno 2940
eicosan 2939
eiettore di vapore 7770
eiezione di particelle di vernici 8276
einsteinio 2941
elastomero 2945, 2946
elastometro 2947
elaterite 2948
elementi di terre rare 6930
elementi transuranici 8376
elementi trascurici 8360
elemento 3010
elemento attivo 166
elemento chimico 1648
elemento di Grove 3881
elemento d'impurezza 4296
elemento di regolazione 3035
elemento di trasporto 8366
elemento leggero 4851
elemento naturale 5660
elemento padre 6124
elemento pesante 3988
elemento pilota 6397
elemento radioattivo 166, 6904
elemento riscaldatore 1324
elemento secondario 7278, 7826
elettrochimica 2962
elettrodecantazione 2964
elettrodo a idrogeno 4196, 7741
elettrodo al calomelano 1399
elettrodo al chinidrone 6865
elettrodo a vetro 3756
elettrodo di riferimento 6997
elettrodo di Soderberg 7472
elettrodo negativo 5672
elettrodo positivo 6586
elettroforesi 2995
elettrolisi 2974
elettrolito 2975
elettrolito colloidale 1968
elettrolito fuso 3641
elettrolito impuro 3580
elettrometallurgia 2989
elettrone 2991
elettroosmosi 2993
elettroraffinazione 3001
elettrospettrogramma 3002
elettrotermica 3005
elettrotipia 3006
eliminazione 7024
elio 3994
eliotropina 3993
elisir 3011
elleborina 3997
elleboro 3996
elutriazione ad aria 255
eluzione 3013
elvite 3999
emanazione 3015
ematimetro 3915
ematina 3920
ematite 3921
ematite da masse reniformi 4631
ematoblasto 3922
ematocita 3923
ematoporfirina 3925

ematossilina 3926
emazia nucleopriva 265
embriologia 3017
emetico 3019
emiaploide 4002
emicellulosa 4000
emicromatidico 4001
emissi 4004
emissione di vapore d'acqua 5525
emizigotico 4005
emoagglutinazione 3917
emoagglutinina 3918
emocitometria 3916
emocoltura 3927
emoglobina 3928
emolisina 3929
emolitico 3930
emolliente 3020
emopoiesi 3924
emulsionamento 3022
emulsionante 3023
emulsione 3025
emulsione adesiva 3026
emulsione autoportante 6161
emulsione reversibile 6242
emulsoide 3028
enantato d'etile 3252
enantiomorfo 3030
enantiotropico 3031
enargite 3032
encaustico 3034
endelionite 3036
endo- 3037
endocrino 3038
endomissi 3041
endomitosi 3040
endoplasma 3042
endopoliploidismo 3043
endosoma 3044
endotermico 3045
eneicosanoato di metile 5387
energia 3050
energia chimica 1649
energia di ionizzazione 4429
energia di legame di particelle alfa
    390
energia intrinseca 4392
enometro 5932
entalpia 3056
entropia 3059
enzima 3068
enzima trasferitore 8362
enzimico 3065
enzimologia 3069
eosina 3070
eosinofilo 3071
eparina 4010
epicloridrina 3073
epidoto 3074
epigenetica 3075
epinefrina 3076
epistasia 3077
epossi- 3079
epsomite 3081
eptabarbital 4011
eptacloro 4012
eptadecanoato di metile 5388
eptadecanol 4013

formazione di schiuma mediante
   ionizzazione 4430
formazione meccanica di schiuma
   5199
formiato di amile 503
formiato di butile 1266
formiato di eptile 4022
formiato di etile 3221
formiato di geraniolo 3735
formiato di magnesio 5062
formiato di metile 5382
formiato di piombo 4779
formiato di sodio 7539
formiato di zinco 8941
formicina 3575
formula 3576
formula chimica 1656
formula del benzene 886
formula di struttura 7864
formula empirica 3021
formula molecolare 5539
fornace a bagno di piombo 4766
fornace a muffola 5609
fornace a tubo di fusione 1997
fornace da calce 4882
fornace per assaggio 694
fornace rotante 7133
fornello essiccatore a vuoto 8669
forno 6028, 7827
forno a bagno 3986
forno a bagno di sale 7189
forno a bagno salino a elettrodi 2968
forno a canale 4819
forno a coke 1948
forno ad arco indiretto 4318
forno ad aria calda 260
forno ad atmosfera controllata 678
forno a induzione a canale 4327
forno a muffola 5608
forno a ricupero di calore 7013
forno a riscaldamento diretto a
   resistenza 2784
forno a riscaldamento indiretto a
   resistenza 4320
forno da cemento 1569
forno di Birkeland e Eyde 974
forno di coppellazione 2224
forno di sinterizzazione 7425
forno fusorio 7461
forno Heroult 4027
forno Keller 4613
forno per piriti 6825
forno per smalto a vetrino 3772
foro di spia 6155
forone 6319
fortificare 3579
forza di legame 1104
forza intramolecolare 4391
forza ionica 4425
fosfati 6322
fosfatide 6326
fosfato bibasico di magnesio 2499
fosfato bibasico di piombo 2498
fosfato bibasico di sodio 2501
fosfato cobaltoso 1919
fosfato cromico 1809
fosfato d'argento 7409
fosfato di ammonio bibasico 2496

fosfato di ammonio e magnesio 5050
fosfato di ammonio monobasico 5565
fosfato di calcio 5566
fosfato di calcio bibasico 2497
fosfato di calcio tribasico 8393
fosfato di istamina 4099
fosfato di magnesio tribasico 8395
fosfato di pirobutammina 6827
fosfato di potassio tribasico 8396
fosfato di primachina 6699
fosfato di sodio ammonico 5475
fosfato di sodio e ammonio 7487
fosfato di triamilfenile 8390
fosfato di tributile 8408
fosfato di tributilfenile 8407
fosfato di tributossietile 8402
fosfato di tricloroetile 8415
fosfato di tricresile 8438
fosfato di tridimetilfenile 8445
fosfato di trietile 8470
fosfato di trifenile 8528
fosfato di triottile 8521
fosfato di zinco 8959
fosfato etilmercurico 3245
fosfato ferrico 3408
fosfato ferroso 3436
fosfato monobasico di magnesio 5567
fosfato monobasico di sodio 5569
fosfato potassico bibasico 2500
fosfato potassico monobasico 5568
fosfato Redonda 6985
fosfato tribasico di sodio 8397
fosfina 6324
fosfiti 6325
fosfito decildifenilico 2750
fosfito di dimetile 2703
fosfito di diottile 2727
fosfito di piombo 4792
fosfito di tributile 8410
fosfito di tricresile 8439
fosfito di tridecile 8444
fosfito di triesile 8481
fosfito di trietile 8471
fosfito di trifenile 8529
fosfito di triisoottile 8488
fosfito di triisopropile 8490
fosfito di trimetile 8515
fosfito di zinco 8961
fosfito ottilfeniletilesilico 3230
fosfolipide 6326
fosfomolibdato di sodio 7588
fosforescenza 6332
fosfori 6336
fosforite 6321
fosforo 6337
fosfotioato di trietile 8472
fosfotungstato di sodio 7580
fosfuranilite 6347
fosfuro 6323
fosfuro di calcio 1374
fosfuro di zinco 8960
fosfuro ferroso 3437
fosgene 6320
fotocatalisi 6348
fotochimica 6349
fotodissociazione 6350
fotoforesi 6355
fotolisi 6353

fotometro 6354
fotopolimero 6356
fotoriattivazione 6357
fotosintesi 6359
fragilità 1170, 3587, 3604
fragilità dovuta all'idrogeno 4197
fragmoplasto 6360
fragmosoma 6361
frammentazione 3588, 7355
frammentazione dei mezzicromatidi
   3936
frammentazione nucleare 5862
frammentazione nucleolare 5873
francio 3589
frangiato 3473
franklinite 3591
frantoio 4629
frantoio a mascella 4593
frantoio a rulli 2190
frantoio per canna da zucchero 7903
frantoio per ossa 1109
frantoio Raymond 6944
frantoio rotativo 7129
frantoio ruotante 3913
frantumatore 5493
frantumatrice per coke 1944
frazionamento 3586
frazionatore di campioni 7201
frazione leggera 4855
freibergite 3601
friabile 6667
fritta 3607
fritta azzurra (per smalti) 1067
fruttosio 3611
ftalammide 6362
ftalato di dicicloesile 2552
ftalato di difenilguanidina 2753
ftalato di glicerile 3800
ftalato di ottildecile 5920
ftalato diottilico 2728
ftalato di tetraidrofurfurile 8143
ftaleine 6363
ftalilsolfacetammide 6369
ftalilsolfatiazolo 6370
ftalimmide 6367
ftalonitrile 6368
fucsina 3612
fuliggine 7666
fulmicotone 3904
fulminati 3619
fulminato di mercurio 5250
fumarato ferroso 3431
fumi 3531
fumiganti 3623
fumo 7463
fungicidi 3628
furano 3631
furfurolo 3633
furoato di metile 5384
fusano 3639
fusiforme 3645
fusione 1499, 3646
fusione in conchiglia a
   rovesciamento 7458
fusioni in tandem 8030
fuso acromatico 101
fuso centrale 1572
fustetto 3649

fusticino 4633
fusticino mezzo 3486
fusto 1494

gadolinio 3651
gadolinite 3650
galattosio 3653
galena 3654
galipot 3655
galleggiabilità 1230
galleggiante di Erdmann 3095
galleria di essicamento 8573
galleria di essicamento a vassoi 8574
gallio 3658
gallocianina 3659
galloni di olio per 1000 litri di resina 5941
galvanizzazione 3660
galvanizzazione preliminare con mercurio 6862
galvanoplastica 2973, 2998
galvanoplastica automatica 3616
galvanoplastica in tamburo 814
galvanoplastica meccanica 5198
galvanoplastica semiautomatica 7306
gametico 3663
gameto 3662
gametocito 3665
gametoide 3666
gamodemo 3670
gangliocita 3671
ganisto 3672
garnierite 3678
garza per imbozzimatura 1094
gas 3679
gas, a prova di ~ 3699
gas blu 1068
gas d'acqua 1068, 8810
gas d'acqua arricchito 1893
gas d'acqua carburato 1463
gas d'altoforno 1026
gas delle paludi 5284
gas d'estinzione 6858
gas di carbone 1890
gas di città 8349
gas di gasogeno 6724
gas di legno 8857
gas di raffinazione 6999
gas di sintesi 7996
gas idrati 3692
gas inerti 4336, 5806
gas Mond 5559
gas naturale 5662
gas naturale acido 7673
gas nobili 5806
gas occluso 5898
gasolina naturale 1493
gasolio 3695
gas perfetto 6203
gas per illuminazione 4856
gas ricco 7096
gas senza additivi 1868
gassificazione del carbone 1891
gassogeno ad aspirazione 7900
gaylussite 3710
gel 3712
gelatina al bicromato 938
gelatina di nitroglicerina 1028

gelatina di nitroglicerina per uso subacqueo 7885
gelatina esplosiva 1028
gelatine 3713
gelatinizzazione 7639
gelatura 1701
gel di silice 7379
gel irreversibile 4454
gel reversibile 7067
gelsemina 3714
gene 3716
gene inibitore 4349
gene maggiore 5095
gene mutatore 5622
generatore di acetilene 91
generatore di aceto 8743
generatore di vapore elettrico 2952
generatore per galvanoplastica 2999
generatore per galvanotecnica 6467
gene restrittivo 7058
geni d'incompatibilità 4305
genoblasto 3726
genoide 3727
genoma 3728
genomio 3728
genosoma 3730
gentisato di sodio 7540
genziana 3731
geranilacetato 3736
geraniolo 3733
germanio 3738
germe 3737
germe cristallino 5880
germicida 3741
germogliare 7731
germoglio 7362
gesso 1620, 3911
gesso di Parigi 6451
gesso francese 3602
gesso in polvere 8846
getto 1499
getto d'aria 249
ghiaccio secco 2882
ghiandola 3751
ghiandole endocrine 3039
ghisa speculare 7709
giada 4583
gialappa 4584
gialli all'ossido di ferro 4450
gialli di ferrite 3416
giallo brillante 4590
giallo di benzidina 891
giallo di cadmio 1316
giallo di calce 4890
giallo di Cassel 1495
giallo di chinolina 6876
giallo di cromo 1800, 4820
giallo di curcuma 2241
giallo di Napoli 5652
giallo d'India 4313
giallo di Parigi 6126
giallo di stronzio 7853
giallo di zinco 8928
giallo d'oro 4591
giallo limone 4821
giallo re 4642
giallo vegetale usato con merdente metallico 5952

giamesonite 4585
gilsonite 3746
ginocardato di sodio 7545
ginogenesi 3909
gipsite 3910
gitossina 3749
glaserite 3752
glauberite 3759
glaucodote 3761
glauconite 3762
gliceraldeide 3786
gliceride 3787
glicerina 3789
glicerofosfato di calcio 1356
glicerofosfato di magnesio 5063
glicerofosfato di manganese 5132
glicerofosfato di sodio 7544
glicerofosfato ferrico 3401
glicerolabietato 3799
glicide 3807
glicidolo 3807
glicinato di rame 2093
glicinato di sodio e teofillina 8200
glicocolla 3808
glicogeno 3810
glicolato di sodio e zirconio 7628
glicolato di zirconio 8989
glicoldietilenico 2579
glicoletilenico 3188
glicoli di poliossipropilene 6549
glicolo di propilene 6763
glicolo dipropilenico 2776
glicolo di trietilene 8455
glicolo di trimetilene 8506
glicolo di tripropilene 8539
glicolonitrile 3813
glicolo polietilenico 6530
glicolsalicilato 3814
glicol titanato di ottilene 5921
glicosecretorio 3815
gliodina 3818
globuline 3766
globulo 3765
glucagone 3774
glucoeptinato di sodio 7541
gluconato di calcio 1355
gluconato di manganese 5131
gluconato di sodio 7542
gluconato ferroso 3432
gluconato potassico 6620
glucosidi 3816
glucuronolattone 3776
glutamato di sodio 7543
glutammina 3780
glutaraldeide 3781
glutaronitrile 3784
glutine 3785
glyossale 3819
gocciolina 2869
goethite 3820
golgiolisi 3837
gomma 7144
gomma acrilica 148
gomma adragante 3903
gomma al butadiene-stirolo 7872
gomma al polisolfuro 6553
gomma al poliuretano 6558
gomma arabica 3898

invariante 4397
invecchiamento 230
invecchiamento naturale 5658
inversione 4398
invertasi 4399
inviluppo d'estrazione 3329
invitalità 4401
in vitro 4402
in vivo 4403
inzuccheramento 3610
iodargirite 4404
iodato di potassio 6622
iodato di sodio 7552
iodeosina 4405
iodio 4407
iodoformio 4414
iodosuccinimmide 4415
ioduro d'argento 7404
ioduro d'etile 3234
ioduro di acetile 94
ioduro di cadmio 1305
ioduro di calcio 1361
ioduro di fosfonio 6329
ioduro di isopropile 4553
ioduro di litio 4960
ioduro di metile 5396
ioduro di metilene 5377
ioduro di metilmagnesio 5408
ioduro di ottile 5923
ioduro di piombo 4782
ioduro di potassio 6623
ioduro di sodio 7553
ioduro di timolo 8283
ioduro di zinco 8943
ioduro ferroso 3433
ioduro mercurico 5235
ioduro mercurico di bario 5231
ioduro mercurico di potassio 5239
ioduro palladioso 6074
ioduro rameoso 2236
ione 4416
ione amfoterico 494
ione complesso 2010
ione negativo 558
ionio 4426
ionizzazione 4427
ionizzazione minima 5505
ionogeno 4435
ionone 4436
ipecacuana 4437
ipercromasia 4256
iperpoliploidismo 4257
ipertelia 4258
ipoaploide 4262
ipoaploidismo 4263
ipoclorito di calcio 1359
ipoclorito di litio 4959
ipoclorito di sodio 7550
ipocromaticità 4261
ipofosfito di calcio 1360
ipofosfito di manganese 5134
ipofosfito di sodio 7551
ipofosfito ferrico 3403
ipopituitarismo 4265
ipostasi 4266
ipostatico 4267
iposurrenalismo 4259
ipotesi di Avogadro 736

iridio 4440
iridosmina 6016
irradiazione di calore 3981
irraggiamento del corpo nero 1008
irritante 4457
isatina 4459
isetionato di stilbamidina 7810
iso- 4463
isoallelo 4464
isoallossazina 4465
isoamilene 4472
isoautopoliploidismo 4475
isoborneol 4476
isobutano 4480
isobuteno 4481
isobutilammina 4484
isobutilundecilenammide 4490
isobutiraldeide 4491
isobutirrato d'etile 3235
isobutirrato di fenlietile 6295
isobutirronitrile 4494
isochinolina 4564
isocianati 4500
isocianato d'etile 3236
isocianato di fenile 6298
isocromatide 4498
isocromosoma 4499
isodecaldeide 4503
isodecanol 4505
isodecilottiladipato 4507
isodurene 4508
isoeptano 4513
isoesano 4514
isoeugenol 4510
isoeugenol di benzile 916
isofenoso 4533
isoforone 4534
isoftalato di dimetile 2695
isogenia 4512
isolamento biologico 955
isolamento cellulare 1549
isolamento del gene dal citoplasma
 3718
isolante 5818, 7825
isolare 7320
isolato 2790
isolecito 4515
isole d'isomeria 4462
isoleucina 4516
isologhi 4517
isomeria nucleare 5864
isomeri ottici 5988
isomerismo dinamico 8054
isomerismo geometrico 3732
isomerizzazione 4519
isomero 4518
isomorfia 4521
isomorfico 4520
isoottano 4523
isoottene 4524
isoottiladipato 4525
isoottilisodecilftalato 4527
isopentaldeide 4530
isopentano 4531
isoploidismo 4537
isopoliploide 4538
isopral 4539
isoprene 4540

isopropanolammina 4541
isopropenilacetilene 4542
isopropilammina 4545
isopropilamminodifenilammina 4546
isopropilamminoetanolo 4547
isopropilato di alluminio 414
isopropilfenolo 4561
isopropilmetilpirazolildimetilcarbammato 4555
isopropilmiristato 4556
isopropiloleato 4557
isopropilpalmitato 4558
isopropilperossibicarbonato 4560
isopropilxantato di sodio 7555
isopulegolo 4563
isosafrolo 4565
isosterismo 4567
isotattico 4568
isotermico 4569
isotiocianato di allile 381
isotipia 4575
isotonico 4571
isotopo 4572
isovaleraldeide 4576
isovalerato d'etile 3237
isovalerianato di bornile 1122
istamina 4098
istidina 4100
istiocita 4101
istochimica 4102
itaconato di dimetile 2697
itterbio 8905
ittrio 8906

kainite 4600
kapok 4603
kermesite 4617
kernite 4618
kerogene 4619
kieselgur 2489
kieserite 4632

labile 4660
labilità 4661
labirinto d'ingresso 259
laboratorio 4662
lacca 4666, 4669, 4691
lacca che si asciuga all'aria 254
lacca del Giappone 4586
lacca di robbia 5045
lacca in bastoni 7805
lacca limpida a tampone 8839
lacca scarlatta 7242
lacmoide 4668
lacrimogeno 4667
lamella 4694
lamellare 4695, 4697
lamina 4696
laminare 4697, 4699
laminaria 4614
laminato di tessuto 4700
laminazione 4701
lampada 4705
lampada a vapori di mercurio 5253
lana di scorie 5504
lanatoside C 4709
lanceolato 4710
langbeinite 4712

livello d'energia d'impurezza 4297
lobelia 4987
lobelina 4988
locale d'imbottigliamento 6887
locale d'infustamento 6887
lotto 840
lubrificante 4998
lubrificazione a immersione 4283
lubrificazione nebulizzata 5942
lucentezza 1165, 3767
lucidatura 3764
lucidatura all'acido 127
lumen 5003
luminescenza 5004
lupanina 5010
luppolo 4131
lupulina 5011
luteina 5016
lutezio 5015
lutidina 5017

macchie da essudazione 7738
macchie saline 2212
macchiettature 7727
macchina ad espansione con
    compressore 1113
macchina masticatrice 5181
macerare 5034, 5167, 7787
macerato 5165
macerazione 5169
macero 7788
macina 1218
macina di pietra 7821
macroasse 5036
macromolecola 5038
macromolecola lineare 4901
macronucleo 5040
macroscopico 5041
macrospora 5042
macrostruttura 5043
maculato a forma di goccia 3908
madreperla 5632
madreperlaceo 5633
maglia 5256
magnesia 5068, 5492
magnesia calcinata 1330
magnesio 5048
magnesite 5047
magnetismo 5085
magnetite 5086
magnetizzazione 5088
magnetometro 5089
magnetostrizione 5090
malachite 5098
malathion 5101
maleato di clorfeniramina 1766
maleato di dibutilstagno 2527
malcato dictilico 2501
maleato di metilergonovina 5378
maleato di pirilammina 6821
malleabile 5108
malonato d'etile 3242
maltare 5111
maltasi 5112
malteni 5113
malteria 5114
malto 5110
malto essiccato 4638

maltosio 5115
malto torrefatto 1016
malto verde 3868, 4990
mancanza di acido fosforico 6330
mandorla amara 1001
maneb 5120
manganato di bario 797
manganato di potassio 6625
manganese 5121
manganite di litio 4961
mangano 5146
manna 5149
mannitolo 5150
mannosio 5152
mano di fondo 822
manometro 5153
manometro a membrana 2483
manometro aneroide 537
manometro a scarica luminosa 3689
manometro a stantuffo 2325
manometro a tubo a U 8664
manometro dell'olio 5945
manometro registratore 6971
mantanite 5141
mantello condriosomico 1779
mantello del fuso 5154
marcassite 5158
marcatori non selezionati 8629
marcia, in ~ 5971
margarina 5159
margarite 5160
marihuana 5162
marmitta dell'agitatore 244
marrone di manganese 5125
marrone di Prussia 6783
marrone di Van Dyke 8702
martensite 5164
martinello a siviera 4683
massa 5171
massa attiva 168
massa molecolare 5541, 5547
massicot 5175
masticazione a snervamento 2323
mastice 5180, 6808
mastice per giunzioni 2927
mastice resiliente 1142
mastleucocita 835
materiale attivo 169
materiale base 823
materiale da strappamento 7846
materiale per inclusioni 3016, 3033
materiale per sviluppo fotografico
    6352
materiale refrattario 7004
materiale refrattario basico 830
materiali fini 3475
materiali refrattari 7005
materia plastica 0450
materia prima 6941
materie plastiche etenoidi 3127
materie plastiche fenoliche 6258
materozza 8364
matraccio 3505
matraccio a fondo piatto 3508
matraccio a fondo rotondo 7139
matraccio conico 3097
matraccio da distillazione 2817
matraccio d'Engler 3052

matraccio di Kjeldahl 4646
matraccio tarato 5194
matrice 5185, 5600
mattone di loppa 7442
mattoni refrattari isolanti 4367
mattoni rotti 3877
maturato 8718
maturazione 5189
mauvina 5191
medicamento 2875
megasporocito 5206
meiocito 5207
meiosoma 5208
melamina 5210
melanina 5213
melanoforina 4384
melasse di olio emulsionato nero
    1018
meldometro 5214
melma 7449
membrana 2480
membrana semipermeabile 7314
mendelevio 5220
mentanediammina 5221
mentolo 5222
mentone 5223
mercaptani 5225
mercaptano di amile 507
mercaptano di esadecile 4057
mercaptano di isopropile 4554
mercaptano di laurile 4746
mercaptano di stearile 7784
mercaptano esilico 4081
mercaptobenzotiazolo 5226
mercaptoetanolo 5227
mercaptotiazolina 5228
mercurio 5248
mescola 7819
mescolanza 1043, 5515
mescola per immersione 2773
mescolare 1040
mescolatore 2191
mescolatore a cilindri 7115
mescolatore a getto 4595
mescolatore a pale in forma di
    griglia 3704
mescolatore a recipiente amovibile
    1628
mescolatore a tamburo 813
mescolatore centrifugatore 1574
mescolatore-chiarificatore a pompe
    6798
mescolatore di pasto opaco 798
mescolatore in serie 843
mescolatore per colori 1632
mescolatore planetario 6429
mescolatrice 2804, 5513
mescolatura a freddo 1955
mesitilene 5257
mesomeria 5259
mesone 5260
mesotorio-I 5261
messaggero RNA 7110
metabisolfito di potassio 6626
metabolismo 5263
metabolismo basale 817
metabolito 5264
metaborato di rame 2096

mobilità elettroforetica 2996
mobilità ionica 4424
mobilometro 5516
modello di danno 6145
modificato chimicamente 1661
modificazione adattiva 180
modificazione genetica criptica 2196
modulo di elasticità 5520
modulo di Young 8904
molasse 5529
molazza 1699
molazza a ruote 2928
molazza a ruote verticali 6092
molecola 5548
molecola attivata 162
molecola isosterica 4566
molecola neutra 5706
molecola omonucleare 4123
molibdato di ammonio 471
molibdato di litio 4963
molibdato di piombo 4784
molibdato di potassio 6628
molibdato di sodio 7566
molibdenite 5550
molibdeno 5551
molibdite 5556
molino 5493
molino a barre 7113
molino a biglie 6152
molino a dischi 2799
molino a due cilindri 8589
molino a martelli 3948
molino a tre rulli 8532
molino a tubo 8563
molino per colloidi 1974
molino per polverizzazione di coke
  1947
momento di legame 1102
monacite 5558
monatomico 5557
mongolismo 5561
monoacetato di glicerina 3793
monoacetato di resorcinolo 7054
monobasico 5563
monobenzoato di resorcinolo 7055
monoblasto 5570
monoborato glicolesilenico di sodio
  7546
monobromuro di iodio 4408
monocarburo di uranio 8643
monoclinico 5572
monocloruro di iodio 4409
monodisperso 5573
monofilamenti di nailon 5888
monofilamento 5575
monoicismo 5574
monolaurato di glicerolo 3794
monomeria 5577
monomero 5576
monometiletere di idrochinone 4215
monomolecolare 8620
monomorfo 5579
mononitroresorcinato di piombo 4785
monooleato di trimetilolpropano 8511
monoricinoleato del glicoletilenico
  3214
monoricinoleato di glicerolo 3795

monoricinoleato di propileneglicolo
  6764
monosaccarosi 5580
monosalicilato di glicol dipropilenico
  2778
monosi 5580
monosolfuro di tetrametiltiuram 8165
monosoma 5581
monossido di carbonio 1453
monossido di dipentene 2739
monossido di litio 4941
monossido di piombo 4940
monossido di sodio 7567
monossido di tallio 8184
monossido di vinilcicloesene 8753
monostearato di alluminio 416
monostearato di diglicol 2615
monostearato di glicerolo 3796
monostearato di glicoltetraetilene
  8125
monostearato di saccarosio 7897
monotropico 5582
monovalente 5583
montaggio a catodo rivestito 772
montaggio con carta impregnata
  6097
montaggio senza rivestimento 5827,
  8631
montaliquidi 109, 1054
montante 1991
montare a schiuma 211
montmorillonite 5585
morchia 3560
mordente 5586, 7737
morfina 5588
morfismo 5589
morfolina 5590
morfologia 5591
morfoplasma 5592
morina 5587
morsetto per treppiede 3506
mortaio 5593
mortaio d'agata 229
mosto 5618, 8864
mosto acidificato 7672
mosto di birra 861
mosto dolce 7976
motilità 5596
movimento anafasico 531
movimento browniano 1197
mucchio 5116
mucillagine 5604
muco 5605
muffo 5620
muffola 5607
mullite 5610
multivalente 5616
muschio d'Irlanda 1475
muscovite 5617
mutamento eutettico 3287
mutante restituibile 7027
mutante sopprimibile 7960
mutazione 5621
mutazione somatica 7663

nafta 5634
naftalene 5635
naftalenesolfonato di piombo 4786

naftalenesolfonato di sodio 7568
naftalina 5635
naftenato cobaltoso 1914
naftenato di calcio 1366
naftenato di cerio 1604
naftenato di cromo 1816
naftenato di manganese 5136
naftenato di mercurio 5251
naftenato di piombo 4787
naftenato di potassio 6629
naftenato di rame 2097
naftenato di sodio 7569
naftenato di zinco 8947
naftenato di zirconio 8993
naftenato fenilmercurico 6303
naftenato ferrico 3404
naftilammina 5643
naftilenediammina 5647
naftiltiourea 5651
naftionato di sodio 7570
naftochinone 5642
naftolo 5640
nagyagite 1019
nailon 5885
narceina 5653
narcotico 5654
narcotina 5655
nauseante 5664
navicella di fusione 1995
nebbia d'acido cromico 1804
nebbia luminosa 4854
nebbia metallica 5267
necrobiosi 5666
necrobiotico 5667
nefelite 5686
nefelometro 5687
negativo 4858
nematicida 5676
neoarsfenammina 5677
neocentromero 5678
neocincofene 5679
neodimio 5680
neoesano 5681
neomicina 5682
neon 5683
neoprene 5684
neostigmina 5685
nero animale 556
nero da canale 1629
nero da gas 3682
nero d'avorio 4579
nero di acetilene 90
nero di alizarina 318
nero di benzolo 896
nero di forno 3636
nero di Francoforte 3590
nero d'ilmenite 4279
nero di magnetite 5087
nero di naftalene 5636
nero di pasta di legno 8859
nero di platino 6472
nero d'ossa 1106
nerofumo 1443, 4706
nerofumo termico 8202
nerol 5689
nerolidolo 5690
nero minerale 5497
nero per cuoio 4812

nero sangue 1050
nero svedese 7972
nero vegetale 8719
nettunio 5688
neurina 5698
neurocheratina 5701
neurogeno 5699
neurolisi 5702
neutralizzare 5705
neutralizzazione 5704
neutro 5703
neutroni C 1882
neutroni epitermici 3078
nichel 5712
nichelatura 5722
nichel carbonile 5718
nichelina 5710
nichetamide 5732
nicols incrociati 2177
nicotina 5728
nicotinammide 5727
nigrosina 5731
niobio 5733
nipplo 5736
nitrare 5737
nitrato 5738
nitrato cobaltoso 1915
nitrato d'argento 7405
nitrato d'etile 3250
nitrato di alluminio 417
nitrato di ammonio 472
nitrato di ammonio cerico 1598
nitrato di ammonio e di calcio 1338
nitrato di bario 799
nitrato di bismuto 985
nitrato di cadmio 1307
nitrato di calcio 1367
nitrato di cellulosa 1557
nitrato di ferro 3405
nitrato di guanidina 3886
nitrato di lantanio 4717
nitrato di litio 4964
nitrato di magnesio 5066
nitrato di metile 5416
nitrato di nichel 5721
nitrato di palladio 6077
nitrato di piombo 4788
nitrato di potassio 6630
nitrato di rame 2098
nitrato di sodio 7571
nitrato di stronzio 7855
nitrato di tallio 8185
nitrato di torio 8271
nitrato di uranile 8652
nitrato di zinco 8948
nitrato di zirconio 8994
nitrato manganoso 5143
nitrato mercurico 5236
nitrato mercuroso 5246
nitratore 5740
nitrazione 5739
nitrificare 5747
nitrile 5748, 5891
nitrile butirrico 1289
nitrili alifatici 3366
nitrito 5749
nitrito d'argento 7406
nitrito d'etile 3251

nitrito di amile 509
nitrito di cobalto e potassio 1923
nitrito di potassio 6631
nitrito di sodio 7572
nitro 6630
nitroacetanilide 5750
nitroamido 5794
nitroamminofenolo 5751
nitroanilina 5752
nitroanisolo 5753
nitrobenzaldeide 5754
nitrobenzene 5755
nitrobenzeneazoresorcinolo 5756
nitrobifenile 5760
nitrocellulosa 5761
nitrocotone 3904
nitrodifenilammina 5762
nitroetano 5763
nitrofenetolo 5781
nitrofenide 5782
nitrofenolo 5783
nitroferricianuro di sodio 7573
nitrofurantoina 5764
nitrofurazone 5765
nitrogenasi 5767
nitroglicerina 5773
nitroguanidina 5774
nitromersolo 5775
nitrometano 5776
nitrometro Lunge 5009
nitron 5777
nitronaftalene 5778
nitroparacresolo 5779
nitroparaffine 5780
nitropropano 5785
nitroprussiato di sodio 7574
nitrosodifenilammina 5789
nitrosodimetilanilina 5788
nitrosofenolo 5792
nitrosoguanidina 5790
nitrosolfatiazolo 5796
nitrosonaftolo 5791
nitrostirene 5795
nitrostirolo 7873
nitrotoluidina 5799
nitrotoluolo 5798
nitrotrifluorometilbenzonitrile 5800
nitrourea 5801
nitroxileno 5803
nitrurazione 5745, 5746
nitruri 5744
nitruro di boro 1126
nitruro di silicio 7389
nitruro di zirconio 8995
nivenite 5804
nobelio 5805
noce vomica 5884
nocività 5808
nocivo 5853
nodale 5809
noduli 5811
nodulo 5810
nonadecano 5813
nonadecanoato di metile 5417
nonanal 5815
nonano 5816
nonanoato di metile 5418
non appiccicoso 8013

non esplosivo 5823
non ferroso 5824
nonilammina 5837
nonilbenzene 5838
nonilene 5839
nonilfenolo 5841
nonillattone 5840
non infiammabile 5826
non irritante 1024
non mescolabile 4285
nonosi 5830
non polare 5831
non poroso 5832
non saturato 8626
non volatile 5834
non vulcanizzato 8609
noradrenalina 5843
norleucina 5844
norma 5845
normale 5846
normalizzazione 5847
normocita 5850
novobiocina 5852
nucleare 5855
nucleina 5868
nucleinazione 5869
nucleo 5879
nucleoalbumina 5870
nucleo definitivo 2356
nucleo di fusione 3647
nucleo di terra grassa 4986
nucleo energico 3048
nucleo generativo 3721
nucleolo 5874
nucleolo satellite 7227
nucleo metabolico 5262
nucleoproteina 5877
nucleotide 5878
nucleo trofico 8555
nullisomico 5882
numero cromosomico del gameto
    3664
numero cromosomico zigotico 8911
numero di acetile 98
numero di acidità 136
numero di Avogadro 735
numero di cetano 1610
numero di coordinazione 2073
numero di ottano 5909
numero di Reynolds 7072
numero di saponificazione 7216
numero di trasporto di ioni 8375
numero di valenza 8679
nylon 5885
nylon elasticizzato 2944

obiettivo 5893, 5894
occludere 5897
occlusione 5899
ocra 5900
ocra francese 3603
oculare 5928
odorifero 5929
odorimetria 5930
olandese imbiancatrice 6658
oleaginoso 5953
oleandomicina 5955
oleato cobaltoso 1916

oleato dell'etere monobutilico del
  glicoletilenico 3201
oleato di alluminio 418
oleato di amile 513
oleato di ammonio 473
oleato di butile 1262
oleato di etile 3253
oleato di magnesio 5067
oleato di manganese 5137
oleato di metile 5422
oleato di piombo 4789
oleato di potassio 6632
oleato di rame 2100
oleato di sodio 7575
oleato di tetraidrofurfurile 8141
oleato di triidrossietilammina 8483
oleato di zinco 8950
oleato mercurico 5237
oleato metossietilico 5295
oleato stannoso 7754
olefine 5957
oleina 5959
oleoammide 5954
oleometro 5960
oleoresina 5961
oli bianchi 8843
oli di acetone 77
oli di assorbimento 35
oli di clornaftalina 1748
oli di lavaggio 35
oli essenziali 3112
oli idrogenati 4187
oli impiegati per il taglio di metalli
  2252
oli leggeri 4859
oli litografici 4973
oli lubrificanti 5000
olio combustibile pesante 3613
olio cotto 1082
olio d'assenzio 8863
olio de comino 2221
olio di andropogon muricatus 8735
olio di arachide 654
olio di balena 8837
olio di cade 1297
olio di cajeput 1321
olio di canapa 4008
olio di canella 1851
olio di canfora 1408
olio di capaiba 2074
olio di cardamomo 1466
olio di cartamo 7171
olio di cassia 1496
olio di cedrina 1860
olio di cocco 1934
olio di colofonia 1053
olio di colza 1992
olio di Dippel 1111
olio diesel 2567
olio di eucalipto 3276
olio di fegato di merluzzo 1936
olio di fegato di pescecane 7353
olio di fegato di rombo 3939
olio di fegato di tonno 8567
olio di flemma 3643
olio di garofano 1878
olio di girasole 7950
olio di lardo 4721

olio di lauro 4738
olio di legno della Cina 8568
olio di legno di cedro 1538
olio di lignite 4873
olio di lino cotto 1081
olio di lino grezzo 6940, 6942
olio di lino raffinato con processo
  alcalinico 332
olio di maggiorana 5163
olio di mirra 5631
olio di noce 8798
olio di noce di palma 6082
olio di ossa 1111
olio di piede di bue 5665
olio di resina 1053, 7125
olio di ricino 1502
olio di ricino solfonato 8581
olio di schisto 7352
olio di semi d'anice 559
olio di semi di cotone 2125
olio di semi di lino 4917
olio di semi di lino soffiato 1060
olio di semi di sego 8026
olio di sempreverdi 8849
olio di senape 5619
olio di sesamo 7335
olio di soia 7675
olio di spermaceti 7703
olio di spigo di lavanda 7710
olio di stilingia 7815
olio di vetriolo 7945
olio d'oliva 5965
olio essenziale de lavanda 4750
olio essenziale di aglio 3675
olio essenziale di Cymbopogon 4822
olio essenziale di labdano 4657
olio essenziale di linaloe 4895
olio essenziale di pino 6402
olio essenziale di ruta 7154
olio essenziale di tiglio 4883
olio essiccante cinese 8568
olio essiccativo 2883
olio iodato 4413
olio lubrificante (di alta viscosità)
  1164
olio non essiccabile 5822
olio per cotone 2124
olio per presse idrauliche 4161
olio pesante 2567
olio semi di gomma 7148
olio solfonato 7930
olio vetriolo 3624
oli per rigenerazione 6965
oli polimerizzati 6541
oli segregati 7290
oli soffiati 1061
oli tessili 8177
olivenite 5964
olivina 5966
olmio 4104
olocrino 4107
oltremare 8600
oltremare di cobalto 1926
omatropina 4108
omeosi 4114
omeostasi 4115
omeotico 4116
omoallelo 4111

omocarion 4112
omociclico 4113
omogametico 4117
omogeneizzazione 4118
omolecitico 4119
omologhi 4121
omomorfi 4122
omoploidismo 4124
omopolare 4125
omopolimero 4126
oncogeno 5967
ondulazione della superficie 7965
oneto 2660
ontogenesi 5972
ontogenia 5972
oosoma 5974
ootidio 5975
opacità 5977, 7447
opacità della superficie 7963
opaco 5013, 5981
opaco alla radiazione 6908
opale nero 1017
opalescente 5979
opalescenza 5978
operazione a lotti 846
operazione di "reforming" 7003
oppiati 5982
oppio 5983
orbita di legame 1103
orbita elettronica comune 1101
ordenina 4132
ordinata 5995
organicometallico 5272
organismo aerobico 212
organismo termofilo 8218
organosolo 6002
orientamento 6004
orientare 6003
orificio 6005
ormone adrenocorticotropo 200
ormone adrenotropo 200
ormone estrogeno 3119
ormoni 4133
ornitina 6006
oro 3821
orpimento 6008
ortocromatico 6009
ortoelio 6010
ortoformato di trietile 8469
ortoidrogeno 6011
ortoidrossido alluminio 402
ortoploidismo 6013
ortorombico 6014
orzo 807
osmiato potassico 6633
osmio 6017
osmiridio 6016
osmometro 6018
osmosi 6019
ospite 4135
ossalacetato di sodio e etile 3265
ossalati 6039
ossalato cobaltoso 1917
ossalato di amile 517
ossalato dibutilico 2519
ossalato di calcio 1368
ossalato di etile 3254
ossalato di ferro 3406

sesquicloruro di alluminio-etile 3140
sesquidiploidismo 7337
sesquiossido di iridio 4442
sesquiossido di molibdeno 5554
sesquiossido di piombo 4796
sesquisilicato di sodio 7601
sesquisolfuro di fosforo 6343
seta artificiale 679
setaccio a fili sagomati 8823
setole di nailon 5886
setosità 7393
settisomico 7326
sezione di rettifica 6979
sezione efficace di urto 2930
sfalerite 1041
sferoma 7705
sferoplasto 7706
sferosoma 7707
sfogliatura 6154
sgocciolatore 2870
sgrassaggio alcalino 326
sgrassaggio anodico 575, 7064
sgrassaggio con solventi 7654
sgrassaggio elettrolitico 2977
sgrassagio alcalino 328
sgrassamento a vapore 8711
sgrassatura 2251, 2367
sgrassatura con emulsione 7655
siderite 7369
sidro 1843
sidro ad alto grado alcoolico 646
siero 7334
sierocoltura 7330
sierosità 7331
sifonare 7426
sifone 8004
silano 7377
silicato 7380
silicato d'etile 3264
silicato di alluminio 422
silicato di calcio 1380
silicato di litio e alluminio 7719
silicato di magnesio 5076
silicato di piombo 4797
silicato di potassio 6645
silicato di sodio 7602
silicato di zirconio e zinco 8978
silicato etilbutilico 3158
silicato glicoletilenico 3215
silicio 7383
silicofluoruro di sodio 7537
silicofluoruro di sodio e alluminio 7481
silicofluoruro di zinco 8967
siliconi 7388
sillimanite 7394
silossano 7395
silvanite 7979
silvinite 7980
silvite 7981
sim- 7982
simaruba 7416
simbiosi 7983
similcuioio 4813
similoro 2899
simmetria 7984
simpatolitico 7985
sinapsi 7986

sinaptico 7987
sincarion 7988
sincito 7989
sindiotattico 7990
sineresi 7991
singoletto 7420
sinterizzare 7422
sinterizzazione 7424
sintesi 7995
sintesi dell'ammoniaca 454
sintesi Diels-Alder 2564
sintesi di Fittig 3491
sintesi di Wurtz 8868
sintesi Fischer-Tropsch 3489
sintetici 8002
sintetico 7997
sintrofo 8003
sirosingopina 8005
sistema cromaffine 1785
sistema in serie 7328
sistema multiplo 5615
sistema periodico 6210
sistema triclino 700
sistemico 8007
sitosterolo 7429
siviera 4682
smaltatura a sale 7192
smaltina 7460
smaltino 7459
smalto 3029
smalto a fuoco 7828
smalto a vetrino 3763
smalto per verniciatrice a rulli 7114
sminuzzatore 2000
smitsonite 7462
sobbollire 6122
soda 553
soda caustica 1535
sodio 7473
sodioacetato di sodio 7603
sodio-butabarbital 1237
sodio di tiamilale 8228
sodio di triiodotironina 8485
soffiatore 1055
soffiatrice per bottiglie 1133
sofisticazione 209
sol 7637
solfacetammide 7905
solfachinossalina 7917
solfadiazina 7906
solfadimetossina 7907
solfaguanidina 7908
solfamato di calcio 1383
solfamerazina 7909
solfametazina 7910
solfametiazolo 7911
solfametossipiridazina 7912
solfanamide di ossidietilenebenzo-
    tiazolo 6053
solfanilammide 7914
solfanilato di zinco 8969
solfanilato sodico 7607
solfapiridina 7916
solfarsfenammina 7918
solfatazione 7922
solfati 7920
solfatiazolo 7921
solfato acido di idrossilammina 4231

solfato ammonico di alluminio 406
solfato cerico 1601
solfato cobaltoso 1921
solfato cromico 1811
solfato d'ammonio cobaltoso 1907
solfato d'argento 7415
solfato di alluminio 424
solfato di ammonio 478
solfato di bario 804
solfato di butacaina 1238
solfato di cadmio 1312
solfato di calcio 1105, 1384, 3911
solfato di chinidina 6869
solfato di chinino 6874
solfato di cromo e ammonio 1815
solfato di cromo e potassio 1818
solfato di dibutolina 2511
solfato di dimetile 2707
solfato di estrone e piperazina 6409
solfato dietilico 2598
solfato di guanile e urea 3893
solfato di idrazina 4166
solfato di idrossichinolina 4249
solfato di idrossilammina 4233
solfato di ittrio 8910
solfato di laurile e sodio 7557
solfato di laurile e trietanolammina
    8447
solfato di magnesio 5078
solfato di magnesio e potassio 6624
solfato di nichel 5725
solfato di nichel e ammonio 5715
solfato di piombo 4800
solfato di piombo basico 828
solfato di potassio 6649
solfato di potassio d'alluminio 421
solfato di radio 6914
solfato di rame 2107
solfato di rame tribasico 8394
solfato di sodio 7608
solfato di sodio commerciale 7190
solfato di sodio di alluminio 423
solfato di sparteina 7681
solfato di stronzio 7858
solfato di tallio 8186
solfato di titanio 8303
solfato di uranile 8653
solfato di vanadile 8700
solfato di zinco 8970
solfato di zirconio 8999
solfato ferrico 3414
solfato ferrico d'ammonio 3394
solfato ferroso 3439
solfato ferroso di ammonio 3427
solfato idrato di alluminio e potassio
    6595
solfato manganoso 5145
solfato mercurico 5241
solfato mercuroso 5247
solfato stannoso 7757
solfato tetradecilsodico 7615
solfato titanoso 8307
solfinpirazone 7924
solfito di calcio 1386
solfito di magnesio 5079
solfito di potassio 6651
solfito di sodio 7610
solfito di zinco 8972

tenifugo 8071
tennantite 8072
tenore di amido 7761
tenore in ceneri 687
tensione del bagno 1565
tensione di cella 8033
tensione di decomposizione 2346
tensione di equilibrio di una
  reazione 3089
tensione di vapore 8715
tensione interfacciale 4380
tensione superficiale 7964
tenuta a vapore 7776
teobromina 8195
teofillina 8198
teofillinametilglucamina 8199
teoria dell'interazione 4371
teoria di separazione capillare 1418
tepido 5002
terbio 8073
terebene 8074
tereftalato di polietilene 6531
terfenile 8080
termite 8212
termochimica 8213
termoindurente 8223
termolabile 8214
termolisi 8216
termoluminescenza 8215
termometro 8217
termometro a dilatazione di gas 3690
termometro a gas 3702
termometro di Beckmann 859
termone 8078
termopila 8219
termoplasticità 8221
termoplastico 8220
termostabile 8225
termostato 8226
terpene 8079
terpineolo 8081
terpinolene 8083
terra alba 8085
terra d'infusori 2489
terra di ombra bruciata 1235
terra di Siena 7370
terra di Siena bruciata 1234
terra di Siena grezza 6943
terra di Verona 3867
terra d'ombra 8606
terra d'ombra turca 8582
terraglia 2913
terra grassa 4985
terra per folloni 3618
terra refrattaria 3878
terra silicea 7378
terra verde 3867
terre coloranti 2912
tessuto di vetro 3754
test cis-trans 1855
testosterone 8088
tetraacetato di piombo 4802
tetraacetilacetonato di zirconio 9000
tetraborato di litio 4969
tetrabromoetilene 8093
tetrabromuro di acetile 92
tetrabutilstagno 8095
tetrabutilurea 8097

tetracarbonile di cobalto 1924
tetracarbossibutano 8100
tetracene 8101
tetracianoetilene 8109
tetraciclina 8110
tetraclorobenzene 8102
tetraclorodifluoroetano 8103
tetracloroetano 8104
tetraclorofenato di sodio 7614
tetraclorofenolo 8105
tetracloruro di carbonio 1455
tetracloruro di silicio 7390
tetracloruro di titanio 8304
tetracloruro di vanadio 8695
tetracloruro di zirconio 9001
tetracosano 8108
tetrade 8111
tetradecano 8112
tetradecene 8113
tetradecilammina 8114
tetradecilmercaptano 8115
tetradimite 8117
tetraedrite 8136
tetraetile di piombo 4803
tetraetileneglicolo 8121
tetraetilenepentammina 8126
tetraetilessiltitanato 8127
tetrafenilsilano 8169
tetrafenilstagno 8170
tetrafluoetilene 8132
tetrafluoroidrazina 8133
tetrafluorometano 8134
tetrafluoruro di silicio 7391
tetrafluoruro di uranio 8647
tetrafosfato di esaetile 4059
tetraidrato cromato di sodio 7517
tetraidrofurano 8137
tetraidrofurfurilpalmitato 8142
tetraidronaftalene 8144
tetraidropiridina 8147
tetraidrossietilenediammina 8149
tetraidrotiofene 8148
tetraiodoetilene 8150
tetralina 8144
tetramero 8154
tetrametilbutanediammina 8156
tetrametildiaminobenzidrolo 8157
tetrametildiaminobenzofenone 8158
tetrametildiaminodifenilmetano 8159
tetrametildiaminodifenilsolfone 8160
tetrametilenediammina 8161
tetrametiletilenediammina 8162
tetramorfo 8166
tetranitrato di pentaeritritolo 6175
tetranitroanilina 8167
tetranitrometano 8168
tetrapropilene 8172
tetrasolfuro di sodio 7616
tetrasomatia 8174
tetravalente 8175
tetrile 8176
tetrossido di vanadio 8696
thiuram 8259
tiazina 8229
tiazolo 8230
tiazosolfone 8231
timochinone 8285
timoftaleina 8284

timolo 8282
tincal 8289
tino di chiarificazione 7341
tino di fermentazione 3386
tino di miscela 5170
tino per infustamento 6888
tinta 8291
tinte acide 108
tintometro Lovibond 4991
tintura 8290
tintura di iodio 4411
tio- 8234
tiocarbanilide 8237
tiocianato di ammonio 480
tiocianato di benzile 918
tiocianato di calcio 1387
tiocianato di piombo 4804
tiocianato di sodio 7017
tiocianato di zinco 8975
tiocianato mercurico 5243
tiocianoacetato di isobornile 4479
tiodiglicolo 8239
tiodipropionato di dilaurile 2658
tiodipropionato di ditridecile 2828
tiodipropionato distearilico 2812
tioetanol d'etile 3269
tioetere di dimiristile 2715
tioeteri 8241
tiofene 8246
tiofenolo 8247
tiofosforammide di trietilene 8468
tiofosgene 8248
tioglicerolo 8242
tioglicolato di isoottile 4529
tioglicolato sodico 7618
tioidantoina 8244
tiomalonato d'oro e sodio 3831
tiosemicarbazide 8252
tiosemicarbazone 8253
tiosolfato di ammonio 481
tiosolfato di bario 806
tiosolfato di sodio 7619
tiouracile 8254
tiourea 8255
tiourea allilica 383
tipolisi 8591
tiraggio forzato 3563
tirammina 8592
tirocidina 8593
tirosina 8594
tirossina 8286
tirotricina 8595
tissotropo 8261
titanato di litio 4970
titanato di piombo 4805
titanato di potassio 6654
titanato di stronzio 7860
titanato di tetrabutile 8096
titanato di tetraisopropile 8152
titanio 8297
titanite 8296
titolazione 8309
titolo 8310
titrimetria 8311
tixotropia 8262
tixotropo 8261
tocoferoli 8313
tolano 8314

DEUTSCH

Aludel 398
Aluminit 403
Aluminium 404
Aluminiumacetat 405
Aluminiumammoniumsulfat 406
Aluminium-blanc fixe 401
Aluminiumborhydrid 407
Aluminiumbromid 408
Aluminiumcarbid 409
Aluminiumchlorid 410
Aluminiumdichloräthylen 3139
Aluminiumdistearat 411
Aluminiumhydrat 402
Aluminiumhydroxyd 412
Aluminiumhydroxystearat 413
Aluminiumisopropylat 414
Aluminiumkaliumsulfat 421
Aluminiummetaphosphat 415
Aluminiummonostearat 416
Aluminiumnatriumsulfat 423
Aluminiumnitrat 417
Aluminiumoleat 418
Aluminiumoxyd 419
Aluminiumpalmitat 420
Aluminiumphosphat 6985
Aluminiumsilikat 422
Aluminiumsulfat 424
Aluminiumtriformiat 425
Aluminothermie 426
Alunit 428
Amalgam 430
Amalgamieren 431
Amalgamzersetzer 2402
Amber 432
Ameiose 436
Ameisensäure 3574
Ameisensäureamylester 503
Ameisensäurebutylester 1266
Americium 437
Amibe 435
Amid 440
Amidine 441
Amidogruppe 442
Amikron 438
amiktisch 439
Amin 443
Aminogruppe 445
Aminoharze 446
Aminosäuren 444
Amitose 447
Amixie 448
Amixis 448
Ammine 449
Ammoniak 450
ammoniakaler Latex 452
Ammoniakbrenner 451
Ammoniaksynthese 454
Ammonium 455
Ammoniumacetat 456
Ammoniumalginat 457
Ammoniumbenzoat 458
Ammoniumbifluorid 459
Ammoniumcarbamat 460
Ammoniumcarbonat 461
Ammoniumchloride 462
Ammoniumchromat 463
Ammoniumdichromat 464
Ammoniumdiphosphat 2496

Ammoniumdiuranat 465
Ammoniumeisenalaun 3394
Ammoniumfluorid 466
Ammoniumhydroxyd 467
Ammoniumlaurat 468
Ammoniumlinoleat 469
Ammoniummetavanadat 470
Ammoniummolybdat 471
Ammoniummonophosphat 5565
Ammoniumnitrat 472
Ammoniumoleat 473
Ammoniumperchlorat 474
Ammoniumpersulfat 475
Ammoniumphosphat 5565
Ammoniumpikrat 476
Ammoniumstearat 477
Ammoniumsulfat 478
Ammoniumsulfid 479
Ammoniumthiocyanat 480
Ammoniumthiosulfat 481
Ammonolyse 482
amorph 483
Amphiaster 484
Amphibole 485
Amphibolit 486
Amphidiploidie 487
Amphigonie 488
Amphimixie 489
Amphimixis 489
Amphogenie 490
Ampholyte 491
Ampholytoid 492
amphoter 493
Amyl 495
Amylacetat 496
Amylalkohol 497
Amylalkoholgärung 510
Amylase 498
Amyläther 502
Amylchloridmischung 499
Amylcitrat 500
Amylen 501
Amylformiat 503
Amylgruppe 504
Amyllactat 505
Amyllaurat 506
Amylmercaptan 507
Amylnaphthalin 508
Amylnitrit 509
Amyloid 511
Amyloidose 512
Amyloleat 513
Amylopsin 515
Amylose 516
Amyloxalat 517
Amylpropionat 518
Amylsalicylat 519
Amylstearat 520
Amyltartrat 521
Ana- 522
Anabasin 523
anabolismusfördernd 3299
Anachromasis 524
Anaerobe 525
Analcim 526
Analyse 528
Analysenwaage 529
Anaphase 530

Anaphasebewegung 531
Anaphorese 532
Andalusit 534
Androsteron 535
Anemometer 536
Aneroidmanometer 537
Anethol 538
Aneuploid 539
Aneuploidie 540
anfeuchten 1072
Anfeuchter 4147
angefeuchtet 5046
angeregtes Molekül 162
angereichertes Lösungsmittel 7097
angereichertes Wasser 3054
Angioblast 541
Anglesit 542
Angriffsstoff 241
Angus-Smith-Verfahren 544
Angusstetzen 8364
Anhydride 545
Anhydrit 546
Anhydrit-Verfahren 547
Anilin 554
Anilinpunkt 555
Anion 558
Anisidine 560
Anisol 561
Anisöl 559
anisomer 562
anisotonisch 563
anisotrop 564
Ankerit 565
Ankerrührwerk 533
Anlaßbatterie 7762
Anlaufen 8049
Annatto-Farbstoff 567
Annidation 569
Anode 570
Anodenreaktion 578
Anodenrest 573
Anodenschlamm 574
anodische Inhibitoren 576
anodische Passivierung 577
anodische Reinigung 575, 7064
anodische Schicht 572
anodische Stromausbeute 571
Anolyt 580
anomale Valenz 581
anorganisch 4356
anorganische Chemie 4357
Anorthosit 582
Anpassung 179
Anreicherung 2017
Anreicherungszone 3055
ansäuern 118, 135
anschwellender Überzug 4395
Ansiedescherben 7260
anstellen (Hefe) 6420
Anstellhefe 6423
Anthophyllit 583
Anthrachinon 588
Anthraflavin 585
Anthranilsäure 586
Anthranol 587
Anthrarobin 589
Anthrazen 584
Antialbumosen 590

Lebensunfähigkeit 4401
Leblanc-Verfahren 4815
Le Chatelier-Prinzip 4816
Lecithin 4817
Leclanché-Element 4818
Lederdichtung 4814
Lederschwarz 4812
Legierungsüberzug 372
Lehm 4985
Lehmkern 4986
Lehre 3705
Leichenwachs 192
Leichtbeton 4867
leichte Fraktion 4855
leichtes Element 4851
leicht färbbar 1824
Leichtfarbe 5008
Leichtgas 4856
Leichtlegierung 4850
Leichtmetall 4857
Leichtöle 4859
leicht siedendes Lösungsmittel 4863
Leim 3777, 7430
Leimen 7432
Leimung 7432, 8566
Leimzucker 3808
Leim zum Verbinden von
  Sperrholzplatten 2927
Leinöl 4917
Leinölfirnis 1081, 1082
Leinsamen 4916
Leipzigergelb 4820
Leistung 6027
Leitsalze 2034
Lemongrasöl 4822
Leonit 4826
Lepidin 4827
Lepidolith 4828
Leptometer 4830
Leptonema 4831
Leptotän 4832
Leptozyt 4829
Leuchten 5004
Leuchtgas 1890, 5662
Leuchtkraft 1166
Leuchtpetroleum 4620
Leuchtschirmsubstanzen 6336
Leuchtstoffe 6333
Leucin 4834
Leucit 4835
Leukobase 4836
Leuko-Derivate 4837
Leuko-Verbindungen 4837
Leukozyt 4838
Leukozytenkörnelung 6445
Lewisit 4842
Lewis-Verfahren 4843
Lichenin 4847
Lichtbeständigkeit 4853
Lichtbeständigkeitsprüfer 3338
Lichtbogenverfahren 657
Lichtdurchlässigkeit 4866
Lichtdurchlässigkeitsmesser 2479
Lichtechtheitsmesser 3615
lichtempfindlich 4862, 6358
Lichtempfindlichkeit 7319
Lichtleitung 4860
Lichtmesser 6354

Lichtnebel 4854
Lichtquelle 4864
Lichtrelais 4861
Lichtschutzmittel 4865
Lichtstrom 5006
lichtunechte Pigmente 3614
lichtwiderständig 4858
Liebig-Kühler 4849
Lignin 4870
Lignocerinsäure 4875
Ligroin 4876
Limonenöl 4883
Limonit 4892
Linaloeöl 4895
Linalool 4896
Linalylacetat 4897
Linde-Verfahren 4898
lineares Chromosom 4899
lineares Makromolekül 4901
lineares Polymer 4903
lineares Superpolymer 4904
Linie mit einem substituierten
  Chromosom 7418
Liniment 4905
linksdrehend 4684
Linneit 4907
Linoleattrockenstoffe 4908
Linolen 4910
Linolensäure 4911
Linolenylalkohol 4912
Linoleylalkohol 4913
Linoleyltrimethylammoniumbromid
  4914
Linolsäure 4909
Linoxyn 4915
Linse 4824
Linsenachse 5986
linsenförmig 4825
Linters 4918
Lipasen 4921
Lipid 193
Lipochrome 4924
Lipoid 4922
Lipoidose 4923
Lipolyse 4926
lipotropisches Mittel 4928
Lipoxydase 4929
Liter 4978
Lithion 4941
Lithionhydrat 4958
Lithium 4942
Lithiumalkoholate 4943
Lithiumaluminat 4944
Lithiumaluminiumhydrid 4945
Lithiumamid 4946
Lithiumborhydrid 4947
Lithiumbromid 4948
Lithiumcarbonat 4949
Lithiumchlorat 4950
Lithiumchlorid 4951
Lithiumchromat 4952
Lithiumcitrat 4953
Lithiumfluophosphat 4955
Lithiumfluorid 4956
Lithiumglimmer 4828
Lithiumhydrat 4958
Lithiumhydrid 4957
Lithiumhydroxyd 4958

Lithiumhypochlorit 4959
Lithiumjodid 4960
Lithiumkobaltit 4954
Lithiummanganit 4961
Lithiummetasilikat 4962
Lithiummolybdat 4963
Lithiumnitrat 4964
Lithiumperoxyd 4965
Lithiumricinoleat 4966
Lithiumsalicylat 4967
Lithiumstearat 4968
Lithiumtetraborat 4969
Lithiumtitanat 4970
Lithiumzirconat 4971
Lithocholsäure 4972
lithographische Öle 4973
lithographischer Lack 4974
Lithopone 4975
Lobelie 4987
Lobelin 4988
Löffel 7254
Lorbeer 4737
Lorbeeröl 4738
löschen 6856
Löschgas 6858
Löschung 6857
Lösemittelaufnahmefähigkeit 7660
Lösemittelbeständigkeit 7659
Lösemittelwiedergewinnung 7658
löslicher Farbstoff 7643
lösliches Nylonharz 7644
Löslichkeitskurve 7642
Losrolle 2313
Lösung 7646
Lösungsgefäß für angereicherte
  Platten 7438
Lösungsmittel 7652
Lösungsmittel mit hohem
  Flammpunkt 4090
Lösungswärme 3979
Lötrohr 5602
Lovibond-Kolorimeter 4991
Luftabscheider 2327
Luftabscheidung 264
Luftäquivalent 256
Luftdosis 3593
luftentzündlich 6837
Luftfaktor 257
Luftkessel 263
Luftkühlung 252
Luftpumpe 262
Luftsauerstoffelement 253
Luftsauger 691
lufttrocknender Lack 254
Luftüberschußfaktor 3297
Luftverdünnung 6932
Luftzirkulation 251
Lumen 5003
Lumineszenz 5004
Lumpenbrei 6920
Lunge-Nitrometer 5009
Lupanin 5010
Lupe 5092
Lupulin 5011
Lutein 5016
Lutetium 5015
Lutidin 5017
Lycopodium 5020

schweres Element 3988
schweres Wasser 3991
Schweröl 2567, 3613
Schwerspat 816
Schwerwasserherstellungsverfahren
  nach Clusius und Starke 1880
Schwesterchromatiden 7428
Schwimmer 7977
Schwimmverfahren 3526
Schwimmvermögen 1230
Schwindwasser 7367
Schwitzen 7971
Schwitzwasser 3532
Scoparium 7255
Scopolamin 7256
Scopolin 7257
Sebacinsäure 7271
Sebaconitril 7272
sechswertig 4071, 7348
Secobarbital 7274
Secobarbitalnatrium 7275
Sedimentationspotential 7284
Seger-Kegel 7287
Segmentallopolyploid 7288
seidenartiges Ansehen 7393
Seidenpapier 8294
Seife 7467
Seifeaufnahmevermögen 7468
Seifenmischer 2191
seifig 7214
Seigerung 4931
Seitenkette 7368
Sek- 7273
Sekretin 6056
sekundäre Alkohole 7276
sekundäre Polyploidie 7279
sekundäres Amid 7277
sekundares Bleiphosphat 2498
Sekundärreaktion 7280
Selbstentladung 7299
Selbstentzündungstemperatur 7302
selbstfertil 7300
Selbstfertilität 7301
selbsttragende Emulsion 6161
Selbstverbrennung 7723
Selektionsfaktor 7292
Selen 7293
Selendiäthyldithiocarbamat 7294
Selendioxyd 7295
selenige Säure 7298
Selenrot 7296
Selensulfid 7297
seltene Erdelemente 6930
seltene Erden 4715
Semiallel 7304
Semicarbazidhydrochlorid 7307
Semichiasma 7309
semiheterotypisch 7311
Semikaryotyp 7312
semiletal 7313
semipolare Bindung 2316
semiunivalent 7315
Senarmontit 7316
Senegawurzel 7317
Senfgas 2536
Senföl 5619
Senkwaage 4208
Sennesblätter 7318

Sepia 7325
Sepiabraun 7325
septisom 7326
Sequestration 7327
Serin 7329
Serokultur 7330
Serosität 7331
Serotonin 7332
Serpentin 7333
Serum 7334
Sesamöl 7335
Sesamolin 7336
Sesquidiploidie 7337
Sicherheitsröhre 7169, 8258
Sicherheitsventil 7170
Sickerwasser 4809
Siderit 7369
Sieb 7263
Siebanalyse 7372
Siebboden 2342
sieben 7371
siebenwertig 4019
Siebplansichter 7264
Siebplatte 7375
Siebrückstand 7373
Siebschüttelvorrichtung 7374
Siebtrommel 8551
Siebwasser 8845
Siedekolben 2817
Sieden 1086, 2916
Siedepunkt 1090
Siederohrkessel 8816
Siennaerde 7370
Sigla-Glas 633
Sikkativ 2866
Silan 7377
Silber 7396
Silberarsphenamin 7397
Silberbromid 7398
Silberchlorid 7399
Silberchloridelement 7400
Silberchromat 7401
Silbercyanid 7402
Silberfahlerz 3601
Silberglanzerz 660
Silberhalogenide 7403
Silberhornerz 1593
Silberjodid 7404
Silberkaliumcyanid 7411
Silbernitrat 7405
Silbernitrit 7406
Silberoxyd 7407
Silberoxydelement 7408
Silberphosphat 7409
Silberpikrat 7410
Silberspat 1593
Silbersulfat 7415
Silber-Zink-Akkumulator 7414
Silikagel 7379
Silikat 7380
Silikatfarben 7381
Silikone 7388
Silikonharze 7386
Silikonkautschuk 7387
Silikonkitt 1142
Silikonlösungen 7385
Silizium 7383
Siliziumcarbid 7384

Siliziumeisen 3423
Siliziumnitrat 7389
Siliziumtetrachlorid 7390
Siliziumtetrafluorid 7391
Sillimanit 7394
Siloxan 7395
Silvanit 7979
Silvin 7981
Simaruba 7416
Simonsches Verfahren 7417
Sinker 7421
Sintermetallurgie 6666
sintern 7422
Sintern 7424
Sinterofen 7425
Sinterplatte 7423
Sisalhanf 7427
Sitosterin 7429
Sitosterol 7429
Skatol 7433
Sklerometer 7253
Skrubber 3700, 7265
Skutterudit 7437
Smalte 1897, 7459
Smaltin 7460
Smaragdgrün 3018
Smithsonit 7462
Soda 7508
Sodaasche 553
Soderberg-Elektrode 7472
Sojabohnenöl 7675
Sol 7637
Solinglas 2183, 7471
solutropische Mischung 7648
Solvat 7649
Solvatisierung 7650
Solvay-Verfahren 7651
Solventnaphtha 7657
somatisch 7662
somatische Mutation 7663
somatische Reduktion 7664
somatoplastisch 7665
Sonde 6713
Sonnenblumenöl 7950
Sonnensalz 7638
Sorbinsäure 7667
Sorbit 7668
sorbose 7669
Sorelzement 7670
Sorption 7671
Sorrelsalze 6643
Sortieren nach der Größe 7431
Spalter 7289
Spaltung 1869, 2345, 7252
spanischer Pfeffer 1424
Spanischrot 7677
Spannungserhöher 5491
Sparteinsulfat 7681
Spat 7678
Spatel 7683
Spatlack 7682
Spearmintöl 7684
Speciation 7685
Specköl 4721
Speckstein 7786
Speisewasservorwärmer 2920
Spektroskop 7692
Spektrum 7693